CRC HANDBOOK OF

Phase Equilibria and Thermodynamic Data of Copolymer Solutions

CRC HANDBOOK OF

Phase Equilibria and Thermodynamic Data of Copolymer Solutions

Christian Wohlfarth

CRC Press
Taylor & Francis Group
Boca Raton London New York

CRC Press is an imprint of the
Taylor & Francis Group, an **informa** business

CRC Press
Taylor & Francis Group
6000 Broken Sound Parkway NW, Suite 300
Boca Raton, FL 33487-2742

First issued in paperback 2019

ISBN-13: 978-1-4398-5038-1 (hbk)
ISBN-13: 978-0-367-38331-2 (pbk)

Visit the Taylor & Francis Web site at
http://www.taylorandfrancis.com

and the CRC Press Web site at
http://www.crcpress.com

CONTENTS

4. HIGH-PRESSURE FLUID PHASE EQUILIBRIUM (HPPE) DATA OF COPOLYMER SOLUTIONS

5. ENTHALPY CHANGES FOR COPOLYMER SOLUTIONS

6. PVT DATA OF COPOLYMERS AND SOLUTIONS

7. SECOND VIRIAL COEFFICIENTS (A_2) OF COPOLYMER SOLUTIONS

APPENDICES

PREFACE

Today, there is still a strong and continuing interest in thermodynamic properties of copolymer solutions. Thus, ten years after the *CRC Handbook of Thermodynamic Data of Copolymer Solutions* was published, necessity as well as desire arises for a supplementary book that includes and provides newly published experimental data from the last decade.

There are about 500 newly published references containing about 150 new vapor-liquid equilibrium datasets, 50 new tables containing classic Henry's coefficients, 250 new liquid-liquid equilibrium datasets, 350 new high-pressure fluid phase equilibrium datasets, 70 new datasets describing PVT-properties, densities, or excess volumes, and 40 new enthalpic datasets. Additionally, a certain amount of data to be included in the supplement was anticipated when the original handbook was made. This last point is especially mentioned for values of second osmotic virial coefficients for which this new supplementary volume gives a much larger data table. In comparison to the original handbook, the new supplementary volume contains a larger amount of data and will be a useful as well as necessary completion of the original handbook.

The new supplementary volume is again divided into the seven chapters as used before: (1) Introduction, (2) Vapor-Liquid Equilibrium (VLE) Data and Gas Solubilities of Copolymer Solutions, (3) Liquid-Liquid Equilibrium (LLE) Data of Copolymer Solutions, (4) High-Pressure Fluid Phase Equilibrium (HPPE) Data of Copolymer Solutions, (5) Enthalpy Changes for Copolymer Solutions, (6) PVT Data of Copolymers and Solutions, and (7) Second Virial Coefficients (A_2) of Copolymer Solutions. Finally, appendices quickly route the user to the desired datasets.

Additionally, tables of systems are provided where results were published only in graphical form in the original literature to lead the reader to further sources. Data are included only if numerical values were published or authors provided their numerical results by personal communication (and I wish to thank all those who did so). No digitized data have been included in this data collection.

The closing day for the data compilation was December 31, 2009. However, the user who is in need of new additional datasets is kindly invited to ask for new information beyond this book via e-mail at christian.wohlfarth@chemie.uni-halle.de. Additionally, the author will be grateful to all users who call his attention to mistakes and make suggestions for improvements.

The new *CRC Handbook of Phase Equilibria and Thermodynamic Data of Copolymer Solutions* will again be useful to researchers, specialists, and engineers working in the fields of polymer science, physical chemistry, chemical engineering, material science, biological science and technology, and those developing computerized predictive packages. The book should also be of use as a data source to Ph.D. students and faculty in chemistry, physics, chemical engineering, biotechnology, and materials science departments at universities.

Christian Wohlfarth
Halle, March 2010

About the Author

Christian Wohlfarth is associate professor for physical chemistry at Martin Luther University Halle-Wittenberg, Germany. He earned his degree in chemistry in 1974 and wrote his Ph.D. thesis in 1977 on investigations of the second dielectric virial coefficient and the intermolecular pair potential, both at Carl Schorlemmer Technical University Merseburg. In 1985, he wrote his habilitation thesis, *Phase Equilibria in Systems with Polymers and Copolymers*, at the Technical University Merseburg.

Since then, Dr. Wohlfarth's main research has been related to polymer systems. Currently, his research topics are molecular thermodynamics, continuous thermodynamics, phase equilibria in polymer mixtures and solutions, polymers in supercritical fluids, PVT behavior and equations of state, and sorption properties of polymers, about which he has published approximately 100 original papers. He has written the following books: *Vapor-Liquid Equilibria of Binary Polymer Solutions*, *CRC Handbook of Thermodynamic Data of Copolymer Solutions*, *CRC Handbook of Thermodynamic Data of Aqueous Polymer Solutions*, *CRC Handbook of Thermodynamic Data of Polymer Solutions at Elevated Pressures*, *CRC Handbook of Enthalpy Data of Polymer-Solvent Systems*, and *CRC Handbook of Liquid-Liquid Equilibrium Data of Polymer Solutions*.

He is working on the evaluation, correlation, and calculation of thermophysical properties of pure compounds and binary mixtures resulting in eleven volumes of the *Landolt-Börnstein New Series*. He is a contributor to the *CRC Handbook of Chemistry and Physics*.

1. INTRODUCTION

1.1. Objectives of the handbook

Knowledge of thermodynamic data of polymer solutions is a necessity for industrial and laboratory processes. Furthermore, such data serve as essential tools for understanding the physical behavior of polymer solutions, for studying intermolecular interactions, and for gaining insights into the molecular nature of mixtures. They also provide the necessary basis for any developments of theoretical thermodynamic models. Scientists and engineers in academic and industrial research need such data and will benefit from a careful collection of existing data. However, the database for polymer solutions is still modest in comparison with the enormous amount of data for low-molecular mixtures, and the specialized database for polymer solutions is even smaller. On the other hand, there is still a strong and continuing interest in thermodynamic properties of copolymer solutions, and during the last ten years after the former *CRC Handbook of Thermodynamic Data of Copolymer Solutions* was published, a large amount of new experimental data has been published, that is now provided with this new *CRC Handbook of Phase Equilibria and Thermodynamic Data of Copolymer Solutions*.

Basic information on polymers can still be found in the *Polymer Handbook* (1999BRA), and there is also a chapter on properties of polymers and polymer solutions in the *CRC Handbook of Chemistry and Physics* (2009LID1). Older data for copolymer solutions can be found in 1992WEN and 1993DAN, but also in the data books written by the author of this *Handbook* (1994WOH, 2001WOH, 2004WOH, 2005WOH, 2006WOH, 2008WOH). The *Handbook* does not present theories and models for polymer solution thermodynamics. Other publications (1971YAM, 1990FUJ, 1990KAM, 1999KLE, 1999PRA, and 2001KON) can serve as starting points for investigating those issues.

The data within this book are divided into six chapters:

- Vapor-Liquid Equilibrium (VLE) Data and Gas Solubilities of Copolymer Solutions
- Liquid-Liquid Equilibrium (LLE) Data of Copolymer Solutions
- High-Pressure Fluid Phase Equilibrium (HPPE) Data of Copolymer Solutions
- Enthalpy Changes for Copolymer Solutions
- PVT Data of Copolymers and Solutions
- Second Virial Coefficients (A_2) of Copolymer Solutions

Data from investigations applying to more than one chapter are divided and appear in the relevant chapters. Data are included only if numerical values were published or authors provided their results by personal communication (and I wish to thank all those who did so). No digitized data have been included in this data collection. However, every data chapter is completed by a table that includes systems and references for data published only in graphical form as phase diagrams or related figures.

1.2. Experimental methods involved

Besides the common progress in instrumentation and computation, no remarkable new developments have been made with respect to the experimental methods involved here. So, a short summary of this chapter should be sufficient for the Handbook. The necessary equations are given together with some short explanations only.

Vapor-liquid equilibrium (VLE) measurements

Investigations on vapor-liquid equilibrium of polymer solutions can be made by various methods:

1. Absolute vapor pressure measurement
2. Differential vapor pressure measurement
3. Isopiestic sorption/desorption methods, i.e., gravimetric sorption, piezoelectric sorption, or isothermal distillation
4. Inverse gas-liquid chromatography (IGC) at infinite dilution, IGC at finite concentrations, and headspace gas chromatography (HSGC)
5. Steady-state vapor-pressure osmometry (VPO)

Experimental techniques for vapor pressure measurements were reviewed in 1975BON and 2000WOH. Methods and results of the application of IGC to polymers and polymer solutions were reviewed in 1976NES, 1988NES, 1989LLO, 1989VIL, and 1991MU1. Reviews on ebulliometry and/or vapor-pressure osmometry can be found in 1974TOM, 1975GLO, 1987COO, 1991MAY, and 1999PET.

VLE-experiments lead either to pressure vs. composition data or to activity vs. composition data. In the low-pressure region, solvent activities, a_A, are commonly determined from measured vapor pressures by:

$$a_A = (P_A / P_A^s) \exp\left[\frac{(B_{AA} - V_A^L)(P - P_A^s)}{RT}\right] \quad (1)$$

where:

a_A	activity of solvent A
B_{AA}	second virial coefficient of the pure solvent A at temperature T
P	system pressure
P_A	partial vapor pressure of the solvent A at temperature T
P_A^s	saturation vapor pressure of the pure liquid solvent A at temperature T
R	gas constant
T	measuring temperature
V_A^L	molar volume of the pure liquid solvent A at temperature T

Quite often, the exponential correction term is neglected, and only the ratio of partial pressure to pure solvent vapor pressure, P_A/P_A^s, is given.

IGC-experiments give specific retention volumina.

$$V_{net} = V_r - V_{dead} \tag{2}$$

where:

V_{net}	net retention volume
V_r	retention volume
V_{dead}	retention volume of the inert marker gas, dead retention, gas holdup

These net retention volumes are reduced to specific retention volumes, $V_g^{\,0}$, by division of equation (1) with the mass of the liquid (here the liquid is the molten copolymer). They are corrected for the pressure difference between column inlet and outlet pressure, and reduced to a temperature $T_0 = 273.15$ K.

$$V_g^{\,0} = \left(\frac{V_{net}}{m_B}\right)\left(\frac{T_0}{T}\right)\frac{3(P_{in}/P_{out})^2 - 1}{2(P_{in}/P_{out})^3 - 1} \tag{3}$$

where:

$V_g^{\,0}$	specific retention volume corrected to 0°C = 273.15 K
m_B	mass of the copolymer in the liquid phase within the column
P_{in}	column inlet pressure
P_{out}	column outlet pressure
T_0	reference temperature = 273.15 K

Theory of GLC provides the relation between $V_g^{\,0}$ and thermodynamic data for the low-molecular component (solvent A) at infinite dilution (superscript ∞):

$$\left(\frac{P_A}{x_A^L}\right)^{\infty} = \frac{RT_0}{V_g^{\,0} M_B} \quad \text{or} \quad \left(\frac{P_A}{w_A^L}\right)^{\infty} = \frac{RT_0}{V_g^{\,0} M_A} \tag{4}$$

where:

M_A	molar mass of the solvent A
M_B	molar mass of the liquid (molten) polymer B (i.e., M_n)
P_A	partial vapor pressure of the solvent A at temperature T
x_A^L	mole fraction of solvent A in the liquid solution
w_A^L	mass fraction of solvent A in the liquid solution

The activity coefficients at infinite dilution read, if we neglect interactions to and between carrier gas molecules (which are normally helium):

$$\gamma_A^{\infty} = \left(\frac{RT_0}{V_g^{\,0} M_B P_A^s}\right)\exp\left[\frac{P^s{}_A(V_A^L - B_{AA})}{RT}\right] \tag{5}$$

$$\Omega_A^{\infty} = \left(\frac{RT_0}{V_g^0 M_A P_A^s}\right) \exp\left[\frac{P_A^s (V_A^L - B_{AA})}{RT}\right] \tag{6}$$

where:

γ_A activity coefficient of the solvent A in the liquid phase with activity $a_A = x_A\gamma_A$

Ω_A mass fraction-based activity coefficient of the solvent A in the liquid phase with activity $a_A = w_A\Omega_A$

One should keep in mind that mole fraction-based activity coefficients γ_A become very small values for common polymer solutions and reach a value of zero for $M_B \to\infty$, which means a limited applicability at least to oligomer solutions. Therefore, the common literature provides only mass fraction-based activity coefficients for (high-molecular) polymer + (low-molecular) solvent pairs. The molar mass M_B of the polymeric liquid is an average value (M_n) according to the usual molar-mass distribution of polymers.

Furthermore, thermodynamic VLE data from GLC measurements are provided in the literature as values for $(P_A/w_A)^{\infty}$, i.e., classical mass fraction-based Henry's constants (if assuming ideal gas phase behavior):

$$H_{A,B} = \left(\frac{P_A}{w_A^L}\right)^{\infty} = \frac{RT_0}{V_g^0 M_A} \tag{7}$$

Since $V_{net} = V_r - V_{dead}$, the marker gas is assumed to not be retained by the polymer stationary phase and will elute at a retention time that is usually very small in comparison with those of the samples investigated. However, for small retention volumes, values for the mass fraction-based Henry's constants should be corrected for the solubility of the marker gas (1976LIU). The apparent Henry's constant is obtained from equation (7) above.

$$H_{A,B} = H_{A,B}^{app}\left[1 + \frac{M_A H_{A,B}^{app}}{M_{ref} H_{A,ref}}\right]^{-1} \tag{8}$$

M_{ref} is the molar mass of the marker gas. The Henry's constant of the marker gas itself, determined by an independent experiment, need not be known very accurately, as it is usually much larger than the apparent Henry's constant of the sample.

VPO-experiments give temperature differences for determining solvent activities:

$$\Delta T^{st} = -k_{VPO}\frac{RT^2}{\Delta_{LV}H_{0A}} \ln a_A \tag{9}$$

where:

k_{VPO}	VPO-specific constant (must be determined separately)
ΔT^{st}	temperature difference between solution and solvent drops in the steady state
$\Delta_{LV}H_{0A}$	molar enthalpy of vaporization of the pure solvent A at temperature T

The steady state must be sufficiently near the vapor-liquid equilibrium and linear non-equilibrium thermodynamics is valid then.

Liquid-liquid equilibrium (LLE) measurements

To understand the results of LLE experiments in polymer solutions, one has to take into account the strong influence of polymer distribution functions on LLE because fractionation occurs during demixing. Fractionation takes place with respect to molar mass distribution as well as to chemical distribution. Fractionation during demixing leads to some special effects by which the LLE phase behavior differs from that of an ordinary, strictly binary mixture because a common polymer solution is a multicomponent system. *Cloud-point curves* are measured instead of binodals; and per each individual feed concentration of the mixture, *two parts of a coexistence curve* occur below (for upper critical solution temperature, UCST, behavior) or above the cloud-point curve (for lower critical solution temperature, LCST, behavior), i.e., produce an infinite number of coexistence data. Distribution functions of the feed polymer belong only to cloud-point data. On the other hand, each pair of coexistence points is characterized by two new and different distribution functions in each coexisting phase. The critical concentration is the only feed concentration where both parts of the coexistence curve meet each other on the cloud-point curve at the critical point that belongs to the feed polymer distribution function. The threshold point (maximum or minimum corresponding to either UCST or LCST behavior) temperature (or pressure) is not equal to the critical point, since the critical point is to be found at a shoulder of the cloud-point curve. *Phase-volume-ratio method* (1968KON) or *coexistence concentration plot* (1969WOL) gives the critical point as the intersection of cloud-point curve and shadow curve. For more details, we refer to 1968KON, 1972KON, 2001KON, 2008WOH.

Treating copolymer solutions with distribution functions by continuous thermodynamics and procedures to measure and calculate liquid-liquid equilibria of such systems is reviewed in 1990RAE.

High-pressure phase equilibrium (HPPE) measurements

Experimental methods for high-pressure fluid phase equilibria in copolymer solutions follow the same lines as for VLE-, LLE-, or VLLE-experiments. The experimental equipment follows on the same techniques, however, extended to high pressure conditions. For more details, we refer to 1994MCH, 1997KIR, 1999KIR, and 2005WOH.

The solvents are in many cases supercritical fluids, i.e., gases/vapors above their critical temperature and pressure. Data were measured mainly for two kinds of solutions: solutions in supercritical CO_2 (and some other fluids) or solutions in supercritical monomers.

Measurement of enthalpy changes in copolymer solutions

Experiments on enthalpy changes in binary copolymer solutions can be made within common microcalorimeters by applying one of the following three methods:

1. Measurement of the enthalpy change caused by solving a given amount of the solute copolymer in an (increasing) amount of solvent, i.e., the solution experiment
2. Measurement of the enthalpy change caused by mixing a given amount of a concentrated copolymer solution with an amount of pure solvent, i.e., the dilution experiment
3. Measurement of the enthalpy change caused by mixing a given amount of a liquid/molten copolymer with an amount of pure solvent, i.e., the mixing experiment

Care must be taken for polymer solutions with respect to the resolution of the instrument, which has to be higher than for common solutions with larger enthalpic effects. Usually employed calorimeters for such purposes are the Calvet-type calorimeters based on heat-flux principle. Details can be found in 1984HEM and 1994MAR.

The (integral) enthalpy of mixing or the (integral) enthalpy of solution of a binary system is the amount of heat which must be supplied when n_A mole of pure solvent A and n_B mole of pure copolymer B are combined to form a homogeneous mixture/solution in order to keep the total system at constant temperature and pressure.

$$\Delta_M h = n_A H_A + n_B H_B - (n_A H_{0A} + n_B H_{0B}) \qquad (10a)$$

$$\Delta_{sol} h = n_A H_A + n_B H_B - (n_A H_{0A} + n_B H_{0B}) \qquad (10b)$$

where:

$\Delta_M h$	(integral) enthalpy of mixing
$\Delta_{sol} h$	(integral) enthalpy of solution
H_A	partial molar enthalpy of solvent A
H_B	partial molar enthalpy of copolymer B
H_{0A}	molar enthalpy of pure solvent A
H_{0B}	molar enthalpy of pure copolymer B
n_A	amount of substance of solvent A
n_B	amount of substance of copolymer B

From thermodynamic reasons follows that the change $\Delta_M H$ of the molar (or specific or segment molar) enthalpy in an isothermal-isobaric mixing process is also the molar (or specific or segment molar) excess enthalpy, H^E, of the mixture. The dependence of H^E upon temperature, T, and pressure, P, permits the correlation of such data with excess heat capacities, $C_p^{\ E}$, and excess volumes, V^E.

$$\left(\partial H^E / \partial T\right)_P = C_P^E \qquad (11)$$

$$\left(\partial H^E / \partial P\right)_T = V^E - T\left(\partial V^E / \partial T\right)_P \qquad (12)$$

Partial molar enthalpies are given by:

$$\Delta_{sol}H_B = (\partial\Delta_{sol}h/\partial n_B)_{P,T,n_j} = H_B - H_{0B} \quad (13a)$$

$$\Delta_M H_B = (\partial\Delta_M h/\partial n_B)_{P,T,n_j} = H_B - H_{0B} \quad (13b)$$

Partial specific enthalpies are given by:

$$\Delta_{sol}H_B = (\partial\Delta_{sol}h/\partial m_B)_{P,T,m_j} \quad (13c)$$

$$\Delta_M H_B = (\partial\Delta_M h/\partial m_B)_{P,T,m_j} \quad (13d)$$

Intermediary enthalpy of dilution are often measured instead of excess enthalpies, i.e., the enthalpy effect obtained if solvent A is added to an existing homogeneous polymer solution. The extensive intermediary enthalpy of dilution is the difference between two values of the enthalpy of the polymer solution corresponding to the concentrations of the polymer solution at the beginning and at the end of the dilution process:

$$\Delta_{dil}H^{12} = H^{(2)} - H^{(1)} \quad (14)$$

with

$$H^{(1)} = n_A^{(1)}H_A^{(1)} + n_B H_B^{(1)} \quad (15a)$$
$$H^{(2)} = n_A^{(2)}H_A^{(2)} + n_B H_B^{(2)} \quad (15b)$$

and

$$n_A^{(2)} = n_A^{(1)} + \Delta n_A \quad (16)$$

where:

$\Delta_{dil}H^{12}$	(extensive) intermediary enthalpy of dilution
$H^{(1)}, H^{(2)}$	enthalpies of the polymer solution before and after the dilution step
$H_A^{(1)}, H_A^{(2)}$	partial molar enthalpies of solvent A before and after the dilution step
$H_B^{(1)}, H_B^{(2)}$	partial molar enthalpies of polymer B before and after the dilution step
$n_A^{(1)}$	amount of solvent in the solution before the dilution step
$n_A^{(2)}$	amount of solvent in the solution after the dilution step
Δn_A	amount of solvent added to solution (1)
n_B	amount of polymer in all solutions.

$\Delta_{dil}H^{12}$ is not directly related to $\Delta_M H$ but to $(\partial\Delta_M H/\partial n_A)_{P,T,n_j}$ by:

$$\Delta_{dil}H^{12} = \int_{n_A^{(1)}}^{n_A^{(2)}} (\partial\Delta_M H/\partial n_A)_{P,T,n_j}\, dn_A \quad (17)$$

Generally, it is known that partial molar enthalpies of mixing (or dilution) of the solvent can also be determined from the temperature dependence of the activity of the solvent, a_A:

$$\Delta_M H_A = R[\partial \ln a_A/\partial(1/T)]_P \quad (18)$$

Enthalpy data from light scattering, osmometry, vapor pressure or vapor sorption measurements, and demixing experiments can be found in the literature. However, agreement between enthalpy changes measured by calorimetry and results determined from the temperature dependence of solvent activity data is often of limited quality. In this *Handbook*, data for $\Delta_M H_A^\infty$ determined by inverse gas-liquid chromatography (IGC) have been included.

$$\Delta_M H_A^\infty = R\,[\partial \ln \Omega_A^\infty / \partial(1/T)]_P \tag{19}$$

Measurement of PVT-behavior of copolymer melt and of excess volume in solution

There are two widely practiced methods for the *PVT* measurement of polymers and copolymers:

1. Piston-die technique
2. Confining fluid technique

which were described in detail by Zoller in papers and books (e.g., 1986ZOL, 1995ZOL).

The tables in Chapter 6 provide specific volumes neither at or below the melting transition of semicrystalline materials nor at or below the glass transition of amorphous samples, since *PVT* data of solid polymer samples are non-equilibrium data and depend on sample history and experimental procedure (which will not be discussed here). Therefore, only equilibrium data for the liquid/molten state are tabulated. Their common accuracy (standard deviation) is about 0.001 cm^3/g in specific volume, 0.1 K in temperature and 0.005*P in pressure (1995ZOL).

Measurement of densities for polymer solutions is usually made today by U-tube vibrating densimeters. Such instruments are commercially available.

Excess volumes at temperature T and pressure p are determined by:

$$V^E_{spec} = V_{spec} - (w_A V_{0A,\,spec} + w_B V_{0B,\,spec}) \tag{20a}$$

or

$$V^E = (x_A M_A + x_B M_B)/\rho - (x_A M_A/\rho_A + x_B M_B/\rho_B) \tag{20b}$$

where:

V_{spec}	specific volume of the polymer solution
V^E, V^E_{spec}	molar or specific excess volume
$V_{0A,\,spec}$, $V_{0B,\,spec}$	specific volume of pure polymer (B) and pure solvent (A)
w_A and w_B	mass fraction of polymer (B) and solvent (A) (definition see below)
x_A and x_B	mole fraction of polymer (B) and solvent (A) (definition see below)
ρ	density of the polymer solution
ρ_A, ρ_B	density of pure polymer (B) or pure solvent (A)

The second term in Eqs. (20a, b) corresponds to the *ideal volume,* V^{id} , of the polymer solution.

Determination of second virial coefficients A_2

There are a couple of methods for the experimental determination of the second virial coefficient: colligative properties (vapor pressure depression, freezing point depression, boiling point increase, membrane osmometry), scattering methods (classical light scattering, X-ray scattering, neutron scattering), sedimentation velocity and sedimentation equilibrium. Details of the special experiments can be found in many textbooks and will not be repeated here (for example, 1972HUG, 1974TOM, 1975CAS, 1975FUJ, 1975GLO, 1987ADA, 1987BER, 1987COO, 1987KRA, 1987WIG, 1991CHU, 1991MAY, 1991MU2, 1992HAR, and 1999PET).

The *vapor pressure depression* of the solvent in a binary copolymer solution, i.e., the difference between the saturation vapor pressure of the pure solvent and the corresponding partial pressure in the solution, $\Delta P_A = P_A^s - P_A$, is expressed as:

$$\frac{\Delta P_A}{P_A} = V_A^L c_B \left[\frac{1}{M_n} + A_2 c_B + A_3 c_B^{\ 2} + \ldots \right] \tag{21}$$

where:

$A_2, A_3, \ldots$	second, third, ... osmotic virial coefficients at temperature T
c_B	(mass/volume) concentration at temperature T
M_n	number-average relative molar mass of the copolymer
ΔP_A	$P_A^s - P_A$, vapor pressure depression of the solvent A at temperature T
V_A^L	molar volume of the pure liquid solvent A at temperature T

The *freezing point depression*, $\Delta_{SL}T_A$, is:

$$\Delta_{SL}T_A = E_{SL} c_B \left[\frac{1}{M_n} + A_2 c_B + A_3 c_B^{\ 2} + \ldots \right] \tag{22}$$

and the *boiling point increase*, $\Delta_{LV}T_A$, is:

$$\Delta_{LV}T_A = E_{LV} c_B \left[\frac{1}{M_n} + A_2 c_B + A_3 c_B^{\ 2} + \ldots \right] \tag{23}$$

where:

E_{LV}	ebullioscopic constant
E_{SL}	cryoscopic constant
$\Delta_{SL}T_A$	freezing point temperature difference between pure solvent and solution, i.e., ${}_{SL}T_A^0 - {}_{SL}T_A$
$\Delta_{LV}T_A$	boiling point temperature difference between solution and pure solvent, i.e., ${}_{LV}T_A - {}_{LV}T_A^0$

The *osmotic pressure*, π, can be described as:

$$\frac{\pi}{c_B} = RT\left[\frac{1}{M_n} + A_2 c_B + A_3 c_B^{\ 2} + \ldots\right] \tag{24}$$

In the *dilute concentration region*, the virial equation is usually truncated after the second virial coefficient which leads to a linear relationship. A linearized relation over a wider concentration range can be constructed if the Stockmayer-Casassa relation between A_2 and A_3 is applied:

$$A_3 M_n = \left(\frac{A_2 M_n}{2}\right)^2 \tag{25}$$

$$\left(\frac{\pi}{c_2}\right)^{0.5} = \left(\frac{RT}{M_n}\right)^{0.5}\left[1 + \frac{A_2 M_n}{2} c_2\right] \tag{26}$$

Scattering methods enable the determination of A_2 via the common relation:

$$\frac{K c_B}{R(q)} = \frac{1}{M_w P_z(q)} + 2A_2 Q(q) c_B + \ldots \tag{27}$$

with

$$q = \frac{4\pi}{\lambda}\sin\frac{\theta}{2} \tag{28}$$

where:

K	a constant that summarizes the optical parameters of a scattering experiment
M_w	mass-average relative molar mass of the copolymer
$P_z(q)$	z-average of the scattering function
q	scattering vector
$Q(q)$	function for the q-dependence of A_2
$R(q)$	excess intensity of the scattered beam at the value q
λ	wavelength
θ	scattering angle

Depending on the chosen experiment (light, X-ray or neutron scattering), the constant K is to be calculated from different relations. For details see the corresponding textbooks (1972HUG, 1975CAS, 1982GLA, 1986HIG, 1987BER, 1987KRA, 1987WIG, and 1991CHU).

1.3. Guide to the data tables

Characterization of the polymers

Copolymers vary by a number of characterization variables. The molar mass and their distribution function are the most important variables, and also the chemical distribution and the average chemical composition have to be given. However, tacticity, sequence distribution, branching, and end groups determine their thermodynamic behavior in solutions too. Unfortunately, much less information is provided with respect to the copolymers that were applied in most of the thermodynamic investigations in the original literature. In many cases, the samples are characterized only by one or two molar mass averages, the average chemical composition, and some additional information (e.g., T_g, T_m, ρ_B, or how and where they were synthesized). Sometimes even this information is missed.

The molar mass averages are defined as follows:

number average M_n

$$M_n = \frac{\sum_i n_{B_i} M_{B_i}}{\sum_i n_{B_i}} = \frac{\sum_i w_{B_i}}{\sum_i w_{B_i} / M_{B_i}} \tag{29}$$

mass average M_w

$$M_w = \frac{\sum_i n_{B_i} M_{B_i}^2}{\sum_i n_{B_i} M_{B_i}} = \frac{\sum_i w_{B_i} M_{B_i}}{\sum_i w_{B_i}} \tag{30}$$

z-average M_z

$$M_z = \frac{\sum_i n_{B_i} M_{B_i}^3}{\sum_i n_{B_i} M_{B_i}^2} = \frac{\sum_i w_{B_i} M_{B_i}^2}{\sum_i w_{B_i} M_{B_i}} \tag{31}$$

viscosity average M_η

$$M_\eta = \left(\frac{\sum_i w_{B_i} M_{B_i}^a}{\sum_i w_{B_i}} \right)^{1/a} \tag{32}$$

where:

a	exponent in the viscosity-molar mass relationship
M_{Bi}	relative molar mass of the polymer species B_i
n_{Bi}	amount of substance of polymer species B_i
w_{Bi}	mass fraction of polymer species B_i

Measures for the polymer concentration

The following concentration measures are used in the tables of this *Handbook* (where B always denotes the main polymer, A denotes the solvent, and in ternary systems C denotes the third component):

mass/volume concentration

$$c_A = m_A/V \qquad c_B = m_B/V \tag{33}$$

mass fraction

$$w_A = m_A/\Sigma\, m_i \qquad w_B = m_B/\Sigma\, m_i \tag{34}$$

mole fraction

$$x_A = n_A/\Sigma\, n_i \qquad x_B = n_B/\Sigma\, n_i \qquad \text{with } n_i = m_i/M_i \text{ and } M_B = M_n \tag{35}$$

volume fraction

$$\varphi_A = (m_A/\rho_A)/\Sigma\,(m_i/\rho_i) \qquad \varphi_B = (m_B/\rho_B)/\Sigma\,(m_i/\rho_i) \tag{36}$$

segment fraction

$$\psi_A = x_A r_A/\Sigma\, x_i r_i \quad \psi_B = x_B r_B/\Sigma\, x_i r \tag{37}$$

base mole fraction

$$z_A = x_A r_A/\Sigma\, x_i r_i \quad z_B = x_B r_B/\Sigma\, x_i r_i \quad \text{with } r_B = M_B/M_0 \text{ and } r_A = 1 \tag{38}$$

where:

c_A	(mass/volume) concentration of solvent A
c_B	(mass/volume) concentration of copolymer B
m_A	mass of solvent A
m_B	mass of copolymer B
M_A	relative molar mass of the solvent A
M_B	relative molar mass of the copolymer B
M_n	number-average relative molar mass of the copolymer B
M_0	molar mass of a basic unit of the copolymer B
n_A	amount of substance of solvent A
n_B	amount of substance of copolymer B
r_A	segment number of the solvent A, usually $r_A = 1$
r_B	segment number of the copolymer B
V	volume of the liquid solution at temperature T
w_A	mass fraction of solvent A
w_B	mass fraction of copolymer B
x_A	mole fraction of solvent A
x_B	mole fraction of copolymer B
z_A	base mole fraction of solvent A
z_B	base mole fraction of copolymer B

φ_A	volume fraction of solvent A
φ_B	volume fraction of copolymer B
ρ_A	density of solvent A
ρ_B	density of copolymer B
ψ_A	segment fraction of solvent A
ψ_B	segment fraction of copolymer B

For high-molecular polymers, a mole fraction is not an appropriate unit to characterize composition. However, for oligomeric products with rather low molar masses, mole fractions were sometimes used. In the common case of a distribution function for the molar mass, $M_B = M_n$ is to be chosen. Mass fraction and volume fraction can be considered as special cases of segment fractions depending on the way by which the segment size is actually determined: $r_i/r_A = M_i/M_A$ or $r_i/r_A = V_i/V_A = (M_i/\rho_i)/(M_A/\rho_A)$, respectively. Classical segment fractions are calculated by applying $r_i/r_A = V_i^{\text{vdW}}/V_A^{\text{vdW}}$ ratios where hard-core van der Waals volumes, V_i^{vdW}, are taken into account. Their special values depend on the chosen equation of state or simply some group contribution schemes (e.g., 1968BON, 1990KRE) and have to be specified. Volume fractions imply a temperature dependence and, as they are defined in the equations above, neglect excess volumes of mixing and, very often, the densities of the polymer in the state of the solution are not known correctly. However, volume fractions can be calculated without the exact knowledge of the polymer molar mass (or its averages). Base mole fractions are seldom applied for copolymer systems. The value for M_0, the molar mass of a basic unit of the copolymer, has to be determined according to the corresponding average chemical composition. Sometimes it is chosen arbitrarily, however, and has to be specified.

Tables of experimental data

The data tables in each chapter are provided in order of the names of the copolymers. In this *Handbook*, usually source-based copolymer names are applied. These names are more common in use, and they are mostly given also in the original sources. For copolymers, their names were built by the two names of the comonomers which are connected by *-co-*, or more specifically by *-alt-* for alternating copolymers, by *-b-* for block copolymers, by *-g-* for graft copolymers, or by *-stat-* for statistical copolymers. Please note: In the original *Handbook* (2001WOH), copolymer names were given by simply separating the comonomers by a slash (/). Since this method of naming was criticized by a number of users, here names are applied as has already been done by the author in the newer *Handbooks* (e.g., 2005WOH, 2006WOH, 2008WOH). Structure-based names, for which details about their nomenclature can be found in the *Polymer Handbook* (1999BRA), are chosen in some single cases only. CAS index names for copolymers are usually not applied here. Latest IUPAC Recommendations for class names of polymers based on chemical structure and molecular architecture are given in (2009BAR). A list of the copolymers in Appendix 1 utilizes the names as given in the chapters of this book.

Within types of copolymers the individual samples are ordered by their increasing average molar mass, and, by increasing average chemical composition. Subsequently, systems are ordered by their solvents. Solvents are listed alphabetically. When necessary, systems are ordered by increasing temperature. In ternary systems, ordering is additionally made subsequently according to the name of the third component in the system. Each dataset begins with the lines for the solution components, e.g., in binary systems

Polymer (B):	**poly(ethylene-*co*-1-octene)**		**2005LE1**
Characterization:	M_n/g.mol^{-1} = 64810, M_w/g.mol^{-1} = 135500, 15.3 mol% 1-octene, T_m/K = 323.2, T_g/K = 214.2, ρ = 0.86 g/cm^3, DuPont Dow Elastomers Corporation		
Solvent (A):	**n-pentane**	C_5H_{12}	**109-66-0**

where the copolymer sample is given in the first line together with the reference. The second line provides then the characterization available for the copolymer sample. The following line gives the solvent's chemical name, molecular formula, and CAS registry number.

In ternary and quaternary systems, the following lines are either for a second solvent or a second polymer, e.g., in ternary systems with two solvents

Polymer (B):	**poly(ethylene oxide-*b*-propylene oxide)**		**1991SA2**
Characterization:	M_n/g.mol^{-1} = 3438, 24.8 mol% ethylene oxide, Polysciences, Inc., Warrington, PA		
Solvent (A):	**water**	H_2O	**7732-18-5**
Solvent (C):	***N*-methylacetamide**	C_3H_7NO	**79-16-3**

or, e.g., in ternary systems with a second polymer

Polymer (B):	**poly(ethylene oxide-*co*-propylene oxide)**		**2004PER**
Characterization:	M_n/g.mol^{-1} = 3900, 50.0 mol% ethylene oxide		
Solvent (A):	**water**	H_2O	**7732-18-5**
Polymer (C):	**hydroxypropylstarch**		
Characterization:	M_n/g.mol^{-1} = 100000, Reppe Glykos AB, Vaxjo, Sweden		

or, e.g., in quaternary (or higher) systems like

Polymer (B):	**poly(ethylene oxide-*b*-propylene oxide-*b*-ethylene oxide)**		**2005SIL**
Characterization:	M_n/g.mol^{-1} = 1706, M_w/g.mol^{-1} = 1945, 50.0 wt% ethylene oxide, (EO)11-(PO)16-(EO)11		
Solvent (A):	**water**	H_2O	**7732-18-5**
Salt (C):	**dipotassium phosphate**	K_2HPO_4	**7758-11-4**
Salt (D):	**potassium hydroxide**	KOH	**1310-58-3**

There are some exceptions from this type of presentation within the tables for the UCST and LCST data. These tables are prepared in the forms as chosen in (2008WOH).

The originally measured data for each single system are sometimes listed together with some comment lines if necessary. The data are usually given as published, but temperatures are always given in K. Pressures are sometimes recalculated into kPa or MPa.

The final day for including data into this *Handbook* was December, 31, 2009.

1.4. List of symbols

a	exponent in the viscosity-molar mass relationship
a_A	activity of solvent A
B	parameter of the Tait equation
B_{AA}	second virial coefficient of the pure solvent A at temperature T
c_A	(mass/volume) concentration of solvent A
c_B	(mass/volume) concentration of copolymer B
C	parameter of the Tait equation
E_{LV}	ebullioscopic constant
E_{SL}	cryoscopic constant
h_D	distance from the center of rotation
H^E	excess enthalpy = $\Delta_M H$ = enthalpy of mixing
H_A	partial molar enthalpy of solvent A
H_B	partial molar (or specific) enthalpy of copolymer B
H_{0A}	molar enthalpy of pure solvent A
H_{0B}	molar (or specific) enthalpy of pure copolymer B
$H_{A,B}$	classical mass fraction Henry's constant of solvent vapor A in the molten copolymer B
$\Delta_{dil}H^{12}$	(integral) intermediary enthalpy of dilution (= $\Delta_M H^{(2)} - \Delta_M H^{(1)}$)
$\Delta_M H$	(integral) enthalpy of mixing
$\Delta_{sol}H$	(integral) enthalpy of solution
${}^{int}\Delta_M H_A$	integral enthalpy of mixing of solvent A (= integral enthalpy of dilution)
$\Delta_M H_A$	partial molar enthalpy of mixing of the solvent A (= differential enthalpy of dilution)
$\Delta_M H_A^\infty$	partial molar enthalpy of mixing at infinite dilution of the solvent A
${}^{int}\Delta_{sol}H_A$	integral enthalpy of solution of solvent A
$\Delta_{sol}H_A$	partial molar enthalpy of solution of the solvent A
$\Delta_{sol}H_A^\infty$	first integral enthalpy of solution of solvent A (= $\Delta_M H_A^\infty$ in the case of liquid/molten copolymers <u>and</u> a liquid solvent, i.e., it is different from the values for solutions of solvent vapors or gases in a liquid/molten copolymer $\Delta_{sol}H_{A(vap)}^\infty$)
$\Delta_{sol}H_{A(vap)}^\infty$	first integral enthalpy of solution of the vapor of solvent A (with $\Delta_{sol}H_{A(vap)}^\infty = \Delta_M H_A^\infty - \Delta_{LV}H_{0A}$)
$\Delta_{LV}H_{0A}$	molar enthalpy of vaporization of the pure solvent A at temperature T
${}^{int}\Delta_M H_B$	integral enthalpy of mixing of copolymer B
$\Delta_M H_B$	partial molar (or specific) enthalpy of mixing of copolymer B
$\Delta_M H_B^\infty$	partial molar (or specific) enthalpy of mixing at infinite dilution of copolymer B
${}^{int}\Delta_{sol}H_B$	integral enthalpy of solution of copolymer B
$\Delta_{sol}H_B$	partial molar (or specific) enthalpy of solution of copolymer B
$\Delta_{sol}H_B^\infty$	first integral enthalpy of solution of copolymer B ($\Delta_M H_B^\infty$ in the case of liquid/molten B)
k_{VPO}	VPO-specific constant (must be determined separately)
K	a constant that summarizes the optical parameters of a scattering experiment

m_A	mass of solvent A
m_B	mass of copolymer B
M	relative molar mass
M_A	molar mass of the solvent A
M_B	molar mass of the copolymer B
M_n	number-average relative molar mass of the copolymer B
M_w	mass-average relative molar mass of the copolymer B
M_η	viscosity-average relative molar mass of the copolymer B
M_z	z-average relative molar mass of the copolymer B
M_0	molar mass of a basic unit of the copolymer B
MI	melting index of the copolymer B
n_A	amount of substance of solvent A
n_B	amount of substance of copolymer B
P	pressure
P_0	standard pressure (= 0.101325 MPa)
P_{crit}	critical pressure
P_A	partial vapor pressure of the solvent A at temperature T
P_A^s	saturation vapor pressure of the pure liquid solvent A at temperature T
ΔP_A	$P_A^s - P_A$, vapor pressure depression of the solvent A at temperature T
P_{in}	column inlet pressure in IGC
P_{out}	column outlet pressure in IGC
$P_z(q)$	z-average of the scattering function
q	scattering vector
$Q(q)$	function for the q-dependence of A_2
R	gas constant
$R(q)$	excess intensity of the scattered beam at the value q
r_A	segment number of the solvent A, usually $r_A = 1$
r_B	segment number of the copolymer B
s	sedimentation coefficient
T	(measuring) temperature
T_{crit}	critical temperature
T_g	glass transition temperature
T_m	melting transition temperature
T_0	reference temperature (= 273.15 K)
ΔT^{st}	temperature difference between solution and solvent drops in VPO
$\Delta_{SL}T_A$	freezing point temperature difference between pure solvent and solution, i.e., ${}_{SL}T_A^0 - {}_{SL}T_A$
$\Delta_{LV}T_A$	boiling point temperature difference between solution and pure solvent, i.e., ${}_{LV}T_A - {}_{LV}T_A^0$
V, V_{spec}	volume or specific volume at temperature T
V_0	reference volume
V^E	excess volume at temperature T
V^{vdW}	hard-core van der Waals volume
V_A^L	molar volume of the pure liquid solvent A at temperature T
V_{net}	net retention volume in IGC
V_r	retention volume in IGC
V_{dead}	retention volume of the (inert) marker gas, dead retention, gas holdup in IGC
V_g^0	specific retention volume corrected to 0°C in IGC

w_A	mass fraction of solvent A
w_B	mass fraction of copolymer B
$w_{B,\ crit}$	mass fraction of the copolymer B at the critical point
x_A	mole fraction of solvent A
x_B	mole fraction of copolymer B
z_A	base mole fraction of solvent A
z_B	base mole fraction of copolymer B
α	critical exponent
γ_A	activity coefficient of the solvent A in the liquid phase with activity $a_A = x_A\gamma_A$
λ	wavelength
φ_A	volume fraction of solvent A
φ_B	volume fraction of copolymer B
$\varphi_{B,\ crit}$	volume fraction of the copolymer B at the critical point
ρ	density (of the mixture) at temperature T
ρ_A	density of solvent A at temperature T
ρ_B	density of copolymer B at temperature T
ψ_A	segment fraction of solvent A
ψ_B	segment fraction of copolymer B
π	osmotic pressure
θ	scattering angle
ω	angular velocity
Ω_A	mass fraction-based activity coefficient of the solvent A in the liquid phase with activity $a_A = w_A\Omega_A$
Ω_A^∞	mass fraction-based activity coefficient of the solvent A at infinite dilution

1.5. References

1968BON Bondi, A., *Physical Properties of Molecular Crystals, Liquids and Glasses*, J. Wiley & Sons, New York, 1968.

1968KON Koningsveld, R. and Staverman, A.J., Liquid-liquid phase separation in multicomponent polymer solutions I and II, *J. Polym. Sci.*, Pt. A-2, 6, 305, 325, 1968.

1969WOL Wolf, B.A., Zur Bestimmung der kritischen Konzentration von Polymerlösungen, *Makromol. Chem.*, 128, 284, 1969.

1971YAM Yamakawa, H., *Modern Theory of Polymer Solutions*, Harper & Row, New York, 1971.

1972HUG Huglin, M.B., Ed., *Light Scattering from Polymer Solutions*, Academic Press, New York, 1972.

1972KON Koningsveld, R., Polymer Solutions and Fractionation, in *Polymer Science*, Jenkins, E.D., Ed., North-Holland, Amsterdam, 1972, 1047.

1974TOM Tombs, M.P. and Peacock, A.R., *The Osmotic Pressure of Macromolecules*, Oxford University Press, London, 1974.

1975BON Bonner, D.C., Vapor-liquid equilibria in concentrated polymer solutions, *Macromol. Sci. Rev. Macromol. Chem.*, C13, 263, 1975.

1975CAS Casassa, E.F. and Berry, G.C., Light scattering from solutions of macromolecules, in *Polymer Molecular Weights*, Marcel Dekker, New York, 1975, Pt. 1, 161.

1975FUJ Fujita, H., *Foundations of Ultracentrifugal Analysis*, J. Wiley & Sons, New York, 1975.

1975GLO Glover, C.A., Absolute colligative property methods, in *Polymer Molecular Weights*, Marcel Dekker, New York, 1975, Pt. 1, 79.

1976LIU Liu, D.D. and Prausnitz, J.M, Solubilities of gases and volatile liquids in polyethylene and in ethylene-vinyl acetate copolymers in the region 125-225 °C, *Ind. Eng. Chem. Fundam.*, 15, 330, 1976.

1976NES Nesterov, A.E. and Lipatov, J.S., *Obrashchennaya Gasovaya Khromatografiya v Termodinamike Polimerov*, Naukova Dumka, Kiev, 1976.

1982GLA Glatter, O. and Kratky, O., Eds., *Small-Angle X-Ray Scattering*, Academic Press, London, 1982.

1984HEM Hemminger, W. and Höhne, G., *Calorimetry: Fundamentals and Practice*, Verlag Chemie, Weinheim, 1984.

1986HIG Higgins, J.S. and Macconachie, A., Neutron scattering from macromolecules in solution, in *Polymer Solutions*, Forsman, W.C., Ed., Plenum Press, New York, 1986, 183.

1986ZOL Zoller, P., Dilatometry, in *Encyclopedia of Polymer Science and Engineering*, Vol. 5, 2nd ed., Mark, H. et al., Eds., J. Wiley & Sons, New York, 1986, 69.

1987ADA Adams, E.T., Osmometry, in *Encyclopedia of Polymer Science and Engineering*, Vol. 10, 2nd ed., Mark, H. et al., Eds., J. Wiley & Sons, New York, 1986, 636.

1987BER Berry, G.C., Light scattering, in *Encyclopedia of Polymer Science and Engineering*, Vol. 8, 2nd ed., Mark, H. et al., Eds., J. Wiley & Sons, New York, 1986, 721.

1987COO Cooper, A.R., Molecular weight determination, in *Encyclopedia of Polymer Science and Engineering*, Vol. 10, 2nd ed., Mark, H. et al., Eds., J. Wiley & Sons, New York, 1986, 1.

1987KRA Kratochvil, P., *Classical Light Scattering from Polymer Solutions*, Elsevier, Amsterdam, 1987.

1987WIG Wignall, G.D., Neutron scattering, in *Encyclopedia of Polymer Science and Engineering*, Vol. 10, 2nd ed., Mark, H. et al., Eds., J. Wiley & Sons, New York, 1986, 112.

1988NES Nesterov, A.E., *Obrashchennaya Gasovaya Khromatografiya Polimerov*, Naukova Dumka, Kiev, 1988.

1989LLO Lloyd, D.R., Ward, T.C., Schreiber, H.P., and Pizana, C.C., Eds., *Inverse Gas Chromato-graphy*, ACS Symposium Series 391, American Chemical Society, Washington, 1989.

1989VIL Vilcu, R. and Leca, M., *Polymer Thermodynamics by Gas Chromatography*, Elsevier, Amsterdam, 1989.

1990BAR Barton, A.F.M., *CRC Handbook of Polymer-Liquid Interaction Parameters and Solubility Parameters*, CRC Press, Boca Raton, 1990.

1990FUJ Fujita, H., *Polymer Solutions*, Elsevier, Amsterdam, 1990.

1990KAM Kamide, K., *Thermodynamics of Polymer Solutions*, Elsevier, Amsterdam, 1990.

1990KRE [Van] Krevelen, D.W., *Properties of Polymers*, 3rd ed., Elsevier, Amsterdam, 1990.

1990RAE Raetzsch, M.D. and Wohlfarth, Ch., Continuous thermodynamics of copolymer systems, *Adv. Polym. Sci.*, 98, 49, 1990.

1991CHU Chu, B., *Laser Light Scattering*, Academic Press, New York, 1991.

1991MAY Mays, J.W. and Hadjichristidis, N., Measurement of molecular weight of polymers by osmometry, in *Modern Methods of Polymer Characterization*, Barth, H.G. and Mays, J.W., Eds., J. Wiley & Sons, New York, 1991, 201.

1991MU1 Munk, P., Polymer characterization using inverse gas chromatography, in *Modern Methods of Polymer Characterization*, Barth, H.G. and Mays, J.W., Eds., J. Wiley & Sons, New York, 1991, 151.

1991MU2 Munk, P., Polymer characterization using the ultracentrifuge, in *Modern Methods of Polymer Characterization*, Barth, H.G. and Mays, J.W., Eds., J. Wiley & Sons, New York, 1991, 271.

1992HAR Harding, S.E., Rowe, A.J., and Horton, J.C., *Analytical Ultracentrifugation in Biochemistry and Polymer Science*, Royal Society of Chemistry, Cambridge, 1992.

1992WEN Wen, H., Elbro, H.S., and Alessi, P., *Polymer Solution Data Collection.* I. Vapor-liquid equilibrium; II. Solvent activity coefficients at infinite dilution; III. Liquid-liquid equlibrium, Chemistry Data Series, Vol. 15, DECHEMA, Frankfurt am Main, 1992.

1993DAN Danner, R.P. and High, M.S., *Handbook of Polymer Solution Thermodynamics*, American Institute of Chemical Engineers, New York, 1993.

1994MAR Marsh, K.N., Ed., *Experimental Thermodynamics, Volume 4, Solution Calorimetry*, Blackwell Science, Oxford, 1994.

1994MCH McHugh, M.A. and Krukonis, V.J., *Supercritical Fluid Extraction: Principles and Practice*, 2nd ed., Butterworth Publishing, Stoneham, 1994.

1994WOH Wohlfarth, Ch., *Vapour-Liquid Equilibrium Data of Binary Polymer Solutions: Physical Science Data*, 44, Elsevier, Amsterdam, 1994.

1995ZOL Zoller, P. and Walsh, D.J., *Standard Pressure-Volume-Temperature Data for Polymers*, Technomic Publishing, Lancaster, 1995.

1997KIR Kiran, E. and Zhuang, W., *Miscibility and Phase Separation of Polymers in Near- and Supercritical Fluids*, ACS Symposium Series 670, 2, 1997.

1997KOL Kolb, B. and Ettre, L.S., *Static Headspace Gas Chromatography: Theory and Practice*, Wiley-VCH, Weinheim, 1997.

1999BRA Brandrup, J., Immergut, E.H., and Grulke, E.A., Eds., *Polymer Handbook*, 4th ed., J. Wiley & Sons, New York, 1999.

1999KIR Kirby, C.F. and McHugh, M.A., Phase behavior of polymers in supercritical fluid solvents, *Chem. Rev.*, 99, 565, 1999.

1999PET Pethrick, R.A. and Dawkins, J.V., Eds., *Modern Techniques for Polymer Characterization*, J. Wiley & Sons, Chichester, 1999.

1999PRA Prausnitz, J.M., Lichtenthaler, R.N., and de Azevedo, E.G., *Molecular Thermodynamics of Fluid Phase Equilibria*, 3rd ed., Prentice Hall, Upper Saddle River, NJ, 1999.

2000WOH Wohlfarth, Ch., Methods for the measurement of solvent activity of polymer solutions, in *Handbook of Solvents*, Wypych, G., Ed., ChemTec Publishing, Toronto, 2000, 146.

2001KON Koningsveld, R., Stockmayer, W.H., and Nies, E., *Polymer Phase Diagrams*, Oxford University Press, Oxford, 2001.

2001WOH Wohlfarth, C., *CRC Handbook of Thermodynamic Data of Copolymer Solutions*, CRC Press, Boca Raton, 2001.

2004WOH Wohlfarth, C., *CRC Handbook of Thermodynamic Data of Aqueous Polymer Solutions*, CRC Press, Boca Raton, 2004.

2005WOH Wohlfarth, C., *CRC Handbook of Thermodynamic Data of Polymer Solutions at Elevated Pressures*, Taylor & Francis, CRC Press, Boca Raton, 2005.

2006WOH Wohlfarth, C., *CRC Handbook of Enthalpy Data of Polymer-Sovent Systems*, Taylor & Francis, CRC Press, Boca Raton, 2006.

2008WOH Wohlfarth, C., *CRC Handbook of Liquid-Liquid Equilibrium Data of Polymer Solutions*, Taylor & Francis, CRC Press, Boca Raton, 2008.

2009BAR Baron, M., Hellwich, K.-H., Hess, M., Horie, K., Jenkins, A.D., Jones, R.G., Kahovec, J., Kratochvil, P., Metanomski, W.V., Mormann, W., Stepto, R.F.T., Vohlidal, J., and Wilks, E.S., Glossary of class names of polymers based on chemical structure and molecular architecture (IUPAC Recommendations 2009), *Pure Appl. Chem.*, 81, 1131, 2008.

2009LID Lide, D.R., Ed., *CRC Handbook of Chemistry and Physics, Section 13: Polymer Properties*, 90th ed., Taylor & Francis, CRC Press, Boca Raton, 2009.

2. VAPOR-LIQUID EQUILIBRIUM (VLE) DATA AND GAS SOLUBILITIES OF COPOLYMER SOLUTIONS

2.1. Binary copolymer solutions

Polymer (B): **poly(acrylonitrile-*co*-butadiene)** **2008SER**
Characterization: 21 % acrylonitrile,
Scientific Polymer Products, Inc., Ontario, NY
Solvent (A): **n-hexane** C_6H_{14} **110-54-3**

Type of data: vapor-liquid equilibrium

T/K = 333.15

w_A	0.026	0.044	0.085	0.109	0.134	0.165	0.228	0.247
P_A/kPa	12.9	25.6	38.8	44.5	50.8	57.5	63.7	64.9

Polymer (B): **poly(acrylonitrile-*co*-butadiene)** **2008SER**
Characterization: 33 % acrylonitrile,
Scientific Polymer Products, Inc., Ontario, NY
Solvent (A): **n-hexane** C_6H_{14} **110-54-3**

Type of data: vapor-liquid equilibrium

T/K = 333.15

w_A	0.003	0.018	0.026	0.038
P_A/kPa	12.4	25.1	36.0	43.3

Polymer (B): **poly(acrylonitrile-*co*-butadiene)** **2008SER**
Characterization: 51 % acrylonitrile,
Scientific Polymer Products, Inc., Ontario, NY
Solvent (A): **n-hexane** C_6H_{14} **110-54-3**

Type of data: vapor-liquid equilibrium

T/K = 333.15

w_A	0.017	0.021	0.029	0.031	0.044	0.044	0.060	0.063
P_A/kPa	12.9	25.6	38.8	44.5	50.8	57.5	63.7	64.9

Polymer (B): **poly(butadiene-*b*-styrene)** **2009XIO**
Characterization: M_n/g.mol^{-1} = 93200, M_w/g.mol^{-1} = 109000, 23.3 mol% styrene, 4.7 mol% vinyl, 5.1 mol% cis, ρ(298.15 K) = 0.956 g/cm^3, synthesized and fractionated in the laboratory
Solvent (A): **tetrahydrofuran** C_4H_8O **109-99-9**

continued

continued

Type of data: vapor-liquid equilibrium

T/K = 298.15

φ_B	0.4730	0.4768	0.4771	0.4806	0.5753	0.6772	0.7727	0.7817	0.8480
P_A/P_{0A}	0.9257	0.9087	0.8952	0.8900	0.8228	0.7124	0.5778	0.5573	0.4214
φ_B	0.8492	0.8527	0.8547	0.8770	0.8834	0.8853	0.8879	0.8901	0.9004
P_A/P_{0A}	0.4165	0.4094	0.4061	0.3628	0.3387	0.3363	0.3162	0.3050	0.2880
φ_B	0.9046	0.9057	0.9406	0.9408	0.9469	0.9524	0.9557	0.9563	0.9581
P_A/P_{0A}	0.2763	0.2749	0.1592	0.1444	0.1454	0.1284	0.1078	0.1158	0.1005

T/K = 313.15

φ_B	0.47505	0.47506	0.57499	0.57514	0.57536	0.57624	0.67634	0.67734	0.67848
P_A/P_{0A}	0.91642	0.91178	0.83399	0.83247	0.82777	0.83288	0.71404	0.71436	0.70812
φ_B	0.67892	0.77766	0.78094	0.78179	0.78204	0.85458	0.85687	0.85688	0.85761
P_A/P_{0A}	0.71526	0.54951	0.54097	0.54052	0.53906	0.39281	0.38865	0.39138	0.37986
φ_B	0.90933	0.91009	0.91167	0.91348	0.96425	0.96436	0.96611	0.96729	
P_A/P_{0A}	0.25899	0.25643	0.25495	0.24684	0.10012	0.09788	0.09351	0.09418	

T/K = 328.15

φ_B	0.47200	0.47358	0.47365	0.57339	0.57403	0.57422	0.57425	0.67467	0.67643
P_A/P_{0A}	0.89922	0.89559	0.88948	0.81774	0.80963	0.83150	0.81156	0.70812	0.71272
φ_B	0.67646	0.67716	0.78169	0.78264	0.78353	0.78420	0.85518	0.85597	0.85887
P_A/P_{0A}	0.70803	0.71284	0.53469	0.53211	0.54045	0.53167	0.38300	0.38292	0.37387
φ_B	0.85927	0.90940	0.91008	0.91012	0.91251	0.96525	0.96584	0.96594	0.96642
P_A/P_{0A}	0.37798	0.24622	0.24883	0.24284	0.24303	0.08874	0.08430	0.07603	0.08752

Polymer (B): **poly(butylene succinate-*co*-butylene adipate)** **2000SA1**
Characterization: M_n/g.mol^{-1} = 53000, M_w/g.mol^{-1} = 180000, 20.0 mol% adipate, T_g/K = 231, T_m/K = 365, 25 wt% crystallinity
Solvent (A): **carbon dioxide** **CO_2** **124-38-9**

Type of data: gas solubility

T/K = 323.15

P/MPa	1.098	2.097	3.065	4.053	6.016	7.870	9.844
m_A/(g CO_2/kg polymer)	12.59	24.18	35.59	47.49	73.64	103.6	150.3

Comments: The isotherm at 323 K is below the melting temperature. Below T_m, only the amorphous parts absorb CO_2. Thus, the solubility can be recalculated with respect to the amount of the amorphous parts of PBS.

P/MPa	1.098	2.097	3.065	4.053	6.016	7.870	9.844
m_A/(g CO_2/kg amorphous)	16.79	32.24	47.45	63.32	98.19	138.1	200.4

continued

continued

T/K = 393.15

P/MPa	2.082	2.581	4.028	4.610	6.200	6.681	8.053
m_A/(g CO_2/kg polymer)	18.39	22.85	35.91	41.22	55.78	60.20	72.80
P/MPa	8.630	11.997	12.460	16.024	16.548	20.036	
m_A/(g CO_2/kg polymer)	78.14	108.4	112.7	142.3	146.7	174.1	

T/K = 423.15

P/MPa	2.102	2.676	4.118	4.684	6.098	6.700	8.177
m_A/(g CO_2/kg polymer)	14.51	18.51	28.63	32.63	42.64	46.92	57.49
P/MPa	8.699	12.314	16.027	16.572	20.074		
m_A/(g CO_2/kg polymer)	61.17	86.49	111.5	115.0	137.4		

T/K = 453.15

P/MPa	2.079	2.600	4.126	4.698	6.070	6.628	8.259
m_A/(g CO_2/kg polymer)	11.84	14.83	23.57	26.89	34.85	38.09	47.63
P/MPa	8.973	12.009	12.766	16.139	20.127		
m_A/(g CO_2/kg polymer)	51.50	69.20	73.49	92.35	113.6		

Polymer (B): **poly(butyl methacrylate-*co*-*N,N*-dimethylamino-ethyl methacrylate)** **2002KAN**

Characterization: M_n/g.mol^{-1} = 22550, M_w/g.mol^{-1} = 26380, 95.13 wt% butyl methacrylate, synthesized in the laboratory

Solvent (A): **benzene** **C_6H_6** **71-43-2**

Type of data: vapor-liquid equilibrium

T/K = 298.15

w_B	0.9484	0.9116	0.8717	0.8411	0.8160	0.7913	0.6624	0.5420	0.5260
P_A/P_{0A}	0.1771	0.3120	0.3894	0.4572	0.5305	0.6024	0.7700	0.8483	0.9056

T/K = 308.15

w_B	0.9422	0.9270	0.8931	0.8741	0.8409	0.7945	0.7645	0.7259	0.6875
P_A/P_{0A}	0.1997	0.3003	0.3938	0.4438	0.4787	0.5960	0.6496	0.7154	0.7815
w_B	0.6252	0.5791							
P_A/P_{0A}	0.8347	0.8752							

Polymer (B): **poly(butyl methacrylate-*co*-*N,N*-dimethylamino-ethyl methacrylate)** **2003WAN**

Characterization: M_w/g.mol^{-1} = 95700, 59.43 mol% butyl methacrylate

Solvent (A): **carbon dioxide** **CO_2** **124-38-9**

Type of data: gas solubility

continued

continued

T/K = 298.15

P/MPa	0.2	0.3	0.4	0.5	0.6	0.7	0.8	0.9	1.0
w_A	0.0054	0.0090	0.0122	0.0154	0.0185	0.0220	0.0255	0.0286	0.0317

Polymer (B): **poly(butyl methacrylate-*co*-*N,N*-dimethylaminoethyl methacrylate)** **2003WAN**
Characterization: M_w/g.mol^{-1} = 87800, 89.54 mol% butyl methacrylate
Solvent (A): **carbon dioxide** CO_2 **124-38-9**

Type of data: gas solubility

T/K = 298.15

P/MPa	0.2	0.3	0.4	0.5	0.6	0.7	0.8	0.9	1.0
w_A	0.0060	0.0091	0.0121	0.0152	0.0183	0.0216	0.0248	0.0278	0.0311

Polymer (B): **poly(butyl methacrylate-*co*-*N,N*-dimethylaminoethyl methacrylate)** **2003WAN**
Characterization: M_w/g.mol^{-1} = 90000, 93.94 mol% butyl methacrylate
Solvent (A): **carbon dioxide** CO_2 **124-38-9**

Type of data: gas solubility

T/K = 298.15

P/MPa	0.2	0.3	0.4	0.5	0.6	0.7	0.8	0.9	1.0
w_A	0.0055	0.0082	0.0115	0.0150	0.0183	0.0223	0.0256	0.0291	0.0326

Polymer (B): **poly(butyl methacrylate-*co*-*N,N*-dimethylaminoethyl methacrylate)** **2003WAN**
Characterization: M_w/g.mol^{-1} = 26400, 95.13 mol% butyl methacrylate
Solvent (A): **carbon dioxide** CO_2 **124-38-9**

Type of data: gas solubility

T/K = 298.15

P/MPa	0.2	0.3	0.4	0.5	0.6	0.7	0.8	0.9	1.0
w_A	0.0091	0.0133	0.0170	0.0213	0.0249	0.0291	0.0329	0.0368	0.0406

Polymer (B): **poly(butyl methacrylate-*co*-*N,N*-dimethylaminoethyl methacrylate)** **2002KAN**
Characterization: M_n/g.mol^{-1} = 84000, M_w/g.mol^{-1} = 96000, 59.43 wt% butyl methacrylate, synthesized in the laboratory
Solvent (A): **toluene** C_7H_8 **108-88-3**

Type of data: vapor-liquid equilibrium

continued

continued

T/K = 308.15

w_B	0.9865	0.9842	0.9774	0.9745	0.9552	0.9288	0.9187
P_A/P_{0A}	0.2710	0.3800	0.4728	0.5783	0.7673	0.8771	0.9541

Polymer (B): **poly(butyl methacrylate-*co*-*N,N*-dimethylaminoethyl methacrylate)** **2002KAN**
Characterization: M_n/g.mol^{-1} = 75000, M_w/g.mol^{-1} = 88000, 89.54 wt% butyl methacrylate, synthesized in the laboratory
Solvent (A): **toluene** **C_7H_8** **108-88-3**

Type of data: vapor-liquid equilibrium

T/K = 308.15

w_B	0.9795	0.9725	0.9598	0.9530	0.9455	0.9253	0.9034
P_A/P_{0A}	0.2710	0.3800	0.4728	0.5783	0.7673	0.8771	0.9541

Polymer (B): **poly(butyl methacrylate-*co*-*N,N*-dimethylaminoethyl methacrylate)** **2002KAN**
Characterization: M_n/g.mol^{-1} = 80500, M_w/g.mol^{-1} = 90000, 93.94 wt% butyl methacrylate, synthesized in the laboratory
Solvent (A): **toluene** **C_7H_8** **108-88-3**

Type of data: vapor-liquid equilibrium

T/K = 308.15

w_B	0.9736	0.9690	0.9645	0.9627	0.9476	0.9297	0.9108
P_A/P_{0A}	0.2710	0.3800	0.4728	0.5783	0.7673	0.8771	0.9541

Polymer (B): **poly(butyl methacrylate-*co*-*N,N*-dimethylaminoethyl methacrylate)** **2002KAN**
Characterization: M_n/g.mol^{-1} = 22550, M_w/g.mol^{-1} = 26380, 95.13 wt% butyl methacrylate, synthesized in the laboratory
Solvent (A): **toluene** **C_7H_8** **108-88-3**

Type of data: vapor-liquid equilibrium

T/K = 308.15

w_B	0.9956	0.9702	0.9395	0.9083	0.9059	0.8280	0.7807	0.6757
P_A/P_{0A}	0.1697	0.2975	0.3693	0.3967	0.4540	0.5655	0.7310	0.9832

Polymer (B): **poly(butyl methacrylate-*b*-perfluoroalkylethyl acrylate)** **2003WAN**
Characterization: M_w/g.mol^{-1} = 18900, 81.99 mol% butyl methacrylate, perfluoroalkyl = $(CF_2)_{8.6}CF_3$
Solvent (A): **carbon dioxide** **CO_2** **124-38-9**

continued

continued

Type of data: gas solubility

T/K = 298.15

P/MPa	0.2	0.3	0.4	0.5	0.6	0.7	0.8	0.9
w_A	0.0054	0.0080	0.0111	0.0151	0.0185	0.0221	0.0249	0.0318

Polymer (B): **poly(butyl methacrylate-*b*-perfluoroalkylethyl acrylate)** **2003WAN**
Characterization: M_w/g.mol^{-1} = 14900, 94.50 mol% butyl methacrylate, perfluoroalkyl = $(CF_2)_{8.6}CF_3$
Solvent (A): **carbon dioxide** **CO_2** **124-38-9**

Type of data: gas solubility

T/K = 298.15

P/MPa	0.2	0.3	0.4	0.5	0.6	0.7	0.8	0.9
w_A	0.0074	0.0111	0.0149	0.0185	0.0221	0.0258	0.0293	0.0328

Polymer (B): **poly(*N,N*-diethylaminoethyl methacrylate-*co*-methyl methacrylate)** **1996COR**
Characterization: 30.0 mol% methyl methacrylate, synthesized in the laboratory
Solvent (A): **water** **H_2O** **7732-18-5**

Type of data: vapor-liquid equilibrium

T/K = 298.15

w_B	0.725	0.826	0.870
a_A	0.984	0.968	0.936

Polymer (B): **poly(*N,N*-diethylaminoethyl methacrylate-*co*-methyl methacrylate)** **1996COR**
Characterization: 40.0 mol% methyl methacrylate, synthesized in the laboratory
Solvent (A): **water** **H_2O** **7732-18-5**

Type of data: vapor-liquid equilibrium

T/K = 298.15

w_B	0.860	0.885	0.920
a_A	0.984	0.968	0.936

Polymer (B): **poly(*N,N*-dimethyl acrylamide-*co*-*tert*-butylacrylamide)** **2009FOR**
Characterization: M_n/g.mol^{-1} = 1350, unspecified comonomer content, synthesized in the laboratory
Solvent (A): **water** **H_2O** **7732-18-5**

continued

continued

Type of data: vapor-liquid equilibrium

T/K = 308.15

w_B	0.45	0.47	0.49	0.51	0.53	0.54	0.57	0.58	0.60
a_A	0.82	0.74	0.72	.62	0.54	0.51	0.38	0.32	0.24
w_B	0.62	0.64	0.67						
a_A	0.17	0.15	0.11						

Polymer (B): **poly(*N,N*-dimethyl acrylamide-*co*-*tert*-butylacrylamide)** **2009FOR**

Characterization: M_n/g.mol^{-1} = 1700, unspecified comonomer content, synthesized in the laboratory

Solvent (A): **water** **H_2O** **7732-18-5**

Type of data: vapor-liquid equilibrium

T/K = 308.15

w_B	0.47	0.49	0.52	0.53	.54	0.56	0.58	0.61	0.62
a_A	0.84	0.77	0.69	0.70	0.69	0.63	0.58	0.55	0.51
w_B	0.63	0.65	0.69						
a_A	0.39	0.30	0.22						

Polymer (B): **poly(*N,N*-dimethyl acrylamide-*co*-*tert*-butylacrylamide)** **2009FOR**

Characterization: M_n/g.mol^{-1} = 2520, unspecified comonomer content, synthesized in the laboratory

Solvent (A): **water** **H_2O** **7732-18-5**

Type of data: vapor-liquid equilibrium

T/K = 308.15

w_B	0.48	0.51	0.53	0.56	0.59	0.61	0.63	0.65	0.67
a_A	0.89	0.83	0.78	0.75	0.71	0.65	0.58	0.53	0.45
w_B	0.69	0.72	0.74						
a_A	0.38	0.35	0.30						

Polymer (B): **poly(ethylene-*co*-1-butene)** **2006NAG**

Characterization: M_n/g.mol^{-1} = 43700, M_w/g.mol^{-1} = 52000, M_z/g.mol^{-1} = 59000, 4.1 mol% 1-butene, 2.05 ethyl branches per 100 backbone C-atomes, hydrogenated polybutadiene PBD 50000, was denoted as LLDPE, DSM, The Netherlands

Solvent (A): **n-hexane** **C_6H_{14}** **110-54-3**

Type of data: vapor-liquid equilibrium

continued

continued

w_B	0.0024	was kept constant						
T/K	411.396	416.30	421.22	426.08	430.92	436.09		
P_A/MPa	0.618	0.703	0.753	0.823	0.903	1.008		
w_B	0.0049	was kept constant						
T/K	410.94	415.91	425.79	430.74	435.63			
P_A/MPa	0.652	0.707	0.827	0.907	0.992			
w_B	0.0078	was kept constant						
T/K	410.94	415.92	425.79	430.74	435.64			
P_A/MPa	0.652	0.707	0.747	0.827	0.992			
w_B	0.0223	was kept constant						
T/K	411.58	416.42	421.41	426.12	431.12			
P_A/MPa	0.613	0.673	0.743	0.808	0.888			
w_B	0.0452	was kept constant						
T/K	411.199	416.162	421.063	425.985				
P_A/MPa	0.614	0.679	0.749	0.819				
w_B	0.0923	was kept constant						
T/K	411.65	416.54	421.46	431.31				
P_A/MPa	0.609	0.674	0.739	0.889				
w_B	0.1946	was kept constant						
T/K	411.092	415.975	420.932	425.915	430.949	435.888		
P_A/MPa	0.613	0.678	0.753	0.828	0.898	0.988		
w_B	0.2435	was kept constant						
T/K	420.924	425.802	430.584	435.601	440.595			
P_A/MPa	0.754	0.814	0.899	0.984	1.069			
w_B	0.3031	was kept constant						
T/K	411.50	416.36	421.27	426.22	431.21	436.10	441.10	446.00
P_A/MPa	0.667	0.752	0.837	0.907	0.972	1.047	1.147	1.252

Polymer (B): **poly(ethylene-*co*-1-butene)** **2002JIN**

Characterization: M_n/g.mol^{-1} = 89300, M_w/g.mol^{-1} = 391000, 26.3 mol% 1-butene, ρ = 0.900 g/cm^3, Ziegler-Natta catalyst $Mg(OEt)_2$/DIBP/$TiCl_4$-TEA

Solvent (A): **1-hexene** C_6H_{12} **592-41-6**

Type of data: vapor solubility

T/K = 323.15

P_A/atm	0.1	0.2	0.3	0.4
w_B	0.97867	0.95807	0.92913	0.88879

continued

continued

T/K = 333.15

P_A/atm	0.1	0.2	0.3	0.4
w_B	0.98406	0.96820	0.94774	0.92617

T/K = 343.15

P_A/atm	0.1	0.2	0.3	0.4
w_B	0.98746	0.97550	0.96300	0.94659

Comments: The copolymer weight fraction belongs to the amorphous fraction of the copolymer.

Polymer (B): **poly(ethylene-*co*-1-butene)** **2007NAG**
Characterization: M_n/g.mol^{-1} = 43700, M_w/g.mol^{-1} = 52000, M_z/g.mol^{-1} = 59000, 4.1 mol% 1-butene, 2.05 ethyl branches per 100 backbone C-atomes, hydrogenated polybutadiene PBD 50000, was denoted as LLDPE, DSM, The Netherlands
Solvent (A): **2-methylpentane** **C_6H_{14}** **107-83-5**

Type of data: vapor-liquid equilibrium

w_B	0.0020	was kept constant			
T/K	391.09	401.08	411.06		
P_A/MPa	0.480	0.595	0.725		
w_B	0.0053	was kept constant			
T/K	390.98	400.93	411.03		
P_A/MPa	0.489	0.604	0.739		
w_B	0.0103	was kept constant			
T/K	391.28	396.13	401.06	406.02	410.89
P_A/MPa	0.520	0.575	0.630	0.700	0.765
w_B	0.0503	was kept constant			
T/K	391.31	396.11	401.19		
P_A/MPa	0.524	0.574	0.639		
w_B	0.0984	was kept constant			
T/K	391.31	396.11	400.97		
P_A/MPa	0.518	0.568	0.623		
w_B	0.1507	was kept constant			
T/K	390.75	395.68	400.46	403.50	
P_A/MPa	0.512	0.567	0.617	0.687	

Polymer (B): **poly(ethylene-*co*-1-hexene)** **2002JIN**
Characterization: M_n/g.mol^{-1} = 1126000, M_w/g.mol^{-1} = 2578000, 1.8 mol% 1-hexene, ρ = 0.923 g/cm^3, metallocene catalyst (2-MeInd)$_2$ZrCl$_2$/MAO

continued

continued

Solvent (A): **1-hexene** C_6H_{12} **592-41-6**

Type of data: vapor solubility

T/K = 323.15

P_A/atm	0.1	0.2	0.3	0.4
w_B	0.98580	0.97049	0.95110	0.92717

T/K = 333.15

P_A/atm	0.1	0.2	0.3	0.4
w_B	0.98952	0.97919	0.96575	0.94878

T/K = 343.15

P_A/atm	0.1	0.2	0.3	0.4
w_B	0.99126	0.98290	0.97242	0.95940

Comments: The copolymer weight fraction belongs to the amorphous fraction of the copolymer.

Polymer (B): **poly(ethylene-*co*-1-hexene)** **2002JIN**
Characterization: M_n/g.mol^{-1} = 59400, M_w/g.mol^{-1} = 243000, 3.5 mol% 1-hexene, ρ = 0.924 g/cm^3, Ziegler-Natta catalyst $Mg(OEt)_2$/DIBP/$TiCl_4$-TEA
Solvent (A): **1-hexene** C_6H_{12} **592-41-6**

Type of data: vapor solubility

T/K = 323.15

P_A/atm	0.1	0.2	0.3	0.4
w_B	0.98270	0.96320	0.94066	0.91568

T/K = 333.15

P_A/atm	0.1	0.2	0.3	0.4
w_B	0.98631	0.97585	0.96177	0.94431

T/K = 343.15

P_A/atm	0.1	0.2	0.3	0.4
w_B	0.98875	0.98023	0.96894	0.95563

Comments: The copolymer weight fraction belongs to the amorphous fraction of the copolymer.

Polymer (B): **poly(ethylene-*co*-1-hexene)** **2002JIN**
Characterization: M_n/g.mol^{-1} = 915000, M_w/g.mol^{-1} = 2323000, 3.5 mol% 1-hexene, ρ = 0.918 g/cm^3, metallocene catalyst $(2\text{-MeInd})_2ZrCl_2$/MAO
Solvent (A): **1-hexene** C_6H_{12} **592-41-6**

continued

continued

Type of data: vapor solubility

T/K = 323.15

P_A/atm	0.1	0.2	0.3	0.4
w_B	0.98598	0.97278	0.95413	0.93251

T/K = 333.15

P_A/atm	0.1	0.2	0.3	0.4
w_B	0.98859	0.97942	0.96627	0.95024

T/K = 343.15

P_A/atm	0.1	0.2	0.3	0.4
w_B	0.99182	0.98310	0.97348	0.96191

Comments: The copolymer weight fraction belongs to the amorphous fraction of the copolymer.

Polymer (B): **poly(ethylene-*co*-1-hexene)** **2002JIN**
Characterization: M_n/g.mol^{-1} = 947000, M_w/g.mol^{-1} = 2566000,
6.6 mol% 1-hexene, ρ = 0.912 g/cm^3,
metallocene catalyst (2-MeInd)$_2$ZrCl$_2$/MAO
Solvent (A): **1-hexene** **C_6H_{12}** **592-41-6**

Type of data: vapor solubility

T/K = 323.15

P_A/atm	0.1	0.2	0.3	0.4
w_B	0.98265	0.96796	0.94612	0.92236

T/K = 333.15

P_A/atm	0.1	0.2	0.3	0.4
w_B	0.98744	0.97474	0.95668	0.93494

T/K = 343.15

P_A/atm	0.1	0.2	0.3	0.4
w_B	0.99871	0.97929	0.96789	0.95462

Comments: The copolymer weight fraction belongs to the amorphous fraction of the copolymer.

Polymer (B): **poly(ethylene-*co*-1-hexene)** **2002JIN**
Characterization: M_n/g.mol^{-1} = 67300, M_w/g.mol^{-1} = 296000,
7.3 mol% 1-hexene, ρ = 0.918 g/cm^3,
Ziegler-Natta catalyst Mg(OEt)$_2$/DIBP/TiCl$_4$-TEA
Solvent (A): **1-hexene** **C_6H_{12}** **592-41-6**

Type of data: vapor solubility

continued

continued

T/K = 323.15

P_A/atm	0.1	0.2	0.3	0.4
w_B	0.98345	0.96474	0.94331	0.91789

T/K = 333.15

P_A/atm	0.1	0.2	0.3	0.4
w_B	0.98717	0.97638	0.96126	0.94339

T/K = 343.15

P_A/atm	0.1	0.2	0.3	0.4
w_B	0.98976	0.98134	0.97046	0.95708

Comments: The copolymer weight fraction belongs to the amorphous fraction of the copolymer.

Polymer (B): **poly(ethylene-*co*-1-hexene)** **2002JIN**
Characterization: M_n/g.mol^{-1} = 875000, M_w/g.mol^{-1} = 2301000, 8.7 mol% 1-hexene, ρ = 0.905 g/cm^3, metallocene catalyst (2-MeInd)$_2$ZrCl$_2$/MAO
Solvent (A): **1-hexene** **C_6H_{12}** **592-41-6**

Type of data: vapor solubility

T/K = 323.15

P_A/atm	0.1	0.2	0.3	0.4
w_B	0.97929	0.95948	0.93482	0.90369

T/K = 333.15

P_A/atm	0.1	0.2	0.3	0.4
w_B	0.98415	0.97095	0.94925	0.92634

T/K = 343.15

P_A/atm	0.1	0.2	0.3	0.4
w_B	0.98606	0.97638	0.96427	0.95042

Comments: The copolymer weight fraction belongs to the amorphous fraction of the copolymer.

Polymer (B): **poly(ethylene-*co*-1-hexene)** **2002JIN**
Characterization: M_n/g.mol^{-1} = 83300, M_w/g.mol^{-1} = 305000, 12.3 mol% 1-hexene, ρ = 0.894 g/cm^3, metallocene catalyst (2-MeInd)$_2$ZrCl$_2$/MAO
Solvent (A): **1-hexene** **C_6H_{12}** **592-41-6**

Type of data: vapor solubility

continued

continued

T/K = 323.15

P_A/atm	0.1	0.2	0.3	0.4
w_B	0.97892	0.95354	0.92306	0.88684

T/K = 333.15

P_A/atm	0.1	0.2	0.3	0.4
w_B	0.98324	0.96362	0.94310	0.91752

T/K = 343.15

P_A/atm	0.1	0.2	0.3	0.4
w_B	0.98671	0.97331	0.95732	0.94221

Comments: The copolymer weight fraction belongs to the amorphous fraction of the copolymer.

Polymer (B): **poly(ethylene-*co*-1-hexene)** **2002JIN**
Characterization: M_n/g.mol^{-1} = 79800, M_w/g.mol^{-1} = 217000, 20.8 mol% 1-hexene, ρ = 0.879 g/cm^3, metallocene catalyst (2-MeInd)$_2$ZrCl$_2$/MAO
Solvent (A): **1-hexene** C_6H_{12} **592-41-6**

Type of data: vapor solubility

T/K = 323.15

P_A/atm	0.1	0.2	0.3	0.4
w_B	0.96920	0.94515	0.90837	0.85802

T/K = 333.15

P_A/atm	0.1	0.2	0.3	0.4
w_B	0.98179	0.95865	0.93221	0.90324

T/K = 343.15

P_A/atm	0.1	0.2	0.3	0.4
w_B	0.98459	0.96925	0.95275	0.93281

Comments: The copolymer weight fraction belongs to the amorphous fraction of the copolymer.

Polymer (B): **poly(ethylene-*co*-1-octadecene)** **2002JIN**
Characterization: M_n/g.mol^{-1} = 800000, M_w/g.mol^{-1} = 2416000, 0.9 mol% 1-octadecene, ρ = 0.919 g/cm^3, metallocene catalyst (2-MeInd)$_2$ZrCl$_2$/MAO
Solvent (A): **1-hexene** C_6H_{12} **592-41-6**

Type of data: vapor solubility

continued

continued

T/K = 323.15

P_A/atm	0.1	0.2	0.3	0.4
w_B	0.98488	0.97154	0.95134	0.93080

T/K = 333.15

P_A/atm	0.1	0.2	0.3	0.4
w_B	0.98770	0.97767	0.96514	0.94860

T/K = 343.15

P_A/atm	0.1	0.2	0.3	0.4
w_B	0.99144	0.98220	0.97247	0.96077

Comments: The copolymer weight fraction belongs to the amorphous fraction of the copolymer.

Polymer (B): **poly(ethylene-*co*-1-octadecene)** **2002JIN**
Characterization: M_n/g.mol^{-1} = 892000, M_w/g.mol^{-1} = 2461000, 1.8 mol% 1-octadecene, ρ = 0.915 g/cm^3, metallocene catalyst $(2\text{-MeInd})_2ZrCl_2$/MAO
Solvent (A): **1-hexene** C_6H_{12} **592-41-6**

Type of data: vapor solubility

T/K = 323.15

P_A/atm	0.1	0.2	0.3	0.4
w_B	0.98307	0.96892	0.94802	0.92355

T/K = 333.15

P_A/atm	0.1	0.2	0.3	0.4
w_B	0.98703	0.97446	0.96079	0.94008

T/K = 343.15

P_A/atm	0.1	0.2	0.3	0.4
w_B	0.98973	0.98010	0.96942	0.95490

Comments: The copolymer weight fraction belongs to the amorphous fraction of the copolymer.

Polymer (B): **poly(ethylene-*co*-1-octadecene)** **2002JIN**
Characterization: M_n/g.mol^{-1} = 908000, M_w/g.mol^{-1} = 2333000, 2.1 mol% 1-octadecene, ρ = 0.912 g/cm^3, metallocene catalyst $(2\text{-MeInd})_2ZrCl_2$/MAO
Solvent (A): **1-hexene** C_6H_{12} **592-41-6**

Type of data: vapor solubility

continued

continued

T/K = 323.15

P_A/atm	0.1	0.2	0.3	0.4
w_B	0.98097	0.96426	0.94154	0.91781

T/K = 333.15

P_A/atm	0.1	0.2	0.3	0.4
w_B	0.98605	0.97218	0.95170	0.93307

T/K = 343.15

P_A/atm	0.1	0.2	0.3	0.4
w_B	0.98737	0.97666	0.96563	0.94965

Comments: The copolymer weight fraction belongs to the amorphous fraction of the copolymer.

Polymer (B): **poly(ethylene-*co*-1-octadecene)** **2002JIN**
Characterization: M_n/g.mol^{-1} = 782000, M_w/g.mol^{-1} = 2018000, 2.3 mol% 1-octadecene, ρ = 0.904 g/cm^3, metallocene catalyst (2-MeInd)$_2$ZrCl$_2$/MAO
Solvent (A): **1-hexene** **C_6H_{12}** **592-41-6**

Type of data: vapor solubility

T/K = 323.15

P_A/atm	0.1	0.2	0.3	0.4
w_B	0.97783	0.95473	0.92578	0.88963

T/K = 333.15

P_A/atm	0.1	0.2	0.3	0.4
w_B	0.98343	0.96662	0.94323	0.91868

T/K = 343.15

P_A/atm	0.1	0.2	0.3	0.4
w_B	0.98477	0.97335	0.95972	0.94178

Comments: The copolymer weight fraction belongs to the amorphous fraction of the copolymer.

Polymer (B): **poly(ethylene-*co*-1-octene)** **2005LEE**
Characterization: M_n/g.mol^{-1} = 64810, M_w/g.mol^{-1} = 135500, 15.3 mol% 1-octene, T_m/K = 323.2, T_g/K = 214.2, ρ = 0.86 g/cm^3, DuPont Dow Elastomers Corporation
Solvent (A): **cyclohexane** **C_6H_{12}** **110-82-7**

Type of data: vapor-liquid equilibrium

continued

continued

w_B 0.0501 was kept constant

T/K	322.85	348.45	373.55	397.95	423.05
P_A/bar	1.601	2.302	3.282	4.752	7.231

Polymer (B): **poly(ethylene-*co*-1-octene)** **2005LEE**
Characterization: M_n/g.mol^{-1} = 64810, M_w/g.mol^{-1} = 135500, 15.3 mol% 1-octene, T_m/K = 323.2, T_g/K = 214.2, ρ = 0.86 g/cm^3, DuPont Dow Elastomers Corporation
Solvent (A): **cyclopentane** **C_5H_{10}** **287-92-3**

Type of data: vapor-liquid equilibrium

w_B 0.0501 was kept constant

T/K	322.85	348.75	373.15	398.35	423.15
P_A/bar	1.870	3.399	5.531	8.809	13.537

Polymer (B): **poly(ethylene-*co*-1-octene)** **2005LEE**
Characterization: M_n/g.mol^{-1} = 64810, M_w/g.mol^{-1} = 135500, 15.3 mol% 1-octene, T_m/K = 323.2, T_g/K = 214.2, ρ = 0.86 g/cm^3, DuPont Dow Elastomers Corporation
Solvent (A): **n-heptane** **C_7H_{16}** **142-82-5**

Type of data: vapor-liquid equilibrium

w_B 0.0500 was kept constant

T/K	323.35	348.25	373.45	398.15	422.85
P_A/bar	1.572	2.101	2.797	3.855	5.354

Polymer (B): **poly(ethylene-*co*-1-octene)** **2005LEE**
Characterization: M_n/g.mol^{-1} = 64810, M_w/g.mol^{-1} = 135500, 15.3 mol% 1-octene, T_m/K = 323.2, T_g/K = 214.2, ρ = 0.86 g/cm^3, DuPont Dow Elastomers Corporation
Solvent (A): **n-hexane** **C_6H_{14}** **110-54-3**

Type of data: vapor-liquid equilibrium

w_B 0.0500 was kept constant

T/K	322.45	353.15	383.45	412.55	442.75
P_A/bar	1.699	3.056	5.472	8.623	13.371

Polymer (B): **poly(ethylene-*co*-1-octene)** **2005LEE**
Characterization: M_n/g.mol^{-1} = 73800, M_w/g.mol^{-1} = 324000, 5.7 mol% 1-octene, ρ = 0.919 g/cm^3, Ziegler-Natta catalyst Mg(OEt)$_2$/DIBP/TiCl$_4$-TEA
Solvent (A): **1-hexene** **C_6H_{12}** **592-41-6**

continued

continued

Type of data: vapor solubility

T/K = 323.15

P_A/atm	0.1	0.2	0.3	0.4
w_B	0.98029	0.96125	0.93855	0.91316

T/K = 333.15

P_A/atm	0.1	0.2	0.3	0.4
w_B	0.98596	0.97380	0.95876	0.94157

T/K = 343.15

P_A/atm	0.1	0.2	0.3	0.4
w_B	0.98797	0.98000	0.96856	0.95430

Comments: The copolymer weight fraction belongs to the amorphous fraction of the copolymer.

Polymer (B): **poly(ethylene-*co*-1-octene)** **2005LEE**
Characterization: M_n/g.mol^{-1} = 64810, M_w/g.mol^{-1} = 135500, 15.3 mol% 1-octene, T_m/K = 323.2, T_g/K = 214.2, ρ = 0.86 g/cm^3, DuPont Dow Elastomers Corporation
Solvent (A): **n-octane** C_8H_{18} **111-65-9**

Type of data: vapor-liquid equilibrium

w_B 0.0499 was kept constant

T/K	323.05	348.65	373.15	398.05	421.85
P_A/bar	1.121	1.513	2.027	2.630	3.375

Polymer (B): **poly(ethylene-*co*-1-octene)** **2005LEE**
Characterization: M_n/g.mol^{-1} = 64810, M_w/g.mol^{-1} = 135500, 15.3 mol% 1-octene, T_m/K = 323.2, T_g/K = 214.2, ρ = 0.86 g/cm^3, DuPont Dow Elastomers Corporation
Solvent (A): **n-pentane** C_5H_{12} **109-66-0**

Type of data: vapor-liquid equilibrium

w_B 0.0099 was kept constant

T/K	323.05	352.55	378.15	402.95
P_A/bar	2.385	4.757	8.049	12.626

w_B 0.0202 was kept constant

T/K	323.05	342.95	368.35	393.15
P_A/bar	2.125	3.581	6.383	10.421

continued

continued

w_B 0.0504 was kept constant

T/K	322.95	343.45	368.25	394.15
P_A/bar	2.248	3.875	6.692	10.955

w_B 0.1133 was kept constant

T/K	322.95	342.85	368.05	393.05
P_A/bar	2.189	3.644	6.359	10.411

w_B 0.2000 was kept constant

T/K	327.95	352.55	377.95	403.85
P_A/bar	3.081	5.217	8.500	13.341

w_B 0.3940 was kept constant

T/K	325.65	347.85	376.55	400.35	424.25
P_A/bar	2.728	4.463	7.976	12.283	18.339

Comments: Some LLE data are given in the corresponding chapter of this book.

Polymer (B): **poly(ethylene-*b*-propylene)** **2000SA2**
Characterization: M_w/g.mol^{-1} = 240000, 4.9 wt% ethene, 90% isotactic, 58.1 wt% crystallinity, impact propylene copolymer, Mitsubishi Chemical, Kurashiki, Japan
Solvent (A): **ethene** **C_2H_4** **74-85-1**

Type of data: gas solubility

T/K = 323.15

P/MPa	0.3800	0.8987	1.4021	1.9014	2.4131	3.0362
w_A	0.00252	0.00580	0.00898	0.01213	0.01535	0.01927

T/K = 343.15

P/MPa	0.3738	0.8009	1.2862	1.7985	2.3093	2.8260
w_A	0.00197	0.00416	0.00664	0.00926	0.01185	0.01447

T/K = 363.15

P/MPa	0.3836	0.8533	1.3355	1.8383	2.3135	2.8231
w_A	0.00169	0.00374	0.00583	0.00800	0.01003	0.01219

Polymer (B): **poly(ethylene-*b*-propylene)** **2000SA2**
Characterization: M_w/g.mol^{-1} = 240000, 4.9 wt% ethene, 90% isotactic, 58.1 wt% crystallinity, impact propylene copolymer, Mitsubishi Chemical, Kurashiki, Japan
Solvent (A): **propene** **C_3H_6** **115-07-1**

Type of data: gas solubility

continued

continued

T/K = 323.15

P/MPa	0.2143	0.6207	1.0232	1.3323	1.6264	1.7866
w_A	0.00664	0.02097	0.03989	0.05663	0.07448	0.08682

T/K = 343.15

P/MPa	0.3024	0.7554	1.4385	1.8452	2.2823	2.6448
w_A	0.00736	0.01898	0.03884	0.05417	0.07091	0.08703

T/K = 363.15

P/MPa	0.3659	0.9153	1.4844	2.0138	2.5029	3.0951
w_A	0.00628	0.01671	0.02847	0.04058	0.05306	0.07089

Polymer (B): **poly(ethylene oxide)-poly(butylene terephthalate) multiblock copolymer** **2003MET**
Characterization: see comments, samples by Isotsis b.v., The Netherlands
Solvent (A): **water** **H_2O** **7732-18-5**

Type of data: sorption of water vapor into block copolymer films

T/K = 293.15

w_B	0.9943	0.9938	0.9916	0.9887	0.9818	0.9733	0.9730
P_A/P_{0A}	0.41	0.53	0.64	0.76	0.87	0.93	0.93

Comments: PEO: M_w/(g/mol) = 300, PBT: T_m/K = 451.45, crystallinity degree = 0.24, mass ratio of PEO/PBT = 32/68.

T/K = 293.15

w_B	0.9949	0.9932	0.9889	0.9883	0.9840	0.9823	0.9765	0.9714	0.9606
P_A/P_{0A}	0.30	0.40	0.50	0.53	0.60	0.64	0.70	0.76	0.80
w_B	0.9411	0.9045	0.8744						
P_A/P_{0A}	0.87	0.93	0.96						

Comments: PEO: M_w/(g/mol) = 600, T_g/K = 224.15, PBT: T_m/K = 451.95, crystallinity degree = 0.18, mass ratio of PEO/PBT = 41/49.

T/K = 293.15

w_B	0.9976	0.9939	0.9881	0.9880	0.9822	0.9807	0.9731	0.9678	0.9565
P_A/P_{0A}	0.18	0.30	0.40	0.41	0.50	0.53	0.60	0.64	0.70
w_B	0.9376	0.9173	0.9078	0.8654	0.8425	0.8114			
P_A/P_{0A}	0.76	0.80	0.81	0.87	0.90	0.93			

Comments: PEO: M_w/(g/mol) = 1000, T_g/K = 224.15, T_m/K = 262.75, crystallinity degree = 0.05, PBT: T_m/K = 436.15, crystallinity degree = 0.12, mass ratio of PEO/PBT = 52/48.

continued

continued

T/K = 293.15

w_B	0.9913	0.9865	0.9844	0.9794	0.9748	0.9676	0.9556	0.9427	0.9053
P_A/P_{0A}	0.30	0.40	0.41	0.50	0.53	0.60	0.64	0.70	0.76
w_B	0.8894	0.8067	0.7358						
P_A/P_{0A}	0.80	0.87	0.93						

Comments: PEO: M_w/(g/mol) = 2000, T_g/K = 220.15, T_m/K = 280.05, crystallinity degree = 0.20, PBT: T_m/K = 455.85, crystallinity degree = 0.15, mass ratio of PEO/PBT = 52/48.

T/K = 293.15

w_B	0.9900	0.9841	0.9846	0.9831	0.9748	0.9589	0.9275	0.9261	0.9264
P_A/P_{0A}	0.30	0.40	0.41	0.41	0.50	0.60	0.70	0.70	0.70
w_B	0.8607	0.8396	0.6969						
P_A/P_{0A}	0.80	0.81	0.93						

Comments: PEO: M_w/(g/mol) = 3000, T_g/K = 221.15, T_m/K = 292.65, crystallinity degree = 0.26, PBT: T_m/K = 466.65, crystallinity degree = 0.15, mass ratio of PEO/PBT = 52/48.

Polymer (B): **poly(ethylene oxide-*co*-propylene oxide)** **2000LIW**
Characterization: M_n/g.mol^{-1} = 780, M_w/g.mol^{-1} = 865, 50.0 mol% ethylene oxide, Zhejiang Univ. Chem. Factory, PR China
Solvent (A): **water** **H_2O** **7732-18-5**

Type of data: vapor-liquid equilibrium

T/K = 298.15

w_B	0.0455	0.0553	0.0817	0.1138	0.1178	0.1288	0.1453	0.2190	0.2472
a_A	0.9987	0.9983	0.9975	0.9963	0.9962	0.99567	0.9952	0.9905	0.9882
w_B	0.3040	0.3850	0.4073	0.5410					
a_A	0.9825	0.9716	0.9675	0.9375					

Polymer (B): **poly(ethylene oxide-*co*-propylene oxide)** **2000LIW**
Characterization: M_n/g.mol^{-1} = 2340, M_w/g.mol^{-1} = 2480, 50.0 mol% ethylene Oxide, Zhejiang Univ. Chem. Factory, PR China
Solvent (A): **water** **H_2O** **7732-18-5**

Type of data: vapor-liquid equilibrium

T/K = 298.15

w_B	0.0715	0.0836	0.1414	0.1558	0.1637	0.1912	0.2677	0.4338	0.6113
a_A	0.9987	0.9983	0.9975	0.9963	0.9962	0.9952	0.9905	0.9716	0.9375

Polymer (B): **poly(ethylene oxide-*co*-propylene oxide)** **2000LIW**
Characterization: M_n/g.mol^{-1} = 3640, M_w/g.mol^{-1} = 4040, 50.0 mol% ethylene Oxide, Zhejiang Univ. Chem. Factory, PR China
Solvent (A): **water** H_2O **7732-18-5**

Type of data: vapor-liquid equilibrium

T/K = 298.15

w_B	0.0970	0.1130	0.1261	0.1522	0.1903	0.1965	0.2065	0.2213	0.2630
a_A	0.9987	0.9983	0.99824	0.9975	0.9963	0.9962	0.99539	0.9952	0.99322
w_B	0.3028	0.3771	0.4716	0.6612					
a_A	0.9905	0.9840	0.9716	0.9375					

Polymer (B): **poly(DL-lactic acid-*co*-glycolic acid)** **2008PIN**
Characterization: M_n/g.mol^{-1} = 45300, M_w/g.mol^{-1} = 77000, T_g/K = 321.75, 15 mol% glycolide, ρ = 1.293 g/cm^3
Solvent (A): **carbon dioxide** CO_2 **124-38-9**

Type of data: gas solubility

T/K = 308.15

P/bar	22	52	78	100	121	151	200
c_A/(g CO_2/g polymer)	0.068	0.190	0.343	0.407	0.431	0.459	0.499

Polymer (B): **poly(DL-lactic acid-*co*-glycolic acid)** **2008KAS**
Characterization: M_w/g.mol^{-1} = 50000-75000, 15 mol% glycolide, Aldrich Chem. Co., Inc., Milwaukee, WI
Solvent (A): **carbon dioxide** CO_2 **124-38-9**

Type of data: gas solubility

T/K = 313.15

P/MPa	1.448	2.427	3.786	4.482	5.055	5.662	6.600	7.700
w_A	0.0236	0.0614	0.1138	0.1407	0.1628	0.1862	0.2224	0.2649

T/K = 333.15

P/MPa	2.248	3.136	3.931	4.669	5.343	6.137	7.800
w_A	0.0348	0.0586	0.0799	0.0997	0.1178	0.1391	0.1836

T/K = 344.15

P/MPa	2.317	3.448	4.669	5.593	6.496	8.103
w_A	0.0344	0.0627	0.0932	0.1163	0.1389	0.1790

Polymer (B): **poly(DL-lactic acid-*co*-glycolic acid)** **2008PIN**
Characterization: M_n/g.mol^{-1} = 41000, M_w/g.mol^{-1} = 72000, T_g/K = 323.55, 25 mol% glycolide, ρ = 1.353 g/cm^3
Solvent (A): **carbon dioxide** CO_2 **124-38-9**

continued

continued

Type of data: gas solubility

T/K = 308.15

P/bar	24	52	77	100	120	151	200
c_A/(g CO_2/g polymer)	0.070	0.176	0.314	0.393	0.420	0.455	0.518

Polymer (B): **poly(DL-lactic acid-*co*-glycolic acid)** **2008PIN**
Characterization: M_n/g.mol^{-1} = 30800, M_w/g.mol^{-1} = 52000, T_g/K = 322.25, 35 mol% glycolide, ρ = 1.346 g/cm^3
Solvent (A): **carbon dioxide** **CO_2** **124-38-9**

Type of data: gas solubility

T/K = 308.15

P/bar	21	50	74	101	120	152	201
c_A/(g CO_2/g polymer)	0.056	0.160	0.268	0.336	0.357	0.377	0.397

Polymer (B): **poly(DL-lactic acid-*co*-glycolic acid)** **2008KAS**
Characterization: M_w/g.mol^{-1} = 40000-75000, 35 mol% glycolide, Sigma Chemical Co., Inc., St. Louis, MO
Solvent (A): **carbon dioxide** **CO_2** **124-38-9**

Type of data: gas solubility

T/K = 313.15

P/MPa	1.380	2.240	3.130	3.896	4.848	5.696	6.351	7.900
w_A	0.0178	0.0487	0.0804	0.1074	0.1413	0.1715	0.1948	0.2499

T/K = 333.15

P/MPa	1.213	1.848	2.069	2.503	3.172	4.241	4.813	5.627	6.500
w_A	0.0054	0.0175	0.0217	0.0299	0.0426	0.0629	0.0738	0.0893	0.1059
P/MPa	7.900								
w_A	0.1325								

T/K = 344.15

P/MPa	2.393	3.206	3.793	4.703	5.593	6.206	6.862	7.689
w_A	0.0235	0.0348	0.0431	0.0558	0.0683	0.0768	0.0860	0.0976

Polymer (B): **poly(DL-lactic acid-*co*-glycolic acid)** **2005ELV**
Characterization: M_n/g.mol^{-1} = 15300, M_w/g.mol^{-1} = 26000, T_g/K = 317.61, 46 mol% glycolide, Alkermes Inc., Cincinnati, OH
Solvent (A): **carbon dioxide** **CO_2** **124-38-9**

Type of data: gas solubility

T/K = 313.15

P/MPa	1.0	2.0	2.4	3.0	3.9
w_A	0.0169	0.0368	0.0442	0.0626	0.0776

Polymer (B): **poly(DL-lactic acid-*co*-glycolic acid)** **2005ELV**
Characterization: M_n/g.mol^{-1} = 15000, M_w/g.mol^{-1} = 27000, T_g/K = 314.14, 47 mol% glycolide, Alkermes Inc., Cincinnati, OH
Solvent (A): **carbon dioxide** **CO_2** **124-38-9**

Type of data: gas solubility

T/K = 313.15

P/MPa	0.6	2.0	3.0
w_A	0.0109	0.0334	0.0508

Polymer (B): **poly(DL-lactic acid-*co*-glycolic acid)** **2005ELV**
Characterization: M_n/g.mol^{-1} = 26100, M_w/g.mol^{-1} = 47000, T_g/K = 316.75, 47 mol% glycolide, Alkermes Inc., Cincinnati, OH
Solvent (A): **carbon dioxide** **CO_2** **124-38-9**

Type of data: gas solubility

T/K = 313.15

P/MPa	1.0	2.0	3.0
w_A	0.0143	0.0306	0.0502

Polymer (B): **poly(DL-lactic acid-*co*-glycolic acid)** **2008PIN**
Characterization: M_n/g.mol^{-1} = 33300, M_w/g.mol^{-1} = 53000, T_g/K = 320.15, 50 mol% glycolide, ρ = 1.407 g/cm^3
Solvent (A): **carbon dioxide** **CO_2** **124-38-9**

Type of data: gas solubility

T/K = 308.15

P/bar	22	53	80	103	123	156	200
c_A/(g CO_2/g polymer)	0.054	0.153	0.249	0.287	0.309	0.330	0.348

Polymer (B): **poly(DL-lactic acid-*co*-glycolic acid)** **2008AI2**
Characterization: M_w/g.mol^{-1} = 70000, 50 mol% glycolide, Boehringer Ingelheim, Germany
Solvent (A): **carbon dioxide** **CO_2** **124-38-9**

Type of data: gas solubility

T/K = 308.15

P/MPa	10.14	11.92	15.31	18.40	21.44	23.11	26.21	29.49	31.47
w_A	0.1464	0.1968	0.2380	0.2577	0.2754	0.2794	0.2880	0.2929	0.2963

T/K = 313.15

P/MPa	10.22	12.09	15.30	18.19	20.61	22.99	25.59	28.46	30.70
w_A	0.0945	0.1666	0.2188	0.2407	0.2555	0.2636	0.2730	0.2823	0.2876

continued

continued

T/K = 323.15

P/MPa	12.43	15.06	19.30	21.02	22.97	24.07	26.44	28.10	30.29
w_A	0.0903	0.1793	0.2236	0.1339	0.2445	0.2492	0.2588	0.2638	0.2702

Polymer (B): **poly(DL-lactic acid-*co*-glycolic acid)** **2005ELV**
Characterization: M_w/g.mol^{-1} = 14000, T_g/K = 309.45,
52 mol% glycolide, Alkermes Inc., Cincinnati, OH
Solvent (A): **carbon dioxide** **CO_2** **124-38-9**

Type of data: gas solubility

T/K = 313.15

P/MPa	1.0	2.0	3.0
w_A	0.0169	0.0288	0.0462

Polymer (B): **poly(styrene-*co*-butadiene)** **2002KAN**
Characterization: M_w/g.mol^{-1} = 100000, 4.1 wt% styrene,
synthesized in the laboratory
Solvent (A): **benzene** **C_6H_6** **71-43-2**

Type of data: vapor-liquid equilibrium

T/K = 298.15

w_B	0.9541	0.9200	0.8883	0.8636	0.8344	0.8136	0.6986	0.6300	0.5725
P_A/P_{0A}	0.1771	0.3120	0.3894	0.4572	0.5305	0.6024	0.7700	0.8483	0.9056

T/K = 308.15

w_B	0.9529	0.9244	0.8956	0.8750	0.8558	0.8041	0.7711	0.7363	0.6902
P_A/P_{0A}	0.1997	0.3003	0.3938	0.4438	0.4787	0.5960	0.6496	0.7154	0.7815

w_B	0.6258	0.5703
P_A/P_{0A}	0.8347	0.8752

Polymer (B): **poly(styrene-*co*-butadiene)** **2008SER**
Characterization: 23 % styrene, Scientific Polymer Products, Inc., Ontario, NY
Solvent (A): **n-hexane** **C_6H_{14}** **110-54-3**

Type of data: vapor-liquid equilibrium

T/K = 343.15

w_A	0.013	0.045	0.060	0.107	0.162	0.255
P_A/kPa	13.9	27.3	38.5	48.8	61.5	75.2

Polymer (B): **poly(styrene-*co*-butadiene)** **2008SER**
Characterization: 45 % styrene, Scientific Polymer Products, Inc., Ontario, NY
Solvent (A): **n-hexane** **C_6H_{14}** **110-54-3**

continued

continued

Type of data: vapor-liquid equilibrium

T/K = 343.15

w_A	0.015	0.036	0.062	0.098	0.155	0.243
P_A/kPa	13.9	27.3	38.5	48.8	61.5	75.2

Polymer (B): **poly(styrene-*co*-butadiene)** **2002KAN**
Characterization: M_w/g.mol^{-1} = 100000, 4.1 wt% styrene, synthesized in the laboratory
Solvent (A): **toluene** C_7H_8 **108-88-3**

Type of data: vapor-liquid equilibrium

T/K = 308.15

w_B	0.9614	0.9367	0.9087	0.8903	0.8762	0.8137	0.7583	0.6622
P_A/P_{0A}	0.1697	0.2975	0.3693	0.3967	0.4540	0.5655	0.7310	0.9832

Polymer (B): **poly(styrene-*b*-butadiene) (four-arm star-shaped with polybutadiene as inner blocks (SB)4)** **2009XIO**
Characterization: M_n/g.mol^{-1} = 198000, M_w/g.mol^{-1} = 220000, 25.6 mol% styrene, 7.4 mol% vinyl, 7.3 mol% cis, ρ(298.15 K) = 0.965 g/cm^3, synthesized and fractionated in the laboratory
Solvent (A): **tetrahydrofuran** C_4H_8O **109-99-9**

Type of data: vapor-liquid equilibrium

T/K = 298.15

φ_B	0.4764	0.4798	0.5577	0.5627	0.5638	0.6724	0.6785	0.6849	0.7834
P_A/P_{0A}	0.8750	0.8636	0.8358	0.8227	0.8174	0.7211	0.6980	0.6914	0.5540
φ_B	0.7855	0.8353	0.8371	0.8386	0.8393	0.9044	0.9070	0.9094	0.9117
P_A/P_{0A}	0.5543	0.4400	0.4341	0.4417	0.4558	0.2873	0.2896	0.2651	0.2715
φ_B	0.9121	0.9127	0.9153	0.9576	0.9589	0.9630	0.9653	0.9662	0.9664
P_A/P_{0A}	0.2686	0.2577	0.2479	0.1143	0.1104	0.0986	0.0998	0.0990	0.0967

T/K = 313.15

φ_B	0.47188	0.47202	0.47376	0.47474	0.56349	0.57177	0.57333	0.57495	0.67927
P_A/P_{0A}	0.89579	0.89268	0.89338	0.88331	0.80101	0.79543	0.80884	0.80710	0.66076
φ_B	0.68038	0.68058	0.68529	0.77658	0.77970	0.77989	0.78095	0.83734	0.83736
P_A/P_{0A}	0.66817	0.69057	0.66102	0.52510	0.53829	0.53850	0.53801	0.41566	0.41027
φ_B	0.84383	0.90299	0.90356	0.90369	0.91008	0.94718	0.95993	0.96485	
P_A/P_{0A}	0.40115	0.25225	0.25407	0.25333	0.23604	0.08563	0.08618	0.08602	

continued

continued

T/K = 328.15

φ_B	0.46901	0.46920	0.46946	0.57025	0.67271	0.77981	0.78096	0.78122	0.85436
P_A/P_{0A}	0.90594	0.89563	0.91862	0.81725	0.67805	0.53219	0.51451	0.50378	0.35032
φ_B	0.85504	0.85554	0.91015	0.91081	0.91195	0.96459	0.96604		
P_A/P_{0A}	0.34492	0.35112	0.22505	0.23535	0.22478	0.08140	0.07830		

Polymer (B): **poly(styrene-*b*-butadiene-*b*-styrene)** **2003WAN**
Characterization: M_w/g.mol^{-1} = 100000, 70 mol% butadiene
Solvent (A): **carbon dioxide** CO_2 **124-38-9**

Type of data: gas solubility

T/K = 298.15

P/MPa	0.2	0.3	0.4	0.5	0.6	0.7	0.8	0.9	1.0
w_A	0.0043	0.0064	0.0081	0.0105	0.0120	0.0140	0.0159	0.0180	0.0202

Polymer (B): **poly(styrene-*b*-butadiene-*b*-styrene)** **2009XIO**
Characterization: M_n/g.mol^{-1} = 76300, M_w/g.mol^{-1} = 85500, 25.8 mol% styrene, 7.3 mol% vinyl, 7.4 mol% cis, ρ(298.15 K) = 0.979 g/cm^3, synthesized and fractionated in the laboratory
Solvent (A): **tetrahydrofuran** C_4H_8O **109-99-9**

Type of data: vapor-liquid equilibrium

T/K = 298.15

φ_B	0.4719	0.4725	0.4749	0.4758	0.5714	0.5717	0.5722	0.5750	0.6736
P_A/P_{0A}	0.8852	0.8763	0.8634	0.8618	0.8178	0.8185	0.8079	0.8102	0.7123
φ_B	0.6736	0.6748	0.6751	0.7779	0.7782	0.7786	0.7788	0.8494	0.8494
P_A/P_{0A}	0.7083	0.7082	0.7089	0.5568	0.5629	0.5564	0.5595	0.4138	0.4148
φ_B	0.8516	0.8729	0.8994	0.9007	0.9024	0.9034	0.9502	0.9545	0.9550
P_A/P_{0A}	0.4191	0.3490	0.2882	0.2983	0.2870	0.2847	0.1253	0.1247	0.1213

T/K = 313.15

φ_B	0.46778	0.46864	0.46865	0.46968	0.56707	0.56756	0.56821	0.57014	0.66887
P_A/P_{0A}	0.87370	0.87160	0.88191	0.87016	0.81272	0.81035	0.81704	0.81837	0.69232
φ_B	0.67084	0.67220	0.67321	0.77565	0.77585	0.77793	0.78126	0.90385	0.90460
P_A/P_{0A}	0.68966	0.68785	0.70205	0.53314	0.53722	0.52803	0.52084	0.25058	0.25601
φ_B	0.90519	0.90596	0.96107	0.96132	0.96247	0.96699			
P_A/P_{0A}	0.25030	0.25929	0.10413	0.10043	0.09741	0.08640			

T/K = 328.15

φ_B	0.46498	0.46515	0.56502	0.56553	0.56573	0.56577	0.66908	0.66961	0.66986
P_A/P_{0A}	0.87860	0.88371	0.78586	0.81550	0.81623	0.80300	0.67255	0.66656	0.68317

continued

continued

φ_B	0.67104	0.77301	0.77688	0.77706	0.77833	0.85157	0.85166	0.85394	0.90691
P_A/P_{0A}	0.65681	0.52022	0.50097	0.49501	0.49328	0.35780	0.35479	0.34361	0.22761
φ_B	0.90720	0.90738	0.90907	0.96360	0.96391	0.96410	0.96587		
P_A/P_{0A}	0.22385	0.21789	0.21292	0.06589	0.06845	0.07566	0.07040		

Polymer (B): **poly(styrene-*b*-isoprene)** **1997ZHA**
Characterization: M_η/g.mol^{-1} = 100000, 25 mol% isoprene, synthesized in the laboratory
Solvent (A): **carbon dioxide** **CO_2** **124-38-9**

Type of data: gas solubility and swelling

T/K = 308.15

P/MPa	1.379	2.758	4.137	5.516	6.895	8.274	10.342
w_A	0.0202	0.0349	0.0534	0.0646	0.0763	0.0858	0.0934
$(\Delta V/V)$/%	1.75	2.87	4.39	6.18	7.72	9.11	11.0

Polymer (B): **poly(styrene-*b*-isoprene)** **1997ZHA**
Characterization: M_η/g.mol^{-1} = 100000, 50 mol% isoprene, synthesized in the laboratory
Solvent (A): **carbon dioxide** **CO_2** **124-38-9**

Type of data: gas solubility and swelling

T/K = 308.15

P/MPa	1.379	2.758	4.137	5.516	6.895	8.274	10.342
w_A	0.0210	0.0397	0.0592	0.0777	0.0848	0.0958	0.1119
$(\Delta V/V)$/%	1.84	3.23	5.21	7.01	8.54	11.1	12.0

Polymer (B): **poly(styrene-*b*-isoprene)** **1997ZHA**
Characterization: M_η/g.mol^{-1} = 100000, 75 mol% isoprene, synthesized in the laboratory
Solvent (A): **carbon dioxide** **CO_2** **124-38-9**

Type of data: gas solubility and swelling

T/K = 308.15

P/MPa	1.379	2.758	4.137	5.516	6.895	8.274	10.342
w_A	0.0208	0.0384	0.0652	0.0839	0.1007	0.1197	0.1416
$(\Delta V/V)$/%	2.39	3.44	5.45	7.84	9.84	12.4	15.1

Polymer (B): **poly(styrene-*b*-methyl methacrylate)** **1997ZHA**
Characterization: M_η/g.mol^{-1} = 75000, 18 mol% methyl methacrylate, synthesized in the laboratory
Solvent (A): **carbon dioxide** **CO_2** **124-38-9**

continued

continued

Type of data: gas solubility and swelling

T/K = 308.15

P/MPa	1.379	2.758	4.137	5.516	6.895	8.274	10.342
w_A	0.0221	0.0388	0.0561	0.0667	0.0733	0.0767	0.0784
$(\Delta V/V)$/%	1.89	3.34	5.01	6.61	8.40	9.48	10.3

Polymer (B): **poly(styrene-*b*-methyl methacrylate)** **1997ZHA**
Characterization: M_η/g.mol^{-1} = 75000, 46 mol% methyl methacrylate, synthesized in the laboratory
Solvent (A): **carbon dioxide** **CO_2** **124-38-9**

Type of data: gas solubility and swelling

T/K = 308.15

P/MPa	1.379	2.758	4.137	5.516	6.895	8.274	10.342
w_A	0.0277	0.0481	0.0666	0.0774	0.0908	0.0950	0.1031
$(\Delta V/V)$/%	2.78	4.44	5.85	7.68	9.68	12.1	13.2

Polymer (B): **poly(styrene-*b*-methyl methacrylate)** **1997ZHA**
Characterization: M_η/g.mol^{-1} = 81000, 56 mol% methyl methacrylate, synthesized in the laboratory
Solvent (A): **carbon dioxide** **CO_2** **124-38-9**

Type of data: gas solubility and swelling

T/K = 308.15

P/MPa	1.379	2.758	4.137	5.516	6.895	8.274	10.342
w_A	0.0300	0.0522	0.0638	0.0850	0.1047	0.1150	0.1319
$(\Delta V/V)$/%	2.52	4.49	6.76	8.86	11.0	12.6	13.7

Polymer (B): **poly(styrene-*b*-methyl methacrylate)** **2003WAN**
Characterization: M_w/g.mol^{-1} = 12100, 58.55 mol% methyl methacrylate
Solvent (A): **carbon dioxide** **CO_2** **124-38-9**

Type of data: gas solubility

T/K = 298.15

P/MPa	0.2	0.3	0.4
w_A	0.0224	0.0329	0.0424

Polymer (B): **poly(styrene-*co*-methyl methacrylate)** **2008SER**
Characterization: 44 % styrene, Scientific Polymer Products, Inc., Ontario, NY
Solvent (A): **benzene** **C_6H_6** **71-43-2**

continued

continued

Type of data: vapor-liquid equilibrium

T/K = 343.15

w_A	0.179	0.231	0.287	0.347	0.380	0.417	0.481
P_A/kPa	40.1	46.3	52.4	56.9	59.3	61.9	65.2

Polymer (B): **poly(styrene-*co*-methyl methacrylate)** **2002KAN**
Characterization: M_w/g.mol^{-1} = 12100, 41.45 wt% styrene, synthesized in the laboratory
Solvent (A): **benzene** C_6H_6 **71-43-2**

Type of data: vapor-liquid equilibrium

T/K = 298.15

w_B	0.9803	0.9707	0.9599	0.9424	0.8863	0.8441	0.7458	0.6704	0.6054
P_A/P_{0A}	0.1771	0.3120	0.3894	0.4572	0.5305	0.6024	0.7700	0.8483	0.9056

T/K = 308.15

w_B	0.9862	0.9776	0.9435	0.9173	0.8890	0.8420	0.8156	0.7803	0.7386
P_A/P_{0A}	0.1997	0.3003	0.3938	0.4438	0.4787	0.5960	0.6496	0.7154	0.7815
w_B	0.6786	0.6253							
P_A/P_{0A}	0.8347	0.8752							

Polymer (B): **poly(styrene-*ran*-methyl methacrylate)** **2008BER**
Characterization: M_n/g.mol^{-1} = 84800, M_w/g.mol^{-1} = 133000, 10 wt% styrene, Polymer Standard Service, Mainz, Germany
Solvent (A): **toluene** C_7H_8 **108-88-3**

Type of data: vapor-liquid equilibrium

T/K = 323.15

w_B	0.4339	0.4429	0.5050	0.5864	0.5968	0.6493	0.6946	0.7014	0.7146
φ_B	0.3560	0.3644	0.4239	0.5056	0.5163	0.5718	0.6213	0.6288	0.6436
P_A/P_{0A}	0.9690	0.9656	0.9590	0.9447	0.9582	0.9413	0.9558	0.9262	0.9216
w_B	0.7670	0.7926							
φ_B	0.7036	0.7337							
P_A/P_{0A}	0.9109	0.8669							

Polymer (B): **poly(styrene-*ran*-methyl methacrylate)** **2008BER**
Characterization: M_n/g.mol^{-1} = 80500, M_w/g.mol^{-1} = 103000, 20 wt% styrene, Polymer Standard Service, Mainz, Germany
Solvent (A): **toluene** C_7H_8 **108-88-3**

Type of data: vapor-liquid equilibrium

continued

continued

T/K = 323.15

w_B	0.5690	0.6229	0.6724	0.7288	0.7533
φ_B	0.4907	0.5466	0.5997	0.6623	0.6903
P_A/P_{0A}	0.9600	0.9418	0.9244	0.8816	0.8707

Polymer (B): **poly(styrene-*ran*-methyl methacrylate)** **2008BER**
Characterization: M_n/g.mol^{-1} = 75000, M_w/g.mol^{-1} = 90000, 50 wt% styrene, Polymer Standard Service, Mainz, Germany
Solvent (A): **toluene** C_7H_8 **108-88-3**

Type of data: vapor-liquid equilibrium

T/K = 323.15

w_B	0.4290	0.5823	0.6312	0.6852	0.6975	0.7367	0.7984
φ_B	0.3625	0.5134	–	0.6223	0.6358	0.6793	0.7498
P_A/P_{0A}	0.9842	0.9505	0.9210	0.8793	0.8708	0.8272	0.7237

Polymer (B): **poly(styrene-*ran*-methyl methacrylate)** **2008BER**
Characterization: M_n/g.mol^{-1} = 34200, M_w/g.mol^{-1} = 50700, 83 wt% styrene,1 Polymer Standard Service, Mainz, Germany
Solvent (A): **toluene** C_7H_8 **108-88-3**

Type of data: vapor-liquid equilibrium

T/K = 323.15

w_B	0.4532	0.5033	0.5759	0.6468	0.6984	0.7299	0.7692	0.8282
φ_B	0.3954	0.4443	0.5172	0.5910	0.6462	0.6808	0.7244	0.7918
P_A/P_{0A}	0.9494	0.9388	0.9027	0.8837	0.8398	0.8070	0.7479	0.6404

Polymer (B): **poly(styrene-*b*-vinyl pyridine)** **1997ZHA**
Characterization: M_η/g.mol^{-1} = 70000, 44 mol% vinyl pyridine, synthesized in the laboratory
Solvent (A): **carbon dioxide** CO_2 **124-38-9**

Type of data: gas solubility and swelling

T/K = 308.15

P/MPa	1.379	2.758	4.137	5.516	6.895	8.274	10.342
w_A	0.0207	0.0350	0.0513	0.0658	0.0777	0.0830	0.0950
$(\Delta V/V)$/%	2.62	4.49	5.81	6.91	8.43	9.05	10.5

Polymer (B): **poly(styrene-*b*-vinyl pyridine)** **1997ZHA**
Characterization: M_η/g.mol^{-1} = 90000, 77 mol% vinyl pyridine, synthesized in the laboratory
Solvent (A): **carbon dioxide** CO_2 **124-38-9**

continued

continued

Type of data: gas solubility and swelling

T/K = 308.15

P/MPa	1.379	2.758	4.137	5.516	6.895	8.274	10.342
w_A	0.0224	0.0368	0.0515	0.0691	0.0832	0.1023	0.1150
$(\Delta V/V)$/%	3.09	5.21	6.68	7.73	8.94	9.78	13.4

Polymer (B): **poly(vinyl acetate-*co*-vinyl alcohol)** **2004PAL**
Characterization: M_n/g.mol^{-1} = 57000, M_w/g.mol^{-1} = 103000, 12 mol% vinyl acetate, semicrystalline sample with 39% crystallinity (by MDSC), Air Products and Chemicals, Inc.
Solvent (A): **methanol** **CH_4O** **67-56-1**

Type of data: vapor-liquid equilibrium

T/K = 363.15

w_A	0.0180	0.0183	0.0185	0.0367	0.0406	0.0526	0.0573
P_A/P_{0A}	0.147	0.145	0.148	0.295	0.297	0.389	0.385

T/K = 373.15

w_A	0.0141	0.0278	0.0384
P_A/P_{0A}	0.106	0.208	0.278

T/K = 383.15

w_A	0.0090	0.0106	0.0119	0.0163	0.0191	0.0207	0.0245	0.0271	0.0315
P_A/P_{0A}	0.083	0.091	0.089	0.164	0.164	0.160	0.179	0.207	0.224

Comments: The concentration of solvent A is related to the 61% amorphous region of the polymer, i.e., the crystalline part is not influenced by sorption and is subtracted when determining w_B.

Polymer (B): **poly(vinyl acetate-*co*-vinyl alcohol)** **2004PAL**
Characterization: M_n/g.mol^{-1} = 57000, M_w/g.mol^{-1} = 103000, 12 mol% vinyl acetate, semicrystalline sample with 39% crystallinity (by MDSC), Air Products and Chemicals, Inc.
Solvent (A): **methyl acetate** **$C_3H_6O_2$** **79-20-9**

Type of data: vapor-liquid equilibrium

T/K = 363.15

w_A	0.0081	0.0178
P_A/P_{0A}	0.118	0.231

T/K = 373.15

w_A	0.0081	0.0152
P_A/P_{0A}	0.087	0.177

continued

continued

T/K = 383.15

w_A	0.0065	0.0143
P_A/P_{0A}	0.088	0.163

Comments: The concentration of solvent A is related to the 61% amorphous region of the polymer, i.e., the crystalline part is not influenced by sorption and is subtracted when determining w_B.

Polymer (B): **poly(vinyl acetate-*co*-vinyl alcohol)** **2004PAL**
Characterization: M_n/g.mol^{-1} = 57000, M_w/g.mol^{-1} = 103000, 12 mol% vinyl acetate, semicrystalline sample with 39% crystallinity (by MDSC), Air Products and Chemicals, Inc.
Solvent (A): **water** **H_2O** **7732-18-5**

Type of data: vapor-liquid equilibrium

T/K = 363.15

w_A	0.0237	0.0273	0.0373	0.0523	0.0701	0.1170	0.1037	0.189	0.219
P_A/P_{0A}	0.222	0.224	0.295	0.461	0.467	0.623	0.701	0.891	0.909

T/K = 373.15

w_A	0.0228	0.0623	0.112	0.208
P_A/P_{0A}	0.221	0.490	0.675	0.849

T/K = 383.15

w_A	0.0222	0.0506	0.0930
P_A/P_{0A}	0.244	0.442	0.621

Comments: The concentration of solvent A is related to the 61% amorphous region of the polymer, i.e., the crystalline part is not influenced by sorption and is subtracted when determining w_B.

Polymer (B): **poly(vinyl acetate-*co*-vinyl chloride)** **2005TOC**
Characterization: M_w/g.mol^{-1} = 300000, 10 wt% vinyl acetate, Aldrich Chem. Co., Inc., Milwaukee, WI
Solvent (A): **benzene** **C_6H_6** **71-43-2**

Type of data: vapor-liquid equilibrium

T/K = 413.15

w_A	0.0017	0.0025	0.0039	0.0058	0.0081	0.0114
a_A	0.011	0.015	0.022	0.032	0.042	0.055

Polymer (B): **poly(vinyl acetate-*co*-vinyl chloride)** **2005TOC**
Characterization: M_w/g.mol^{-1} = 300000, 10 wt% vinyl acetate, Aldrich Chem. Co., Inc., Milwaukee, WI
Solvent (A): **2-butanone** **C_4H_8O** **78-93-3**

Type of data: vapor-liquid equilibrium

T/K = 413.15

w_A	0.0019	0.0030	0.0038	0.0051	0.0086	0.0097
a_A	0.013	0.020	0.026	0.034	0.052	0.057

Polymer (B): **poly(vinyl acetate-*co*-vinyl chloride)** **2005TOC**
Characterization: M_w/g.mol^{-1} = 300000, 10 wt% vinyl acetate, Aldrich Chem. Co., Inc., Milwaukee, WI
Solvent (A): **ethanol** **C_2H_6O** **64-17-5**

Type of data: vapor-liquid equilibrium

T/K = 413.15

w_A	0.0003	0.0006	0.0012	0.0028	0.0036	0.0046
a_A	0.004	0.008	0.015	0.029	0.037	0.045

Polymer (B): **poly(vinyl acetate-*co*-vinyl chloride)** **2005TOC**
Characterization: M_w/g.mol^{-1} = 300000, 10 wt% vinyl acetate, Aldrich Chem. Co., Inc., Milwaukee, WI
Solvent (A): **ethyl acetate** **$C_4H_8O_2$** **141-78-6**

Type of data: vapor-liquid equilibrium

T/K = 413.15

w_A	0.0012	0.0021	0.0035	0.0056	0.0090	0.0125
a_A	0.009	0.015	0.023	0.033	0.047	0.059

Polymer (B): **poly(vinyl acetate-*co*-vinyl chloride)** **2005TOC**
Characterization: M_w/g.mol^{-1} = 300000, 10 wt% vinyl acetate, Aldrich Chem. Co., Inc., Milwaukee, WI
Solvent (A): **n-hexane** **C_6H_{14}** **110-54-3**

Type of data: vapor-liquid equilibrium

T/K = 413.15

w_A	0.0005	0.0008	0.0012	0.0015	0.0019	0.0022
a_A	0.013	0.020	0.030	0.036	0.043	0.050

Polymer (B): **poly(vinyl acetate-*co*-vinyl chloride)** **2005TOC**
Characterization: M_w/g.mol^{-1} = 300000, 10 wt% vinyl acetate, Aldrich Chem. Co., Inc., Milwaukee, WI
Solvent (A): **toluene** **C_7H_8** **108-88-3**

Type of data: vapor-liquid equilibrium

T/K = 413.15

w_A	0.0015	0.0024	0.0049	0.0079	0.0131	0.0187
a_A	0.009	0.015	0.030	0.047	0.074	0.100

Polymer (B): **poly(vinyl acetate-*co*-vinyl chloride)** **2005TOC**
Characterization: M_w/g.mol^{-1} = 300000, 10 wt% vinyl acetate, Aldrich Chem. Co., Inc., Milwaukee, WI
Solvent (A): **vinyl acetate** **$C_4H_6O_2$** **108-05-4**

Type of data: vapor-liquid equilibrium

T/K = 413.15

w_A	0.0013	0.0021	0.0032	0.0049	0.0075	0.0113
a_A	0.010	0.015	0.023	0.032	0.043	0.057

Polymer (B): **poly(vinyl alcohol-*co*-vinyl acetal)** **2005CSA**
Characterization: M_n/g.mol^{-1} = 75600, 7.0 mol% vinyl acetal, prepared and fractionated in the laboratory from Poval 420, Kuraray Co., Japan
Solvent (A): **water** **H_2O** **7732-18-5**

Type of data: vapor-liquid equilibrium

T/K = 298.15

φ_B	0.00574	0.01256	0.01563	0.02079	0.02461	0.03028	0.03538
ln(a_A)	$-2.52\ 10^{-6}$	$-8.36\ 10^{-6}$	$-1.02\ 10^{-5}$	$-1.47\ 10^{-5}$	$-1.91\ 10^{-5}$	$-3.04\ 10^{-5}$	$-3.86\ 10^{-5}$
φ_B	0.04022	0.04962					
ln(a_A)	$-4.68\ 10^{-5}$	$-6.70\ 10^{-5}$					

Polymer (B): **poly(vinyl alcohol-*co*-vinyl butyral)** **2005CSA**
Characterization: M_n/g.mol^{-1} = 77260, 7.0 mol% vinyl butyral, prepared and fractionated in the laboratory from Poval 420, Kuraray Co., Japan
Solvent (A): **water** **H_2O** **7732-18-5**

Type of data: vapor-liquid equilibrium

continued

continued

T/K = 298.15

φ_B	0.00586	0.00970	0.01527	0.02008	0.02416	0.02902	0.03089
$\ln(a_A)$	$-2.73\ 10^{-6}$	$-3.24\ 10^{-6}$	$-4.82\ 10^{-6}$	$-6.90\ 10^{-6}$	$-1.14\ 10^{-5}$	$-1.20\ 10^{-5}$	$-1.36\ 10^{-5}$
φ_B	0.03807	0.04490					
$\ln(a_A)$	$-1.69\ 10^{-5}$	$-2.22\ 10^{-5}$					

Polymer (B): **poly(vinyl alcohol-*co*-vinyl propional)** **2005CSA**
Characterization: M_n/g.mol^{-1} = 76430, 7.0 mol% vinyl propional, prepared and fractionated in the laboratory from Poval 420, Kuraray Co., Japan
Solvent (A): **water** **H_2O** **7732-18-5**

Type of data: vapor-liquid equilibrium

T/K = 298.15

φ_B	0.00409	0.01112	0.01383	0.02045	0.02209	0.03115	0.0350
$\ln(a_A)$	$-1.39\ 10^{-6}$	$-4.30\ 10^{-6}$	$-6.54\ 10^{-6}$	$-1.02\ 10^{-5}$	$-1.13\ 10^{-5}$	$-2.15\ 10^{-5}$	$-2.75\ 10^{-5}$
φ_B	0.04621						
$\ln(a_A)$	$-4.31\ 10^{-5}$						

Polymer (B): **poly(vinyl alcohol-*co*-vinyl propional)** **2005CSA**
Characterization: M_n/g.mol^{-1} = 77270, 9.5 mol% vinyl propional, prepared and fractionated in the laboratory from Poval 420, Kuraray Co., Japan
Solvent (A): **water** **H_2O** **7732-18-5**

Type of data: vapor-liquid equilibrium

T/K = 298.15

φ_B	0.00475	0.00806	0.01122	0.02801	0.02675	0.04273	0.04160
$\ln(a_A)$	$-1.76\ 10^{-6}$	$-3.48\ 10^{-6}$	$-4.46\ 10^{-6}$	$-1.56\ 10^{-5}$	$-1.42\ 10^{-5}$	$-3.19\ 10^{-5}$	$-3.20\ 10^{-5}$
φ_B	0.05267						
$\ln(a_A)$	$-4.49\ 10^{-5}$						

Polymer (B): **poly(4-vinylpyridine-*b*-styrene)** **2000FOR**
Characterization: M_n/g.mol^{-1} = 42100 (20700-*b*-21400), M_w/g.mol^{-1} = 47600, 50.0 wt% styrene, Polymer Source
Solvent (A): **methanol** **CH_4O** **67-56-1**

Type of data: vapor-liquid equilibrium

T/K = 343.15

w_B	0.987	0.981	0.972	0.957	0.944	0.923	0.891	0.861	0.793
a_A	0.081	0.165	0.267	0.368	0.486	0.581	0.680	0.756	0.834

Polymer (B): **poly(4-vinylpyridine-*co*-styrene)** **2000FOR**
Characterization: M_w/g.mol^{-1} = 1200000-1500000, 10.0 wt% styrene,
Aldrich Chem. Co., Inc., Milwaukee, WI
Solvent (A): **methanol** **CH_4O** **67-56-1**

Type of data: vapor-liquid equilibrium

T/K = 343.15

w_B	0.957	0.944	0.922	0.898	0.862	0.806	0.745	0.676	0.595
a_A	0.171	0.255	0.344	0.431	0.533	0.652	0.753	0.840	0.891

Polymer (B): **poly(4-vinylpyridine-*co*-styrene)** **2000FOR**
Characterization: M_w/g.mol^{-1} = 60000, 50.0 wt% styrene,
Scientific Polymer Products, Inc., Ontario, NY
Solvent (A): **methanol** **CH_4O** **67-56-1**

Type of data: vapor-liquid equilibrium

T/K = 343.15

w_B	0.987	0.977	0.966	0.951	0.929	0.906	0.873	0.841	0.776
a_A	0.083	0.168	0.275	0.371	0.487	0.582	0.698	0.772	0.854

Polymer (B): **poly(4-vinylpyridine-*co*-styrene)** **2000FOR**
Characterization: M_w/g.mol^{-1} = 1200000-1500000, 10.0 wt% styrene,
Aldrich Chem. Co., Inc., Milwaukee, WI
Solvent (A): **2-propanol** **C_3H_8O** **67-63-0**

Type of data: vapor-liquid equilibrium

T/K = 343.15

w_B	0.984	0.954	0.934	0.903	0.873	0.821	0.761	0.725
a_A	0.111	0.243	0.368	0.482	0.614	0.747	0.862	0.911

Polymer (B): **poly(4-vinylpyridine-*co*-styrene)** **2000FOR**
Characterization: M_w/g.mol^{-1} = 60000, 50.0 wt% styrene,
Scientific Polymer Products, Inc., Ontario, NY
Solvent (A): **2-propanol** **C_3H_8O** **67-63-0**

Type of data: vapor-liquid equilibrium

T/K = 343.15

w_B	0.985	0.966	0.951	0.927	0.897	0.853	0.800	0.767
a_A	0.111	0.250	0.374	0.501	0.621	0.758	0.874	0.928

2.2. Table of binary systems where data were published only in graphical form as phase diagrams or related figures

Polymer (B)	Solvent (A)	Ref.
Poly(acrylonitrile-*co*-butadiene-*co*-styrene)		
	carbon dioxide	2007NAW
Poly(bisphenol A carbonate-*co*-4,40-(3,3,5-trimethylcyclohexylidene) diphenol carbonate)		
	nitrogen	2005LOP
	oxygen	2005LOP
Poly(2,2-bistrifluoromethyl-4,5-difluoro-1,3-dioxole-*co*-tetrafluoroethylene)		
	n-butane	2002DEA
	carbon dioxide	1999BON
	carbon dioxide	2002DEA
	ethane	1999BON
	ethane	2002DEA
	helium	1999BON
	methane	1999BON
	methane	2002DEA
	nitrogen	1999BON
	nitrogen	2002DEA
	oxygen	1999BON
	oxygen	2002DEA
	propane	1999BON
	propane	2002DEA
	tetrafluoroethane	1999BON
	tetrafluoroethane	2002DEA
	tetrafluoromethane	1999BON
	tetrafluoromethane	2002DEA
Poly(ethylene-*co*-ethyl acrylate)		
	carbon dioxide	2004ARE

Polymer (B)	Solvent (A)	Ref.
Poly(ethylene-*co*-1-hexene)		
	ethene	2006NOV
	1-hexene	2006NOV
Poly(ethylene-*co*-1-octene)		
	carbon dioxide	2007LIG
	n-heptane	2005LEE
	n-hexane	2005LEE
	nitrogen	2007LIG
	n-octane	2005LEE
	n-pentane	2005LEE
Poly(ethylene-*co*-propylene)		
	ethane	2001TSU
	ethene	1994YOO
	ethene	2001TSU
	propane	2001TSU
	propene	1994YOO
	propene	2001TSU
Poly(ethylene-*co*-propylene-*co*-norbornene)		
	argon	2005RUT
	carbon dioxide	2005RUT
	methane	2005RUT
	oxygen	2005RUT
Poly(ethylene-*co*-vinyl acetate)		
	carbon dioxide	2002SHI
	carbon dioxide	1994KAM
	methane	1994KAM
	nitrogen	1994KAM
Poly(ethylene oxide-*b*-1,1'-dihydroperflurooctyl methacrylate)		
	carbon dioxide	2007LIY
Poly(3-hydroxybutyrate-*co*-3-hydroxyvalerate)		
	carbon dioxide	2007CRA

Polymer (B)	Solvent (A)	Ref.
Poly(DL-lactic acid-*co*-glycolic acid)		
	carbon dioxide	2006LIU
	carbon dioxide	2007PIN
	duchloromethane	2009FOS
Poly(methyl methacrylate-*co*-ethylhexyl acrylate)		
	carbon dioxide	2006DU1
Poly(methyl methacrylate-*co*-ethylhexyl acrylate-*co*-ethylene glycol dimethacrylate)		
	carbon dioxide	2005DUA
	carbon dioxide	2006DU1
	carbon dioxide	2006DU2
Poly(styrene-*co*-butadiene)		
	carbon dioxide	2003KOG
Poly[tetrafluoroethylene-*co*-perfluoro(methyl vinyl ether)]		
	carbon dioxide	2005PRA
	carbon dioxide	2006BON
	carbon dioxide	2007FOS
	ethane	2005PRA
	hexfluoroethane	2005PRA
	methane	2005PRA
	nitrogen	2005PRA
	octafluoropropane	2005PRA
	propane	2005PRA
	tetrafluoromethane	2005PRA
Poly(tetrafluoroethylene-*co*-2,2,4-trifluoro-5-trifluoromethoxy-1,3-dioxole)		
	carbon dioxide	2004PRA
	ethane	2004PRA
	methane	2004PRA
	nitrogen	2004PRA
	octafluoropropane	2004PRA
	ozone	2008DIN
	propane	2004PRA

Polymer (B)	**Solvent (A)**	**Ref.**
Poly(tetrahydropyranyl methacrylate-*co*-1H,1H-perfluorooctyl methacrylate)	carbon dioxide	2004PHA
Poly(vinyl acetate-*co*-1-vinyl-2-pyrrolidinone)	carbon dioxide	2003KIK

2.3. Ternary and quaternary copolymer solutions

Polymer (B): **poly(ethylene-*co*-1-butene)** **2006NAG**
Characterization: M_n/g.mol^{-1} = 43700, M_w/g.mol^{-1} = 52000, M_z/g.mol^{-1} = 59000, 4.1 mol% 1-butene, 2.05 ethyl branches per 100 backbone C-atomes, hydrogenated polybutadiene PBD 50000, was denoted as LLDPE, DSM, The Netherlands
Solvent (A): **ethene** C_2H_4 **74-85-1**
Solvent (C): **n-hexane** C_6H_{14} **110-54-3**

Type of data: vapor-liquid equilibrium data

w_A	0.0118	0.0118	0.0118	0.0205	0.0205	0.0298	0.0298	0.0298	0.0099
w_B	0.0502	0.0502	0.0502	0.0501	0.0501	0.0499	0.0499	0.0499	0.0998
w_C	0.9380	0.9380	0.9380	0.9294	0.9294	0.9203	0.9203	0.9203	0.8903
T/K	411.46	416.37	421.20	406.00	410.93	396.49	401.10	406.30	406.11
P/MPa	1.052	1.116	1.181	1.354	1.434	1.529	1.604	1.699	0.969
w_A	0.0099	0.0099	0.0099	0.0196	0.0196	0.0196	0.0196	0.0296	0.0296
w_B	0.0998	0.0998	0.0998	0.1007	0.1007	0.1007	0.1007	0.1002	0.1002
w_C	0.8903	0.8903	0.8903	0.8797	0.8797	0.8797	0.8797	0.8702	0.8702
T/K	411.11	416.06	421.13	401.65	406.43	411.45	416.30	396.26	401.23
P/MPa	1.039	1.109	1.189	1.309	1.374	1.479	1.549	1.580	1.660
w_A	0.0296	0.0098	0.0098	0.0098	0.0210	0.0210	0.0210	0.0210	0.0293
w_B	0.1002	0.1508	0.1508	0.1508	0.1503	0.1503	0.1503	0.1503	0.1506
w_C	0.8702	0.8394	0.8394	0.8394	0.8287	0.8287	0.8287	0.8287	0.8201
T/K	406.21	416.15	421.11	426.07	400.89	405.73	410.70	415.66	396.32
P/MPa	1.750	1.139	1.219	1.299	1.354	1.429	1.504	1.589	1.664
w_A	0.0293	0.0293							
w_B	0.1506	0.1506							
w_C	0.8201	0.8201							
T/K	401.21	406.10							
P/MPa	1.744	1.834							

Polymer (B): **poly(ethylene-*co*-norbornene)** **2003LEE**
Characterization: M_n/g.mol^{-1} = 34800, M_w/g.mol^{-1} = 83300, 50 mol% ethylene, T_g/K = 408.15, Union Chemical Laboratories, Taiwan
Solvent (A): **ethene** C_2H_4 **74-85-1**
Solvent (C): **toluene** C_7H_8 **108-88-3**
Solvent (D): **bicyclo[2,2,1]-2-heptene (norbornene)** C_7H_{10} **498-66-8**

Type of data: gas solubility

continued

continued

Comments: The mass fractions of B/C/D belong to the feed liquid copolymer solution.

T/K = 323.15

$w_B/w_C/w_D$ = 0.04/0.40/0.56

P/bar	5.1	9.9	14.6	19.2	23.8
m_A/(g/g feed liquid copolymer solution)	0.0160	0.0321	0.0502	0.0698	0.0915

$w_B/w_C/w_D$ = 0.16/0.35/0.49

P/bar	6.4	10.8	14.6	19.1	25.2
m_A/(g/g feed liquid copolymer solution)	0.0157	0.0299	0.0436	0.0590	0.0836

$w_B/w_C/w_D$ = 0.28/0.30/0.42

P/bar	5.7	9.4	16.0	20.6	23.3
m_A/(g/g feed liquid copolymer solution)	0.0112	0.0230	0.0376	0.0518	0.0661

T/K = 373.15

$w_B/w_C/w_D$ = 0.04/0.40/0.56

P/bar	6.1	10.7	15.4	20.5	23.7
m_A/(g/g feed liquid copolymer solution)	0.0111	0.0224	0.0333	0.0465	0.0563

$w_B/w_C/w_D$ = 0.16/0.35/0.49

P/bar	5.6	9.9	13.7	19.4	25.6
m_A/(g/g feed liquid copolymer solution)	0.0088	0.0169	0.0251	0.0380	0.0530

T/K = 423.15

$w_B/w_C/w_D$ = 0.04/0.40/0.56

P/bar	7.0	10.5	14.8	19.9	25.3
m_A/(g/g feed liquid copolymer solution)	0.0055	0.0118	0.0179	0.0288	0.0385

$w_B/w_C/w_D$ = 0.16/0.35/0.49

P/bar	6.8	10.7	15.4	20.0	23.9
m_A/(g/g feed liquid copolymer solution)	0.0047	0.0100	0.0171	0.0250	0.0313

Polymer (B): **poly(ethylene-*co*-norbornene)** **2003LEE**
Characterization: M_n/g.mol^{-1} = 34800, M_w/g.mol^{-1} = 83300, 50 mol% ethylene, T_g/K = 408.15, Union Chemical Laboratories, Taiwan
Solvent (A): **ethene** C_2H_4 **74-85-1**
Solvent (C): **toluene** C_7H_8 **108-88-3**
Solvent (D): **bicyclo[2,2,1]-2-heptene (norbornene)** C_7H_{10} **498-66-8**

Type of data: vapor-liquid equilibrium

continued

continued

T/K = 323.15

	liquid phase				vapor phase		
P/bar	x_A	x_B	x_C	x_D	y_A	y_C	y_D
5.1	0.0526	0.0001	0.3995	0.5478	0.9625	0.0088	0.0287
9.9	0.1001	0.0001	0.3794	0.5204	0.9767	0.0057	0.0176
14.6	0.1481	0.0001	0.3592	0.4926	0.9860	0.0037	0.0103
19.2	0.1948	0.0001	0.3395	0.4656	0.9876	0.0030	0.0094
23.8	0.2406	0.0001	0.3202	0.4391	0.9893	0.0029	0.0078
6.4	0.0583	0.0005	0.3968	0.5444	0.9660	0.0093	0.0247
10.8	0.1057	0.0004	0.3768	0.5171	0.9852	0.0034	0.0114
14.6	0.1471	0.0004	0.3594	0.4931	0.9867	0.0033	0.0100
19.1	0.1894	0.0004	0.3416	0.4686	0.9900	0.0021	0.0079
25.2	0.2486	0.0004	0.3166	0.4344	0.9908	0.0024	0.0068
5.7	0.0494	0.0010	0.4003	0.5493	0.9622	0.0122	0.0256
9.4	0.0958	0.0010	0.3807	0.5225	0.9757	0.0072	0.0171
16.0	0.1479	0.0009	0.3588	0.4924	0.9855	0.0043	0.0102
20.6	0.1928	0.0008	0.3399	0.4665	0.9886	0.0032	0.0082
23.3	0.2339	0.0008	0.3227	0.4426	0.9902	0.0032	0.0066
T/K = 373.15							
6.1	0.0371	0.0001	0.4147	0.5481	0.8305	0.0681	0.1014
10.7	0.0719	0.0001	0.3997	0.5283	0.9012	0.0276	0.0712
15.4	0.1032	0.0001	0.3863	0.5104	0.9257	0.0208	0.0535
20.5	0.1385	0.0001	0.3710	0.4904	0.9411	0.0167	0.0422
23.7	0.1630	0.0001	0.3605	0.4764	0.9476	0.0153	0.0371
5.6	0.0337	0.0005	0.4090	0.5568	0.8305	0.0762	0.0933
9.9	0.0627	0.0005	0.3968	0.5400	0.8919	0.0614	0.0467
13.7	0.0904	0.0005	0.3850	0.5241	0.9210	0.0371	0.0419
19.4	0.1306	0.0004	0.3680	0.5010	0.9402	0.0258	0.0340
25.6	0.1734	0.0004	0.3499	0.4763	0.9500	0.0238	0.0262
T/K = 423.15							
7.0	0.0185	0.0001	0.4149	0.5665	0.4610	0.1940	0.3450
10.5	0.0393	0.0001	0.4061	0.5545	0.6322	0.1223	0.2455
14.8	0.0583	0.0001	0.3980	0.5436	0.7084	0.0942	0.1974
19.9	0.0906	0.0001	0.3844	0.5249	0.7833	0.0692	0.1475
25.3	0.1177	0.0001	0.3729	0.5093	0.8162	0.0469	0.1369
6.8	0.0181	0.0005	0.4140	0.5674	0.4725	0.1784	0.3491
10.7	0.0382	0.0005	0.4055	0.5558	0.6479	0.1113	0.2408
15.4	0.0632	0.0005	0.3950	0.5413	0.7411	0.0820	0.1769
20.0	0.0866	0.0005	0.3851	0.5278	0.7878	0.0655	0.1467
23.9	0.1102	0.0005	0.3752	0.5141	0.8170	0.0559	0.1271

Polymer (B):	**poly(ethylene-*b*-propylene)**		**2000SA2**
Characterization:	M_w/g.mol^{-1} = 240000, 4.9 wt% ethene, 90% isotactic, 58.1 wt% crystallinity, impact propylene copolymer, Mitsubishi Chemical, Kurashiki, Japan		
Solvent (A):	**ethene**	C_2H_4	**74-85-1**
Solvent (C):	**n-hexane**	C_6H_{14}	**110-54-3**

Type of data: weight fraction-based partition coefficient at infinite dilution, $K^\infty = w_C^{gas}/w_C^{liq}$

T/K = 323.15

P/MPa	0.6	1.2	1.8
K^∞	30.6	15.2	12.0

T/K = 343.15

P/MPa	0.6	1.2	1.8	2.4
K^∞	43.9	20.7	15.9	13.1

T/K = 363.15

P/MPa	0.6	1.2	1.8	2.4
K^∞	59.7	31.9	23.2	17.4

Comments: The weight fraction-based partition coefficients of n-hexane at infinite dilution are given for the equilibrated system of C_2H_4 + polymer at the binary system equilibrium temperature, pressure, and concentration (see Chapter 2.1).

Polymer (B):	**poly(ethylene-*b*-propylene)**		**2000SA2**
Characterization:	M_w/g.mol^{-1} = 240000, 4.9 wt% ethene, 90% isotactic, 58.1 wt% crystallinity, impact propylene copolymer, Mitsubishi Chemical, Kurashiki, Japan		
Solvent (A):	**propene**	C_3H_6	**115-07-1**
Solvent (C):	**n-hexane**	C_6H_{14}	**110-54-3**

Type of data: weight fraction-based partition coefficient at infinite dilution, $K^\infty = w_C^{gas}/w_C^{liq}$

T/K = 323.15

P/MPa	1.2	1.6	1.8
K^∞	12.4	9.4	8.6

T/K = 343.15

P/MPa	1.2	1.6	1.8	2.4
K^∞	21.2	15.7	14.1	11.2

T/K = 363.15

P/MPa	1.2	1.8	2.4
K^∞	38.2	20.4	14.7

Comments: The weight fraction-based partition coefficients of n-hexane at infinite dilution are given for the equilibrated system of C_2H_4 + polymer at the binary system equilibrium temperature, pressure, and concentration (see Chapter 2.1).

Polymer (B):	**poly(ethylene oxide-*co*-propylene oxide)**		**2000LIW**
Characterization:	M_n/g.mol^{-1} = 780, M_w/g.mol^{-1} = 865, 50.0 mol% ethylene oxide, Zhejiang Univ. Chem. Factory, PR China		
Solvent (A):	**water**	**H_2O**	**7732-18-5**
Salt (C):	**ammonium sulfate**	**$(NH_4)_2SO_4$**	**7783-20-2**

Type of data: vapor-liquid equilibrium data

T/K = 298.15

w_A	0.8965	0.9237	0.9417	0.9658	0.7882	0.8324	0.8639	0.9116	0.7331
w_B	0.1011	0.0710	0.0510	0.0239	0.2068	0.1560	0.1191	0.0618	0.2606
w_C	0.00244	0.00530	0.00728	0.0103	0.00498	0.0116	0.0170	0.0266	0.00628
a_A	0.99567	0.99567	0.99567	0.99567	0.9882	0.9882	0.9882	0.9882	0.9825

w_A	0.7828	0.8191	0.8774	0.6329	0.6874	0.7306	0.8014
w_B	0.2021	0.1583	0.0859	0.3585	0.2909	0.2358	0.1388
w_C	0.0151	0.0226	0.0367	0.00863	0.0217	0.0336	0.0598
a_A	0.9825	0.9825	0.9825	0.9675	0.9675	0.9675	0.9675

Polymer (B):	**poly(ethylene oxide-*co*-propylene oxide)**		**2000LIW**
Characterization:	M_n/g.mol^{-1} = 3640, M_w/g.mol^{-1} = 4040, 50.0 mol% ethylene Oxide, Zhejiang Univ. Chem. Factory, PR China		
Solvent (A):	**water**	**H_2O**	**7732-18-5**
Salt (C):	**ammonium sulfate**	**$(NH_4)_2SO_4$**	**7783-20-2**

Type of data: vapor-liquid equilibrium data

T/K = 298.15

w_A	0.8974	0.9222	0.9454	0.9703	0.8129	0.8491	0.8830	0.9263	0.7653
w_B	0.1017	0.0758	0.0517	0.0259	0.1855	0.1471	0.1109	0.0643	0.2326
w_C	0.000904	0.00196	0.00286	0.00379	0.00165	0.00381	0.00615	0.00938	0.00207
a_A	0.99824	0.99824	0.99824	0.99824	0.99539	0.99539	0.99539	0.99539	0.99322

w_A	0.8026	0.8416	0.8958	0.6520	0.6914	0.7358	0.8063
w_B	0.1924	0.1501	0.0909	0.3449	0.3008	0.2503	0.1690
w_C	0.00499	0.00832	0.0133	0.00306	0.00779	0.0139	0.0247
a_A	0.99322	0.99322	0.99322	0.9840	0.9840	0.9840	0.9840

Polymer (B):	**poly(styrene-*co*-butadiene)**		**2005WUH**
Characterization:	M_w/g.mol^{-1} = 160000, 18 wt% styrene, SBR, Petrofina, Inc., Brussels, Belgium		
Solvent (A):	**ethene**	**C_2H_4**	**74-85-1**
Solvent (C):	**toluene**	**C_7H_8**	**108-88-3**

Type of data: gas solubility

Comments: w_B/w_C = 10/90 was kept constant.

T/K = 293.15

continued

continued

P/bar	3.45	5.52	6.89	10.34	12.41
w_A	0.0142	0.0259	0.0313	0.0452	0.0596

T/K = 303.15

P/bar	3.45	5.52	6.89	10.34	12.41
w_A	0.0118	0.0213	0.0255	0.0411	0.0545

T/K = 313.15

P/bar	3.45	5.52	6.89	10.34	12.41
w_A	0.0119	0.0143	0.0231	0.0322	0.0434

T/K = 323.15

P/bar	3.45	5.52	6.89	10.34	12.41
w_A	0.0093	0.0124	0.0202	0.0247	0.0389

T/K = 333.15

P/bar	3.45	5.52	6.89	10.34	12.41
w_A	0.0063	0.0123	0.0182	0.0253	0.0319

T/K = 343.15

P/bar	3.45	5.52	6.89	10.34	12.41
w_A	0.0054	0.0128	0.0180	0.0259	0.0293

Polymer (B): **poly(vinyl acetate-*co*-vinyl alcohol)** **2002SEI**
Characterization: M_w/g.mol^{-1} = 6000, 20 mol% vinyl acetate, Polysciences, Inc., Warrington, PA, USA
Solvent (A): **water** **H_2O** **7732-18-5**
Solvent (C): **ethanol** **C_2H_6O** **64-17-5**

Type of data: vapor-liquid equilibrium

T/K = 363.15 w_B = 0.040 was kept constant

x_C	0.1996	0.3998	0.5997	0.7997	0.8999
y_C	0.5335	0.6093	0.6925	0.8211	0.8986

Polymer (B): **poly(vinyl alcohol-*co*-sodium acrylate)** **2004MAT**
Characterization: 40 mol% sodium acrylate, Sumitomo Chemical Co., Ltd., Japan
Solvent (A): **water** **H_2O** **7732-18-5**
Solvent (C): **ethanol** **C_2H_6O** **64-17-5**

Type of data: vapor-liquid equilibrium

Comments: The mole fraction of ethanol in the liquid mixture, x_C, is given for the polymer-free solvent mixture. The copolymer was considered to be a physical gel, i.e., cross-linked by hydrogen bonds.

continued

continued

w_B	0.05	0.05	0.05	0.05	0.05	0.10	0.10	0.10	0.10
x_C	0.400	0.500	0.600	0.700	0.800	0.400	0.500	0.600	0.700
y_C	0.661	0.699	0.735	0.785	0.839	0.676	0.721	0.760	0.796
P_A/kPa	2.39	2.24	2.07	1.71	1.30	2.39	2.04	1.86	1.60
P_C/kPa	4.65	5.21	5.75	6.26	6.77	4.99	5.26	5.89	6.23

w_B	0.10	0.15	0.15	0.15	0.15	0.15
x_C	0.800	0.400	0.500	0.600	0.700	0.800
y_C	0.858	0.697	0.733	0.782	0.822	0.860
P_A/kPa	1.15	2.20	2.03	1.68	1.41	1.08
P_C/kPa	6.93	5.07	5.57	6.01	6.51	6.64

Polymer (B): **poly(vinyl alcohol-*co*-sodium acrylate)** **1996MIS**
Characterization: 40 mol% sodium acrylate, Sumitomo Chemical Co., Ltd., Japan
Solvent (A): **water** **H_2O** **7732-18-5**
Solvent (C): **1-propanol** **C_3H_8O** **71-23-8**

Type of data: vapor-liquid equilibrium

Comments: The copolymer was considered to be a physical gel, i.e., cross-linked by hydrogen bonds.

w_A	0.5182	0.3912	0.2948	0.2192	0.1589	0.1082	0.0667	0.4909	0.3706
w_B	0.0500	0.0500	0.0500	0.0500	0.0500	0.0500	0.0500	0.1000	0.1000
w_C	0.4318	0.5588	0.6552	0.7308	0.7911	0.8418	0.8833	0.4091	0.5294
y_C	0.368	0.384	0.404	0.420	0.451	0.504	0.608	0.375	0.393
P_A/kPa	2.87	2.81	2.74	2.69	2.43	2.09	1.54	2.87	2.79
P_C/kPa	1.67	1.75	1.86	1.95	2.00	2.12	2.39	1.72	1.81

w_A	0.2833	0.2090	0.1526	0.1025	0.0628	0.4623	0.3510	0.2638	0.1962
w_B	0.1000	0.1000	0.1000	0.1000	0.1000	0.1500	0.1500	0.1500	0.1500
w_C	0.6167	0.6910	0.7474	0.7975	0.8372	0.3877	0.4990	0.5862	0.6538
y_C	0.400	0.438	0.489	0.547	0.629	0.379	0.403	0.421	0.481
P_A/kPa	2.85	2.30	2.13	1.92	1.37	2.85	2.75	2.63	2.19
P_C/kPa	1.90	1.79	2.04	2.32	2.32	1.76	1.86	1.91	2.03

w_A	0.1417	0.0977	0.0590
w_B	0.1500	0.1500	0.1500
w_C	0.7083	0.7523	0.7910
y_C	0.526	0.592	0.653
P_A/kPa	1.97	1.59	1.31
P_C/kPa	2.19	2.31	2.46

Polymer (B): **poly(vinyl alcohol-*co*-sodium acrylate)** **2004MAT**
Characterization: 40 mol% sodium acrylate, Sumitomo Chemical Co., Ltd., Japan
Solvent (A): **water** **H_2O** **7732-18-5**
Solvent (C): **2-propanol** **C_3H_8O** **67-63-0**

Type of data: vapor-liquid equilibrium

continued

continued

Comments: The mole fraction of 2-propanol in the liquid mixture, x_C, is given for the polymer-free solvent mixture. The copolymer was considered to be a physical gel, i.e., cross-linked by hydrogen bonds.

w_B	0.05	0.05	0.05	0.05	0.05	0.05	0.05	0.10	0.10
x_C	0.200	0.299	0.399	0.500	0.599	0.700	0.797	0.200	0.300
y_C	0.558	0.599	0.610	0.632	0.669	0.727	0.784	0.560	0.591
P_A/kPa	2.72	2.63	2.44	2.40	2.23	1.81	1.47	2.76	2.57
P_C/kPa	3.44	3.68	3.82	4.13	4.50	4.81	5.35	3.51	3.72

w_B	0.10	0.10	0.10	0.10	0.10	0.15	0.15	0.15	0.15
x_C	0.400	0.500	0.599	0.700	0.799	0.200	0.299	0.400	0.500
y_C	0.618	0.650	0.691	0.740	0.804	0.579	0.603	0.630	0.689
P_A/kPa	2.50	2.34	1.99	1.73	1.28	2.69	2.62	2.42	2.14
P_C/kPa	4.04	4.34	4.46	4.93	5.24	3.70	3.98	4.12	4.73

w_B	0.15	0.15	0.15
x_C	0.600	0.700	0.800
y_C	0.718	0.761	0.821
P_A/kPa	1.94	1.57	1.15
P_C/kPa	4.94	4.99	5.28

Polymer (B): **poly(vinyl alcohol-*co*-vinyl propional)** **2005CSA**
Characterization: M_n/g.mol^{-1} = 76430, 7.0 mol% vinyl propional, prepared and fractionated in the laboratory from Poval 420, Kuraray Co., Japan
Solvent (A): **water** **H_2O** **7732-18-5**
Polymer (C): **poly(1-vinyl-2-pyrrolidinone)**
Characterization: M_n/g.mol^{-1} = 117000, fractionated in the laboratory from Fluka K90 PVP, Fluka AG, Buchs, Switzerland

Type of data: vapor-liquid equilibrium

Comments: w_B/w_C = 1/1 was kept constant.

T/K = 298.15

φ_{B+C}	0.01067	0.01413	0.01863	0.02328	0.02824	0.03641	0.04602
ln(a_A)	$-5.74\ 10^{-6}$	$-6.32\ 10^{-6}$	$-1.09\ 10^{-5}$	$-1.74\ 10^{-5}$	$-2.23\ 10^{-5}$	$-3.77\ 10^{-5}$	$-5.45\ 10^{-5}$

Polymer (B): **poly(vinylidene fluoride-*co*-chlorotrifluoro-ethylene)** **2005SOL**
Characterization: see comments
Solvent (A): **carbon dioxide** **CO_2** **124-38-9**
Polymer (C): **poly(vinylidene fluoride)**
Characterization: see comments

continued

continued

Comments: The polymer mixture is a commercial polymer SOLEF VF2-CTFE copolymer grade 60512 consiting of 67% (B) and 33% (C) plus a small amount of high-density polyethylene. The density of this alloy is 1.77 g/cm^3, its crystallinity is 45% (by DSC).

Type of data: gas solubility

T/K = 353.15

P/bar	19.1	19.4	31.2	31.0	38.9	38.78
w_A	0.00633	0.00675	0.00848	0.00883	0.01031	0.01388

T/K = 373.15

P/bar	20.1	20.1	28.1	28.1	37.6	37.6
w_A	0.00712	0.00614	0.00890	0.00867	0.01098	0.01106

T/K = 393.15

P/bar	18.6	18.6	29.0	28.5	28.5	29.0	38.4	38.4
w_A	0.00570	0.00411	0.00759	0.00743	0.00599	0.00538	0.00932	0.00854

Polymer (B): **poly(vinylidene fluoride-*co*-chlorotrifluoro-ethylene)** **2005SOL**
Characterization: see comments
Solvent (A): **methane** **CH_4** **74-82-8**
Polymer (C): **poly(vinylidene fluoride)**
Characterization: see comments

Comments: The polymer mixture is a commercial polymer SOLEF VF2-CTFE copolymer grade 60512 consiting of 67% (B) and 33% (C) plus a small amount of high-density polyethylene. The density of this alloy is 1.77 g/cm^3, its crystallinity is 45% (by DSC).

Type of data: gas solubility

T/K = 353.15

P/bar	53.7	53.5	104.8	104.8	158.0	158.3
w_A	0.00148	0.00147	0.00387	0.00220	0.00474	0.00336

T/K = 373.15

P/bar	51.4	51.4	50.0	50.0	100.1	100.1	158.2	158.2
w_A	0.00131	0.00051	0.00157	0.00069	0.00255	0.00120	0.00341	0.00276

T/K = 393.15

P/bar	42.4	54.2	54.2	71.9	97.9	97.9	158.5	158.3
w_A	0.00116	0.00111	0.00061	0.00140	0.00217	0.00139	0.00331	0.00250

2.4. Table of ternary or quaternary systems where data were published only in graphical form as phase diagrams or related figures

Polymer (B)	Second and third component	Ref.
Poly(ethylene-*co*-1-hexene)		
	ethene and 1-hexene	2006NOV
Poly(ethylene-*co*-propylene)		
	ethene and propene	1994YOO
Poly(ethylene oxide-*b*-propylene oxide-*b*-ethylene oxide)		
	carbon dioxide and p-xylene	2002ZHA
	ethene and p-xylene	2005ZHA
	ethene and water/p-xylene	2005ZHA

2.5. Classical mass-fraction Henry's constants of solvent vapors in molten copolymers

Polymer (B): **poly(*tert*-butyl acrylate-*b*-methyl methacrylate)** **2008AYD**
Characterization: M_n/g.mol^{-1} = 69000, M_w/g.mol^{-1} = 80700,
5 mol% tert-butyl acrylate, synthesized in the laboratory

Solvent (A)	T/ K	$H_{A,B}$/ MPa	Solvent (A)	T/ K	$H_{A,B}$/ MPa
ethyl acetate	423.15	5.064	n-decane	443.15	2.125
ethyl acetate	433.15	4.891	toluene	423.15	1.838
ethyl acetate	443.15	5.653	toluene	433.15	2.179
propyl acetate	423.15	2.515	toluene	443.15	2.525
propyl acetate	433.15	2.949	ethylbenzene	423.15	1.177
propyl acetate	443.15	3.369	ethylbenzene	433.15	1.259
3-methylbutyl acetate	423.15	0.9850	ethylbenzene	443.15	1.616
3-methylbutyl acetate	433.15	1.203	chlorobenzene	423.15	0.7418
3-methylbutyl acetate	443.15	1.590	chlorobenzene	433.15	0.8731
n-octane	423.15	4.498	chlorobenzene	443.15	1.177
n-octane	433.15	4.983	propylbenzene	423.15	0.6777
n-octane	443.15	6.174	propylbenzene	433.15	0.8064
n-nonane	423.15	2.784	propylbenzene	443.15	1.025
n-nonane	433.15	2.976	isopropylbenzene	423.15	0.8120
n-nonane	443.15	3.145	isopropylbenzene	433.15	0.9382
n-decane	423.15	1.698	isopropylbenzene	443.15	1.157
n-decane	433.15	1.820			

Polymer (B): **poly(*tert*-butyl acrylate-*b*-methyl methacrylate)** **2006SAK**
Characterization: M_n/g.mol^{-1} = 69000 (9000-*b*-60000),
5 mol% *tert*-butyl acrylate, synthesized in the laboratory

Solvent (A)	T/ K	$H_{A,B}$/ MPa	Solvent (A)	T/ K	$H_{A,B}$/ MPa
benzene	413.15	3.007	isobutyl acetate	433.15	2.199
benzene	423.15	3.576	isobutyl acetate	443.15	2.607
benzene	433.15	4.038	*tert*-butyl acetate	413.15	1.145
benzene	443.15	4.674	*tert*-butyl acetate	423.15	1.388

continued

continued

Solvent (A)	T/ K	$H_{A,B}$/ MPa	Solvent (A)	T/ K	$H_{A,B}$/ MPa
butyl acetate	413.15	1.220	*tert*-butyl acetate	433.15	1.655
butyl acetate	423.15	1.420	*tert*-butyl acetate	443.15	2.207
butyl acetate	433.15	1.750	3-methylbutyl acetate	413.15	0.8518
butyl acetate	443.15	2.007	3-methylbutyl acetate	423.15	0.9850
chlorobenzene	413.15	0.5600	3-methylbutyl acetate	433.15	1.203
chlorobenzene	423.15	0.7418	3-methylbutyl acetate	443.15	1.590
chlorobenzene	433.15	0.8731	propyl acetate	413.15	2.094
chlorobenzene	443.15	1.177	propyl acetate	423.15	2.515
cumene	413.15	0.7827	propyl acetate	433.15	2.949
cumene	423.15	0.8120	propyl acetate	443.15	3.369
cumene	433.15	0.9382	2-propyl acetate	413.15	3.163
cumene	443.15	1.157	2-propyl acetate	423.15	3.453
ethyl acetate	413.15	3.646	2-propyl acetate	433.15	4.557
ethyl acetate	423.15	5.064	2-propyl acetate	443.15	5.358
ethyl acetate	433.15	4.891	propylbenzene	413.15	0.5412
ethyl acetate	443.15	5.653	propylbenzene	423.15	0.6777
ethylbenzene	413.15	0.9469	propylbenzene	433.15	0.8064
ethylbenzene	423.15	1.177	propylbenzene	443.15	1.025
ethylbenzene	433.15	1.259	toluene	413.15	1.456
ethylbenzene	443.15	1.616	toluene	423.15	1.838
isobutyl acetate	413.15	1.665	toluene	433.15	2.179
isobutyl acetate	423.15	1.989	toluene	443.15	2.525

Polymer (B): **poly(*tert*-butyl acrylate-*b*-methyl methacrylate)** **2008AYD**
Characterization: M_n/g.mol^{-1} = 62000, M_w/g.mol^{-1} = 45000,
16 mol% *tert*-butyl acrylate, synthesized in the laboratory

Solvent (A)	T/ K	$H_{A,B}$/ MPa	Solvent (A)	T/ K	$H_{A,B}$/ MPa
ethyl acetate	418.15	3.383	n-decane	433.15	1.709
ethyl acetate	423.15	3.757	n-decane	438.15	1.752
ethyl acetate	428.15	4.021	n-decane	443.15	1.793
ethyl acetate	433.15	4.184	toluene	418.15	2.014
ethyl acetate	438.15	4.361	toluene	423.15	2.044
ethyl acetate	443.15	4.603	toluene	428.15	2.092
propyl acetate	418.15	2.285	toluene	433.15	2.160
propyl acetate	423.15	2.366	toluene	438.15	2.343
propyl acetate	428.15	2.574	toluene	443.15	2.450

continued

continued

Solvent (A)	T/ K	$H_{A,B}$/ MPa	Solvent (A)	T/ K	$H_{A,B}$/ MPa
propyl acetate	433.15	2.752	ethylbenzene	418.15	1.260
propyl acetate	438.15	2.914	ethylbenzene	423.15	1.273
propyl acetate	443.15	3.084	ethylbenzene	428.15	1.309
3-methylbutyl acetate	418.15	1.203	ethylbenzene	433.15	1.352
3-methylbutyl acetate	423.15	1.214	ethylbenzene	438.15	1.441
3-methylbutyl acetate	428.15	1.238	ethylbenzene	443.15	1.543
3-methylbutyl acetate	433.15	1.261	chlorobenzene	418.15	0.8169
3-methylbutyl acetate	438.15	1.302	chlorobenzene	423.15	0.8637
3-methylbutyl acetate	443.15	1.398	chlorobenzene	428.15	0.9251
n-octane	418.15	3.875	chlorobenzene	433.15	1.004
n-octane	423.15	3.906	chlorobenzene	438.15	1.114
n-octane	428.15	3.976	chlorobenzene	443.15	1.144
n-octane	433.15	3.992	propylbenzene	418.15	0.8130
n-octane	438.15	4.016	propylbenzene	423.15	0.8405
n-octane	443.15	4.194	propylbenzene	428.15	0.8707
n-nonane	418.15	2.452	propylbenzene	433.15	0.9154
n-nonane	423.15	2.544	propylbenzene	438.15	0.9795
n-nonane	428.15	2.551	propylbenzene	443.15	1.036
n-nonane	433.15	2.551	isopropylbenzene	418.15	1.136
n-nonane	438.15	2.589	isopropylbenzene	423.15	1.144
n-nonane	443.15	2.754	isopropylbenzene	428.15	1.145
n-decane	418.15	1.659	isopropylbenzene	433.15	1.151
n-decane	423.15	1.670	isopropylbenzene	438.15	1.165
n-decane	428.15	1.696	isopropylbenzene	443.15	1.196

Polymer (B): **poly(3,4-dichlorobenzyl methacrylate-*co*-ethyl methacrylate)** **2001DEM**

Characterization: 13 mol% ethyl methacrylate, synthesized in the laboratory

Solvent (A)	T/ K	$H_{A,B}$/ MPa	Solvent (A)	T/ K	$H_{A,B}$/ MPa
methanol	403.15	23.47	toluene	403.15	4.668
methanol	413.15	25.59	toluene	413.15	5.233
methanol	423.15	29.17	toluene	423.15	6.086
ethanol	403.15	16.32	1,2-dimethylbenzene	403.15	2.417
ethanol	413.15	16.54	1,2-dimethylbenzene	413.15	3.000
ethanol	423.15	18.46	1,2-dimethylbenzene	423.15	3.542

continued

continued

Solvent (A)	T/ K	$H_{A,B}$/ MPa	Solvent (A)	T/ K	$H_{A,B}$/ MPa
2-propanone	403.15	14.32	n-octane	403.15	5.746
2-propanone	413.15	15.58	n-octane	413.15	6.174
2-propanone	423.15	16.78	n-octane	423.15	6.627
2-butanone	403.15	8.605	n-nonane	403.15	4.061
2-butanone	413.15	9.573	n-nonane	413.15	4.351
2-butanone	423.15	10.68	n-nonane	423.15	4.799
methyl acetate	403.15	10.79	n-decane	403.15	2.433
methyl acetate	413.15	11.66	n-decane	413.15	2.820
methyl acetate	423.15	13.05	n-decane	423.15	3.579
ethyl acetate	403.15	7.471	n-undecane	403.15	1.823
ethyl acetate	413.15	8.055	n-undecane	413.15	1.961
ethyl acetate	423.15	8.888	n-undecane	423.15	2.475
benzene	403.15	6.955	n-dodecane	403.15	0.8818
benzene	413.15	7.532	n-dodecane	413.15	1.081
benzene	423.15	8.551	n-dodecane	423.15	1.362

Polymer (B): **poly(3,4-dichlorobenzyl methacrylate-*co*-ethyl methacrylate)** **2001KA1**

Characterization: 27 mol% ethyl methacrylate, T_g/K = 336.2, ρ (298 K) = 1.09 g/cm^3, synthesized in the laboratory

Solvent (A)	T/ K	$H_{A,B}$/ MPa	Solvent (A)	T/ K	$H_{A,B}$/ MPa
methanol	403.15	24.11	toluene	403.15	5.072
methanol	413.15	27.05	toluene	413.15	5.799
methanol	423.15	29.05	toluene	423.15	6.320
ethanol	403.15	15.70	1,2-dimethylbenzene	403.15	2.899
ethanol	413.15	17.12	1,2-dimethylbenzene	413.15	3.301
ethanol	423.15	19.33	1,2-dimethylbenzene	423.15	3.766
2-propanone	403.15	13.30	n-octane	403.15	6.194
2-propanone	413.15	14.70	n-octane	413.15	6.272
2-propanone	423.15	15.83	n-octane	423.15	6.694
2-butanone	403.15	8.897	n-nonane	403.15	4.540
2-butanone	413.15	9.573	n-nonane	413.15	4.946
2-butanone	423.15	10.79	n-nonane	423.15	5.398
methyl acetate	403.15	10.91	n-decane	403.15	2.478

continued

continued

Solvent (A)	T/ K	$H_{A,B}$/ MPa	Solvent (A)	T/ K	$H_{A,B}$/ MPa
methyl acetate	413.15	11.93	n-decane	413.15	3.244
methyl acetate	423.15	12.83	n-decane	423.15	3.961
ethyl acetate	403.15	7.980	n-undecane	403.15	2.038
ethyl acetate	413.15	8.592	n-undecane	413.15	2.292
ethyl acetate	423.15	9.239	n-undecane	423.15	2.711
benzene	403.15	7.631	n-dodecane	403.15	1.122
benzene	413.15	8.705	n-dodecane	413.15	1.333
benzene	423.15	9.409	n-dodecane	423.15	1.718

Polymer (B): **poly(3,4-dichlorobenzyl methacrylate-*co*-ethyl methacrylate)** **2001KA1**

Characterization: 60 mol% ethyl methacrylate, T_g/K = 334.2, ρ (298 K) = 1.14 g/cm^3, synthesized in the laboratory

Solvent (A)	T/ K	$H_{A,B}$/ MPa	Solvent (A)	T/ K	$H_{A,B}$/ MPa
methanol	403.15	22.15	toluene	403.15	4.871
methanol	413.15	25.59	toluene	413.15	5.453
methanol	423.15	27.69	toluene	423.15	5.925
ethanol	403.15	13.77	1,2-dimethylbenzene	403.15	2.818
ethanol	413.15	14.94	1,2-dimethylbenzene	413.15	3.174
ethanol	423.15	16.06	1,2-dimethylbenzene	423.15	3.727
2-propanone	403.15	12.74	n-octane	403.15	5.507
2-propanone	413.15	15.46	n-octane	413.15	5.970
2-propanone	423.15	17.30	n-octane	423.15	6.476
2-butanone	403.15	8.181	n-nonane	403.15	4.166
2-butanone	413.15	8.872	n-nonane	413.15	4.506
2-butanone	423.15	9.602	n-nonane	423.15	4.905
methyl acetate	403.15	9.986	n-decane	403.15	3.099
methyl acetate	413.15	11.07	n-decane	413.15	3.396
methyl acetate	423.15	11.84	n-decane	423.15	3.695
ethyl acetate	403.15	7.062	n-undecane	403.15	2.052
ethyl acetate	413.15	7.672	n-undecane	413.15	2.362
ethyl acetate	423.15	7.220	n-undecane	423.15	2.656
benzene	403.15	6.989	n-dodecane	403.15	1.277
benzene	413.15	7.493	n-dodecane	413.15	1.540
benzene	423.15	8.283	n-dodecane	423.15	1.831

Polymer (B): **poly(3,4-dichlorobenzyl methacrylate-*co*-ethyl methacrylate)** **2001KA1**

Characterization: 70 mol% ethyl methacrylate, T_g/K = 332.2, ρ(298 K) = 1.14 g/cm^3, synthesized in the laboratory

Solvent (A)	T/ K	$H_{A,B}$/ MPa	Solvent (A)	T/ K	$H_{A,B}$/ MPa
methanol	403.15	15.68	toluene	403.15	3.668
methanol	413.15	16.88	toluene	413.15	4.081
methanol	423.15	18.27	toluene	423.15	4.506
ethanol	403.15	10.16	1,2-dimethylbenzene	403.15	2.097
ethanol	413.15	10.93	1,2-dimethylbenzene	413.15	2.482
ethanol	423.15	11.68	1,2-dimethylbenzene	423.15	2.830
2-propanone	403.15	9.377	n-octane	403.15	3.780
2-propanone	413.15	10.21	n-octane	413.15	4.082
2-propanone	423.15	11.53	n-octane	423.15	4.379
2-butanone	403.15	5.920	n-nonane	403.15	2.908
2-butanone	413.15	6.402	n-nonane	413.15	3.151
2-butanone	423.15	7.207	n-nonane	423.15	3.472
methyl acetate	403.15	7.130	n-decane	403.15	2.157
methyl acetate	413.15	7.742	n-decane	413.15	2.351
methyl acetate	423.15	8.331	n-decane	423.15	2.634
ethyl acetate	403.15	5.166	n-undecane	403.15	1.453
ethyl acetate	413.15	5.591	n-undecane	413.15	1.664
ethyl acetate	423.15	6.181	n-undecane	423.15	1.929
benzene	403.15	5.305	n-dodecane	403.15	0.9483
benzene	413.15	5.757	n-dodecane	413.15	1.143
benzene	423.15	6.307	n-dodecane	423.15	1.393

Polymer (B): **poly(3,4-dichlorobenzyl methacrylate-*co*-ethyl methacrylate)** **2001KA1**

Characterization: 82 mol% ethyl methacrylate, T_g/K = 331.2, ρ(298 K) = 1.17 g/cm^3, synthesized in the laboratory

Solvent (A)	T/ K	$H_{A,B}$/ MPa	Solvent (A)	T/ K	$H_{A,B}$/ MPa
methanol	403.15	20.60	toluene	403.15	3.827
methanol	413.15	20.79	toluene	413.15	4.457
methanol	423.15	23.16	toluene	423.15	5.312

continued

continued

Solvent (A)	T/ K	$H_{A,B}$/ MPa	Solvent (A)	T/ K	$H_{A,B}$/ MPa
ethanol	403.15	11.97	1,2-dimethylbenzene	403.15	2.133
ethanol	413.15	12.23	1,2-dimethylbenzene	413.15	2.587
ethanol	423.15	14.00	1,2-dimethylbenzene	423.15	3.073
2-propanone	403.15	10.80	n-octane	403.15	4.667
2-propanone	413.15	11.01	n-octane	413.15	4.995
2-propanone	423.15	12.37	n-octane	423.15	5.388
2-butanone	403.15	6.832	n-nonane	403.15	3.621
2-butanone	413.15	7.376	n-nonane	413.15	4.043
2-butanone	423.15	8.535	n-nonane	423.15	4.157
methyl acetate	403.15	8.286	n-decane	403.15	2.437
methyl acetate	413.15	8.636	n-decane	413.15	3.006
methyl acetate	423.15	9.347	n-decane	423.15	3.124
ethyl acetate	403.15	5.994	n-undecane	403.15	1.518
ethyl acetate	413.15	6.837	n-undecane	413.15	1.899
ethyl acetate	423.15	7.428	n-undecane	423.15	2.463
benzene	403.15	5.946	n-dodecane	403.15	0.9126
benzene	413.15	6.519	n-dodecane	413.15	1.239
benzene	423.15	7.091	n-dodecane	423.15	1.660

Polymer (B): **poly(3,4-dichlorobenzyl methacrylate-*co*-ethyl methacrylate)** **2001DEM**

Characterization: 93 mol% ethyl methacrylate, synthesized in the laboratory

Solvent (A)	T/ K	$H_{A,B}$/ MPa	Solvent (A)	T/ K	$H_{A,B}$/ MPa
methanol	403.15	24.70	toluene	403.15	5.222
methanol	413.15	24.87	toluene	413.15	6.385
methanol	423.15	28.70	toluene	423.15	7.228
ethanol	403.15	14.37	1,2-dimethylbenzene	403.15	3.256
ethanol	413.15	14.58	1,2-dimethylbenzene	413.15	3.484
ethanol	423.15	16.16	1,2-dimethylbenzene	423.15	4.420
2-propanone	403.15	14.87	n-octane	403.15	6.561
2-propanone	413.15	14.43	n-octane	413.15	6.809
2-propanone	423.15	15.70	n-octane	423.15	7.336
2-butanone	403.15	9.291	n-nonane	403.15	5.132
2-butanone	413.15	9.936	n-nonane	413.15	5.382
2-butanone	423.15	10.60	n-nonane	423.15	5.787

continued

continued

Solvent (A)	T/ K	$H_{A,B}$/ MPa	Solvent (A)	T/ K	$H_{A,B}$/ MPa
methyl acetate	403.15	12.99	n-decane	403.15	3.563
methyl acetate	413.15	12.07	n-decane	413.15	3.652
methyl acetate	423.15	13.21	n-decane	423.15	4.178
ethyl acetate	403.15	8.157	n-undecane	403.15	2.090
ethyl acetate	413.15	8.738	n-undecane	413.15	2.671
ethyl acetate	423.15	9.618	n-undecane	423.15	3.403
benzene	403.15	8.121	n-dodecane	403.15	1.239
benzene	413.15	8.551	n-dodecane	413.15	2.045
benzene	423.15	9.259	n-dodecane	423.15	2.323

Polymer (B): **poly[2-(*N,N*-dimethylamino)ethyl methacrylate-*b*-2-(*N*-morpholino)ethyl methacrylate]** **2008YAZ**

Characterization: M_n/g.mol^{-1} = 33500, M_w/g.mol^{-1} = 38200, 50 mol%/50 mol%, T_g/K = 332.5, ρ(298 K) = 1.0892 g/cm^3, synthesized in the laboratory

Solvent (A)	T/ K	$H_{A,B}$/ MPa	Solvent (A)	T/ K	$H_{A,B}$/ MPa
n-hexane	383.15	6.381	cyclopentane	383.15	6.636
n-hexane	393.15	7.820	cyclopentane	393.15	7.244
n-hexane	403.15	8.903	cyclopentane	403.15	8.728
n-hexane	413.15	10.18	cyclopentane	413.15	9.753
n-heptane	383.15	2.876	cyclohexane	383.15	2.995
n-heptane	393.15	3.803	cyclohexane	393.15	3.622
n-heptane	403.15	4.375	cyclohexane	403.15	4.290
n-heptane	413.15	5.105	cyclohexane	413.15	5.219
n-octane	383.15	1.393	cycloheptane	383.15	0.9338
n-octane	393.15	1.778	cycloheptane	393.15	1.221
n-octane	403.15	2.239	cycloheptane	403.15	1.563
n-octane	413.15	2.683	cycloheptane	413.15	1.958
n-nonane	383.15	0.6735	benzene	383.15	1.290
n-nonane	393.15	0.8796	benzene	393.15	1.660
n-nonane	403.15	1.167	benzene	403.15	2.068
n-nonane	413.15	1.499	benzene	413.15	2.626
n-decane	383.15	0.3245	toluene	383.15	0.6044
n-decane	393.15	0.4415	toluene	393.15	0.7902
n-decane	403.15	0.6074	toluene	403.15	1.021
n-decane	413.15	0.8009	toluene	413.15	1.310

Polymer (B): **poly(ethylene-*co*-1-octene)** **2005LIU**
Characterization: M_n/g.mol^{-1} = 34600, M_w/g.mol^{-1} = 69200, LLDPE, 1.2 mol% 1-octene, ρ = 0.938 g/cm^3, 3 branches/1000CH_2, single-site catalyst, Nova Chemicals Corp., Calgary, Canada

Comments: Henry constants are corrected for solubility of the marker gas CH_4.

Solvent (A)	T/ K	$H_{A,B}$/ MPa	Solvent (A)	T/ K	$H_{A,B}$/ MPa
benzene	443.15	3.74	1-hexene	483.15	10.6
benzene	463.15	5.15	1-hexene	503.15	13.4
benzene	483.15	6.84	n-nonane	443.15	0.839
benzene	503.15	8.38	n-nonane	463.15	1.26
cyclohexane	443.15	3.32	n-nonane	483.15	1.81
cyclohexane	463.15	4.65	n-nonane	503.15	2.66
cyclohexane	483.15	6.13	n-octane	443.15	1.57
cyclohexane	503.15	7.78	n-octane	463.15	2.27
n-dodecane	443.15	0.140	n-octane	483.15	3.03
n-dodecane	463.15	0.233	n-octane	503.15	4.43
n-dodecane	483.15	0.321	1-octene	443.15	1.69
n-dodecane	503.15	0.592	1-octene	463.15	2.40
n-heptane	443.15	2.98	1-octene	483.15	3.23
n-heptane	463.15	4.14	1-octene	503.15	4.66
n-heptane	483.15	5.75	n-pentadecane	443.15	0.025
n-heptane	503.15	7.47	n-pentadecane	463.15	0.046
n-hexane	443.15	5.71	n-pentadecane	483.15	0.077
n-hexane	463.15	7.52	n-pentadecane	503.15	0.137
n-hexane	483.15	9.74	toluene	443.15	1.82
n-hexane	503.15	12.2	toluene	463.15	2.57
1-hexene	443.15	6.32	toluene	483.15	3.43
1-hexene	463.15	8.50	toluene	503.15	4.86

Polymer (B): **poly(ethylene-*co*-1-octene)** **2004ZHA**
Characterization: M_n/g.mol^{-1} = 33200, M_w/g.mol^{-1} = 93000, LLDPE, 4.2 mol% 1-octene, ρ = 0.917 g/cm^3, 10.6 branches/1000CH$_2$, single-site catalyst, Nova Chemicals Corp., Calgary, Canada

Solvent (A)	T/ K	$H_{A,B}$/ MPa	Solvent (A)	T/ K	$H_{A,B}$/ MPa
n-hexane	443.15	5.7	1-hexene	483.15	10.1
n-hexane	463.15	7.6	1-hexene	503.15	12.6
n-hexane	483.15	8.9	1-octene	443.15	1.7
n-hexane	503.15	11.5	1-octene	463.15	2.4
n-octane	443.15	1.58	1-octene	483.15	3.2
n-octane	463.15	2.20	1-octene	503.15	4.3
n-octane	483.15	3.01	benzene	443.15	3.8
n-octane	503.15	4.2	benzene	463.15	4.9
1-hexene	443.15	6.4	benzene	483.15	6.3
1-hexene	463.15	8.1	benzene	503.15	8.1

Polymer (B): **poly(ethylene-*co*-1-octene)** **2004ZHA**
Characterization: M_n/g.mol^{-1} = 34800, M_w/g.mol^{-1} = 1044000, LLDPE, 4.5 mol% 1-octene, ρ = 0.920 g/cm^3, 11.3 branches/1000CH$_2$, single-site catalyst, Nova Chemicals Corp., Calgary, Canada

Solvent (A)	T/ K	$H_{A,B}$/ MPa	Solvent (A)	T/ K	$H_{A,B}$/ MPa
n-hexane	443.15	5.3	1-hexene	483.15	9.8
n-hexane	463.15	7.0	1-hexene	503.15	11.4
n-hexane	483.15	8.5	1-octene	443.15	1.7
n-hexane	503.15	9.8	1-octene	463.15	2.3
n-octane	443.15	1.5	1-octene	483.15	3.1
n-octane	463.15	2.1	1-octene	503.15	4.1
n-octane	483.15	2.9	benzene	443.15	3.5
n-octane	503.15	3.8	benzene	463.15	4.8
1-hexene	443.15	6.1	benzene	483.15	6.1
1-hexene	463.15	7.9	benzene	503.15	7.6

Polymer (B): **poly(ethylene-*co*-1-octene)** **2004ZHA**
Characterization: M_n/g.mol^{-1} = 38700, M_w/g.mol^{-1} = 77400, LLDPE, 4.6 mol% 1-octene, ρ = 0.922 g/cm^3, 11.4 branches/1000CH_2, single-site catalyst, Nova Chemicals Corp., Calgary, Canada

Solvent (A)	T/ K	$H_{A,B}$/ MPa	Solvent (A)	T/ K	$H_{A,B}$/ MPa
n-hexane	443.15	5.7	1-hexene	483.15	9.82
n-hexane	463.15	8.0	1-hexene	503.15	13.3
n-hexane	483.15	9.6	1-octene	443.15	1.7
n-hexane	503.15	12.1	1-octene	463.15	2.4
n-octane	443.15	1.57	1-octene	483.15	3.2
n-octane	463.15	2.28	1-octene	503.15	4.8
n-octane	483.15	3.26	benzene	443.15	3.8
n-octane	503.15	4.30	benzene	463.15	5.0
1-hexene	443.15	6.07	benzene	483.15	6.3
1-hexene	463.15	8.66	benzene	503.15	8.4

Polymer (B): **poly(ethylene-*co*-1-octene)** **2005LIU**
Characterization: M_n/g.mol^{-1} = 38700, M_w/g.mol^{-1} = 77400, LLDPE, 4.6 mol% 1-octene, ρ = 0.922 g/cm^3, 11.4 branches/1000CH_2, single-site catalyst, Nova Chemicals Corp., Calgary, Canada

Solvent (A)	T/ K	$H_{A,B}$/ MPa	Solvent (A)	T/ K	$H_{A,B}$/ MPa
benzene	443.15	3.72	1-hexene	483.15	9.05
benzene	463.15	4.80	1-hexene	503.15	12.1
benzene	483.15	6.02	n-nonane	443.15	0.857
benzene	503.15	7.92	n-nonane	463.15	1.28
cyclohexane	443.15	3.37	n-nonane	483.15	1.89
cyclohexane	463.15	4.45	n-nonane	503.15	2.64
cyclohexane	483.15	5.59	octane	443.15	1.55
cyclohexane	503.15	7.04	octane	463.15	2.22
n-dodecane	443.15	0.143	octane	483.15	3.14
n-dodecane	463.15	0.232	octane	503.15	4.15
n-dodecane	483.15	0.367	1-octene	443.15	1.68
n-dodecane	503.15	0.593	1-octene	463.15	2.35
n-heptane	443.15	2.90	1-octene	483.15	3.11
n-heptane	463.15	4.11	1-octene	503.15	4.53

continued

continued

Solvent (A)	T/ K	$H_{A,B}$/ MPa	Solvent (A)	T/ K	$H_{A,B}$/ MPa
n-heptane	483.15	5.24	n-pentadecane	443.15	0.026
n-heptane	503.15	7.10	n-pentadecane	463.15	0.046
n-hexane	443.15	5.54	n-pentadecane	483.15	0.076
n-hexane	463.15	7.48	n-pentadecane	503.15	0.134
n-hexane	483.15	8.84	toluene	443.15	1.82
n-hexane	503.15	11.0	toluene	463.15	2.54
1-hexene	443.15	5.91	toluene	483.15	3.52
1-hexene	463.15	8.08	toluene	503.15	4.86

Polymer (B): **poly(ethylene-*co*-1-octene)** **2004ZHA**

Characterization: M_n/g.mol^{-1} = 18800, M_w/g.mol^{-1} = 122200, LLDPE, 4.8 mol% 1-octene, ρ = 0.924 g/cm^3, 11.9 branches/1000CH_2, Ziegler-Natta catalyst, Nova Chemicals Corp., Calgary, Canada

Solvent (A)	T/ K	$H_{A,B}$/ MPa	Solvent (A)	T/ K	$H_{A,B}$/ MPa
n-hexane	443.15	6.1	1-hexene	483.15	11.0
n-hexane	463.15	7.9	1-hexene	503.15	14.6
n-hexane	483.15	10.1	1-octene	443.15	1.8
n-hexane	503.15	12.5	1-octene	463.15	2.6
n-octane	443.15	1.7	1-octene	483.15	3.4
n-octane	463.15	2.3	1-octene	503.15	4.7
n-octane	483.15	3.2	benzene	443.15	4.0
n-octane	503.15	4.6	benzene	463.15	5.4
1-hexene	443.15	6.8	benzene	483.15	7.0
1-hexene	463.15	9.4	benzene	503.15	8.8

Polymer (B): **poly(ethylene-*co*-1-octene)** **2004ZHA**
Characterization: M_n/g.mol^{-1} = 26400, M_w/g.mol^{-1} = 116200, LLDPE, 5.1 mol% 1-octene, ρ = 0.923 g/cm^3, 12.7 branches/1000CH$_2$, Ziegler-Natta catalyst, Nova Chemicals Corp., Calgary, Canada

Solvent (A)	T/ K	$H_{A,B}$/ MPa	Solvent (A)	T/ K	$H_{A,B}$/ MPa
n-hexane	443.15	6.1	1-hexene	483.15	10.4
n-hexane	463.15	7.4	1-hexene	503.15	12.6
n-hexane	483.15	9.7	1-octene	443.15	1.7
n-hexane	503.15	11.7	1-octene	463.15	2.4
n-octane	443.15	1.6	1-octene	483.15	3.2
n-octane	463.15	2.2	1-octene	503.15	4.4
n-octane	483.15	3.1	benzene	443.15	3.8
n-octane	503.15	4.2	benzene	463.15	4.8
1-hexene	443.15	6.7	benzene	483.15	6.7
1-hexene	463.15	8.2	benzene	503.15	8.7

Polymer (B): **poly(ethylene-*co*-1-octene)** **2004ZHA**
Characterization: M_n/g.mol^{-1} = 20500, M_w/g.mol^{-1} = 102500, LLDPE, 5.2 mol% 1-octene, ρ = 0.923 g/cm^3, 12.9 branches/1000CH$_2$, Ziegler-Natta catalyst, Nova Chemicals Corp., Calgary, Canada

Solvent (A)	T/ K	$H_{A,B}$/ MPa	Solvent (A)	T/ K	$H_{A,B}$/ MPa
n-hexane	443.15	6.3	1-hexene	483.15	11.5
n-hexane	463.15	7.6	1-hexene	503.15	13.6
n-hexane	483.15	10.7	1-octene	443.15	1.9
n-hexane	503.15	13.3	1-octene	463.15	2.6
n-octane	443.15	1.7	1-octene	483.15	3.6
n-octane	463.15	2.5	1-octene	503.15	4.8
n-octane	483.15	3.3	benzene	443.15	4.0
n-octane	503.15	4.8	benzene	463.15	5.4
1-hexene	443.15	6.8	benzene	483.15	7.0
1-hexene	463.15	8.6	benzene	503.15	8.8

Polymer (B): **poly(ethylene-*co*-1-octene)** **2004ZHA**
Characterization: M_n/g.mol^{-1} = 15300, M_w/g.mol^{-1} = 47400, LLDPE, 6.1 mol% 1-octene, ρ = 0.919 g/cm^3, 15.3 branches/1000CH$_2$, single-site catalyst, Nova Chemicals Corp., Calgary, Canada

Solvent (A)	T/ K	$H_{A,B}$/ MPa	Solvent (A)	T/ K	$H_{A,B}$/ MPa
n-hexane	443.15	5.9	1-hexene	483.15	10.6
n-hexane	463.15	7.4	1-hexene	503.15	12.4
n-hexane	483.15	9.8	1-octene	443.15	1.7
n-hexane	503.15	11.4	1-octene	463.15	2.4
n-octane	443.15	1.5	1-octene	483.15	3.4
n-octane	463.15	2.3	1-octene	503.15	4.4
n-octane	483.15	3.1	benzene	443.15	3.8
n-octane	503.15	4.1	benzene	463.15	4.9
1-hexene	443.15	6.3	benzene	483.15	6.6
1-hexene	463.15	8.2	benzene	503.15	8.2

Polymer (B): **poly(ethylene-*co*-1-octene)** **2005LIU**
Characterization: M_n/g.mol^{-1} = 20300, M_w/g.mol^{-1} = 69000, LLDPE, 7.2 mol% 1-octene, ρ = 0.914 g/cm^3, 18.1 branches/1000CH$_2$, single-site catalyst, Nova Chemicals Corp., Calgary, Canada

Solvent (A)	T/ K	$H_{A,B}$/ MPa	Solvent (A)	T/ K	$H_{A,B}$/ MPa
benzene	443.15	3.85	1-hexene	483.15	13.2
benzene	463.15	5.28	1-hexene	503.15	15.8
benzene	483.15	7.55	n-nonane	443.15	0.847
benzene	503.15	9.49	n-nonane	463.15	1.27
cyclohexane	443.15	3.39	n-nonane	483.15	1.91
cyclohexane	463.15	4.90	n-nonane	503.15	2.74
cyclohexane	483.15	6.79	octane	443.15	1.57
cyclohexane	503.15	8.80	octane	463.15	2.28
n-dodecane	443.15	0.141	octane	483.15	3.37
n-dodecane	463.15	0.235	octane	503.15	4.61
n-dodecane	483.15	0.374	1-octene	443.15	1.75
n-dodecane	503.15	0.625	1-octene	463.15	2.51
n-heptane	443.15	3.06	1-octene	483.15	3.74

continued

continued

Solvent (A)	T/ K	$H_{A,B}$/ MPa	Solvent (A)	T/ K	$H_{A,B}$/ MPa
n-heptane	463.15	4.21	1-octene	503.15	5.09
n-heptane	483.15	6.07	n-pentadecane	443.15	0.025
n-heptane	503.15	8.32	n-pentadecane	463.15	0.047
n-hexane	443.15	6.11	n-pentadecane	483.15	0.081
n-hexane	463.15	8.23	n-pentadecane	503.15	0.143
n-hexane	483.15	11.5	toluene	443.15	1.83
n-hexane	503.15	13.7	toluene	463.15	2.60
1-hexene	443.15	6.67	toluene	483.15	3.79
1-hexene	463.15	9.58	toluene	503.15	5.05

Polymer (B): **poly(ethylene-*co*-1-octene)** **2004ZHA**
Characterization: M_n/g.mol^{-1} = 10500, M_w/g.mol^{-1} = 99800, LLDPE, 7.8 mol% 1-octene, ρ = 0.917 g/cm^3, 19.6 branches/1000CH$_2$, single-site catalyst, Nova Chemicals Corp., Calgary, Canada

Solvent (A)	T/ K	$H_{A,B}$/ MPa	Solvent (A)	T/ K	$H_{A,B}$/ MPa
n-hexane	443.15	5.6	1-hexene	483.15	10.9
n-hexane	463.15	7.5	1-hexene	503.15	12.8
n-hexane	483.15	9.5	1-octene	443.15	1.7
n-hexane	503.15	11.0	1-octene	463.15	2.4
n-octane	443.15	1.5	1-octene	483.15	3.2
n-octane	463.15	2.2	1-octene	503.15	4.2
n-octane	483.15	3.0	benzene	443.15	3.7
n-octane	503.15	4.1	benzene	463.15	5.0
1-hexene	443.15	6.2	benzene	483.15	6.6
1-hexene	463.15	8.4	benzene	503.15	8.2

Polymer (B): **poly(ethylene-*co*-1-octene)** **2004ZHA**
Characterization: M_n/g.mol^{-1} = 25700, M_w/g.mol^{-1} = 69400, LLDPE, 12.2 mol% 1-octene, ρ = 0.902 g/cm^3, 30.4 branches/1000CH$_2$, single-site catalyst, Nova Chemicals Corp., Calgary, Canada

Solvent (A)	T/ K	$H_{A,B}$/ MPa	Solvent (A)	T/ K	$H_{A,B}$/ MPa
n-hexane	443.15	5.9	1-hexene	483.15	11.0
n-hexane	463.15	7.5	1-hexene	503.15	13.9
n-hexane	483.15	9.8	1-octene	443.15	1.8
n-hexane	503.15	12.2	1-octene	463.15	2.5
n-octane	443.15	1.6	1-octene	483.15	3.6
n-octane	463.15	2.4	1-octene	503.15	4.8
n-octane	483.15	3.2	benzene	443.15	3.9
n-octane	503.15	4.5	benzene	463.15	5.2
1-hexene	443.15	6.4	benzene	483.15	6.9
1-hexene	463.15	8.5	benzene	503.15	8.8

Polymer (B): **poly(ethylene-*co*-1-octene)** **2004ZHA**
Characterization: M_n/g.mol^{-1} = 17300, M_w/g.mol^{-1} = 105300, LLDPE, 14.0 mol% 1-octene, ρ = 0.902 g/cm^3, 35.0 branches/1000CH$_2$, Ziegler-Natta catalyst, Nova Chemicals Corp., Calgary, Canada

Solvent (A)	T/ K	$H_{A,B}$/ MPa	Solvent (A)	T/ K	$H_{A,B}$/ MPa
n-hexane	443.15	5.9	1-hexene	483.15	9.4
n-hexane	463.15	7.9	1-hexene	503.15	11.6
n-hexane	483.15	8.3	1-octene	443.15	1.7
n-hexane	503.15	10.7	1-octene	463.15	2.3
n-octane	443.15	1.5	1-octene	483.15	3.0
n-octane	463.15	2.2	1-octene	503.15	4.1
n-octane	483.15	2.9	benzene	443.15	3.7
n-octane	503.15	3.8	benzene	463.15	5.2
1-hexene	443.15	6.6	benzene	483.15	6.1
1-hexene	463.15	8.5	benzene	503.15	8.0

Polymer (B): **poly(ethylene-*co*-1-octene)** **2005LIU**
Characterization: M_n/g.mol^{-1} = 53800, M_w/g.mol^{-1} = 96900, LLDPE, 19.9 mol% 1-octene, ρ = 0.874 g/cm^3, 49.7 branches/1000CH_2, single-site catalyst, Exxon Mobile Chemicals Corporation

Solvent (A)	T/ K	$H_{A,B}$/ MPa	Solvent (A)	T/ K	$H_{A,B}$/ MPa
benzene	443.15	4.42	1-hexene	483.15	12.2
benzene	463.15	5.78	1-hexene	503.15	16.0
benzene	483.15	7.84	n-nonane	443.15	0.984
benzene	503.15	10.3	n-nonane	463.15	1.48
cyclohexane	443.15	3.86	n-nonane	483.15	2.10
cyclohexane	463.15	5.09	n-nonane	503.15	3.11
cyclohexane	483.15	6.82	octane	443.15	1.84
cyclohexane	503.15	9.05	octane	463.15	2.58
n-dodecane	443.15	0.163	octane	483.15	3.65
n-dodecane	463.15	0.278	octane	503.15	5.07
n-dodecane	483.15	0.426	1-octene	443.15	2.00
n-dodecane	503.15	0.706	1-octene	463.15	2.82
n-heptane	443.15	3.52	1-octene	483.15	3.99
n-heptane	463.15	4.63	1-octene	503.15	5.32
n-heptane	483.15	6.32	n-pentadecane	443.15	0.030
n-heptane	503.15	8.26	n-pentadecane	463.15	0.055
n-hexane	443.15	6.31	n-pentadecane	483.15	0.091
n-hexane	463.15	8.34	n-pentadecane	503.15	0.161
n-hexane	483.15	11.2	toluene	443.15	2.20
n-hexane	503.15	14.6	toluene	463.15	2.99
1-hexene	443.15	7.18	toluene	483.15	4.15
1-hexene	463.15	8.92	toluene	503.15	5.61

Polymer (B): **poly(ethylene-*co*-1-octene)** **2005LIU**
Characterization: M_n/g.mol^{-1} = 52000, M_w/g.mol^{-1} = 104000, LLDPE, 34.9 mol% 1-octene, ρ = 0.865 g/cm^3, 87.2 branches/1000CH_2, single-site catalyst, Exxon Mobile Chemicals Corporation

Solvent (A)	T/ K	$H_{A,B}$/ MPa	Solvent (A)	T/ K	$H_{A,B}$/ MPa
benzene	443.15	4.21	1-hexene	483.15	10.6
benzene	463.15	5.67	1-hexene	503.15	14.3
benzene	483.15	7.48	n-nonane	443.15	0.939

continued

continued

Solvent (A)	T/ K	$H_{A,B}$/ MPa	Solvent (A)	T/ K	$H_{A,B}$/ MPa
benzene	503.15	9.23	n-nonane	463.15	1.43
cyclohexane	443.15	3.69	n-nonane	483.15	1.97
cyclohexane	463.15	5.01	n-nonane	503.15	2.78
cyclohexane	483.15	6.54	octane	443.15	1.73
cyclohexane	503.15	8.26	octane	463.15	2.53
n-dodecane	443.15	0.155	octane	483.15	3.34
n-dodecane	463.15	0.264	octane	503.15	4.62
n-dodecane	483.15	0.403	1-octene	443.15	1.90
n-dodecane	503.15	0.635	1-octene	463.15	2.71
n-heptane	443.15	3.22	1-octene	483.15	3.70
n-heptane	463.15	4.53	1-octene	503.15	5.12
n-heptane	483.15	6.02	n-pentadecane	443.15	0.028
n-heptane	503.15	7.76	n-pentadecane	463.15	0.052
n-hexane	443.15	6.39	n-pentadecane	483.15	0.085
n-hexane	463.15	8.61	n-pentadecane	503.15	0.150
n-hexane	483.15	10.4	toluene	443.15	2.02
n-hexane	503.15	12.5	toluene	463.15	2.90
1-hexene	443.15	6.77	toluene	483.15	3.76
1-hexene	463.15	9.20	toluene	503.15	5.18

Polymer (B): **poly(glycidyl methacrylate-*co*-butyl methacrylate)** **2002KAY**

Characterization: M_n/g.mol^{-1} = 530000, M_w/g.mol^{-1} = 738000,
41 mol% butyl methacrylate, T_g/K = 334,
ρ(298 K) = 1.178 g/cm^3, synthesized in the laboratory

Solvent (A)	T/ K	$H_{A,B}$/ MPa	Solvent (A)	T/ K	$H_{A,B}$/ MPa
benzene	383.15	2.135	n-pentane	383.15	5.381
benzene	388.15	2.462	n-pentane	388.15	5.723
benzene	393.15	2.643	n-pentane	393.15	5.786
benzene	403.15	3.157	n-pentane	403.15	6.270
benzene	413.15	3.598	n-pentane	413.15	6.726
benzene	423.15	4.078	n-pentane	423.15	7.058
n-decane	383.15	0.755	1,4-dimethylbenzene	383.15	0.6418
n-decane	388.15	0.892	1,4-dimethylbenzene	388.15	0.7934
n-decane	393.15	0.990	1,4-dimethylbenzene	393.15	0.8895

continued

continued

Solvent (A)	T/ K	$H_{A,B}$/ MPa	Solvent (A)	T/ K	$H_{A,B}$/ MPa
n-decane	403.15	1.271	1,4-dimethylbenzene	403.15	1.174
n-decane	413.15	1.524	1,4-dimethylbenzene	413.15	1.459
n-decane	423.15	1.785	1,4-dimethylbenzene	423.15	1.772
n-heptane	383.15	3.014	tetrachloromethane	383.15	1.455
n-heptane	388.15	3.188	tetrachloromethane	388.15	1.589
n-heptane	393.15	3.323	tetrachloromethane	393.15	1.648
n-heptane	403.15	3.673	tetrachloromethane	403.15	1.905
n-heptane	413.15	4.062	tetrachloromethane	413.15	2.146
n-heptane	423.15	4.334	tetrachloromethane	423.15	2.389
n-hexane	383.15	4.048	toluene	383.15	1.179
n-hexane	388.15	4.217	toluene	388.15	1.396
n-hexane	393.15	4.385	toluene	393.15	1.530
n-hexane	403.15	4.783	toluene	403.15	1.945
n-hexane	413.15	5.188	toluene	413.15	2.310
n-hexane	423.15	5.548	toluene	423.15	2.700
n-nonane	383.15	1.289	1-chlorobutane	383.15	2.417
n-nonane	388.15	1.471	1-chlorobutane	388.15	2.641
n-nonane	393.15	1.577	1-chlorobutane	393.15	2.810
n-nonane	403.15	1.923	1-chlorobutane	403.15	3.254
n-nonane	413.15	2.219	1-chlorobutane	413.15	3.619
n-nonane	423.15	2.519	1-chlorobutane	423.15	4.035
n-octane	383.15	2.056	1-chloropropane	383.15	3.726
n-octane	388.15	2.252	1-chloropropane	388.15	4.005
n-octane	393.15	2.336	1-chloropropane	393.15	4.179
n-octane	403.15	2.754	1-chloropropane	403.15	4.686
n-octane	413.15	3.068	1-chloropropane	413.15	5.073
n-octane	423.15	3.375	1-chloropropane	423.15	5.529

Polymer (B): **poly(glycidyl methacrylate-*co*-ethyl methacrylate)** **2002KAY**

Characterization: M_n/g.mol^{-1} = 441000, M_w/g.mol^{-1} = 700000, 44 mol% ethyl methacrylate, T_g/K = 355, ρ(298 K) = 1.186 g/cm^3, synthesized in the laboratory

Solvent (A)	T/ K	$H_{A,B}$/ MPa	Solvent (A)	T/ K	$H_{A,B}$/ MPa
benzene	403.15	1.962	n-pentane	403.15	3.388
benzene	413.15	2.105	n-pentane	413.15	3.635
benzene	423.15	2.884	n-pentane	423.15	3.876
benzene	433.15	3.248	n-pentane	433.15	4.175
n-decane	403.15	0.943	1,4-dimethylbenzene	403.15	0.8066
n-decane	413.15	1.038	1,4-dimethylbenzene	413.15	0.9177
n-decane	423.15	1.103	1,4-dimethylbenzene	423.15	1.160
n-decane	433.15	1.410	1,4-dimethylbenzene	433.15	1.283
n-heptane	403.15	2.205	tetrachloromethane	403.15	1.241
n-heptane	413.15	2.351	tetrachloromethane	413.15	1.300
n-heptane	423.15	2.504	tetrachloromethane	423.15	0.9338
n-heptane	433.15	2.721	tetrachloromethane	433.15	1.530
n-hexane	403.15	2.760	toluene	403.15	1.256
n-hexane	413.15	2.909	toluene	413.15	1.422
n-hexane	423.15	3.104	toluene	423.15	1.703
n-hexane	433.15	3.357	toluene	433.15	1.830
n-nonane	403.15	1.330	1-chlorobutane	403.15	1.967
n-nonane	413.15	1.391	1-chlorobutane	413.15	2.133
n-nonane	423.15	1.724	1-chlorobutane	423.15	44.61
n-nonane	433.15	2.083	1-chlorobutane	433.15	2.717
n-octane	403.15	1.747	1-chloropropane	403.15	2.677
n-octane	413.15	1.863	1-chloropropane	413.15	2.658
n-octane	423.15	2.088	1-chloropropane	423.15	2.936
n-octane	433.15	2.267	1-chloropropane	433.15	3.297

Polymer (B): **poly(glycidyl methacrylate-*co*-methyl methacrylate)** **2002KAY**

Characterization: M_n/g.mol^{-1} = 555000, M_w/g.mol^{-1} = 710000, 38 mol% methyl methacrylate, T_g/K = 373, ρ(298 K) = 1.204 g/cm^3, synthesized in the laboratory

Solvent (A)	T/ K	$H_{A,B}$/ MPa	Solvent (A)	T/ K	$H_{A,B}$/ MPa
benzene	423.15	2.539	n-pentane	443.15	4.663
benzene	433.15	2.798	1,4-dimethylbenzene	423.15	2.219
benzene	443.15	3.054	1,4-dimethylbenzene	433.15	1.384
n-decane	423.15	1.438	1,4-dimethylbenzene	443.15	1.468
n-decane	433.15	1.562	tetrachloromethane	423.15	1.561
n-decane	443.15	1.677	tetrachloromethane	433.15	1.725
n-heptane	423.15	2.757	tetrachloromethane	443.15	1.859
n-heptane	433.15	2.998	toluene	423.15	1.781
n-heptane	443.15	3.285	toluene	433.15	1.914
n-hexane	423.15	3.349	toluene	443.15	2.049
n-hexane	433.15	3.645	1-chlorobutane	423.15	2.446
n-hexane	443.15	3.904	1-chlorobutane	433.15	2.661
n-nonane	423.15	1.837	1-chlorobutane	443.15	2.890
n-nonane	433.15	2.049	1-chloropropane	423.15	3.116
n-nonane	443.15	2.208	1-chloropropane	433.15	3.480
n-octane	423.15	2.272	1-chloropropane	443.15	3.755
n-octane	433.15	2.467	1,4-dioxane	423.15	1.606
n-octane	443.15	2.665	1,4-dioxane	433.15	1.723
n-pentane	423.15	4.093	1,4-dioxane	443.15	1.824
n-pentane	433.15	4.402			

Polymer (B): **poly(L-lactic acid-*co*-glycolic acid)** **2006ESE**

Characterization: M_n/g.mol^{-1} = 57500, 35 wt% glycolide, T_g/K = 315, ρ = 1.24 g/cm^3, Aldrich Chem. Co., Inc., Milwaukee, WI

Solvent (A)	T/ K	$H_{A,B}$/ MPa	Solvent (A)	T/ K	$H_{A,B}$/ MPa
dichloromethane	453.15	1.355	2-propanone	473.15	3.192
dichloromethane	463.15	1.820	2-propanone	493.15	5.539
dichloromethane	473.15	2.321	tetrahydrofuran	453.15	2.330
dichloromethane	493.15	3.820	tetrahydrofuran	463.15	2.607
ethyl acetate	453.15	1.701	tetrahydrofuran	473.15	3.088

continued

continued

Solvent (A)	T/ K	$H_{A,B}$/ MPa	Solvent (A)	T/ K	$H_{A,B}$/ MPa
ethyl acetate	463.15	1.762	tetrahydrofuran	493.15	4.578
ethyl acetate	473.15	2.085	trichloromethane	453.15	0.5532
ethyl acetate	493.15	3.736	trichloromethane	463.15	0.8520
ethanol	453.15	2.863	trichloromethane	473.15	1.051
ethanol	463.15	3.649	trichloromethane	493.15	1.880
ethanol	473.15	4.581	water	453.15	4.928
ethanol	493.15	11.94	water	463.15	6.877
2-propanone	453.15	1.976	water	473.15	9.359
2-propanone	463.15	2.433	water	493.15	16.33

Polymer (B): **poly(3-mesityl-2-hydroxypropyl methacrylate-*co*-1-vinyl-2-pyrrolidinone)** **2005ACI**

Characterization: M_n/g.mol^{-1} = 58000, M_w/g.mol^{-1} = 481000, 45.75 mol% 1-vinyl-2-pyrrolidinone, T_g/K = 380, ρ(298 K) = 1.040 g/cm^3, synthesized in the laboratory

Solvent (A)	T/ K	$H_{A,B}$/ MPa	Solvent (A)	T/ K	$H_{A,B}$/ MPa
n-hexane	323.15	7.717	methanol	403.15	30.32
n-hexane	333.15	8.626	methanol	413.15	31.31
n-hexane	343.15	9.717	methanol	423.15	32.31
n-hexane	353.15	10.93	methanol	433.15	33.24
n-hexane	363.15	11.75	methanol	443.15	34.73
n-hexane	373.15	12.36	methanol	453.15	35.05
n-hexane	383.15	11.49	methanol	463.15	34.85
n-hexane	393.15	13.47	ethanol	323.15	11.04
n-hexane	403.15	13.91	ethanol	333.15	12.39
n-hexane	413.15	14.28	ethanol	343.15	13.74
n-hexane	423.15	14.39	ethanol	353.15	14.52
n-hexane	433.15	14.54	ethanol	363.15	16.26
n-hexane	443.15	14.60	ethanol	373.15	19.70
n-hexane	453.15	14.16	ethanol	383.15	18.19
n-hexane	463.15	14.49	ethanol	393.15	20.40
n-heptane	323.15	6.299	ethanol	403.15	21.83
n-heptane	333.15	6.899	ethanol	413.15	22.14
n-heptane	343.15	7.883	ethanol	423.15	22.47
n-heptane	353.15	9.198	ethanol	433.15	22.76

continued

continued

Solvent (A)	T/ K	$H_{A,B}$/ MPa	Solvent (A)	T/ K	$H_{A,B}$/ MPa
n-heptane	363.15	10.10	ethanol	443.15	23.40
n-heptane	373.15	10.32	ethanol	453.15	23.29
n-heptane	383.15	9.308	ethanol	463.15	23.71
n-heptane	393.15	11.03	1-propanol	323.15	8.579
n-heptane	403.15	11.26	1-propanol	333.15	10.08
n-heptane	413.15	11.36	1-propanol	343.15	11.98
n-heptane	423.15	11.68	1-propanol	353.15	12.89
n-heptane	433.15	12.04	1-propanol	363.15	13.80
n-heptane	443.15	12.11	1-propanol	373.15	15.33
n-heptane	453.15	12.08	1-propanol	383.15	14.42
n-heptane	463.15	12.06	1-propanol	393.15	16.50
n-octane	323.15	5.175	1-propanol	403.15	17.03
n-octane	333.15	6.052	1-propanol	413.15	17.26
n-octane	343.15	6.918	1-propanol	423.15	17.81
n-octane	353.15	7.905	1-propanol	433.15	17.15
n-octane	363.15	8.674	1-propanol	443.15	17.39
n-octane	373.15	9.025	1-propanol	453.15	17.85
n-octane	383.15	8.581	1-propanol	463.15	18.05
n-octane	393.15	9.554	1-butanol	323.15	6.677
n-octane	403.15	9.774	1-butanol	333.15	7.278
n-octane	413.15	9.966	1-butanol	343.15	7.938
n-octane	423.15	10.06	1-butanol	353.15	8.761
n-octane	433.15	10.37	1-butanol	363.15	9.495
n-octane	443.15	10.40	1-butanol	373.15	12.27
n-octane	453.15	10.68	1-butanol	383.15	11.59
n-octane	463.15	10.71	1-butanol	393.15	12.90
n-decane	323.15	3.585	1-butanol	403.15	13.33
n-decane	333.15	3.957	1-butanol	413.15	13.53
n-decane	343.15	4.328	1-butanol	423.15	13.40
n-decane	353.15	4.785	1-butanol	433.15	13.50
n-decane	363.15	5.548	1-butanol	443.15	13.69
n-decane	373.15	6.284	1-butanol	453.15	13.29
n-decane	383.15	5.978	1-butanol	463.15	13.64
n-decane	393.15	7.232	1-pentanol	323.15	5.691
n-decane	403.15	7.775	1-pentanol	333.15	6.120
n-decane	413.15	7.711	1-pentanol	343.15	6.867
n-decane	423.15	7.794	1-pentanol	353.15	7.367
n-decane	433.15	8.041	1-pentanol	363.15	7.864
n-decane	443.15	7.843	1-pentanol	373.15	9.338
n-decane	453.15	7.863	1-pentanol	383.15	8.689
n-decane	463.15	7.965	1-pentanol	393.15	9.678

continued

continued

Solvent (A)	T/ K	$H_{A,B}$/ MPa	Solvent (A)	T/ K	$H_{A,B}$/ MPa
methanol	323.15	15.66	1-pentanol	403.15	10.49
methanol	333.15	17.56	1-pentanol	413.15	10.72
methanol	343.15	19.46	1-pentanol	423.15	11.02
methanol	353.15	21.21	1-pentanol	433.15	11.10
methanol	363.15	23.38	1-pentanol	443.15	10.87
methanol	373.15	29.37	1-pentanol	453.15	11.02
methanol	383.15	26.31	1-pentanol	463.15	11.17
methanol	393.15	29.13			

Polymer (B): **poly(methyl methacrylate-*co*-butyl methacrylate)** **2005ESE**
Characterization: M_n/g.mol^{-1} = 75000, 15 wt% butyl methacrylate,
T_g/K = 378, ρ (298 K) = 1.115 g/cm^3,
Aldrich Chem. Co., Inc., Milwaukee, WI

Solvent (A)	T/ K	$H_{A,B}$/ MPa	Solvent (A)	T/ K	$H_{A,B}$/ MPa
1-butanol	423.15	2.547	methyl acetate	423.15	7.441
1-butanol	443.15	3.040	methyl acetate	443.15	8.587
1-butanol	453.15	4.157	methyl acetate	453.15	10.32
1-butanol	473.15	5.836	methyl acetate	473.15	15.41
butyl methacrylate	423.15	0.7481	methyl methacrylate	423.15	2.923
butyl methacrylate	443.15	0.7783	methyl methacrylate	443.15	3.137
butyl methacrylate	453.15	0.9484	methyl methacrylate	453.15	3.584
butyl methacrylate	473.15	1.439	methyl methacrylate	473.15	5.203
dichloromethane	423.15	4.051	1-propanol	423.15	4.895
dichloromethane	443.15	5.402	1-propanol	443.15	6.008
dichloromethane	453.15	6.787	1-propanol	453.15	6.809
dichloromethane	473.15	9.584	1-propanol	473.15	11.25
ethanol	423.15	9.444	2-propanone	423.15	8.689
ethanol	443.15	12.77	2-propanone	443.15	10.71
ethanol	453.15	14.72	2-propanone	453.15	12.61
ethanol	473.15	23.70	2-propanone	473.15	17.85
ethyl acetate	423.15	5.228	propyl acetate	423.15	3.246
ethyl acetate	443.15	5.832	propyl acetate	443.15	3.633
ethyl acetate	453.15	6.661	propyl acetate	453.15	3.943
ethyl acetate	473.15	10.79	propyl acetate	473.15	5.821

continued

continued

Solvent (A)	T/ K	$H_{A,B}$/ MPa	Solvent (A)	T/ K	$H_{A,B}$/ MPa
methanol	423.15	16.68	trichloromethane	423.15	2.510
methanol	443.15	23.39	trichloromethane	443.15	2.510
methanol	453.15	26.35	trichloromethane	453.15	3.044
methanol	473.15	44.02	trichloromethane	473.15	4.424

Polymer (B): **poly[(2-phenyl-1,3-dioxolane-4-yl)methyl methacrylate-*co*-butyl methacrylate]** **2003ACI**

Characterization: M_n/g.mol^{-1} = 301100, M_w/g.mol^{-1} = 880200, 55.0 mol% butyl methacrylate, T_g/K = 370, ρ(298 K) = 1.217 g/cm^3, could be a block copolymer, synthesized in the laboratory

Solvent (A)	T/ K	$H_{A,B}$/ MPa	Solvent (A)	T/ K	$H_{A,B}$/ MPa
methanol	413.15	36.92	1-propanol	443.15	22.36
methanol	423.15	40.04	1-propanol	453.15	23.62
methanol	433.15	40.97	1-butanol	413.15	14.12
methanol	443.15	42.70	1-butanol	423.15	15.40
methanol	453.15	45.14	1-butanol	433.15	16.47
ethanol	413.15	25.95	1-butanol	443.15	17.92
ethanol	423.15	27.24	1-butanol	453.15	18.24
ethanol	433.15	28.01	1-pentanol	413.15	10.26
ethanol	443.15	29.17	1-pentanol	423.15	11.66
ethanol	453.15	32.86	1-pentanol	433.15	12.45
1-propanol	413.15	18.90	1-pentanol	443.15	13.78
1-propanol	423.15	20.54	1-pentanol	453.15	14.47
1-propanol	433.15	21.47			

Polymer (B): **poly[(2-phenyl-1,3-dioxolane-4-yl)methyl methacrylate-*co*-ethyl methacrylate]** **2009KAR**

Characterization: M_n/g.mol^{-1} = 365400, M_w/g.mol^{-1} = 1215000, 52 mol% ethyl methacrylate, ρ (298 K) = 1.21 g/cm^3, T_g/K = 363, synthesized in the laboratory

Solvent (A)	T/ K	$H_{A,B}$/ MPa	Solvent (A)	T/ K	$H_{A,B}$/ MPa
ethanol	393.15	15.08	n-hexane	393.15	7.745
ethanol	403.15	14.68	n-hexane	403.15	8.224
ethanol	413.15	15.65	n-hexane	413.15	8.499
ethanol	423.15	15.68	n-hexane	423.15	8.840
ethanol	433.15	16.79	n-hexane	433.15	9.181
1-propanol	393.15	10.68	n-heptane	393.15	6.266
1-propanol	403.15	11.61	n-heptane	403.15	6.751
1-propanol	413.15	12.19	n-heptane	413.15	6.723
1-propanol	423.15	12.20	n-heptane	423.15	7.208
1-propanol	433.15	12.49	n-heptane	433.15	7.332
1-butanol	393.15	6.930	n-octane	393.15	5.189
1-butanol	403.15	7.926	n-octane	403.15	5.665
1-butanol	413.15	8.470	n-octane	413.15	5.690
1-butanol	423.15	9.200	n-octane	423.15	5.928
1-butanol	433.15	9.773	n-octane	433.15	6.525

Polymer (B): **poly[(2-phenyl-1,3-dioxolane-4-yl)methyl methacrylate-*co*-glycidyl methacrylate]** **2004ILT**

Characterization: M_n/g.mol^{-1} = 244000, M_w/g.mol^{-1} = 623400, 60.0 mol% glycidyl methacrylate, T_g/K = 367, ρ (298 K) = 1.229 g/cm^3, could be a block copolymer, synthesized in the laboratory

Solvent (A)	T/ K	$H_{A,B}$/ MPa	Solvent (A)	T/ K	$H_{A,B}$/ MPa
ethanol	413.15	28.50	n-hexane	413.15	15.59
ethanol	423.15	29.17	n-hexane	423.15	17.45
ethanol	433.15	29.88	n-hexane	433.15	17.93
ethanol	443.15	36.25	n-hexane	443.15	20.75
ethanol	453.15	37.07	n-hexane	453.15	20.43
ethanol	463.15	42.50	n-hexane	463.15	24.18

continued

continued

Solvent (A)	T/ K	$H_{A,B}$/ MPa	Solvent (A)	T/ K	$H_{A,B}$/ MPa
1-propanol	413.15	21.23	n-heptane	413.15	13.49
1-propanol	423.15	22.49	n-heptane	423.15	14.25
1-propanol	433.15	23.04	n-heptane	433.15	15.42
1-propanol	443.15	27.99	n-heptane	443.15	16.66
1-propanol	453.15	29.29	n-heptane	453.15	17.57
1-propanol	463.15	30.72	n-heptane	463.15	20.79
1-butanol	413.15	15.88	n-octane	413.15	11.17
1-butanol	423.15	17.21	n-octane	423.15	12.50
1-butanol	433.15	17.71	n-octane	433.15	12.74
1-butanol	443.15	20.16	n-octane	443.15	14.73
1-butanol	453.15	22.36	n-octane	453.15	15.41
1-butanol	463.15	24.91	n-octane	463.15	17.14
1-pentanol	413.15	11.45	n-decane	413.15	8.102
1-pentanol	423.15	11.93	n-decane	423.15	8.581
1-pentanol	433.15	12.63	n-decane	433.15	9.226
1-pentanol	443.15	16.20	n-decane	443.15	10.50
1-pentanol	453.15	17.77	n-decane	453.15	11.01
1-pentanol	463.15	19.82	n-decane	463.15	12.98

Polymer (B): **poly[(2-phenyl-1,3-dioxolane-4-yl)methyl methacrylate-*co*-styrene]** **2006KAR**

Characterization: M_n/g.mol^{-1} = 181000, M_w/g.mol^{-1} = 436500, unknown styrene content, T_g/K = 373-383, synthesized in the laboratory

Solvent (A)	T/ K	$H_{A,B}$/ MPa	Solvent (A)	T/ K	$H_{A,B}$/ MPa
ethanol	413.15	2.581	n-hexane	413.15	1.370
ethanol	423.15	2.556	n-hexane	423.15	1.418
ethanol	433.15	2.653	n-hexane	433.15	1.415
1-propanol	413.15	1.969	n-heptane	413.15	1.125
1-propanol	423.15	1.964	n-heptane	423.15	1.165
1-propanol	433.15	1.979	n-heptane	433.15	1.168
1-butanol	413.15	1.237	n-octane	413.15	0.9467
1-butanol	423.15	1.343	n-octane	423.15	0.9689
1-butanol	433.15	1.413	n-octane	433.15	1.001

Polymer (B): **poly[2-(3-phenyl-3-methylcyclobutyl)-2-hydroxyethyl methacrylate-*co*-methacrylic acid]** **2001KA2**

Characterization: M_n/g.mol^{-1} = 17200, M_w/g.mol^{-1} = 45000, 45 mol% methacrylic acid, synthesized in the laboratory

Solvent (A)	T/ K	$H_{A,B}$/ MPa	Solvent (A)	T/ K	$H_{A,B}$/ MPa
methanol	423.15	10.53	ethyl acetate	423.15	2.864
methanol	433.15	10.87	ethyl acetate	433.15	3.405
methanol	443.15	11.81	ethyl acetate	443.15	3.720
methanol	453.15	12.31	ethyl acetate	453.15	3.785
ethanol	423.15	7.063	benzene	423.15	2.175
ethanol	433.15	7.413	benzene	433.15	2.364
ethanol	443.15	8.135	benzene	443.15	4.016
ethanol	453.15	8.081	benzene	453.15	4.320
2-propanone	423.15	5.683	toluene	423.15	1.521
2-propanone	433.15	6.317	toluene	433.15	1.628
2-propanone	443.15	6.062	toluene	443.15	1.811
2-propanone	453.15	6.119	toluene	453.15	2.191
2-butanone	423.15	3.523	1,2-dimethylbenzene	423.15	0.5281
2-butanone	433.15	4.233	1,2-dimethylbenzene	433.15	0.5954
2-butanone	443.15	4.512	1,2-dimethylbenzene	443.15	0.6649
2-butanone	453.15	4.680	1,2-dimethylbenzene	453.15	0.8202
methyl acetate	423.15	3.484	n-dodecane	423.15	0.1621
methyl acetate	433.15	4.775	n-dodecane	433.15	0.1803
methyl acetate	443.15	5.059	n-dodecane	443.15	0.2269
methyl acetate	453.15	5.205	n-dodecane	453.15	0.2996

Polymer (B): **poly[styrene-*b*-(1-butene-*co*-ethylene)-*b*-styrene]** **2009OVE**

Characterization: M_η/g.mol^{-1} = 90000, 32 wt% styren, REPSOL-YPF, Madrid, Spain

Solvent (A)	T/ K	$H_{A,B}$/ MPa	Solvent (A)	T/ K	$H_{A,B}$/ MPa
benzene	303.15	0.08792	n-hexane	323.15	0.4049
benzene	313.15	0.1355	n-hexane	333.15	0.5844
benzene	323.15	0.2085	1-hexene	303.15	0.2094
benzene	333.15	0.3111	1-hexene	313.15	0.3125
cyclohexane	303.15	0.08834	1-hexene	323.15	0.4765
cyclohexane	313.15	0.1357	1-hexene	333.15	0.7314

continued

continued

Solvent (A)	T/ K	$H_{A,B}$/ MPa	Solvent (A)	T/ K	$H_{A,B}$/ MPa
cyclohexane	323.15	0.1971	methylcyclohexane	303.15	0.04081
cyclohexane	333.15	0.2992	methylcyclohexane	313.15	0.06564
cyclopentane	303.15	0.2875	methylcyclohexane	323.15	0.09950
cyclopentane	313.15	0.4219	methylcyclohexane	333.15	0.1485
cyclopentane	323.15	0.6493	n-octane	303.15	0.01639
1,2-dimethylbenzene	303.15	0.00675	n-octane	313.15	0.02871
1,2-dimethylbenzene	313.15	0.01224	n-octane	323.15	0.04771
1,2-dimethylbenzene	323.15	0.02033	n-octane	333.15	0.07578
1,2-dimethylbenzene	333.15	0.03201	n-pentane	303.15	0.5822
ethylbenzene	303.15	0.00993	n-pentane	313.15	0.9253
ethylbenzene	313.15	0.01709	n-pentane	323.15	1.284
ethylbenzene	323.15	0.02847	tetrahydrofuran	303.15	0.1582
ethylbenzene	333.15	0.04400	tetrahydrofuran	313.15	0.2300
n-heptane	303.15	0.05363	tetrahydrofuran	323.15	0.3653
n-heptane	313.15	0.08632	tetrahydrofuran	333.15	0.5567
n-heptane	323.15	0.1389	toluene	303.15	0.02571
n-heptane	333.15	0.2110	toluene	313.15	0.04048
n-hexane	303.15	0.1762	toluene	323.15	0.06466
n-hexane	313.15	0.2933	toluene	333.15	0.1037

Polymer (B): **poly(styrene-*b*-ethylene oxide-*b*-styrene)** **2007ZOU**

Characterization: M_n/g.mol^{-1} = 11900, M_w/g.mol^{-1} = 16700,
42 wt% ethylene oxide, T_m/K = 328.3,
Polymeric Science Department, Hubei University, China

Solvent (A)	T/ K	$H_{A,B}$/ MPa	Solvent (A)	T/ K	$H_{A,B}$/ MPa
n-hexane	343.15	2.479	propyl acetate	373.15	0.5908
n-hexane	353.15	2.720	propyl acetate	383.15	0.7407
n-hexane	363.15	2.951	propyl acetate	393.15	0.9250
n-hexane	373.15	2.988	butyl acetate	343.15	0.1180
n-hexane	383.15	3.400	butyl acetate	353.15	0.1662
n-hexane	393.15	3.500	butyl acetate	363.15	0.2298
n-heptane	343.15	1.318	butyl acetate	373.15	0.3051
n-heptane	353.15	1.547	butyl acetate	383.15	0.3967
n-heptane	363.15	1.785	butyl acetate	393.15	0.5123
n-heptane	373.15	1.921	pentyl acetate	343.15	0.05342

continued

continued

Solvent (A)	T/ K	$H_{A,B}$/ MPa	Solvent (A)	T/ K	$H_{A,B}$/ MPa
n-heptane	383.15	2.235	pentyl acetate	353.15	0.07808
n-heptane	393.15	2.378	pentyl acetate	363.15	0.1121
n-octane	343.15	0.6605	pentyl acetate	373.15	0.1552
n-octane	353.15	0.8145	pentyl acetate	383.15	0.2099
n-octane	363.15	0.9901	pentyl acetate	393.15	0.2804
n-octane	373.15	1.145	methanol	343.15	1.379
n-octane	383.15	1.352	methanol	353.15	1.879
n-octane	393.15	1.541	methanol	363.15	2.363
n-nonane	343.15	0.3142	methanol	373.15	2.940
n-nonane	353.15	0.4091	methanol	383.15	3.676
n-nonane	363.15	0.5240	methanol	393.15	4.639
n-nonane	373.15	0.6395	ethanol	343.15	0.8038
n-nonane	383.15	0.7887	ethanol	353.15	1.109
n-nonane	393.15	0.9389	ethanol	363.15	1.476
n-decane	343.15	0.1461	ethanol	373.15	1.887
n-decane	353.15	0.1998	ethanol	383.15	2.361
n-decane	363.15	0.2677	ethanol	393.15	2.790
n-decane	373.15	0.3456	1-propanol	343.15	0.3176
n-decane	383.15	0.4436	1-propanol	353.15	0.4539
n-decane	393.15	0.5504	1-propanol	363.15	0.6274
methyl acetate	343.15	0.9206	1-propanol	373.15	0.8326
methyl acetate	353.15	1.170	1-propanol	383.15	1.086
methyl acetate	363.15	1.456	1-propanol	393.15	1.378
methyl acetate	373.15	1.689	1-butanol	343.15	0.1254
methyl acetate	383.15	2.005	1-butanol	353.15	0.1868
methyl acetate	393.15	2.338	1-butanol	363.15	0.2694
ethyl acetate	343.15	0.5347	1-butanol	373.15	0.3747
ethyl acetate	353.15	0.6978	1-butanol	383.15	0.5086
ethyl acetate	363.15	0.8947	1-butanol	393.15	0.6732
ethyl acetate	373.15	1.078	1-pentanol	343.15	0.05105
ethyl acetate	383.15	1.310	1-pentanol	353.15	0.07869
ethyl acetate	393.15	1.564	1-pentanol	363.15	0.1181
propyl acetate	343.15	0.2581	1-pentanol	373.15	0.1711
propyl acetate	353.15	0.3498	1-pentanol	383.15	0.2409
propyl acetate	363.15	0.4662	1-pentanol	393.15	0.3315

Polymer (B):	**poly(styrene-*b*-ethylene oxide-*b*-styrene)**	**2007ZOU**
Characterization:	M_n/g.mol^{-1} = 11200, M_w/g.mol^{-1} = 14300, 60 wt% ethylene oxide, T_m/K = 335., Polymeric Science Department, Hubei University, China	

Solvent (A)	T/ K	$H_{A,B}$/ MPa	Solvent (A)	T/ K	$H_{A,B}$/ MPa
n-hexane	343.15	2.035	propyl acetate	373.15	0.5275
n-hexane	353.15	2.276	propyl acetate	383.15	0.6931
n-hexane	363.15	2.536	propyl acetate	393.15	0.8300
n-hexane	373.15	2.651	butyl acetate	343.15	0.1021
n-hexane	383.15	3.141	butyl acetate	353.15	0.1449
n-hexane	393.15	3.104	butyl acetate	363.15	0.2003
n-heptane	343.15	1.046	butyl acetate	373.15	0.2702
n-heptane	353.15	1.241	butyl acetate	383.15	0.3635
n-heptane	363.15	1.456	butyl acetate	393.15	0.4594
n-heptane	373.15	1.633	pentyl acetate	343.15	0.04562
n-heptane	383.15	1.961	pentyl acetate	353.15	0.06735
n-heptane	393.15	2.047	pentyl acetate	363.15	0.09698
n-octane	343.15	0.4990	pentyl acetate	373.15	0.1362
n-octane	353.15	0.6311	pentyl acetate	383.15	0.1886
n-octane	363.15	0.7827	pentyl acetate	393.15	0.2490
n-octane	373.15	0.9343	methanol	343.15	1.672
n-octane	383.15	1.150	methanol	353.15	2.037
n-octane	393.15	1.280	methanol	363.15	2.618
n-nonane	343.15	0.2321	methanol	373.15	3.290
n-nonane	353.15	0.3089	methanol	383.15	4.043
n-nonane	363.15	0.4028	methanol	393.15	4.543
n-nonane	373.15	0.5066	ethanol	343.15	0.8783
n-nonane	383.15	0.6496	ethanol	353.15	1.106
n-nonane	393.15	0.7606	ethanol	363.15	1.509
n-decane	343.15	0.1062	ethanol	373.15	1.887
n-decane	353.15	0.1483	ethanol	383.15	2.481
n-decane	363.15	0.2021	ethanol	393.15	2.815
n-decane	373.15	0.2676	1-propanol	343.15	0.3247
n-decane	383.15	0.3522	1-propanol	353.15	0.4618
n-decane	393.15	0.4385	1-propanol	363.15	0.6380
methyl acetate	343.15	0.8308	1-propanol	373.15	0.8394
methyl acetate	353.15	1.045	1-propanol	383.15	1.117
methyl acetate	363.15	1.298	1-propanol	393.15	1.385
methyl acetate	373.15	1.530	1-butanol	343.15	0.1298
methyl acetate	383.15	1.954	1-butanol	353.15	0.1917
methyl acetate	393.15	2.122	1-butanol	363.15	0.2747

continued

continued

Solvent (A)	T/ K	$H_{A,B}$/ MPa	Solvent (A)	T/ K	$H_{A,B}$/ MPa
ethyl acetate	343.15	0.4744	1-butanol	373.15	0.3799
ethyl acetate	353.15	0.6198	1-butanol	383.15	0.5206
ethyl acetate	363.15	0.7912	1-butanol	393.15	0.6764
ethyl acetate	373.15	0.9734	1-pentanol	343.15	0.05321
ethyl acetate	383.15	1.252	1-pentanol	353.15	0.08144
ethyl acetate	393.15	1.420	1-pentanol	363.15	0.1210
propyl acetate	343.15	0.2261	1-pentanol	373.15	0.1738
propyl acetate	353.15	0.3086	1-pentanol	383.15	0.2459
propyl acetate	363.15	0.4098	1-pentanol	393.15	0.3332

Polymer (B): **poly(styrene-*g*-ethyl methacrylate)** **2006TEM**
Characterization: 68 units EMA/graft, 54% of styrene-units are grafted, T_g/K = 358, synthesized in the laboratory

Solvent (A)	T/ K	$H_{A,B}$/ MPa	Solvent (A)	T/ K	$H_{A,B}$/ MPa
methanol	413.15	38.94	n-hexane	443.15	18.56
methanol	423.15	42.44	n-hexane	453.15	19.10
methanol	433.15	43.48	n-hexane	463.15	21.08
methanol	443.15	48.22	n-heptane	413.15	12.06
methanol	453.15	50.27	n-heptane	423.15	13.57
methanol	463.15	55.81	n-heptane	433.15	14.34
ethanol	413.15	26.65	n-heptane	443.15	15.52
ethanol	423.15	29.17	n-heptane	453.15	16.07
ethanol	433.15	30.24	n-heptane	463.15	17.85
ethanol	443.15	33.09	n-octane	413.15	10.25
ethanol	453.15	34.47	n-octane	423.15	11.56
ethanol	463.15	38.21	n-octane	433.15	11.98
1-propanol	413.15	19.48	n-octane	443.15	13.08
1-propanol	423.15	21.84	n-octane	453.15	13.62
1-propanol	433.15	22.36	n-octane	463.15	15.41
1-propanol	443.15	24.54	n-nonane	413.15	8.898
1-propanol	453.15	25.88	n-nonane	423.15	9.783
1-propanol	463.15	28.63	n-nonane	433.15	10.00
1-butanol	413.15	14.87	n-nonane	443.15	11.14
1-butanol	423.15	16.65	n-nonane	453.15	11.73
1-butanol	433.15	17.12	n-nonane	463.15	12.92

continued

continued

Solvent (A)	T/ K	$H_{A,B}$/ MPa	Solvent (A)	T/ K	$H_{A,B}$/ MPa
1-butanol	443.15	19.27	n-decane	413.15	7.222
1-butanol	453.15	20.03	n-decane	423.15	8.185
1-butanol	463.15	22.36	n-decane	433.15	8.357
1-pentanol	413.15	11.25	n-decane	443.15	9.445
1-pentanol	423.15	12.69	n-decane	453.15	10.10
1-pentanol	433.15	13.28	n-decane	463.15	11.08
1-pentanol	443.15	14.98	tetrachloromethane	413.15	7.981
1-pentanol	453.15	15.24	tetrachloromethane	423.15	8.736
1-pentanol	463.15	17.29	tetrachloromethane	433.15	8.788
n-hexane	413.15	14.72	tetrachloromethane	443.15	9.587
n-hexane	423.15	16.37	tetrachloromethane	453.15	10.11
n-hexane	433.15	17.00	tetrachloromethane	463.15	11.18

Polymer (B): **poly(tetrafluoroethylene-*co*-hexafluoropropylene)** **2003HUT**

Characterization: M_n/g.mol^{-1} = 104000, M_w/g.mol^{-1} = 292000, 63.0 wt% hexafluoropropylene, T_g/K = 311.5, 15 % HFP as diads, DuPont, Wilmington, Delaware

Solvent (A)	T/ K	$H_{A,B}$/ MPa	Solvent (A)	T/ K	$H_{A,B}$/ MPa
n-hexane	321.5	7.91	hexafluoropropylene	411.9	54.0
n-hexane	357.2	14.6	hexafluoropropylene	466.4	76.9
n-hexane	411.9	43.9	hexafluoropropylene	511.1	95.1
n-hexane	466.5	76.5	octafluorocyclobutane	357.1	12.2
n-hexane	511.1	114.0	octafluorocyclobutane	412.0	23.3
benzene	332.5	5.26	octafluorocyclobutane	466.4	43.0
benzene	357.2	9.25	octafluorocyclobutane	511.1	58.5
benzene	411.9	30.0	toluene	321.5	1.68
benzene	466.5	44.6	toluene	357.2	4.37
benzene	511.2	70.4	toluene	411.8	15.8
1,1-difluoroethane	321.5	39.0	toluene	466.4	26.2
1,1-difluoroethane	357.1	80.4	toluene	511.2	43.3
1,1-difluoroethane	411.8	198.0	tetrafluoroethene	321.5	47.2
1,1-difluoroethane	466.3	180.0	tetrafluoroethene	332.2	55.4
1,1-difluoroethane	511.1	278.0	tetrafluoroethene	357.2	83.5

continued

continued

Solvent (A)	*T*/ K	$H_{A,B}$/ MPa	Solvent (A)	*T*/ K	$H_{A,B}$/ MPa
hexafluoroethane	332.5	67.6	tetrafluoroethene	411.9	145.9
hexafluoroethane	357.2	94.2	tetrafluoroethene	466.5	193.5
hexafluoroethane	412.0	133.0	tetrafluoroethene	511.1	228.0
hexafluoroethane	466.5	180.0	trifluoromethane	332.3	179.0
hexafluoroethane	511.1	210.0	trifluoromethane	357.2	194.0
hexafluoropropylene	321.5	17.3	trifluoromethane	412.0	274.0
hexafluoropropylene	332.3	15.5	trifluoromethane	466.4	381.0
hexafluoropropylene	357.1	25.8	trifluoromethane	511.0	397.0
2H,3H-decafluoropentane				332.4	1.58
2H,3H-decafluoropentane				357.2	3.29
2H,3H-decafluoropentane				411.8	8.72
2H,3H-decafluoropentane				466.5	18.0
2H,3H-decafluoropentane				511.1	29.9
hexafluoro-3,4-bis(trifluoromethyl)cyclobutane				357.1	3.23
hexafluoro-3,4-bis(trifluoromethyl)cyclobutane				411.8	7.08
hexafluoro-3,4-bis(trifluoromethyl)cyclobutane				466.5	13.2
hexafluoro-3,4-bis(trifluoromethyl)cyclobutane				511.2	22.4
1,1,2-trichloro-1,2,2-trifluoroethane				332.4	2.06
1,1,2-trichloro-1,2,2-trifluoroethane				357.1	3.81
1,1,2-trichloro-1,2,2-trifluoroethane				412.0	8.84
1,1,2-trichloro-1,2,2-trifluoroethane				466.5	17.2
1,1,2-trichloro-1,2,2-trifluoroethane				511.2	27.1
2,2,4-trimethylpentane				332.5	3.37
2,2,4-trimethylpentane				357.2	5.78
2,2,4-trimethylpentane				412.0	14.7
2,2,4-trimethylpentane				466.5	28.1
2,2,4-trimethylpentane				511.1	48.7

2.6. References

1994KAM Kamiya, Y., Naito, Y., and Borubon, D., Sorption and partial molar volumes of gases in poly(ethylene-*co*-vinyl acetate), *J. Polym. Sci.: Part B: Polym. Phys.*, 32, 281, 1994.

1994YOO Yoon, J.S., Chung, C.Y., and Lee, I.H., Solubility and diffusion coefficient of gaseous ethylene and α-olefin in ethylene/α-olefin random copolymers, *Eur. Polym. J.*, 30, 1209, 1994.

1996COR Cornejo-Bravo, J.M. and Siegel, R.A., Water vapour sorption behaviour of copolymers of *N,N*-diethylaminoethyl methacrylate and methyl methacrylate, *Biomaterials*, 17, 1187, 1996.

1996MIS Mishima, K., Matsuyama, K., Kutsumi, M., Komorita, N., Tokuyasu, T., Miyake, Y., and Taylor, F., Effect of vinyl alcohol + sodium acrylate copolymer gel on the vapor-liquid equilibrium of 1-propanol + water, *J. Chem. Eng. Data*, 41, 953, 1996.

1997ZHA Zhang, Y., Gangwani, K.K., and Lemert, R.M., Sorption and swelling of block copolymers in the presence of supercritical carbon dioxide, *J. Supercrit. Fluids*, 11, 115, 1997.

1999BON Bondar, V.I., Freeman, B.D., and Yampolskii, Yu.P., Sorption of gases and vapors in an amorphous glassy perfluorodioxole copolymer, *Macromolecules*, 32, 6163, 1999.

2000FOR Fornasiero, F., Halim, M., and Prausnitz, J.M., Vapor-sorption equilibria for 4-vinylpyridine-based copolymer and cross-linked polymer/alcohol systems. Effect of 'intramolecular repulsion', *Macromolecules*, 33, 8435, 2000.

2000LIW Li, W., Lin, D.-Q., and Zhu, Z.-Q., Measurement of water activities and prediction of liquid-liquid equilibria for water + ethylene oxide-propylene oxide random copolymer + ammonium sulfate systems, *Fluid Phase Equil.*, 175, 7, 2000.

2000SA1 Sato, Y., Takikawa, T., Sorakubo, A., Takishima, S., Masuoka, H., and Imaizumi, M., Solubility and diffusion coefficient of carbon dioxide in biodegradable polymers, *Ind. Eng. Chem. Res.*, 39, 4813, 2000.

2000SA2 Sato, Y., Tsuboi, A., Sorakubo, A., Takishima, S., Masuoka, H., and Ishikawa, T., Vapor-liquid equilibrium ratios for hexane at infinite dilution in ethylene + impact polypropylene copolymer and propylene + impact polypropylene copolymer, *Fluid Phase Equil.*, 170, 49, 2000.

2001DEM Demirelli, K., Kaya, I., and Coskun, M., 3,4-Dichlorobenzyl methacrylate and ethyl methacrylate system. Monomer reactivity ratios and determination of thermodynamic properties at infinite dilution using inverse gas chromatography, *Polymer*, 42, 5181, 2001.

2001KA1 Kaya, I. and Demirelli, K., Study of some thermodynamic properties of poly(3,4-dichlorobenzyl methacrylate-*co*-ethyl methacrylate) using inverse gas chromatography, *J. Polym. Eng.*, 21, 1, 2001.

2001KA2 Kaya, I. and Demirelli, K., Determination of the thermodynamic properties of poly[2-(3-phenyl-3-methylcyclobutyl)-2-hydroxyethyl methacrylate-*co*-methacrylic acid] at infinite dilution by inverse gas chromatography, *Turk. J. Chem.*, 25, 11, 2001.

2001TSU Tsuboi, A., Kolar, P., Ishikawa, T., Kamiya, Y., and Masuoka, H., Sorption and partial molar volumes of C_2 and C_3 hydrocarbons in polypropylene copolymers, *J. Polym. Sci.: Part B: Polym. Phys.*, 39, 1255, 2001.

2002DEA DeAngelis, M.G., Merkel, T.C., Bondar, V.I., Freeman, B.D., Doghieri, F., and Sarti, G.C., Gas sorption and dilation in poly(2,2-bistrifluoromethyl-4,5-difluoro-1,3-dioxole-*co*-tetrafluoroethylene): Comparison of experimental data with predictions of the nonequilibrium lattice fluid model, *Macromolecules*, 35, 1276, 2002.

2002JIN Jin, H.-J., Kim, S., and Yoon, J.-S., Solubility of 1-hexene in LLDPE synthesized by $(2\text{-MeInd})_2ZrCl_2$/MAO and by $Mg(OEt)_2$/DIBP/$TiCl_4$-TEA, *J. Appl. Polym. Sci.*, 84, 1566, 2002.

2002KAN Kang, S., Huang, Y., Fu, J., Liu, H., and Hu, Y., Vapor-liquid equilibria of several copolymer + solvent systems, *J. Chem. Eng. Data*, 47, 788, 2002.

2002KAY Kaya, I., Ilter, Z., and Senol, D., Thermodynamic interactions and characterisation of poly[(glycidyl methacrylate-*co*-methyl, ethyl, butyl) methacrylate] by inverse gas chromatography, *Polymer*, 43, 6455, 2002.

2002SEI Seiler, M., Arlt, W., Kautz, H., and Frey, H., Experimental data and theoretical considerations on vapor-liquid and liquid-liquid equilibria of hyperbranched polyglycerol and PVA solutions, *Fluid Phase Equil.*, 201, 359, 2002.

2002SHI Shieh, Y.-T. and Lin, Y.-G., Equilibrium solubility of CO_2 in rubbery EVA over a wide pressure range: Effects of carbonyl group content and crystallinity, *Polymer*, 43, 1849, 2002.

2002ZHA Zhang, R., Liu, J., He, J., Han, B., Zhang, X, Liu, Z., Jiang, T., and Hu, G., Compressed CO_2-assisted formation of reverse micelles of PEO-PPO-PEO copolymer, *Macromolecules*, 35, 7869, 2002.

2003ACI Acikses, A., Kaya, I., and Ilter, Z., Study of some thermodynamic properties of poly[(2-phenyl-1,3-dioxolane-4-yl)methyl methacrylate-*co*-butyl methacrylate] by inverse gas chromatography, *Polym.-Plast. Technol. Eng.*, 42, 431, 2003.

2003HUT Hutchenson, K.W., Henry's law coefficients for monomers and selected solvents in amorphous tetrafluoroethylene-hexafluoropropylene copolymers, *J. Chem. Eng. Data*, 48, 1028, 2003.

2003KIK Kikic, I., Vecchione, F., Alessi, P., Cortesi, A., Eva, F., and Elvassore, N., Polymer plasticization using supercritical carbon dioxide: Experiment and modeling, *Ind. Eng. Chem. Res.*, 42, 3022, 2003.

2003KOG Koga, T., Seo, Y.-S., Shin, K., Zhang, Y., Rafailovich, M.H., Sokolov, J.C., Chu, B., and Satija, S.K., The role of elasticity in the anomalous swelling of polymer thin films in density fluctuating supercritical fluids, *Macromolecules*, 36, 5236, 2003.

2003LEE Lee, L.-S., Shih, R.-F., Ou, H.-J., and Lee, T.-S., Solubility of ethylene in mixtures of toluene, norbornene, and cyclic olefin copolymer at various temperatures and pressures, *Ind. Eng. Chem. Res.*, 42, 6977, 2003.

2003MET Metz, S.J., van der Vegt, N.F.A., Mulder, M.H.V., and Wessling, M., Thermodynamics of water vapor sorption in poly(ethylene oxide) poly(butylene terephthalate) block copolymers (experimental data by S.J. Metz and M. Wessling), *J. Phys. Chem. B*, 107, 13629, 2003.

2003WAN Wang, S., Peng, C., Li, K., Wang, J., Shi, J., and Liu, H., Measurement of solubilities of CO_2 in polymers by quartz crystal micobalance (Chin.), *J. Chem. Ind. Eng. (China)*, 54, 141, 2003.

2004ARE Areerat, S., Funami, E., Hayata, Y., Nakagawa, D., and Ohshima, M., Measurement and prediction of diffusion coefficients of supercritical CO_2 in molten polymers, *Polym. Eng. Sci.*, 44, 1915, 2004.

2004ILT Ilter, Z., Kaya, I., and Acikses, A., Determination of poly[(2-phenyl-1,3-dioxolane-4-yl)methyl methacrylate-*co*-glycidyl methacrylate]-probe interactions by inverse gas chromatography, *Polym.-Plast. Technol. Eng.*, 43, 229, 2004.

2004MAT Matsuyama, K. and Mishima, K., Effect of vinyl alcohol + sodium acrylate copolymer gel on the vapor-liquid equilibrium compositions of ethanol + water and 2-propanol + water systems, *J. Chem. Eng. Data*, 49, 1688, 2004.

2004PAL Palamara, J.E., Zielinski, J.M., Hamedi, M., Duda, J.L., and Danner, R.P., Vapor-liquid equilibria of water, methanol, and methyl acetate in poly(vinyl acetate) and partially and fully hydrolyzed poly(vinyl alcohol), *Macromolecules*, 37, 6189, 2004.

2004PHA Pham, V.Q., Rao, N., and Ober, C.K., Swelling and dissolution rate measurements of polymer thin films in supercritical carbon dioxide, *J. Supercrit. Fluids*, 31, 323, 2004.

2004PRA Prabhakar, R.S., Freeman, B.D., and Roman, I., Gas and vapor sorption and permeation in poly(2,2,4-trifluoro-5-trifluoromethoxy-1,3-dioxole-*co*-tetrafluoroethylene), *Macromolecules*, 37, 7688, 2004.

2004ZHA Zhao, L. and Choi, P., Differences between Ziegler-Natta and single-site linear low-density polyethylenes as characterized by inverse gas chromatography, *Macromol. Rapid Commun.*, 25, 535, 2004.

2005ACI Acikses, A., Kaya, I., Sezek, U., and Kirilmis, C., Synthesis, characterization and thermodynamic properties of poly(3-mesityl-2-hydroxypropyl methacrylate-*co*-*N*-vinyl-2-pyrrolidone), *Polymer*, 46, 11322, 2005.

2005CSA Csaki, K.F., Nagy, M., and Csempesz, F., Influence of the chain composition on the thermodynamic properties of binary and ternary polymer solutions (experimental data by F. Csempesz), *Langmuir*, 21, 761, 2005.

2005DUA Duarte, A.R.C., Anderson, L.E., Duarte, C.M.M., and Kazarian, S.G., A comparison between gravimetric and in situ spectroscopic methods to measure the sorption of CO_2 in a biocompatible polymer, *J. Supercrit. Fluids*, 36, 160, 2005.

2005ELV Elvassore, N., Vezzu, K., and Bertucco, A., Measurement and modeling of CO_2 absorption in poly(lactic-*co*-glycolic acid), *J. Supercrit. Fluids*, 33, 1, 2005.

2005ESE Eser, H. and Tihminlioglu, F., Solubility and diffusivity of solvents and nonsolvents in poly(methyl methacrylate-*co*-butyl methacrylate), *Fluid Phase Equil.*, 237, 68, 2005.

2005LEE Lee, S.-H., Phase behavior of binary and ternary mixtures of poly(ethylene-*co*-octene)–hydrocarbons (experimental data by S.-H. Lee), *J. Appl. Polym. Sci.*, 95, 161, 2005.

2005LIU Liu, Z.H., Zhang, M., Zhao, L., and Choi, P., Molecular origin of the anomalous thermodynamic behavior of single-site ethylene-1-octene copolymer liquids with different branch contents, *Macromolecules*, 38, 4512, 2005.

2005LOP Lopez-Gonzalez, M.M., Saiz, E., and Riande, E., Experimental and simulation studies of gas sorption processes in polycarbonate films, *Polymer*, 46, 4322, 2005.

2005PRA Prabhakar, R.S., DeAngelis, M.G., Sarti, G.C., Freeman, B.D., and Coughlin, M.C., Gas and vapor sorption, permeation, and diffusion in poly(tetrafluoroethylene-*co*-perfluoromethyl vinyl ether), *Macromolecules*, 38, 7043, 2005.

2005RUT Rutherford, S.W., Kurtz, R.E., Smith, M.G., Honnell, K.G., and Coons, J.E., Measurement and correlation of sorption and transport properties of ethylene-propylene-diene monomer (EPDM) elastomers, *J. Membrane Sci.*, 263, 57, 2005.

2005SOL Solms, N. von, Zecchin, N., Rubin, A., Andersen, S.I., and Stenby, E.H., Direct measurement of gas solubility and diffusivity in poly(vinylidene fluoride) with a high-pressure microbalance, *Eur. Polym. J.*, 41, 341, 2005.

2005TOC Tochigi, K., Kurita, S., Okitsu, Y., Kurihara, K., and Ochi, K., Measurement and prediction of activity coefficients of solvents in polymer solutions using gas chromatography and a cubic-perturbed equation of state with group contribution, *Fluid Phase Equil.*, 228-229, 527, 2005.

2005WUH Wu, J., Pan, Q., and Rempel, G.L., Solubility of ethylene in toluene and toluene/styrene–butadiene rubber solutions, *J. Appl. Polym. Sci.*, 96, 645, 2005.

2005ZHA Zhang, R., Liu, J., Han, B., Wang, B., Sun, D., and He, J., Effect of PEO–PPO–PEO structure on the compressed ethylene-induced reverse micelle formation and water solubilization, *Polymer*, 46, 3936, 2005.

2006BON Bonavoglia, B., Storti, G., Morbidelli, M., Rajendran, A., and Mazotti, M., Sorption and swelling of semicrystalline polymers in supercritical CO_2, *J. Polym. Sci.: Part B: Polym. Phys.*, 44, 1531, 2006.

2006DU1 Duarte, R.C., Sampaio de Sousa, A.R., de Sousa, H.C., Gil, M.H.M., Jespersen, H.T., and Duarte, C.M.M., Solubility of dense CO_2 in two biocompatible acrylate copolymers, *Brazil. J. Chem. Eng.*, 23, 191, 2006.

2006DU2 Duarte, A.R.C., Martins, C., Coimbra, P., Gil, M.H.M., de Sousa, H.C., and Duarte, C.M.M., Sorption and diffusion of dense carbon dioxide in a biocompatible polymer, *J. Supercrit. Fluids*, 38, 392, 2006.

2006ESE Eser, H. and Tihminlioglu, F., Determination of thermodynamic and transport properties of solvents and non solvents in poly(L-lactide-*co*-glycolide), *J. Appl. Polym. Sci.*, 102, 2426, 2006.

2006KAR Karagöz, M.H., Zorer, O.S., and Ilter, Z., Analysis of physical and thermodynamic properties of poly(2-phenyl-1,3-dioxolane-4-yl-methyl-methacrylate-*co*-styrene) with inverse gas chromatography, *Polym.-Plast. Technol. Eng.*, 45, 785, 2006.

2006LIU Liu, D. and Tomasko, D.L., Carbon dioxide sorption and dilation of poly(lactide-*co*-glycolide), *J. Supercrit. Fluids*, 39, 416, 2006.

2006NAG Nagy, I., Loos, Th.W.de, Krenz, R.A., and Heidemann, R.A., High pressure phase equilibria in the systems linear low density polyethylene + n-hexane and linear low density polyethylene + n-hexane + ethylene: Experimental results and modelling with the Sanchez-Lacombe equation of state, *J. Supercrit. Fluids*, 37, 115, 2006.

2006NOV Novak, A., Bobak, M., Kosek, J., Banaszak, B.J., Lo, D., Widya, T., Ray, W.H., Pablo, J.J.de, Ethylene and 1-hexene sorption in LLDPE under typical gas-phase reactor conditions: Experiments, *J. Appl. Polym. Sci.*, 100, 1124, 2006.

2006PAR Park, H.E. and Dealy, J.M., Effects of pressure and supercritical fluids on the viscosity of polyethylene, *Macromolecules*, 39, 5438, 2006.

2006SAK Sakar, D., Erdogan, T., Cankurtaran, O., Hizal, G., Karaman, F., and Tunca, U., Physicochemical characterization of poly(*tert*-butyl acrylate-*b*-methyl methacrylate) prepared with atom transfer radical polymerization by inverse gas chromatography, *Polymer*, 47, 132, 2006.

2006TEM Temüz, M.M., Coskun, M., and Acikses, A., Determination of some thermodynamic parameters of poly(styrene-*graft*-ethyl methacrylate) using inverse gas chromatography, *J. Macromol. Sci. Part A: Pure Appl. Chem.*, 43, 609, 2006.

2007CRA Cravo, C., Duarte, A.R.C., and Duarte, C.M.M., Solubility of carbon dioxide in a natural biodegradable polymer: Determination of diffusion coefficients, *J. Supercrit. Fluids*, 40, 194, 2007.

2007FOS	Fossati, P., Sanguineti, A, DeAngelis, M.G., Baschetti, M.G., Doghieri, F., and Sarti, G.C., Gas solubility and permeability in MFA, *J. Polym. Sci.: Part B: Polym. Phys.*, 45, 1637, 2007.
2007LIG	Li, G., Gunkel, F., Wang, J., Park, C.B., and Altstaedt, V., Solubility measurements of N_2 and CO_2 in polypropylene and ethene/octene copolymer, *J. Appl. Polym. Sci.*, 103, 2945, 2007.
2007LIY	Li, Y., Wang, X., Sanchez, I.C., Johnston, K.P., and Green, P.F, Ordering in asymmetric block copolymer films by a compressible fluid, *J. Phys. Chem. B*, 111, 16, 2007.
2007NAG	Nagy, I., Krenz, R.A., Heidemann, R.A., and de Loos, Th.W., High-pressure phase equilibria in the system linear low density polyethylene + isohexane: Experimental results and modelling, *J. Supercrit. Fluids*, 40, 125, 2007.
2007NAW	Nawaby, A.V., Handa, Y.P., Liao, X., Yoshitaka, Y., and Tomohiro, M., Polymer-CO_2 systems exhibiting retrograde behavior and formation of nanofoams, *Polym. Int.*, 56, 67, 2007.
2007PIN	Pini, R., Storti, G., Mazzotti, M., Tai, H., Shakesheff, K.M., and Howdle, S.M., Sorption and swelling of poly(DL-lactic acid) and poly(lactic-*co*-glycolic acid) in supercritical CO_2, *Macromol. Symp.*, 259, 197, 2007.
2007ZOU	Zou, Q.-C. and WU, L.-M., Inverse gas chromatographic characterization of triblock copolymer of polystyrene-*b*-poly(ethylene oxide)-*b*-polystyrene., *J. Polym. Sci.: Part B: Polym. Phys.*, 45, 2015, 2007.
2008AI2	Aionicesei, E., Skerget, M., and Knez, Z., Measurement of CO_2 solubility and diffusivity in poly(L-lactide) and poly(DL-lactide-*co*-glycolide) by magnetic suspension balance, *J. Supercrit. Fluids*, 47, 296, 2008.
2008AYD	Aydin, S., Erdogan, T., Sakar, D., Hizal, G., Cankurtaran, O., Tunca, U., and Karaman, F., Detection of microphase separation in poly(*tert*-butyl acrylate-*b*-methyl methacrylate) synthesized via atom transfer radical polymerization by inverse gas chromatography, *Eur. Polym. J.*, 44, 2115, 2008.
2008BER	Bercea, M., Eckelt, J., and Wolf, B.A., Random copolymers: Their solution thermodynamics as compared with that of the corresponding homopolymers (experimental data by M. Bercea), *Ind. Eng. Chem. Res.*, 47, 2434, 2008.
2008DIN	Dingemans, M., Dewulf, J., Van Hecke, W., and Van Langenhove, H., Determination of ozone solubility in polymeric materials, *Chem. Eng. J.*, 138, 172, 2008.
2008KAS	Kasturirangan, A., Grant, C., and Teja, A.S., Compressible lattice model for phase equilibria in CO_2 + polymer systems, *Ind. Eng. Chem. Res.*, 47, 645, 2008.
2008PIN	Pini, R., Storti, G., Mazzotti, M., Tai, H., Shakesheff, K.M., and Howdle, S.M., Sorption and swelling of poly(DL-lactic acid) and poly(lactic-*co*-glycolic acid) in supercritical CO_2: An experimental and modeling study, *J. Polym. Sci.: Part B: Polym. Phys.*, 46, 483, 2008.
2008SER	Se, R.A.G. and Aznar, M., Vapor-liquid equilibrium of copolymer + solvent systems: Experimental data and thermodynamic modeling with new UNIFAC groups, *Chin. J. Chem. Eng.*, 16, 605, 2008.
2008YAZ	Yazici, D.T., Askin, A., and Bütün, V., Thermodynamic interactions of water-soluble homopolymers and double-hydrophilic diblock copolymer, *J. Chem. Thermodyn.*, 40, 353, 2008.

2009FOR	Foroutan, M. and Khomami, M.H., Influence of copolymer molar mass on the thermodynamic properties of aqueous solution of an amphiphilic copolymer, *J. Chem. Eng. Data*, 54, 861, 2009.
2009FOS	Foss, W.R., Anderl, J.N., Clausi, A.L., and Burke, P.A., Diffusivities of dichloromethane in poly(lactide-*co*-glycolide), *J. Appl. Polym. Sci.*, 112, 1622, 2009.
2009KAR	Karagöz, M.H., Erge, H., and Ilter, Z., Physical and thermodynamic properties of poly(2-phenyl-1,3-dioxolane-4-yl-methyl-methacrylate-*co*-ethyl methacrylate) polymer with inverse gas chromatography, *Asian J. Chem.*, 21, 4032, 2009.
2009OVE	Ovejero, G., Perez, P., Romero, M.D., Diaz, I., and Diez, E., SEBS triblock copolymer-solvent interaction parameters from inverse gas chromatography measurements, *Eur. Polym. J.*, 45, 590, 2009.
2009XIO	Xiong, X., Eckelt, J., Zhang, L., and Wolf, B.A., Thermodynamics of block copolymer solutions as compared with the corresponding homopolymer solutions: Experiment and theory (exp. data by J. Eckelt and B.A. Wolf), *Macromolecules*, 42, 8398, 2009.

3. LIQUID-LIQUID EQUILIBRIUM (LLE) DATA OF POLYMER SOLUTIONS

3.1. Cloud-point and/or coexistence curves of quasibinary solutions

Polymer (B): **poly(*N,N*-diethylacrylamide-*co*-acrylic acid)** **2001CAI**
Characterization: M_w/g.mol^{-1} = 319000, 5.98 mol% acrylic acid
Solvent (A): **water** **H_2O** **7732-18-5**

Type of data: cloud points (LCST-behavior)

w_B	0.005	T/K	305.05	
w_B	0.005	T/K	301.25	(in a solution of 0.05 M NaCl)

Polymer (B): **poly(*N,N*-diethylacrylamide-*co*-acrylic acid)** **2001CAI**
Characterization: M_w/g.mol^{-1} = 306000, 13.22 mol% acrylic acid
Solvent (A): **water** **H_2O** **7732-18-5**

Type of data: cloud points (LCST-behavior)

w_B	0.005	T/K	304.15

Polymer (B): **poly(*N,N*-diethylacrylamide-*co*-acrylic acid)** **2001CAI**
Characterization: M_w/g.mol^{-1} = 308000, 20.64 mol% acrylic acid
Solvent (A): **water** **H_2O** **7732-18-5**

Type of data: cloud points (LCST-behavior)

w_B	0.005	T/K	300.15

Polymer (B): **poly(*N,N*-dimethylacrylamide-*co*-allyl methacrylate)** **2005YI2**
Characterization: M_n/g.mol^{-1} = 9200, M_w/M_n = 1.9, 14.0 mol% allyl methacrylate
Solvent (A): **water** **H_2O** **7732-18-5**

Type of data: cloud points (LCST-behavior)

w_B	0.005	T/K	345.15

Polymer (B): **poly(*N,N*-dimethylacrylamide-*co*-allyl methacrylate)** **2005YI2**
Characterization: M_n/g.mol^{-1} = 10000, M_w/M_n = 2.2, 19.0 mol% allyl methacrylate
Solvent (A): **water** **H_2O** **7732-18-5**

Type of data: cloud points (LCST-behavior)

w_B	0.005	T/K	327.15

Polymer (B): **poly(*N,N*-dimethylacrylamide-*co*-allyl methacrylate)** **2005YI2**
Characterization: M_n/g.mol^{-1} = 12000, M_w/M_n = 2.3, 21.0 mol% allyl methacrylate
Solvent (A): **water** **H_2O** **7732-18-5**

Type of data: cloud points (LCST-behavior)

w_B	0.005	T/K	313.75

Polymer (B): **poly(*N,N*-dimethylacrylamide-*co*-allyl methacrylate)** **2005YI2**
Characterization: M_n/g.mol^{-1} = 13000, M_w/M_n = 2.2, 23.0 mol% allyl methacrylate
Solvent (A): **water** **H_2O** **7732-18-5**

Type of data: cloud points (LCST-behavior)

w_B	0.005	T/K	302.55

Polymer (B): **poly(*N,N*-dimethylacrylamide-*co*-allyl methacrylate)** **2005YI2**
Characterization: M_n/g.mol^{-1} = 13000, M_w/M_n = 2.3, 28.0 mol% allyl methacrylate
Solvent (A): **water** **H_2O** **7732-18-5**

Type of data: cloud points (LCST-behavior)

w_B	0.005	T/K	296.25

Polymer (B): **poly(*N,N*-dimethylacrylamide-*co*-allyl methacrylate)** **2005YI2**
Characterization: M_n/g.mol^{-1} = 12000, M_w/M_n = 2.1, 30.0 mol% allyl methacrylate
Solvent (A): **water** **H_2O** **7732-18-5**

Type of data: cloud points (LCST-behavior)

w_B	0.005	T/K	288.85

Polymer (B): **poly(*N,N*-dimethylacrylamide-*co*-*N*-phenyl-acrylamide)** **2005YI1**
Characterization: M_n/g.mol^{-1} = 9700, M_w/M_n = 1.09, 12.0 mol% *N*-phenylacrylamide, synthesized in the laboratory
Solvent (A): **water** **H_2O** **7732-18-5**

Type of data: cloud points (LCST-behavior)

w_B	0.01	T/K	343.25

Polymer (B): **poly(*N,N*-dimethylacrylamide-*co*-*N*-phenyl-acrylamide)** **2005YI1**
Characterization: M_n/g.mol^{-1} = 2000, M_w/M_n = 1.08, 15.9 mol% *N*-phenylacrylamide, synthesized in the laboratory
Solvent (A): **water** **H_2O** **7732-18-5**

Type of data: cloud points (LCST-behavior)

w_B	0.01	T/K	312.45

Polymer (B): **poly(*N,N*-dimethylacrylamide-*co*-*N*-phenyl-acrylamide)** **2005YI1**
Characterization: M_n/g.mol^{-1} = 10200, M_w/M_n = 1.07, 15.9 mol% *N*-phenylacrylamide, synthesized in the laboratory
Solvent (A): **water** **H_2O** **7732-18-5**

Type of data: cloud points (LCST-behavior)

w_B	0.01	T/K	316.65

Polymer (B): **poly(*N,N*-dimethylacrylamide-*co*-*N*-phenyl-acrylamide)** **2005YI1**
Characterization: M_n/g.mol^{-1} = 3500, M_w/M_n = 1.06, 16.2 mol% *N*-phenylacrylamide, synthesized in the laboratory
Solvent (A): **water** **H_2O** **7732-18-5**

Type of data: cloud points (LCST-behavior)

w_B	0.01	T/K	313.35

Polymer (B): **poly(*N,N*-dimethylacrylamide-*co*-*N*-phenyl-acrylamide)** **2005YI1**
Characterization: M_n/g.mol^{-1} = 4700, M_w/M_n = 1.05, 16.3 mol% *N*-phenylacrylamide, synthesized in the laboratory
Solvent (A): **water** **H_2O** **7732-18-5**

Type of data: cloud points (LCST-behavior)

w_B	0.01	T/K	312.25

Polymer (B): **poly(*N,N*-dimethylacrylamide-*co*-*N*-phenyl-acrylamide)** **2005YI1**
Characterization: M_n/g.mol^{-1} = 4600, M_w/M_n = 1.07,
21.7 mol% *N*-phenylacrylamide, synthesized in the laboratory
Solvent (A): **water** **H_2O** **7732-18-5**

Type of data: cloud points (LCST-behavior)

w_B	0.01	T/K	290.35

Polymer (B): **poly(*N,N*-dimethylacrylamide-*co*-*N*-phenyl-acrylamide)** **2005YI1**
Characterization: M_n/g.mol^{-1} = 8600, M_w/M_n = 1.07,
21.7 mol% *N*-phenylacrylamide, synthesized in the laboratory
Solvent (A): **water** **H_2O** **7732-18-5**

Type of data: cloud points (LCST-behavior)

w_B	0.01	T/K	293.25

Polymer (B): **poly(*N,N*-dimethylacrylamide-*co*-*N*-phenyl-acrylamide)** **2005YI1**
Characterization: M_n/g.mol^{-1} = 3200, M_w/M_n = 1.07,
21.8 mol% *N*-phenylacrylamide, synthesized in the laboratory
Solvent (A): **water** **H_2O** **7732-18-5**

Type of data: cloud points (LCST-behavior)

w_B	0.01	T/K	289.45

Polymer (B): **poly(*N,N*-dimethylacrylamide-*co*-*N*-phenyl-acrylamide)** **2005YI1**
Characterization: M_n/g.mol^{-1} = 10600, M_w/M_n = 1.07,
22.0 mol% *N*-phenylacrylamide, synthesized in the laboratory
Solvent (A): **water** **H_2O** **7732-18-5**

Type of data: cloud points (LCST-behavior)

w_B	0.01	T/K	293.25

Polymer (B): **poly(ethylene-*co*-1-butene)** **2006NAG**
Characterization: M_n/g.mol^{-1} = 43700, M_w/g.mol^{-1} = 52000, M_z/g.mol^{-1} = 59000,
4.1 mol% 1-butene, 2.05 ethyl branches per 100 backbone C-atomes, hydrogenated polybutadiene PBD 50000, was denoted as LLDPE, DSM, The Netherlands
Solvent (A): **n-hexane** **C_6H_{14}** **110-54-3**

continued

continued

Type of data: cloud points

w_B	0.0005	0.0005	0.0005	0.0005	0.0005	0.0005	0.0005	0.0005	0.0005
T/K	450.65	455.60	460.52	465.47	470.26	475.12	480.14	485.02	490.13
P/MPa	1.706	2.386	3.051	3.746	4.336	4.911	5.611	6.136	6.806
w_B	0.0005	0.0024	0.0024	0.0024	0.0024	0.0024	0.0024	0.0024	0.0024
T/K	495.03	440.92	445.88	450.92	455.86	460.76	466.02	470.62	475.56
P/MPa	7.281	1.198	1.898	2.668	3.408	4.048	4.703	5.333	5.948
w_B	0.0024	0.0024	0.0024	0.0024	0.0049	0.0049	0.0049	0.0049	0.0049
T/K	480.46	485.21	490.07	494.94	440.48	445.50	450.36	455.27	460.43
P/MPa	6.498	7.068	7.643	8.223	1.572	2.322	3.017	3.662	4.402
w_B	0.0049	0.0049	0.0049	0.0049	0.0049	0.0049	0.0049	0.0078	0.0078
T/K	465.40	470.32	475.12	480.08	485.17	490.14	494.91	435.55	440.60
P/MPa	5.052	5.652	6.242	6.867	7.437	8.017	8.547	1.099	1.874
w_B	0.0078	0.0078	0.0078	0.0078	0.0078	0.0078	0.0078	0.0078	0.0078
T/K	445.56	450.47	455.40	460.33	465.27	470.19	475.15	480.05	484.91
P/MPa	2.599	3.274	3.964	4.719	5.269	5.899	6.494	7.099	7.699
w_B	0.0078	0.0078	0.0100	0.0100	0.0100	0.0100	0.0100	0.0100	0.0100
T/K	490.19	495.08	435.95	440.86	445.93	450.76	455.68	460.64	465.65
P/MPa	8.274	8.809	1.223	1.973	2.738	3.453	4.103	4.798	5.428
w_B	0.0100	0.0100	0.0100	0.0100	0.0100	0.0100	0.0223	0.0223	0.0223
T/K	470.57	475.52	480.43	485.28	490.18	495.20	435.86	440.94	445.88
P/MPa	6.023	6.648	7.228	7.788	8.353	8.913	1.563	2.348	3.048
w_B	0.0223	0.0223	0.0223	0.0223	0.0223	0.0223	0.0223	0.0223	0.0223
T/K	450.77	455.64	460.68	465.64	470.62	475.53	480.52	485.32	490.24
P/MPa	3.748	4.408	5.098	5.763	6.388	6.988	7.588	8.168	8.738
w_B	0.0223	0.0452	0.0452	0.0452	0.0452	0.0452	0.0452	0.0452	0.0452
T/K	495.24	435.79	440.73	445.66	450.62	455.59	460.51	465.56	470.53
P/MPa	9.298	1.750	2.490	3.200	3.900	4.590	5.260	5.920	6.560
w_B	0.0452	0.0452	0.0452	0.0452	0.0452	0.0585	0.0585	0.0585	0.0585
T/K	475.45	480.39	485.32	490.13	494.97	435.70	440.62	445.52	450.36
P/MPa	7.165	7.765	8.345	8.905	9.445	2.024	2.719	3.394	4.119
w_B	0.0585	0.0585	0.0585	0.0585	0.0585	0.0585	0.0585	0.0585	0.0606
T/K	460.25	465.20	470.10	475.03	480.03	485.07	490.13	495.02	435.90
P/MPa	5.449	6.084	6.674	7.334	7.969	8.569	9.124	9.699	1.964
w_B	0.0606	0.0606	0.0606	0.0606	0.0606	0.0606	0.0606	0.0606	0.0606
T/K	440.70	445.68	450.67	455.37	460.64	465.45	470.40	475.34	480.24
P/MPa	2.614	3.369	4.099	4.724	5.459	6.094	6.724	7.344	7.944
w_B	0.0606	0.0606	0.0606	0.06242	0.06242	0.06242	0.06242	0.06242	0.06242
T/K	485.23	490.26	495.15	435.72	440.65	445.61	450.46	455.52	460.47
P/MPa	8.519	9.084	9.619	1.894	2.624	3.384	4.119	4.789	5.494

continued

continued

w_B	0.06242	0.06242	0.06242	0.06242	0.06242	0.06242	0.06242	0.065445	0.065445
T/K	465.71	470.67	475.56	480.50	485.43	490.46	495.72	435.89	440.81
P/MPa	6.144	6.734	7.374	7.994	8.559	9.134	9.709	1.924	2.749
w_B	0.065445	0.065445	0.065445	0.065445	0.065445	0.065445	0.065445	0.065445	0.065445
T/K	445.79	450.61	455.61	460.50	465.30	470.25	475.23	480.14	484.88
P/MPa	3.399	4.069	4.774	5.474	6.044	6.724	7.319	7.899	8.469
w_B	0.065445	0.065445	0.06690	0.06690	0.06690	0.06690	0.06690	0.06690	0.06690
T/K	489.85	494.89	435.93	440.80	445.73	450.66	455.55	460.52	465.40
P/MPa	9.069	9.624	1.969	2.594	3.349	4.034	4.684	5.394	5.994
w_B	0.06690	0.06690	0.06690	0.06690	0.06690	0.06690	0.0706	0.0706	0.0706
T/K	470.31	475.21	480.28	485.25	490.00	495.00	436.14	441.05	445.86
P/MPa	6.644	7.244	7.869	8.454	9.014	9.589	1.909	2.684	3.374
w_B	0.0706	0.0706	0.0706	0.0706	0.0706	0.0706	0.0706	0.0706	0.0706
T/K	450.805	455.70	460.81	465.77	470.61	475.57	480.38	485.36	490.26
P/MPa	4.084	4.794	5.434	6.029	6.709	7.324	7.899	8.469	9.029
w_B	0.0706	0.0734	0.0734	0.0734	0.0734	0.0734	0.0734	0.0734	0.0734
T/K	495.323	430.71	435.74	440.62	445.60	450.52	455.46	460.35	465.27
P/MPa	9.629	0.973	1.748	2.503	3.223	3.918	4.603	5.263	5.913
w_B	0.0734	0.0734	0.0734	0.0734	0.0734	0.0734	0.0826	0.0826	0.0826
T/K	470.15	475.21	480.11	485.29	490.16	495.18	436.21	441.16	446.06
P/MPa	6.538	7.188	7.768	8.443	9.018	9.543	1.814	2.564	3.279
w_B	0.0826	0.0826	0.0826	0.0826	0.0826	0.0826	0.0826	0.0826	0.0826
T/K	450.95	455.85	460.80	465.70	470.83	475.71	480.57	485.49	490.59
P/MPa	3.989	4.664	5.329	5.969	6.624	7.229	7.834	8.414	8.989
w_B	0.0826	0.0923	0.0923	0.0923	0.0923	0.0923	0.0923	0.0923	0.0923
T/K	495.31	431.30	436.40	441.11	446.17	451.05	456.00	458.55	460.95
P/MPa	9.529	0.985	1.790	2.475	3.210	3.905	4.590	4.940	5.255
w_B	0.0923	0.0923	0.0923	0.0923	0.0923	0.0923	0.0923	0.0923	0.1534
T/K	463.34	465.91	473.36	475.73	482.45	487.53	491.14	495.73	435.805
P/MPa	5.575	5.905	6.900	7.175	7.990	8.585	8.980	9.485	1.363
w_B	0.1534	0.1534	0.1534	0.1534	0.1534	0.1534	0.1534	0.1534	0.1534
T/K	440.700	445.574	450.492	455.459	460.303	465.374	470.264	475.266	480.228
P/MPa	2.103	2.798	3.513	4.233	4.898	5.563	6.213	6.843	7.438
w_B	0.1534	0.1534	0.1534	0.19463	0.19463	0.19463	0.19463	0.19463	0.19463
T/K	485.118	489.966	495.032	435.888	440.916	446.062	451.113	455.543	460.414
P/MPa	8.028	8.598	9.178	0.989	1.624	2.449	3.149	3.774	4.424
w_B	0.19463	0.19463	0.19463	0.19463	0.19463	0.19463	0.19463	0.2435	0.2435
T/K	465.513	470.451	475.400	480.519	485.453	490.548	495.514	445.479	450.554
P/MPa	5.144	5.799	6.419	7.074	7.649	8.244	8.849	1.644	2.374

continued

continued

w_B	0.2435	0.2435	0.2435	0.2435	0.2435	0.2435	0.2435	0.2435	0.2435
T/K	455.445	460.460	465.535	470.421	475.462	480.029	484.864	489.887	494.885
P/MPa	3.094	3.849	4.499	5.104	5.774	6.199	6.894	7.499	8.099
w_B	0.3031	0.3031	0.3031	0.3031	0.3031	0.3031	0.3031	0.3031	0.3031
T/K	451.011	455.705	460.899	465.863	470.829	475.762	480.675	485.650	490.540
P/MPa	1.597	2.297	3.067	3.847	4.547	4.897	5.642	6.247	6.847
w_B	0.3031								
T/K	495.435								
P/MPa	7.497								

Type of data: coexistence data (liquid-liquid-vapor three phase equilibrium)

w_B	0.0005	0.0005	0.0005	0.0005	0.0005	0.0024	0.0024	0.0024	0.0024
T/K	450.58	460.52	470.32	480.30	490.07	440.94	450.82	460.74	470.75
P/MPa	1.251	1.481	1.736	2.026	2.341	1.068	1.278	1.603	1.758
w_B	0.0024	0.0024	0.0049	0.0049	0.0049	0.0049	0.0049	0.0049	0.0078
T/K	480.45	490.02	440.62	450.62	460.54	470.46	480.31	490.20	440.61
P/MPa	2.053	2.363	1.087	1.292	1.517	1.762	2.052	2.367	1.081
w_B	0.0078	0.0078	0.0078	0.0078	0.0078	0.0223	0.0223	0.0223	0.0223
T/K	450.62	460.54	470.46	480.32	490.20	440.970	450.959	460.669	470.571
P/MPa	1.320	1.519	1.782	2.057	2.371	1.068	1.273	1.538	1.753
w_B	0.0223	0.0223	0.0452	0.0452	0.0452	0.0452	0.0452	0.0452	0.0452
T/K	480.380	490.182	431.016	440.959	450.757	460.601	470.530	480.378	490.215
P/MPa	2.128	2.353	0.899	1.079	1.274	1.499	1.759	2.049	2.364
w_B	0.0923	0.0923	0.0923	0.0923	0.0923	0.0923	0.0923	0.1946	0.1946
T/K	431.31	441.38	451.33	460.96	470.67	480.46	490.32	441.080	450.976
P/MPa	0.890	1.075	1.275	1.490	1.740	2.030	2.355	1.098	1.288
w_B	0.1946	0.1946	0.1946	0.1946	0.2435	0.2435	0.2435	0.3031	0.3031
T/K	460.856	470.696	480.547	491.112	445.488	450.430	460.276	460.793	470.728
P/MPa	1.513	1.773	2.058	2.418	1.164	1.269	1.494	1.521	1.748
w_B	0.3031								
T/K	480.391								
P/MPa	2.087								

Polymer (B): **poly(ethylene-*co*-1-butene)** **2007NAG**

Characterization: M_n/g.mol^{-1} = 43700, M_w/g.mol^{-1} = 52000, M_z/g.mol^{-1} = 59000, 4.1 mol% 1-butene, 2.05 ethyl branches per 100 backbone C-atomes, hydrogenated polybutadiene PBD 50000, was denoted as LLDPE, DSM, The Netherlands

Solvent (A): **2-methylpentane** C_6H_{14} **107-83-5**

Type of data: cloud points

continued

continued

w_B	0.0020	0.0020	0.0020	0.0020	0.0020	0.0020	0.0020	0.0020	0.0020
T/K	420.92	425.84	430.77	435.89	440.81	445.84	450.72	455.58	460.45
P/MPa	1.060	1.765	2.350	3.460	4.185	4.795	5.410	6.040	6.645
w_B	0.0020	0.0020	0.0020	0.0020	0.0020	0.0020	0.0020	0.0053	0.0053
T/K	465.35	470.40	475.24	480.25	485.13	490.17	495.10	421.03	425.96
P/MPa	7.220	7.820	8.345	8.890	9.410	9.935	10.450	1.975	2.729
w_B	0.0053	0.0053	0.0053	0.0053	0.0053	0.0053	0.0053	0.0053	0.0053
T/K	430.96	435.86	440.86	445.74	450.66	455.60	460.58	465.43	470.34
P/MPa	3.434	4.119	4.784	5.444	6.059	6.664	7.274	7.894	8.419
w_B	0.0053	0.0053	0.0053	0.0053	0.0053	0.0103	0.0103	0.0103	0.0103
T/K	475.28	480.17	484.99	489.91	494.85	415.69	420.61	425.57	430.83
P/MPa	8.969	9.499	10.019	10.519	11.039	1.570	2.385	3.060	3.940
w_B	0.0103	0.0103	0.0103	0.0103	0.0103	0.0103	0.0103	0.0103	0.0103
T/K	435.61	440.61	445.54	450.46	455.16	460.09	465.04	470.01	474.83
P/MPa	4.640	5.275	5.900	6.535	7.130	7.750	8.285	8.645	9.450
w_B	0.0103	0.0103	0.0103	0.0103	0.0288	0.0288	0.0288	0.0288	0.0288
T/K	479.72	484.62	489.48	494.39	410.72	416.06	420.97	425.91	430.79
P/MPa	9.935	10.465	10.965	11.465	1.519	2.284	3.009	3.694	4.374
w_B	0.0288	0.0288	0.0288	0.0288	0.0288	0.0288	0.0288	0.0288	0.0288
T/K	435.70	440.68	445.60	450.57	455.50	460.40	465.21	470.09	475.03
P/MPa	5.049	5.699	6.334	6.969	7.584	8.174	8.724	9.284	9.844
w_B	0.0288	0.0288	0.0288	0.0288	0.0503	0.0503	0.0503	0.0503	0.0503
T/K	479.98	484.97	489.97	494.89	411.02	415.89	420.79	425.72	430.65
P/MPa	10.364	10.894	11.419	11.894	1.874	2.569	3.344	4.049	4.774
w_B	0.0503	0.0503	0.0503	0.0503	0.0503	0.0503	0.0503	0.0503	0.0503
T/K	435.61	440.59	445.52	450.34	455.49	460.44	465.38	470.35	475.32
P/MPa	5.394	6.069	6.844	7.449	8.099	8.669	9.224	9.799	10.294
w_B	0.0503	0.0503	0.0503	0.0503	0.0604	0.0604	0.0604	0.0604	0.0604
T/K	480.11	485.09	490.01	494.96	410.57	415.12	420.77	425.77	430.71
P/MPa	10.809	11.339	11.769	12.309	1.497	2.347	3.072	3.767	4.467
w_B	0.0604	0.0604	0.0604	0.0604	0.0604	0.0604	0.0604	0.0604	0.0604
T/K	435.64	440.68	445.55	450.57	455.68	460.69	465.38	470.22	475.23
P/MPa	5.132	5.787	6.417	7.067	7.667	8.377	8.867	9.397	9.937
w_B	0.0604	0.0604	0.0604	0.0604	0.0638	0.0638	0.0638	0.0638	0.0638
T/K	480.11	485.07	489.96	494.86	411.71	416.61	421.60	426.50	431.40
P/MPa	10.462	11.057	11.492	11.992	1.893	2.598	3.368	4.033	4.743
w_B	0.0638	0.0638	0.0638	0.0638	0.0638	0.0638	0.0638	0.0638	0.0638
T/K	436.35	441.27	446.24	451.19	456.18	461.06	465.84	470.83	475.70
P/MPa	5.443	6.048	6.718	7.323	7.968	8.548	9.043	9.693	10.173

continued

continued

w_B	0.0638	0.0638	0.0638	0.0638	0.0647	0.0647	0.0647	0.0647	0.0647
T/K	480.63	485.60	490.54	495.43	410.78	415.70	420.63	425.71	430.72
P/MPa	10.798	11.273	11.818	12.273	1.718	2.523	3.208	3.983	4.718
w_B	0.0647	0.0647	0.0647	0.0647	0.0647	0.0647	0.0647	0.0647	0.0647
T/K	435.77	440.66	445.56	450.62	455.58	460.40	465.39	470.27	475.19
P/MPa	5.448	6.118	6.673	7.323	7.933	8.393	8.973	9.543	10.068
w_B	0.0647	0.0647	0.0647	0.0647	0.0984	0.0984	0.0984	0.0984	0.0984
T/K	480.25	485.08	489.96	494.84	405.92	410.92	415.77	420.68	425.67
P/MPa	10.623	11.143	11.633	12.143	0.873	1.698	2.498	3.143	3.968
w_B	0.0984	0.0984	0.0984	0.0984	0.0984	0.0984	0.0984	0.0984	0.0984
T/K	435.76	440.72	445.69	450.51	455.36	460.31	465.21	470.15	475.06
P/MPa	5.298	5.933	6.643	7.243	7.828	8.408	9.068	9.718	10.263
w_B	0.0984	0.0984	0.0984	0.0984	0.1507	0.1507	0.1507	0.1507	0.1507
T/K	479.97	484.85	489.77	494.64	411.08	416.02	420.98	425.94	430.88
P/MPa	10.758	11.198	11.743	12.218	1.197	1.957	2.682	3.402	4.107
w_B	0.1507	0.1507	0.1507	0.1507	0.1507	0.1507	0.1507	0.1507	0.1507
T/K	435.94	440.72	445.60	450.41	455.48	460.49	465.40	470.20	474.94
P/MPa	4.807	5.452	6.097	6.717	7.342	7.937	8.542	9.087	9.717
w_B	0.1507	0.1507	0.1507	0.1507	0.2494	0.2494	0.2494	0.2494	0.2494
T/K	480.04	485.01	489.92	494.85	425.78	430.67	435.64	440.57	445.51
P/MPa	10.207	10.797	11.322	11.772	2.047	2.797	3.417	4.212	4.867
w_B	0.2494	0.2494	0.2494	0.2494	0.2494	0.2494	0.2494	0.2494	0.2494
T/K	450.39	455.39	460.34	465.24	470.14	475.12	480.10	484.96	489.82
P/MPa	5.542	6.132	6.707	7.272	7.942	8.442	9.017	9.542	10.022
w_B	0.2494								
T/K	494.77								
P/MPa	10.642								

Type of data: coexistence data (liquid-liquid-vapor three phase equilibrium)

w_B	0.0020	0.0020	0.0020	0.0020	0.0020	0.0020	0.0020	0.0020	0.0053
T/K	420.92	430.65	440.88	450.74	460.55	470.39	480.26	490.13	420.75
P/MPa	0.875	1.045	1.250	1.480	1.730	2.020	2.340	2.710	0.884
w_B	0.0053	0.0053	0.0053	0.0053	0.0053	0.0053	0.0053	0.0103	0.0103
T/K	430.70	440.52	450.50	460.48	470.21	480.12	489.90	420.61	430.71
P/MPa	1.064	1.264	1.489	1.749	2.029	2.354	2.719	0.920	1.100
w_B	0.0103	0.0103	0.0103	0.0103	0.0103	0.0103	0.0288	0.0288	0.0288
T/K	440.50	450.27	460.19	469.97	479.83	489.62	411.02	421.01	430.77
P/MPa	1.295	1.520	1.775	2.060	2.385	2.740	0.729	0.884	1.065
w_B	0.0288	0.0288	0.0288	0.0288	0.0288	0.0288	0.0503	0.0503	0.0503
T/K	440.72	450.46	460.45	470.31	480.24	490.00	411.00	420.86	430.94
P/MPa	1.269	1.479	1.739	2.029	2.349	2.714	0.774	0.924	1.104

continued

continued

w_B	0.0503	0.0503	0.0503	0.0503	0.0503	0.0503	0.0984	0.0984	0.0984
T/K	440.81	450.71	460.47	470.34	480.11	489.94	410.89	420.75	440.55
P/MPa	1.304	1.534	1.784	2.074	2.394	2.774	0.753	0.908	1.283
w_B	0.0984	0.0984	0.0984	0.0984	0.0984	0.1507	0.1507	0.1507	0.1507
T/K	450.39	460.24	470.15	479.86	489.69	410.42	420.34	430.45	440.21
P/MPa	1.508	1.763	2.048	2.373	2.728	0.752	0.902	1.077	1.277
w_B	0.1507	0.1507	0.1507	0.1507	0.1507				
T/K	450.09	459.98	469.86	479.81	489.73				
P/MPa	1.502	1.757	2.047	2.367	2.667				

Polymer (B): **poly(ethylene-*co*-1-butene), deuterated** **2008KOS**
Characterization: M_n/g.mol^{-1} = 222700, M_w/g.mol^{-1} = 245000,
20.2 mol% 1-butene, 10 ethyl branches per 100 backbone C-atomes, completely deuterated polybutadiene, synthesized in the laboratory
Solvent (A): **2,2-dimethylpropane** C_5H_{12} **463-82-1**

Type of data: cloud points

w_B 0.003 was kept constant

T/K	285.95	289.15	292.75	296.65	299.45	304.15	306.55	310.35	314.75
P/MPa	35.3	32.4	30.3	27.1	26.3	25.0	23.8	22.6	21.7
T/K	317.85	322.65	327.35	329.65	333.15	337.05	339.55	341.35	345.45
P/MPa	21.3	20.8	20.4	20.2	20.0	19.6	19.5	19.4	19.4
T/K	349.45	352.75	358.55	361.65	364.65	366.95	369.35	372.75	375.45
P/MPa	19.4	19.5	19.8	19.9	20.0	20.1	20.2	20.4	20.6
T/K	379.15	381.75	385.35	389.55	391.35	394.85	398.15	401.85	406.45
P/MPa	20.8	20.9	21.1	21.4	21.5	21.7	22.0	22.3	22.6
T/K	409.05	412.95	415.55	418.75	423.05	425.75	428.75	431.25	434.85
P/MPa	22.9	23.1	23.3	23.5	23.9	24.1	24.3	24.5	24.7
T/K	435.85	440.05	441.25	445.65	450.15	454.35	456.05	460.65	461.35
P/MPa	24.7	25.0	25.1	25.3	25.7	26.2	26.4	27.1	27.2

Comments: Solid-liquid equilibrium data are additionally given in 2008KOS.

Polymer (B): **poly(ethylene-*co*-1-butene), deuterated** **2008KOS**
Characterization: M_n/g.mol^{-1} = 222700, M_w/g.mol^{-1} = 245000,
20.2 mol% 1-butene, 10 ethyl branches per 100 backbone C-atomes, completely deuterated polybutadiene, synthesized in the laboratory
Solvent (A): **2-methylbutane** C_5H_{12} **78-78-4**

Type of data: cloud points

continued

continued

w_B 0.005 was kept constant

T/K	371.15	391.75	414.45	437.05
P/MPa	1.5	4.6	8.6	12.0

w_B 0.019 was kept constant

T/K	367.95	378.25	386.25	410.85	426.95	448.15
P/MPa	1.6	3.4	5.1	8.8	11.1	14.3

Comments: Solid-liquid equilibrium data are additionally given in 2008KOS.

Polymer (B): **poly(ethylene-*co*-1-butene), deuterated** **2008KOS**
Characterization: M_n/g.mol^{-1} = 222700, M_w/g.mol^{-1} = 245000,
20.2 mol% 1-butene, 10 ethyl branches per 100 backbone C-atomes, completely deuterated polybutadiene, synthesized in the laboratory
Solvent (A): **n-pentane** C_5H_{12} **109-66-0**

Type of data: cloud points

w_B	0.005	0.005	0.005	0.005	0.018	0.018
T/K	395.65	408.45	420.95	439.55	389.85	408.45
P/MPa	2.6	4.6	6.8	9.5	1.8	5.0

Comments: Solid-liquid equilibrium data are additionally given in 2008KOS.

Polymer (B): **poly(ethylene-*co*-1-octene)** **2005LE1**
Characterization: M_n/g.mol^{-1} = 64810, M_w/g.mol^{-1} = 135500,
15.3 mol% 1-octene, T_m/K = 323.2, T_g/K = 214.2,
ρ = 0.86 g/cm^3, DuPont Dow Elastomers Corporation
Solvent (A): **n-pentane** C_5H_{12} **109-66-0**

Type of data: cloud points

w_B	0.0099	0.0202	0.0202	0.0504	0.0504	0.1133	0.1133	0.2000
T/K	413.55	404.35	413.45	403.35	416.05	403.15	412.95	414.95
P/bar	23.553	25.268	40.458	26.542	46.730	21.348	37.763	26.297

Polymer (B): **poly(ethylene-*co*-vinyl acetate)** **1999BEY**
Characterization: M_n/g.mol^{-1} = 74000, M_w/g.mol^{-1} = 285000, 70.0 wt% vinyl acetate, Scientific Polymer Products, Inc., Ontario, NY
Solvent (A): **cyclopentane** C_5H_{10} **287-92-3**

Type of data: cloud points (LCST-behavior)

w_B	0.06	0.06	0.06	0.06	0.06	0.06
T/K	462.55	480.05	494.45	507.05	518.35	530.35
P/bar	34.6	60.7	80.9	97.4	113.3	125.3

Polymer (B): **poly(ethylene-*co*-vinyl acetate)** **1999BEY**
Characterization: M_n/g.mol^{-1} = 74000, M_w/g.mol^{-1} = 285000, 70.0 wt% vinyl acetate, Scientific Polymer Products, Inc., Ontario, NY
Solvent (A): **cyclopentene** **C_5H_8** **142-29-0**

Type of data: cloud points (LCST-behavior)

w_B	0.06	0.06	0.06	0.06	0.06	0.06	0.06	0.06
T/K	475.75	475.95	485.25	486.35	496.75	497.95	505.65	507.15
P/bar	37.3	37.6	51.9	53.4	68.4	70.1	80.7	83.2

Polymer (B): **poly(ethylene oxide-*b*-propylene fumarate-*b*-ethylene oxide) dimethyl ether** **2002BEH**
Characterization: M_n/g.mol^{-1} = 2730, about 40 mol% ethylene oxide in 1.9 blocks, M_n/g.mol^{-1} = 570 PEG monomethyl ether block, M_n/g.mol^{-1} = 1660 poly(propylene fumarate) block
Solvent (A): **water** **H_2O** **7732-18-5**

Type of data: cloud points (LCST-behavior)

w_B	0.05	0.15	0.25
T/K	313.15	318.15	318.15

Polymer (B): **poly(ethylene oxide-*b*-propylene fumarate-*b*-ethylene oxide) dimethyl ether** **2002BEH**
Characterization: M_n/g.mol^{-1} = 3120, about 50 mol% ethylene oxide in 1.8 blocks, M_n/g.mol^{-1} = 800 PEG monomethyl ether block, M_n/g.mol^{-1} = 1660 poly(propylene fumarate) block
Solvent (A): **water** **H_2O** **7732-18-5**

Type of data: cloud points (LCST-behavior)

w_B	0.05	0.15	0.25
T/K	328.15	333.15	333.15

Polymer (B): **poly(ethylene oxide-*co*-propylene oxide)** **2000PE2**
Characterization: M_n/g.mol^{-1} = 3400, 20.0 mol% ethylene oxide, Shearwater Polymers, Huntsville, AL
Solvent (A): **water** **H_2O** **7732-18-5**

Type of data: cloud points (LCST-behavior)

w_B 0.10 T/K 303.15

Polymer (B): **poly(ethylene oxide-*co*-propylene oxide)** **2000PE1**
Characterization: M_n/g.mol^{-1} = 3000, 27.0 mol% ethylene oxide, Shearwater Polymers, Huntsville, AL
Solvent (A): **water** **H_2O** **7732-18-5**

Type of data: cloud points (LCST-behavior)

w_B	0.01	T/K	309.15

Polymer (B): **poly(ethylene oxide-*co*-propylene oxide)** **2000PE2**
Characterization: M_n/g.mol^{-1} = 5400, 30.0 mol% ethylene oxide, Shearwater Polymers, Huntsville, AL
Solvent (A): **water** **H_2O** **7732-18-5**

Type of data: cloud points (LCST-behavior)

w_B	0.10	T/K	313.15

Polymer (B): **poly(ethylene oxide-*co*-propylene oxide)** **1997LIM**
Characterization: M_n/g.mol^{-1} = 3000-3500, 33.3 mol% ethylene oxide, Zhejiang Univ.Chem.Factory, PR China
Solvent (A): **water** **H_2O** **7732-18-5**

Type of data: cloud points (LCST-behavior)

w_B	0.10	T/K	340.15

Polymer (B): **poly(ethylene oxide-*co*-propylene oxide)** **2000PE1**
Characterization: M_n/g.mol^{-1} = 5000, 38.5 mol% ethylene oxide, Shearwater Polymers, Huntsville, AL
Solvent (A): **water** **H_2O** **7732-18-5**

Type of data: cloud points (LCST-behavior)

w_B	0.01	T/K	309.15

Polymer (B): **poly(ethylene oxide-*co*-propylene oxide)** **2000LIW**
Characterization: M_n/g.mol^{-1} = 2340, M_w/g.mol^{-1} = 2480, 50.0 mol% ethylene oxide
Solvent (A): **water** **H_2O** **7732-18-5**

Type of data: cloud points (LCST-behavior)

w_B	0.2000	T/K	338.25

Type of data: coexistence data (tie lines, LCST-behavior)

Comments: The total feed concentration of the polymer is w_B = 0.2000.

T/K	341.35	345.25	350.65	357.95	364.65	373.15
w_B (top phase)	0.1870	0.1087	0.0710	0.0424	0.0315	0.0216
w_B (bottom phase)	0.6124	0.6592	0.6942	0.7173	0.7528	0.7654

Polymer (B): **poly(ethylene oxide-*co*-propylene oxide)** **1997LIM**
Characterization: M_n/g.mol^{-1} = 3000-3500, 50.0 mol% ethylene oxide, Zhejiang Univ.Chem.Factory, PR China
Solvent (A): **water** **H_2O** **7732-18-5**

Type of data: cloud points (LCST-behavior)

w_B 0.10 T/K 343.15

Polymer (B): **poly(ethylene oxide-*co*-propylene oxide)** **2000LIW**
Characterization: M_n/g.mol^{-1} = 3640, M_w/g.mol^{-1} = 4040, 50.0 mol% ethylene oxide
Solvent (A): **water** **H_2O** **7732-18-5**

Type of data: cloud points (LCST-behavior)

w_B 0.2000 T/K 324.85

Type of data: coexistence data (tie lines, LCST-behavior)

Comments: The total feed concentration of the polymer is w_B = 0.2000.

T/K	328.85	332.55	339.05	343.30	347.20	352.15	363.35	373.15
w_B (top phase)	0.1876	0.1275	0.0530	0.0320	0.0210	0.0140	0.0079	0.0084
w_B (bottom phase)	0.6151	0.6531	0.6992	0.7394	0.7685	0.7881	0.8315	0.8461

Polymer (B): **poly(ethylene oxide-*co*-propylene oxide)** **2000PE2**
Characterization: M_n/g.mol^{-1} = 3900, 50.0 mol% ethylene oxide, Breox PAG 50A 1000, Int. Speciality Chemicals, Southampton, UK
Solvent (A): **water** **H_2O** **7732-18-5**

Type of data: cloud points (LCST-behavior)

w_B 0.10 T/K 323.15

Polymer (B): **poly(ethylene oxide-*co*-propylene oxide)** **2000PE1**
Characterization: M_n/g.mol^{-1} = 3000, 58.8 mol% ethylene oxide, Shearwater Polymers, Huntsville, AL
Solvent (A): **water** **H_2O** **7732-18-5**

Type of data: cloud points (LCST-behavior)

w_B 0.01 T/K 326.65

Polymer (B): **poly(ethylene oxide-*co*-propylene oxide)** **1997LIM**
Characterization: M_n/g.mol^{-1} = 3000-3500, 66.7 mol% ethylene oxide, Zhejiang Univ.Chem.Factory, PR China
Solvent (A): **water** **H_2O** **7732-18-5**

Type of data: cloud points (LCST-behavior)

w_B 0.10 T/K 359.15

Polymer (B): **poly(ethylene oxide-*b*-propylene oxide-*b*-ethylene oxide)** **2000LAM**
Characterization: M_w/g.mol^{-1} = 4400, 30 mol% ethylene oxide, ρ(298.15 K) = 1.036 g/cm^3
Solvent (A): **water** **H_2O** **7732-18-5**

Type of data: cloud points (LCST-behavior)

w_B	0.05	T/K	286.65

Polymer (B): **poly(ethylene oxide-*b*-propylene oxide-*b*-ethylene oxide)** **2005ZH2**
Characterization: M_n/g.mol^{-1} = 4950, 30 wt% ethylene oxide, EO17PO60EO17
Solvent (A): **water** **H_2O** **7732-18-5**

Type of data: cloud points (LCST-behavior)

w_B	0.10	T/K	325.15

Polymer (B): **poly(ethylene oxide-*b*-propylene oxide-*b*-ethylene oxide)** **2005ZH2**
Characterization: M_n/g.mol^{-1} = 2200, 40 wt% ethylene oxide, EO10PO23EO10
Solvent (A): **water** **H_2O** **7732-18-5**

Type of data: cloud points (LCST-behavior)

w_B	0.10	T/K	346.15

Polymer (B): **poly(ethylene oxide-*b*-propylene oxide-*b*-ethylene oxide)** **2005ZH2**
Characterization: M_n/g.mol^{-1} = 2900, 40 wt% ethylene oxide, EO13PO30EO13
Solvent (A): **water** **H_2O** **7732-18-5**

Type of data: cloud points (LCST-behavior)

w_B	0.10	T/K	333.15

Polymer (B): **poly(ethylene oxide-*b*-propylene oxide-*b*-ethylene oxide)** **2005ZH2**
Characterization: M_n/g.mol^{-1} = 5900, 40 wt% ethylene oxide, EO27PO61EO27
Solvent (A): **water** **H_2O** **7732-18-5**

Type of data: cloud points (LCST-behavior)

w_B	0.10	T/K	351.15

Polymer (B): **poly(ethylene oxide-*b*-propylene oxide-*b*-ethylene oxide)** **2005ZH2**
Characterization: M_n/g.mol^{-1} = 4600, 50 wt% ethylene oxide, EO26PO40EO26
Solvent (A): **water** H_2O **7732-18-5**

Type of data: cloud points (LCST-behavior)

w_B 0.10 T/K 359.15

Polymer (B): **poly(ethylene oxide-*b*-propylene oxide-*b*-ethylene oxide)** **2005ZH2**
Characterization: M_n/g.mol^{-1} = 6500, 50 wt% ethylene oxide, EO37PO56EO37
Solvent (A): **water** H_2O **7732-18-5**

Type of data: cloud points (LCST-behavior)

w_B 0.10 T/K 367.15

Polymer (B): **poly(2-hydroxypropyl acrylate-*co*-*N*-acryloylmorpholine)** **2008EGG**
Characterization: see table, synthesized in the laboratory
Solvent (A): **water** H_2O **7732-18-5**

Type of data: cloud points (LCST-behavior)

M_n/ g.mol^{-1}	M_w/M_n	composition ratio wt% (NAM/HPA)	w_B	T/ K
8500	1.23	49/51	0.005	352.65
8500	1.23	49/51	0.010	339.05
8300	1.20	40/60	0.005	335.85
8300	1.20	40/60	0.010	326.15
8800	1.20	30/70	0.005	322.35
8800	1.20	30/70	0.010	319.45
8400	1.20	19/82	0.005	314.65
8400	1.20	19/82	0.010	304.05
8100	1.16	08/92	0.005	337.05
8100	1.16	08/92	0.010	298.45
8200	1.16	0/100	0.005	299.85
8200	1.16	0/100	0.010	294.55

Polymer (B): **poly(2-hydroxypropyl acrylate-*co*-*N,N*-dimethylacrylamide)** **2008EGG**
Characterization: see table, synthesized in the laboratory
Solvent (A): **water** **H_2O** **7732-18-5**

Type of data: cloud points (LCST-behavior)

M_n/ g.mol^{-1}	M_w/M_n	composition ratio wt% (NAM/HPA)	w_B	T/ K
9800	1.27	43/57	0.010	356.05
9600	1.26	34/66	0.005	344.75
9600	1.26	34/66	0.010	335.45
10900	1.24	26/74	0.005	328.95
10900	1.24	26/74	0.010	321.85
10700	1.22	18/82	0.005	319.85
10700	1.22	18/82	0.010	311.75
10500	1.20	10/90	0.005	308.45
10500	1.20	10/90	0.010	301.65
11100	1.21	0/100	0.005	299.85
11100	1.21	0/100	0.010	294.55

Polymer (B): **poly(*N*-isopropylacrylamide-*co*-acrylamide)** **2006SHE**
Characterization: 2.2 mol% acrylamide, synthesized in the laboratory
Solvent (A): **water** **H_2O** **7732-18-5**

Type of data: cloud points (LCST-behavior)

c_B/(g/cm^3) 0.005 T/K 307.55

Polymer (B): **poly(*N*-isopropylacrylamide-*co*-acrylamide)** **2006SHE**
Characterization: 4.5 mol% acrylamide, synthesized in the laboratory
Solvent (A): **water** **H_2O** **7732-18-5**

Type of data: cloud points (LCST-behavior)

c_B/(g/cm^3) 0.005 T/K 309.45

Polymer (B): **poly(*N*-isopropylacrylamide-*co*-acrylamide)** **2006SHE**
Characterization: 6.6 mol% acrylamide, synthesized in the laboratory
Solvent (A): **water** **H_2O** **7732-18-5**

Type of data: cloud points (LCST-behavior)

c_B/(g/cm^3) 0.005 T/K 311.15

Polymer (B): **poly(*N*-isopropylacrylamide-*co*-acrylamide)** **2006SHE**
Characterization: M_η/g.mol^{-1} = 15600, 7.4 mol% acrylamide, synthesized in the laboratory
Solvent (A): **water** **H_2O** **7732-18-5**

Type of data: cloud points (LCST-behavior)

c_B/(g/cm^3) 0.005 *T*/K 311.35

Polymer (B): **poly(*N*-isopropylacrylamide-*co*-acrylamide)** **2006SHE**
Characterization: M_η/g.mol^{-1} = 16900, 8.1 mol% acrylamide, synthesized in the laboratory
Solvent (A): **water** **H_2O** **7732-18-5**

Type of data: cloud points (LCST-behavior)

c_B/(g/cm^3) 0.005 *T*/K 312.15

Polymer (B): **poly(*N*-isopropylacrylamide-*co*-acrylamide)** **2006SHE**
Characterization: M_η/g.mol^{-1} = 9700, 14.4 mol% acrylamide, synthesized in the laboratory
Solvent (A): **water** **H_2O** **7732-18-5**

Type of data: cloud points (LCST-behavior)

c_B/(g/cm^3) 0.005 *T*/K 316.85

Polymer (B): **poly(*N*-isopropylacrylamide-*co*-acrylic acid)** **2000YAM**
Characterization: M_n/g.mol^{-1} = 330000, M_w/g.mol^{-1} = 924000
5.4 mol% acrylic acid, synthesized in the laboratory
Solvent (A): **water** **H_2O** **7732-18-5**

Type of data: cloud points (LCST-behavior)

w_B	0.001	0.001	0.001	0.001	0.001
pH	2.0	3.0	4.0	5.0	6.0
T/K	307.35	307.45	309.35	311.15	312.55

Comments: The pH of the solution was adjusted by adding the appropriate amount of NaOH solution.

Polymer (B): **poly(*N*-isopropylacrylamide-*co*-acrylic acid)** **1999JON**
Characterization: 32 mol% acrylic acid, synthesized in the laboratory
Solvent (A): **water** **H_2O** **7732-18-5**

Type of data: cloud points (LCST-behavior)

w_B	0.005	0.005	0.005	0.005	0.005	0.005	0.005
pH	1.0	1.5	2.0	2.5	3.0	3.5	4.0
T/K	<273.15	284.15	287.15	291.15	295.15	301.15	316.15

Polymer (B): **poly(*N*-isopropylacrylamide-*co*-acrylic acid)** **2000YAM**
Characterization: M_n/g.mol^{-1} = 340000, M_w/g.mol^{-1} = 1020000
37.9 mol% acrylic acid, synthesized in the laboratory
Solvent (A): **water** **H_2O** **7732-18-5**

Type of data: cloud points (LCST-behavior)

w_B 0.001 pH 5.0 T/K 315.25

Comments: The pH of the solution was adjusted by adding the appropriate amount of NaOH solution.

Polymer (B): **poly(*N*-isopropylacrylamide-*co*-acrylic acid)** **1999JON**
Characterization: 54 mol% acrylic acid, synthesized in the laboratory
Solvent (A): **water** **H_2O** **7732-18-5**

Type of data: cloud points (LCST-behavior)

w_B	0.005	0.005	0.005	0.005	0.005
pH	1.0	1.5	2.0	2.5	3.0
T/K	<273.15	283.15	288.15	299.15	318.15

Polymer (B): **poly(*N*-isopropylacrylamide-*co*-acrylic acid)** **2000YAM**
Characterization: M_n/g.mol^{-1} = 94000, M_w/g.mol^{-1} = 169200,
75.0 mol% acrylic acid, synthesized in the laboratory
Solvent (A): **water** **H_2O** **7732-18-5**

Type of data: cloud points (LCST-behavior)

w_B 0.001 pH 5.0 T/K 320.25

Comments: The pH of the solution was adjusted by adding appropriate amounts of NaOH solution.

Polymer (B): **poly(*N*-isopropylacrylamide-*co*-acrylic acid)** **1999JON**
Characterization: 80 mol% acrylic acid, synthesized in the laboratory
Solvent (A): **water** **H_2O** **7732-18-5**

Type of data: cloud points (LCST-behavior)

w_B	0.005	0.005	0.005	0.005	0.005
pH	1.0	1.5	2.0	2.5	3.0
T/K	<273.15	276.15	289.15	299.15	324.15

Polymer (B): **poly(*N*-isopropylacrylamide-*b*-ε-caprolactone-*b*-*N*-isopropylacrylamide)** **2008LOH**
Characterization: M_n/g.mol^{-1} = 17420, M_w/M_n = 1.54,
14.0 wt% ε-caprolactone, synthesized in the laboratory
Solvent (A): **water** **H_2O** **7732-18-5**

Type of data: cloud points (LCST-behavior)

c_B/(g/dm^3) 0.250 T/K 310.05

Polymer (B): **poly(*N*-isopropylacrylamide-*b*-ε-caprolactone-*b*-*N*-isopropylacrylamide)** **2008LOH**
Characterization: M_n/g.mol^{-1} = 6830, M_w/M_n = 1.36, 24.4 wt% ε-caprolactone, synthesized in the laboratory
Solvent (A): **water** **H_2O** **7732-18-5**

Type of data: cloud points (LCST-behavior)

c_B/(g/dm^3) 0.250 *T*/K 304.95

Polymer (B): **poly(*N*-isopropylacrylamide-*b*-ε-caprolactone-*b*-*N*-isopropylacrylamide)** **2008LOH**
Characterization: M_n/g.mol^{-1} = 5210, M_w/M_n = 1.68, 34.4 wt% ε-caprolactone, synthesized in the laboratory
Solvent (A): **water** **H_2O** **7732-18-5**

Type of data: cloud points (LCST-behavior)

c_B/(g/dm^3) 0.250 *T*/K 303.75

Polymer (B): **poly(*N*-isopropylacrylamide-*co*-1-deoxy-1-methacrylamido-D-glucitol)** **2002REB**
Characterization: M_n/g.mol^{-1} = 78000, M_w/g.mol^{-1} = 170000, 12.9 mol% glucitol synthesized in the laboratory by radical polymerization
Solvent (A): **water** **H_2O** **7732-18-5**

Type of data: cloud points (LCST-behavior)

w_B	0.00500	0.01060	0.02062	0.03010	0.04010	0.05260	0.05440	0.06080	0.07080
T/K	320.1	316.1	314.3	313.6	313.7	312.7	312.8	313.1	313.1
w_B	0.07790	0.08530	0.09720	0.18900					
T/K	313.0	312.8	312.4	310.6					

Polymer (B): **poly(*N*-isopropylacrylamide-*co*-1-deoxy-1-methacrylamido-D-glucitol)** **2002REB**
Characterization: M_n/g.mol^{-1} = 28600, M_w/g.mol^{-1} = 56000, 13.3 mol% glucitol synthesized in the laboratory by radical polymerization
Solvent (A): **water** **H_2O** **7732-18-5**

Type of data: cloud points (LCST-behavior)

w_B	0.0102	0.0174	0.0233	0.0263	0.0397	0.0550	0.0102	0.0102	0.0102
T/K	413.2	384.8	368.5	358.1	353.0	366.0	428.6	450.3	465.0
P/bar	1.00	1.00	1.00	1.00	1.00	1.00	2.90	3.87	4.48
w_B	0.0174	0.0174	0.0233	0.0233	0.0233	0.0233	0.0233	0.0397	0.0397
T/K	401.6	432.5	391.3	395.0	419.0	433.5	453.0	377.8	400.0
P/bar	2.27	3.87	2.10	2.27	3.58	5.71	8.57	2.57	3.65

continued

continued

w_B	0.0550	0.0550	0.0550	0.0550	0.0550	0.0550	0.0550	0.0550
T/K	373.0	388.5	392.2	403.8	411.5	421.6	440.2	451.6
P/bar	1.62	2.26	2.45	2.90	3.58	4.70	7.43	8.84

Polymer (B): **poly(*N*-isopropylacrylamide-*co*-1-deoxy-1-methacrylamido-D-glucitol)** **2002REB**

Characterization: M_n/g.mol^{-1} = 51600, M_w/g.mol^{-1} = 110000, 13.7 mol% glucitol synthesized in the laboratory by radical polymerization

Solvent (A): **water** **H_2O** **7732-18-5**

Type of data: cloud points (LCST-behavior)

w_B	0.01001	0.03914	0.0400	0.08810	0.0070	0.0070	0.0070	0.0070	0.0070
T/K	321.1	315.6	313.7	315.1	320.9	321.8	322.7	323.4	323.8
P/bar	1.0	1.0	1.0	1.0	20.0	100.0	200.0	300.0	500.0
w_B	0.0070	0.0230	0.0230	0.0230	0.0230	0.0230	0.0230		
T/K	324.0	317.6	319.0	320.3	320.7	321.9	322.4		
P/bar	600.0	20.0	200.0	300.0	400.0	600.0	700.0		

Polymer (B): **poly(*N*-isopropylacrylamide-*co*-1-deoxy-1-methacrylamido-D-glucitol)** **2002REB**

Characterization: M_n/g.mol^{-1} = 145000, M_w/g.mol^{-1} = 432000, 14.0 mol% glucitol synthesized in the laboratory by radical polymerization

Solvent (A): **water** **H_2O** **7732-18-5**

Type of data: cloud points (LCST-behavior)

w_B	0.0050	0.0100	0.0305	0.0432	0.0530	0.0983	0.1103	0.1294	0.0038
T/K	313.1	311.3	310.1	309.8	309.3	308.6	308.0	307.3	311.7
P/bar	1.0	1.0	1.0	1.0	1.0	1.0	1.0	1.0	3.5
w_B	0.0038	0.0038	0.0038	0.0038	0.0038	0.0038	0.0038	0.0038	0.0432
T/K	312.0	312.4	313.3	313.6	313.7	314.3	314.3	314.0	310.0
P/bar	62.0	110.0	200.0	300.0	400.0	500.0	600.0	700.0	10.0
w_B	0.0432	0.0432	0.0432	0.0432	0.0432	0.0432	0.0432	0.0432	0.0432
T/K	310.1	310.3	310.5	310.8	311.1	311.4	311.8	311.8	312.0
P/bar	40.0	70.0	100.0	150.0	200.0	250.0	300.0	350.0	400.0
w_B	0.0432	0.0432	0.0432	0.0432					
T/K	312.1	312.3	312.3	312.0					
P/bar	450.0	550.0	600.0	700.0					

Polymer (B): **poly(*N*-isopropylacrylamide-*co*-*N,N*-dimethylacrylamide)** **2006SHE**
Characterization: 10.7 mol% *N,N*-dimethylacrylamide
Solvent (A): **water** **H_2O** **7732-18-5**

Type of data: cloud points (LCST-behavior)

c_B/(g/cm^3)	0.005	T/K	310.05

Polymer (B): **poly(*N*-isopropylacrylamide-*co*-*N,N*-dimethylacrylamide)** **2003BAR**
Characterization: 13.0 mol% *N,N*-dimethylacrylamide
Solvent (A): **water** **H_2O** **7732-18-5**

Type of data: cloud points (LCST-behavior)

w_B	0.01	T/K	309.15

Polymer (B): **poly(*N*-isopropylacrylamide-*co*-*N,N*-dimethylacrylamide)** **2006SHE**
Characterization: 14.7 mol% *N,N*-dimethylacrylamide
Solvent (A): **water** **H_2O** **7732-18-5**

Type of data: cloud points (LCST-behavior)

c_B/(g/cm^3)	0.005	T/K	311.25

Polymer (B): **poly(*N*-isopropylacrylamide-*co*-*N,N*-dimethylacrylamide)** **2006SHE**
Characterization: 17.3 mol% *N,N*-dimethylacrylamide
Solvent (A): **water** **H_2O** **7732-18-5**

Type of data: cloud points (LCST-behavior)

c_B/(g/cm^3)	0.005	T/K	312.95

Polymer (B): **poly(*N*-isopropylacrylamide-*co*-*N,N*-dimethylacrylamide)** **2006SHE**
Characterization: 21.2 mol% *N,N*-dimethylacrylamide
Solvent (A): **water** **H_2O** **7732-18-5**

Type of data: cloud points (LCST-behavior)

c_B/(g/cm^3)	0.005	T/K	314.25

Polymer (B): **poly(*N*-isopropylacrylamide-*co*-*N,N*-dimethylacrylamide)** **2003BAR**
Characterization: 27.0 mol% *N,N*-dimethylacrylamide
Solvent (A): **water** **H_2O** **7732-18-5**

Type of data: cloud points (LCST-behavior)

w_B	0.01	T/K	312.15

Polymer (B): **poly(*N*-isopropylacrylamide-*co*-*N,N*-dimethylacrylamide)** **2003BAR**
Characterization: 30.0 mol% *N,N*-dimethylacrylamide
Solvent (A): **water** H_2O **7732-18-5**

Type of data: cloud points (LCST-behavior)

w_B	0.01	T/K	315.15

Polymer (B): **poly(*N*-isopropylacrylamide-*co*-*N,N*-dimethylacrylamide)** **2006SHE**
Characterization: 31.4 mol% *N,N*-dimethylacrylamide
Solvent (A): **water** H_2O **7732-18-5**

Type of data: cloud points (LCST-behavior)

c_B/(g/cm^3)	0.005	T/K	319.15

Polymer (B): **poly(*N*-isopropylacrylamide-*co*-*N,N*-dimethylacrylamide)** **2003BAR**
Characterization: 50.0 mol% *N,N*-dimethylacrylamide
Solvent (A): **water** H_2O **7732-18-5**

Type of data: cloud points (LCST-behavior)

w_B	0.01	T/K	323.15

Polymer (B): **poly(*N*-isopropylacrylamide-*co*-*N,N*-dimethylacrylamide)** **2003BAR**
Characterization: 60.0 mol% *N,N*-dimethylacrylamide
Solvent (A): **water** H_2O **7732-18-5**

Type of data: cloud points (LCST-behavior)

w_B	0.01	T/K	336.15

Polymer (B): **poly(*N*-isopropylacrylamide-*co*-*N,N*-dimethylacrylamide)** **2003BAR**
Characterization: 66.0 mol% *N,N*-dimethylacrylamide
Solvent (A): **water** H_2O **7732-18-5**

Type of data: cloud points (LCST-behavior)

w_B	0.01	T/K	345.15

Polymer (B): **poly(*N*-isopropylacrylamide-*co*-*N*-glycineacrylamide)** **2000PRI**
Characterization: M_n/g.mol^{-1} = 30000, M_w/g.mol^{-1} = 77000, 20 mol% *N*-glycine-acrylamide, synthesized in the laboratory
Solvent (A): **water** H_2O **7732-18-5**

Type of data: cloud points (LCST-behavior)

c_B/(g/l)	1.0	1.0	1.0	1.0
pH	2.77	3.83	4.25	5.10
T/K	303.55	305.85	307.55	311.45

Comments: pH values were prepared by buffer solutions from 0.1 M citric acid and 0.1 M sodium hydroxide at constant ionic strength of 0.1 M NaCl.

Polymer (B): **poly(*N*-isopropylacrylamide-*co*-2-hydroxyethyl methacrylate)** **2006SHE**
Characterization: 9.4 mol% 2-hydroxyethyl methacrylate
Solvent (A): **water** H_2O **7732-18-5**

Type of data: cloud points (LCST-behavior)

c_B/(g/cm^3) 0.005 *T*/K 303.45

Polymer (B): **poly(*N*-isopropylacrylamide-*co*-2-hydroxyethyl methacrylate)** **2006SHE**
Characterization: 17.0 mol% 2-hydroxyethyl methacrylate
Solvent (A): **water** H_2O **7732-18-5**

Type of data: cloud points (LCST-behavior)

c_B/(g/cm^3) 0.005 *T*/K 299.75

Polymer (B): **poly(*N*-isopropylacrylamide-*co*-2-hydroxyethyl methacrylate)** **2006SHE**
Characterization: 25.8 mol% 2-hydroxyethyl methacrylate
Solvent (A): **water** H_2O **7732-18-5**

Type of data: cloud points (LCST-behavior)

c_B/(g/cm^3) 0.005 *T*/K 294.55

Polymer (B): **poly(*N*-isopropylacrylamide-*co*-2-hydroxyethyl methacrylate)** **2006SHE**
Characterization: 34.9 mol% 2-hydroxyethyl methacrylate
Solvent (A): **water** H_2O **7732-18-5**

Type of data: cloud points (LCST-behavior)

c_B/(g/cm^3) 0.005 *T*/K 290.15

Polymer (B):	**poly(*N*-isopropylacrylamide-*co*-*N*-isopropylmethacrylamide)**		**2001DJO**
Characterization:	M_n/g.mol^{-1} = 55300, M_w/g.mol^{-1} = 177000, 10.56 mol% *N*-isopropylmethacrylamide, synthesized in the laboratory		
Solvent (A):	**water**	H_2O	**7732-18-5**

Type of data: cloud points (LCST-behavior)

w_B	0.01	T/K	307.15

Polymer (B):	**poly(*N*-isopropylacrylamide-*co*-*N*-isopropylmethacrylamide)**		**2001DJO**
Characterization:	M_n/g.mol^{-1} = 42800, M_w/g.mol^{-1} = 137000, 20.30 mol% *N*-isopropylmethacrylamide, synthesized in the laboratory		
Solvent (A):	**water**	H_2O	**7732-18-5**

Type of data: cloud points (LCST-behavior)

w_B	0.01	T/K	308.45

Polymer (B):	**poly(*N*-isopropylacrylamide-*co*-*N*-isopropylmethacrylamide)**		**2001DJO**
Characterization:	M_n/g.mol^{-1} = 28800, M_w/g.mol^{-1} = 92000, 30.00 mol% *N*-isopropylmethacrylamide, synthesized in the laboratory		
Solvent (A):	**water**	H_2O	**7732-18-5**

Type of data: cloud points (LCST-behavior)

w_B	0.01	T/K	309.75

Polymer (B):	**poly(*N*-isopropylacrylamide-*co*-*N*-isopropylmethacrylamide)**		**2001DJO**
Characterization:	M_n/g.mol^{-1} = 23100, M_w/g.mol^{-1} = 74000, 39.99 mol% *N*-isopropylmethacrylamide, synthesized in the laboratory		
Solvent (A):	**water**	H_2O	**7732-18-5**

Type of data: cloud points (LCST-behavior)

w_B	0.01	T/K	311.05

Polymer (B):	**poly(*N*-isopropylacrylamide-*co*-*N*-isopropylmethacrylamide)**		**2001DJO**
Characterization:	M_n/g.mol^{-1} = 18800, M_w/g.mol^{-1} = 60000, 50.22 mol% *N*-isopropylmethacrylamide, synthesized in the laboratory		
Solvent (A):	**water**	H_2O	**7732-18-5**

Type of data: cloud points (LCST-behavior)

w_B	0.01	T/K	312.95

Polymer (B): **poly(*N*-isopropylacrylamide-*co*-*N*-isopropylmethacrylamide)** **2001DJO**
Characterization: M_n/g.mol^{-1} = 23100, M_w/g.mol^{-1} = 74000, 59.89 mol% *N*-isopropylmethacrylamide, synthesized in the laboratory
Solvent (A): **water** **H_2O** **7732-18-5**

Type of data: cloud points (LCST-behavior)

w_B	0.01	T/K	314.65

Polymer (B): **poly(*N*-isopropylacrylamide-*co*-*N*-isopropylmethacrylamide)** **2001DJO**
Characterization: M_n/g.mol^{-1} = 17500, M_w/g.mol^{-1} = 56000, 69.86 mol% *N*-isopropylmethacrylamide, synthesized in the laboratory
Solvent (A): **water** **H_2O** **7732-18-5**

Type of data: cloud points (LCST-behavior)

w_B	0.01	T/K	315.55

Polymer (B): **poly(*N*-isopropylacrylamide-*co*-*N*-isopropylmethacrylamide)** **2001DJO**
Characterization: M_n/g.mol^{-1} = 16600, M_w/g.mol^{-1} = 53000, 79.81 mol% *N*-isopropylmethacrylamide, synthesized in the laboratory
Solvent (A): **water** **H_2O** **7732-18-5**

Type of data: cloud points (LCST-behavior)

w_B	0.01	T/K	317.35

Polymer (B): **poly(*N*-isopropylacrylamide-*co*-*N*-isopropylmethacrylamide)** **2001DJO**
Characterization: M_n/g.mol^{-1} = 14700, M_w/g.mol^{-1} = 47000, 89.99 mol% *N*-isopropylmethacrylamide, synthesized in the laboratory
Solvent (A): **water** **H_2O** **7732-18-5**

Type of data: cloud points (LCST-behavior)

w_B	0.01	T/K	318.75

Polymer (B): **poly(*N*-isopropylacrylamide-*co*-2-methacryloamidohistidine)** **2003KAL**
Characterization: see table, synthesized in the laboratory
Solvent (A): **water** **H_2O** **7732-18-5**

Type of data: cloud points (LCST-behavior)

continued

continued

w_D 0.005 was kept constant

Mol% MAH	M_η/ g.mol^{-1}	pH = 4.0	*T*/K pH = 7.4	pH = 9.0
1.26	62376	303.05	304.35	303.75
2.48	62790	303.95	305.85	304.45
4.85	62082	304.45	306.55	304.85
9.24	62541	304.75	307.25	305.15
4.85	68142	304.25	305.85	304.65
4.85	77590	304.05	305.65	304.55

Polymer (B): **poly(*N*-isopropylacrylamide-*co*-8-methacryloyloxyoctanoic acid)** **2004SAL**

Characterization: M_w/g.mol^{-1} = 354000, 7.1 mol% 8-methacryloyloxy-octanoic acid, synthesized in the laboratory

Solvent (A): **water** **H_2O** **7732-18-5**

Type of data: cloud points (LCST-behavior)

c_B/(mg/ml)	5.0	5.0	5.0	5.0	5.0
pH	*)	7	8	9	10
T/K	290.95	299.65	307.65	310.75	315.05

Comments: pH values were prepared by buffer solutions at constant ionic strength of 0.1 M. *) is deionized water. Cloud points were detected by using modulated DSC.

Polymer (B): **poly(*N*-isopropylacrylamide-*co*-8-methacryloyloxyoctanoic acid)** **2004SAL**

Characterization: M_w/g.mol^{-1} = 318000, 12.4 mol% 8-methacryloyloxy-octanoic acid, synthesized in the laboratory

Solvent (A): **water** **H_2O** **7732-18-5**

Type of data: cloud points (LCST-behavior)

c_B/(mg/ml)	5.0	5.0
pH	7	8
T/K	294.25	313.35

Comments: pH values were prepared by buffer solutions at constant ionic strength of 0.1 M. Cloud points were detected by using modulated DSC.

Polymer (B): **poly(*N*-isopropylacrylamide-*co*-8-methacryloyloxyoctanoic acid)** **2004SAL**

Characterization: M_w/g.mol^{-1} = 275000, 17.4 mol% 8-methacryloyloxyoctanoic acid, synthesized in the laboratory

Solvent (A): **water** **H_2O** **7732-18-5**

Type of data: cloud points (LCST-behavior)

c_B/(mg/ml)	5.0
pH	8
T/K	314.75

Comments: pH values were prepared by buffer solutions at constant ionic strength of 0.1 M. Cloud points were detected by using modulated DSC.

Polymer (B): **poly(*N*-isopropylacrylamide-*co*-5-methacryloyloxypentanoic acid)** **2004SAL**

Characterization: M_w/g.mol^{-1} = 291000, 6.7 mol% 5-methacryloyloxypentanoic acid, synthesized in the laboratory

Solvent (A): **water** **H_2O** **7732-18-5**

Type of data: cloud points (LCST-behavior)

c_B/(mg/ml)	5.0	5.0	5.0	5.0	5.0
pH	*)	7	8	9	10
T/K	299.45	305.45	307.55	308.15	306.65

Comments: pH values were prepared by buffer solutions at constant ionic strength of 0.1 M. *) is deionized water. Cloud points were detected by using modulated DSC.

Polymer (B): **poly(*N*-isopropylacrylamide-*co*-5-methacryloyloxypentanoic acid)** **2004SAL**

Characterization: M_w/g.mol^{-1} = 340000, 10.8 mol% 5-methacryloyloxypentanoic acid, synthesized in the laboratory

Solvent (A): **water** **H_2O** **7732-18-5**

Type of data: cloud points (LCST-behavior)

c_B/(mg/ml)	5.0	5.0	5.0	5.0
pH	*)	7	8	9
T/K	283.25	307.65	310.25	315.15

Comments: pH values were prepared by buffer solutions at constant ionic strength of 0.1 M. *) is deionized water. Cloud points were detected by using modulated DSC.

Polymer (B): **poly(*N*-isopropylacrylamide-*co*-5-methacryloyloxypentanoic acid)** **2004SAL**
Characterization: M_w/g.mol^{-1} = 241000, 17.9 mol% 5-methacryloyloxypentanoic acid, synthesized in the laboratory
Solvent (A): **water** **H_2O** **7732-18-5**

Type of data: cloud points (LCST-behavior)

c_B/(mg/ml)	5.0	5.0
pH	7	8
T/K	307.55	316.95

Comments: pH values were prepared by buffer solutions at constant ionic strength of 0.1 M. Cloud points were detected by using modulated DSC.

Polymer (B): **poly(*N*-isopropylacrylamide-*co*-5-methacryloyloxypentanoic acid)** **2004SAL**
Characterization: M_w/g.mol^{-1} = 310000, 23.7 mol% 5-methacryloyloxypentanoic acid, synthesized in the laboratory
Solvent (A): **water** **H_2O** **7732-18-5**

Type of data: cloud points (LCST-behavior)

c_B/(mg/ml)	5.0	5.0
pH	7	8
T/K	304.55	325.95

Comments: pH values were prepared by buffer solutions at constant ionic strength of 0.1 M. Cloud points were detected by using modulated DSC.

Polymer (B): **poly(*N*-isopropylacrylamide-*co*-11-methacryloyloxyundecanoic acid)** **2004SAL**
Characterization: M_w/g.mol^{-1} = 287000, 6.5 mol% 11-methacryloyloxyundecanoic acid, synthesized in the laboratory
Solvent (A): **water** **H_2O** **7732-18-5**

Type of data: cloud points (LCST-behavior)

c_B/(mg/ml)	5.0	5.0	5.0	5.0	5.0
pH	*)	7	8	9	10
T/K	279.15	290.25	306.65	301.65	305.85

Comments: pH values were prepared by buffer solutions at constant ionic strength of 0.1 M. *) is deionized water. Cloud points were detected by using modulated DSC.

Polymer (B): **poly(*N*-isopropylacrylamide-*co*-11-methacryloyloxyundecanoic acid)** **2004SAL**
Characterization: M_w/g.mol^{-1} = 289000, 11.2 mol% 11-methacryloyloxyundecanoic acid, synthesized in the laboratory
Solvent (A): **water** **H_2O** **7732-18-5**

Type of data: cloud points (LCST-behavior)

c_B/(mg/ml)	5.0	5.0
pH	8	9
T/K	306.35	308.85

Comments: pH values were prepared by buffer solutions at constant ionic strength of 0.1 M. Cloud points were detected by using modulated DSC.

Polymer (B): **poly(*N*-isopropylacrylamide-*co*-11-methacryloyloxyundecanoic acid)** **2004SAL**
Characterization: M_w/g.mol^{-1} = 231000, 17.2 mol% 11-methacryloyloxyundecanoic acid, synthesized in the laboratory
Solvent (A): **water** **H_2O** **7732-18-5**

Type of data: cloud points (LCST-behavior)

c_B/(mg/ml)	5.0
pH	8
T/K	306.85

Comments: pH values were prepared by buffer solutions at constant ionic strength of 0.1 M. Cloud points were detected by using modulated DSC.

Polymer (B): **poly(*N*-isopropylacrylamide-*co*-4-pentenoic acid)** **1995CHE**
Characterization: M_w/g.mol^{-1} = 240000, 0.7 mol% 4-pentenoic acid
Solvent (A): **water** **H_2O** **7732-18-5**

Type of data: cloud points (LCST-behavior)

c_B/(mg/ml)	2.0	2.0
pH	4.0	7.5
T/K	303.05	306.15

Comments: pH values were prepared by phosphate buffer solutions.

Polymer (B): **poly(*N*-isopropylacrylamide-*co*-4-pentenoic acid)** **1995CHE**
Characterization: M_w/g.mol^{-1} = 280000, 1.3 mol% 4-pentenoic acid
Solvent (A): **water** **H_2O** **7732-18-5**

Type of data: cloud points (LCST-behavior)

c_B/(mg/ml)	2.0	2.0
pH	4.0	7.5
T/K	302.85	307.45

Comments: pH values were prepared by phosphate buffer solutions.

Polymer (B): **poly(*N*-isopropylacrylamide-*co*-4-pentenoic acid) 1995CHE**
Characterization: M_w/g.mol^{-1} = 370000, 2.3 mol% 4-pentenoic acid
Solvent (A): **water** H_2O **7732-18-5**

Type of data: cloud points (LCST-behavior)

c_B/(mg/ml)	2.0	2.0
pH	4.0	7.5
T/K	302.05	309.65

Comments: pH values were prepared by phosphate buffer solutions.

Polymer (B): **poly(*N*-isopropylacrylamide-*co*-4-pentenoic acid) 1995CHE**
Characterization: M_w/g.mol^{-1} = 400000, 6.5 mol% 4-pentenoic acid
Solvent (A): **water** H_2O **7732-18-5**

Type of data: cloud points (LCST-behavior)

c_B/(mg/ml)	2.0
pH	4.0
T/K	300.45

Comments: pH values were prepared by phosphate buffer solutions.

Polymer (B): **poly(*N*-isopropylacrylamide-*co*-4-pentenoic acid) 1995CHE**
Characterization: M_w/g.mol^{-1} = 300000, 10.8 mol% 4-pentenoic acid
Solvent (A): **water** H_2O **7732-18-5**

Type of data: cloud points (LCST-behavior)

c_B/(mg/ml)	2.0
pH	4.0
T/K	297.45

Comments: pH values were prepared by phosphate buffer solutions.

Polymer (B): **poly(*N*-isopropylacrylamide-*co*-4-pentenoic acid) 1995CHE**
Characterization: M_w/g.mol^{-1} = 220000, 21.7 mol% 4-pentenoic acid
Solvent (A): **water** H_2O **7732-18-5**

Type of data: cloud points (LCST-behavior)

c_B/(mg/ml)	2.0
pH	4.0
T/K	292.35

Comments: pH values were prepared by phosphate buffer solutions.

Polymer (B): **poly(*N*-isopropylacrylamide-*co*-4-pentenoic acid)** **1995CHE**
Characterization: M_w/g.mol^{-1} = 190000, 28.6 mol% 4-pentenoic acid
Solvent (A): **water** H_2O **7732-18-5**

Type of data: cloud points (LCST-behavior)

c_B/(mg/ml)	2.0
pH	4.0
T/K	277.15

Comments: pH values were prepared by phosphate buffer solutions.

Polymer (B): **poly(*N*-isopropylacrylamide-*co*-sodium 2-acrylamido-2-methyl-1-propanesulfonate-*co*-*N*-*tert*-butylacrylamide)** **2007SH1**
Characterization: synthesized in the laboratory
Solvent (A): **water** H_2O **7732-18-5**

Type of data: cloud points (LCST-behavior)

Comments: molar ratio of NIPAm/AMPS/NTBAm = 100/4.3/0

w_B	0.002	0.004	0.006	0.008	0.010
T/K	323.45	320.05	318.15	316.25	315.65

Comments: molar ratio of NIPAm/AMPS/NTBAm = 100/4.1/3.1

w_B	0.002	0.004	0.006	0.008	0.010
T/K	321.35	317.35	315.25	313.45	312.85

Comments: molar ratio of NIPAm/AMPS/NTBAm = 100/4.0/9.5

w_B	0.002	0.004	0.006	0.008	0.010
T/K	318.35	314.35	312.05	310.55	309.95

Comments: molar ratio of NIPAm/AMPS/NTBAm = 100/3.5/18.4

w_B	0.002	0.004	0.006	0.008	0.010
T/K	314.45	310.65	308.15	307.95	306.55

Polymer (B): **poly(*N*-isopropylacrylamide-*co*-1-vinylimidazole)** **2000PE1**
Characterization: synthesized in the laboratory
Solvent (A): **water** H_2O **7732-18-5**

Type of data: cloud points (LCST-behavior)

w_B	0.01	T/K	309.65	for a copolymer of 4.76 mol% 1-vinylimidazole
w_B	0.01	T/K	318.15	for a copolymer of 33.33 mol% 1-vinylimidazole
w_B	0.01	T/K	329.15	for a copolymer of 50.00 mol% 1-vinylimidazole

Polymer (B): **poly(*N*-isopropylacrylamide-*co*-p-vinylphenylboronic acid)** **2005CIM**
Characterization: 3.2 mol% p-vinylphenylboronic acid
Solvent (A): **water** **H_2O** **7732-18-5**

Type of data: cloud points (LCST-behavior)

pH	4.0	5.0	7.4
T/K	299.75	300.05	300.95

Polymer (B): **poly(*N*-isopropylacrylamide-*co*-p-vinylphenylboronic acid)** **2005CIM**
Characterization: 7.3 mol% p-vinylphenylboronic acid
Solvent (A): **water** **H_2O** **7732-18-5**

Type of data: cloud points (LCST-behavior)

pH	4.0	5.0	7.4
T/K	299.05	299.35	300.65

Polymer (B): **poly(*N*-isopropylacrylamide-*co*-p-vinylphenylboronic acid)** **2005CIM**
Characterization: 31.1 mol% p-vinylphenylboronic acid
Solvent (A): **water** **H_2O** **7732-18-5**

Type of data: cloud points (LCST-behavior)

pH	4.0	5.0	7.4
T/K	298.75	298.95	299.45

Polymer (B): **poly(methoxydiethylene glycol methacrylate-*co*-dodecyl methacrylate)** **2004KIT**
Characterization: M_n/g.mol^{-1} = 32100, M_w/g.mol^{-1} = 52600, 9.43 mol% methoxydiethylene glycol methacrylate, molar ratio = 1/9.61, synthesized in the laboratory
Solvent (A): **water** **H_2O** **7732-18-5**

Type of data: cloud points (LCST-behavior)

c_B/(g/l) 1.0 *T*/K 292.65

Polymer (B): **poly(methoxydiethylene glycol methacrylate-*co*-methoxyoligo(ethylene glycol) methacrylate)** **2004KIT**
Characterization: M_n/g.mol^{-1} = 46600, M_w/g.mol^{-1} = 116000, 2.75 mol% methoxyoligoethylene glycol(9 EO units) methacrylate, molar ratio = 1/35.4, synthesized in the laboratory
Solvent (A): **water** **H_2O** **7732-18-5**

Type of data: cloud points (LCST-behavior)

c_B/(g/l) 1.0 *T*/K 300.85

Polymer (B): **poly(methoxydiethylene glycol methacrylate-*co*-methoxyoligo(ethylene glycol) methacrylate)** **2004KIT**

Characterization: M_n/g.mol^{-1} = 35200, M_w/g.mol^{-1} = 67200, 4.12 mol% methoxyoligoethylene glycol(9 EO units) methacrylate, molar ratio = 1/23.3, synthesized in the laboratory

Solvent (A): **water** **H_2O** **7732-18-5**

Type of data: cloud points (LCST-behavior)

c_B/(g/l)	1.0	T/K	302.85

Polymer (B): **poly(methoxydiethylene glycol methacrylate-*co*-methoxyoligo(ethylene glycol) methacrylate)** **2004KIT**

Characterization: M_n/g.mol^{-1} = 13100, M_w/g.mol^{-1} = 20100, 9.69 mol% methoxyoligoethylene glycol(9 EO units) methacrylate, molar ratio = 1/9.32, synthesized in the laboratory

Solvent (A): **water** **H_2O** **7732-18-5**

Type of data: cloud points (LCST-behavior)

c_B/(g/l)	1.0	T/K	312.45

Polymer (B): **poly(methoxydiethylene glycol methacrylate-*co*-methoxyoligo(ethylene glycol) methacrylate)** **2004KIT**

Characterization: M_n/g.mol^{-1} = 39700, M_w/g.mol^{-1} = 76300, 1.83 mol% methoxyoligoethylene glycol(23 EO units) methacrylate, molar ratio = 1/53.5, synthesized in the laboratory

Solvent (A): **water** **H_2O** **7732-18-5**

Type of data: cloud points (LCST-behavior)

c_B/(g/l)	1.0	T/K	302.75

Polymer (B): **poly(methoxydiethylene glycol methacrylate-*co*-methoxyoligo(ethylene glycol) methacrylate)** **2004KIT**

Characterization: M_n/g.mol^{-1} = 18000, M_w/g.mol^{-1} = 34200, 4.40 mol% methoxyoligoethylene glycol(23 EO units) methacrylate, molar ratio = 1/21.7, synthesized in the laboratory

Solvent (A): **water** **H_2O** **7732-18-5**

Type of data: cloud points (LCST-behavior)

c_B/(g/l)	1.0	T/K	309.15

Polymer (B): **poly(methoxydiethylene glycol methacrylate-*co*-methoxyoligo(ethylene glycol) methacrylate)** **2004KIT**

Characterization: M_n/g.mol^{-1} = 33200, M_w/g.mol^{-1} = 56800, 9.43 mol% methoxyoligoethylene glycol(23 EO units) methacrylate, molar ratio = 1/9.61, synthesized in the laboratory

Solvent (A): **water** **H_2O** **7732-18-5**

Type of data: cloud points (LCST-behavior)

c_B/(g/l) 1.0 T/K 334.65

Polymer (B): **poly[oligo(ethylene glycol) methylacrylate-*co*-oligo(propylene glycol) methacrylate]** **2009WA2**

Characterization: see table, synthesized in the laboratory, oligomers have M_n(OEGM)/g.mol^{-1} = 475, M_n(OPGM)/g.mol^{-1} = 430

Solvent (A): **water** **H_2O** **7732-18-5**

Type of data: cloud points (LCST-behavior)

c_B/(mg/ml) 3.0 was kept constant, phosphate buffer, pH = 7.4

M_n/ g.mol^{-1}	M_w/M_n	composition ratio mol% (OEGM/OPGM)	T/ K
33500	1.43	79:21	343.15
32700	1.37	56:44	332.15
31700	1.33	22:78	309.15
16700	1.39	22:78	308.15
30300	1.69	21:79	307.15
15500	1.61	22:78	309.15
160000	1.72	22:78	309.15

Polymer (B): **poly(styrene-*b*-butadiene-*b*-styrene-*b*-butadiene-*b*-styrene)** **2004NIE**

Characterization: M_n/g.mol^{-1} = 109000, M_w/g.mol^{-1} = 120000, both polybutadiene blocks have M_n/g.mol^{-1} = 32000, the middle polystyrene block has M_n/g.mol^{-1} = 9000, both outside polystyrene blocks have M_n/g.mol^{-1} = 13000

Solvent (A): **n-heptane** **C_7H_{16}** **142-82-5**

Type of data: cloud points (UCST-behavior)

w_B 0.005 T/K 330.15

Polymer (B): **poly(styrene-*co*-methyl methacrylate)** **2006GAR**
Characterization: M_n/g.mol^{-1} = 55200, M_w/g.mol^{-1} = 58500
Solvent (A): **cyclohexanol** $C_6H_{12}O$ **108-93-0**

Type of data: cloud points (UCST-behavior)

w_B	0.0015	0.0062	0.0104	0.0417	0.0602	0.0941	0.1179	0.1589	0.1639
T/K	340.74	346.65	347.14	345.68	345.41	345.61	345.45	345.73	344.80

w_B	0.1830
T/K	345.15

Polymer (B): **poly(vinyl acetate-*co*-vinyl alcohol)** **1991EAG**
Characterization: synthesized in the laboratory
Solvent (A): **water** H_2O **7732-18-5**

Type of data: cloud points (LCST-behavior)

w_B	0.02	T/K	338.35	for a copolymer of 18.0 mol% vinyl acetate
w_B	0.02	T/K	308.75	for a copolymer of 23.7 mol% vinyl acetate
w_B	0.02	T/K	305.35	for a copolymer of 24.4 mol% vinyl acetate
w_B	0.02	T/K	315.05	for a copolymer of 30.0 mol% vinyl acetate

Polymer (B): **poly(*N*-vinylcaprolactam-*co*-methacrylic acid)** **2003OKH**
Characterization: M_w/g.mol^{-1} = 4100, 9 mol% methacrylic acid
Solvent (A): **water** H_2O **7732-18-5**

Type of data: cloud points (LCST-behavior)

c_B/(g/l) 1.0 T/K 303.15 (in 0.05 M NaCl aqueous solution)

Polymer (B): **poly(*N*-vinylcaprolactam-*co*-methacrylic acid)** **2003OKH**
Characterization: 12 mol% methacrylic acid, synthesized in the laboratory
Solvent (A): **water** H_2O **7732-18-5**

Type of data: cloud points (LCST-behavior)

c_B/(g/l) 1.0 T/K 310.65 (in 0.05 M NaCl aqueous solution)

Polymer (B): **poly(*N*-vinylcaprolactam-*co*-methacrylic acid)** **2003OKH**
Characterization: 18 mol% methacrylic acid, synthesized in the laboratory
Solvent (A): **water** H_2O **7732-18-5**

Type of data: cloud points (LCST-behavior)

c_B/(g/l) 1.0 T/K 313.85 (in 0.05 M NaCl aqueous solution)

Polymer (B): **poly(*N*-vinylcaprolactam-*co*-methacrylic acid)** **2003OKH**
Characterization: M_w/g.mol^{-1} = 4100, 37 mol% methacrylic acid
Solvent (A): **water** **H_2O** **7732-18-5**

Type of data: cloud points (LCST-behavior)

c_B/(g/l) 1.0 *T*/K 304.15 (in 0.05 M NaCl aqueous solution)

Polymer (B): **poly[*N*-vinylcaprolactam-*g*-poly(ethyleneoxidoxyalkyl methacrylate)]** **2005LAU**
Characterization: M_w/g.mol^{-1} = 71000, 6.3 wt% poly(ethyleneoxidoxyalkyl methacrylate), MAC11EO42, synthesized in the laboratory
Solvent (A): **water** **H_2O** **7732-18-5**

Type of data: cloud points (LCST-behavior)

c_B/(g/l) 1.0 *T*/K 306.25

Polymer (B): **poly[*N*-vinylcaprolactam-*g*-poly(ethyleneoxidoxyalkyl methacrylate)]** **2005LAU**
Characterization: M_w/g.mol^{-1} = 310000, 13.0 wt% poly(ethyleneoxidoxyalkyl methacrylate), MAC11EO42, synthesized in the laboratory
Solvent (A): **water** **H_2O** **7732-18-5**

Type of data: cloud points (LCST-behavior)

c_B/(g/l) 1.0 *T*/K 306.25

Polymer (B): **poly[*N*-vinylcaprolactam-*g*-poly(ethyleneoxidoxyalkyl methacrylate)]** **2005LAU**
Characterization: M_w/g.mol^{-1} = 250000, 15.8 wt% poly(ethyleneoxidoxyalkyl methacrylate), MAC11EO42, synthesized in the laboratory
Solvent (A): **water** **H_2O** **7732-18-5**

Type of data: cloud points (LCST-behavior)

c_B/(g/l) 1.0 *T*/K 306.55

Polymer (B): **poly[*N*-vinylcaprolactam-*g*-poly(ethyleneoxidoxyalkyl methacrylate)]** **2005LAU**
Characterization: M_w/g.mol^{-1} = 300000, 18.3 wt% poly(ethyleneoxidoxyalkyl methacrylate), MAC11EO42, synthesized in the laboratory
Solvent (A): **water** **H_2O** **7732-18-5**

Type of data: cloud points (LCST-behavior)

c_B/(g/l) 1.0 *T*/K 306.65

Polymer (B): **poly[*N*-vinylcaprolactam-*g*-poly(ethyleneoxidoxyalkyl methacrylate)]** **2005LAU**
Characterization: M_w/g.mol^{-1} = 260000, 34.0 wt% poly(ethyleneoxidoxyalkyl methacrylate), MAC11EO42, synthesized in the laboratory
Solvent (A): **water** **H_2O** **7732-18-5**

Type of data: cloud points (LCST-behavior)

c_B/(g/l)	1.0	*T*/K	306.95

Polymer (B): **poly(*N*-vinylcaprolactam-*co*-1-vinylimidazole)** **2000PE1**
Characterization: synthesized in the laboratory
Solvent (A): **water** **H_2O** **7732-18-5**

Type of data: cloud points (LCST-behavior)

w_B	0.01	*T*/K	310.15	for a copolymer of 9.09 mol% 1-vinylimidazole
w_B	0.01	*T*/K	308.15	for a copolymer of 16.66 mol% 1-vinylimidazole
w_B	0.01	*T*/K	310.15	for a copolymer of 33.33 mol% 1-vinylimidazole
w_B	0.01	*T*/K	312.65	for a copolymer of 50.00 mol% 1-vinylimidazole

Polymer (B): **poly(*N*-vinylisobutyramide-*co*-*N*-vinylamine)** **2000KUN**
Characterization: M_n/g.mol^{-1} = 20000, M_w/g.mol^{-1} = 40000
8.0 mol% *N*-vinylamine, synthesized in the laboratory
Solvent (A): **water** **H_2O** **7732-18-5**

Type of data: cloud points (LCST-behavior)

w_B	0.001	0.001	0.001
pH	13	11	10
T/K	320.35	327.75	341.05

Polymer (B): **poly(*N*-vinylisobutyramide-*co*-*N*-vinylamine)** **2000KUN**
Characterization: M_n/g.mol^{-1} = 34000, M_w/g.mol^{-1} = 95200
18.0 mol% *N*-vinylamine, synthesized in the laboratory
Solvent (A): **water** **H_2O** **7732-18-5**

Type of data: cloud points (LCST-behavior)

w_B	0.001	pH	13	*T*/K	325.85

Polymer (B): **poly(*N*-vinylisobutyramide-*co*-*N*-vinylamine)** **2000KUN**
Characterization: M_n/g.mol^{-1} = 33000, M_w/g.mol^{-1} = 89100
26.0 mol% *N*-vinylamine, synthesized in the laboratory
Solvent (A): **water** **H_2O** **7732-18-5**

Type of data: cloud points (LCST-behavior)

w_B	0.001	pH	13	*T*/K	334.85

Polymer (B): **poly(*N*-vinylisobutyramide-*co*-*N*-vinylamine)** **2000KUN**
Characterization: M_n/g.mol^{-1} = 31000, M_w/g.mol^{-1} = 77500
36.0 mol% *N*-vinylamine, synthesized in the laboratory
Solvent (A): **water** **H_2O** **7732-18-5**

Type of data: cloud points (LCST-behavior)

w_B	0.001	pH	13	T/K	346.75

Polymer (B): **poly(vinyl methyl ether-*b*-vinyl isobutyl ether)** **2005VER**
Characterization: M_n/g.mol^{-1} = 9800, M_n/g.mol^{-1} (PVME-block) = 9300,
M_n/g.mol^{-1} (PVIBE-block) = 500, M_w/g.mol^{-1} = 11270,
3 mol% vinyl isobutyl ether, PVME160-*b*-PVIBE5
Solvent (A): **water** **H_2O** **7732-18-5**

Type of data: cloud points (LCST-behavior)

c_B/(g/l)	1.0	T/K	305.15

Polymer (B): **poly(vinyl methyl ether-*b*-vinyl isobutyl ether)** **2005VER**
Characterization: M_n/g.mol^{-1} = 5900, M_n/g.mol^{-1} (PVME-block) = 5000,
M_n/g.mol^{-1} (PVIBE-block) = 900, M_w/g.mol^{-1} = 6785,
10 mol% vinyl isobutyl ether, PVME85-*b*-PVIBE9
Solvent (A): **water** **H_2O** **7732-18-5**

Type of data: cloud points (LCST-behavior)

c_B/(g/l)	1.0	T/K	312.15

Polymer (B): **poly(vinyl methyl ether-*b*-vinyl isobutyl ether-*b*-vinyl methyl ether)** **2005VER**
Characterization: M_n/g.mol^{-1} = 9500, M_n/g.mol^{-1} (PVME-block) = 4500,
M_n/g.mol^{-1} (PVIBE-block) = 1000, M_w/g.mol^{-1} = 11210,
7 mol% vinyl isobutyl ether, PVME65-*b*-PVIBE10-*b*-PVME65
Solvent (A): **water** **H_2O** **7732-18-5**

Type of data: cloud points (LCST-behavior)

c_B/(g/l)	1.0	T/K	314.15

3.2. Table of binary systems where data were published only in graphical form as phase diagrams or related figures

Polymer (B)	Solvent (A)	Ref.
Chitosan-*g*-poly(*N*-isopropylacrylamide)		
	water	2009LIX
	water	2009REC
Dextran-*g*-poly(*N*-isopropylacrylamide)		
	water	2009PAT
Ethylcellulose-*g*-[poly(ethylene glycol) methyl ether methacrylate]		
	water	2008LIY
Hydroxypropylcellulose-*g*-poly(*N*-isopropyl-acrylamide)-*g*-poly(acrylic acid)		
	water	2009LIX
Poly(acrylamide-*co*-hydroxypropyl acrylate)		
	water	1975TAY
Poly(acrylamide-*co*-*N*-isopropylacrylamide)		
	water	2004EEC
	water	2007ERB
Poly(3-acrylamido-3-deoxy-1,2:5,6-di-O-isopropylidene-α-D-glucofuranose-*co*-*N*-isopropylacrylamide)		
	water	2009SHI
Poly(acrylic acid-*co*-*N*-isopropylacrylamide)		
	water	1997YOO
	water	2000BO1
	water	2001BUL
	water	2000YAM
	water	2004MAE
	water	2006WEN

Polymer (B)	**Solvent (A)**	**Ref.**
Poly(acrylic acid-*g*-*N*-isopropylacrylamide)	water	2008LI1
Poly(acrylic acid-*co*-nonyl acrylate)	acrylic acid	2007MIK
	bisphenol-A diglycidyl ether	2004MIK
	diethylene glycol bis(methacyl-oxyethylene carbonate)	2004MIK
	nonyl acrylate	2007MIK
Poly(acrylonitrile-*co*-butadiene)	ethyl acetate	2002VSH
	ethyl acetate	2004VSH
Poly(*N*-acryloylpyrrolidine-*b*-*N*-isopropyl-acrylamide)	water	2007SKR
Poly(allylamine-*g*-*N*-isopropylacrylamide)	water	2005GAO
Poly(5,6-benzo-2-methylene-1,3-dioxepane-*co*-*N*-isopropylacrylamide)	water	2007REN
Poly{[bis(ethyl glycinat-*N*-yl)phosphazene]-*b*-ethylene oxide}	water	2002CH2
Poly(butadiene-*co*-α-methylstyrene)	butyl acetate	2004VSH
	ethyl acetate	2002VSH
	ethyl acetate	2004VSH
Poly(butadiene-*b*-styrene)	dichlorobenzene	2004ABB
Poly(*N*-*tert*-butylacrylamide-*co*-acrylamide)	water	2009MAH

Polymer (B)	Solvent (A)	Ref.
Poly(*N-tert*-butylacrylamide-*co*-*N,N*-dimethylacrylamide)		
	water	1999LIU
Poly(*N-tert*-butylacrylamide-*co*-*N*-ethylacrylamide)		
	water	1999LIU
Poly(butylene oxide-*b*-ethylene oxide)		
	water	2001HAM
	water	2002CH1
	water	2002SON
	water	2004KEL
	water	2005CHA
Poly(butylene oxide-*b*-ethylene oxide-*b*-butylene oxide)		
	water	2004KEL
Poly(ε-caprolactone-*b*-ethylene glycol-*b*-ε-caprolactone)		
	water	2005BAE
	water	2006LUC
Poly(ε-caprolactone-*b*-*N*-isopropylacrylamide-*b*-ε-caprolactone)		
	water	2008CHA
Poly[*N*-cyclopropylacrylamide-*co*-4-(2-phenyl-diazenyl)benzamido-*N*-(2-aminoethyl)acrylamide]		
	water	2009JOC
Poly(diacetone acrylamide-*co*-acrylamide)		
	water	1975TAY
Poly(diacetone acrylamide-*co*-hydroxyethyl acrylate)		
	water	1975TAY

Polymer (B)	Solvent (A)	Ref.
Poly(*N,N*-diethylacrylamide-*co*-acrylamide)	water	1999LIU
Poly(*N,N*-diethylacrylamide-*co*-acrylic acid)	water	2001CAI
	water	2004MAE
	water	2008FAN
Poly(*N,N*-diethylacrylamide-*co*-*N,N*-dimethyl-acrylamide)	water	1999LIU
Poly(*N,N*-diethylacrylamide-*co*-*N*-ethylacrylamide)	water	1999LIU
Poly(*N,N*-diethylacrylamide-*co*-methacrylic acid)	water	2003LI1
Poly[*N,N*-diethylacrylamide-*co*-4-(2-phenyldiaze-nyl)benzamido-*N*-(2-aminoethyl)acrylamide]	water	2009JOC
Poly[di(ethylene glycol) methyl ether methacrylate-*b*-tri(ethylene glycol) methyl ether methacrylate]	water	2008YAM
Poly[di(ethylene glycol) methyl ether methacrylate-*co*-tri(ethylene glycol) methyl ether methacrylate]	water	2008YAM
Poly(*N,N*-dimethylacrylamide-*co*-glycidyl methacrylate)	water	2003YIN
Poly(*N,N*-dimethylacrylamide-*b*-*N*-isopropyl-acrylamide)	water	2007SKR

Polymer (B)	**Solvent (A)**	**Ref.**
Poly(*N,N*-dimethylacrylamide-*b*-*N*-isopropyl-acrylamide-*b*-*N*-acryloylpyrrolidine)		
	water	2007SKR
Poly[2-(*N,N*-dimethylamino)ethyl methacrylate-*b*-(2,2,3,4,4,4-hexafluorobutyl methacrylate)]		
	water	2009LI1
Poly[2-(*N,N*-dimethylamino)ethyl methacrylate-*b*-*N*-isopropylacrylamide]		
	water	2008LI3
Poly[2-(*N,N*-dimethylamino)ethyl methacrylate-*g*-*N*-isopropylacrylamide]		
	water	2008LI2
	water	2009LI2
Poly[2-(*N,N*-dimethylamino)ethyl methacrylate-*b*-(2,2,3,3,4,4,5,5-octafluoropentyl methacrylate)]		
	water	2009LI1
Poly[2-(*N,N*-dimethylamino)ethyl methacrylate-*stat*-oligo(ethylene glycol) methyl ether methacrylate]		
	water	2007FOU
Poly[2-(*N,N*-dimethylamino)ethyl methacrylate-*b*-(2,2,2-trifluoroethyl methacrylate)]		
	water	2009LI1
Poly[2-(*N,N*-dimethylamino)ethyl methacrylate-*co*-*N*-vinylcaprolactam]		
	water	2006VER
Poly(dimethylsiloxane-*co*-methylphenylsiloxane)		
	anisole	2000SCH
	2-propanone	2000SCH
Poly(dimethylsiloxane-*b*-1,1,3,3-tetramethyl-disiloxanylethylene)		
	ethoxybenzene	2000AZU

Polymer (B)	**Solvent (A)**	**Ref.**
Poly(divinyl ether-*alt*-maleic anhydride)		
	water	2004VOL
	water	2005IZU
Poly[2-(2-ethoxyethoxy)ethyl vinyl ether-*b*-(2-methoxyethyl vinyl ether)]		
	deuterium oxide	2006OSA
	water	2006OSA
Poly[*N*-(2-ethoxyethyl)acrylamide-*co*-*N*-isopropylacrylamide]		
	water	2009MA3
Poly(*N*-ethylacrylamide-*co*-*N*-isopropylacrylamide)		
	water	1975TAY
Poly(ethylene-*co*-acrylic acid)		
	dioctyl phthalate	2005ZHO
	dioctyl phthalate	2006ZHO
Poly(ethylene-*co*-acrylic acid)-*g*-poly(ethylene glycol) monomethyl ether		
	dioctyl phthalate	2005ZHO
	dioctyl phthalate	2006ZHO
Poly(ethylene-*co*-1-octene)		
	n-heptane	2005LE1
	n-hexane	2005HEI
	n-hexane	2005LE1
	n-octane	2005LE1
	n-pentane	2005LE1
Poly(ethylene-*co*-vinyl acetate)		
	cyclopentane	1999BEY
	cyclopentane	2000BEY
	cyclopentene	1999BEY
	cyclopentene	2000BEY
	tetraethoxysilane	2006CHA

Polymer (B)	**Solvent (A)**	**Ref.**
Poly(ethylene-*co*-vinyl alcohol)		
	1,4-butanediol	2007LVR
	1,3-propanediol	2007LVR
	1,2,3-propanetriol	2003SH1
	1,2,3-propanetriol	2003SH2
	1,2,3-propanetriol	2003SH3
	1,2,3-propanetriol	2007LVR
Poly(ethylene glycol-*b*-ε-caprolactone)		
	water	2006LUC
Poly(ethylene glycol-*b*-*N*-isopropylacrylamide)		
	water	2005MOT
	water	2005ZH3
Poly(ethylene glycol-*stat*-propylene glycol) monobutyl ether		
	water	2005AUB
Poly(ethylene glycol-*b*-4-vinylpyridine-*b*-*N*-isopropylacrylamide)		
	water	2005ZH4
Poly[(ethylene glycol) monomethacrylate-*co*-methyl methacrylate]		
	water	2004ALI
Poly(ethylene oxide-*b*-n-alkyl glycidyl carbamate-*b*-ethylene oxide)		
	water	2007DIM
Poly(ethylene oxide-*co*-alkyl glycidyl ether)		
	water	2001LIU
Poly(ethylene oxide-*b*-butylene oxide)		
	water	2001HAM
	water	2002CH1
	water	2002SON
	water	2004KEL
	water	2005CHA

Polymer (B)	**Solvent (A)**	**Ref.**
Poly(ethylene oxide-*co*-1,2-butylene oxide)	water	2000SAH
Poly(ethylene oxide-*b*-butylene oxide-*b*-ethylene oxide)	water	2002SON
Poly[ethylene oxide-*co*-(dialkoxymethyl)-propylglycidyl ether]	water	2001LIU
Poly[ethylene oxide-*b*-(ethylene oxide-*co*-propylene oxide)-*b*-DL-lactide]	water	2005AUB
Poly(ethylene oxide-*b*-*N*-isopropylacrylamide)	water	2004NED
	water	2006QIN
Poly(ethylene oxide-*b*-L-lactide-*b*-ethylene oxide)	water	1999JEO
Poly[ethylene oxide-*b*-(DL-lactide-*co*-glycolide)-*b*-ethylene oxide]	water	2004PA1
Poly(ethylene oxide-*b*-propylene oxide)	water	2007MAN
Poly(ethylene oxide-*co*-propylene oxide)	water	1999JOH
	water	1999PER
	water	2000PE2
	water	2003CAM
	water	2005AUB

Polymer (B)	Solvent (A)	Ref.
Poly(ethylene oxide-*b*-propylene oxide-*b*-ethylene oxide)		
	water	2000LAM
	water	2001DES
	water	2002DES
	water	2004VAR
	water	2008BER
	water	2008SHA
	water	2009ALV
	water	2009KOS
	water	2009NAN
Poly[(ethylene oxide-*co*-propylene oxide)-*b*-DL-lactide]		
	water	2005AUB
Poly(ethylene oxide-*b*-styrene oxide)		
	water	2003YAN
Poly(ethyl ethylene phosphate-*b*-propylene oxide-*b*-ethyl ethylene phosphate)		
	water	2009WA3
Poly[*N*-ethyl-*N*-methylacrylamide-*co*-4-(2-phenyl-diazenyl)benzamido-*N*-(2-aminoethyl)acrylamide]		
	water	2009JOC
Poly(2-ethyl-2-oxazoline-*b*-ε-caprolactone)		
	water	2000KIM
Poly(2-ethyl-2-oxazoline-*b*-2-nonyl-2-oxazoline)		
	water	2009LAM
Poly(2-ethyl-2-oxazoline-*co*-2-nonyl-2-oxazoline)		
	water	2009LAM
Poly(3-ethyl-1-vinyl-2-pyrrolidinone-*co*-1-vinyl-2-pyrrolidinone)		
	water	2009TRE

Polymer (B)	Solvent (A)	Ref.
Poly(L-glutamic acid-*g*-*N*-isopropylacrylamide)	water	2008HE2
Poly(glycidol-*b*-propylene oxide-*b*-glycidol)	water	2006HAL
Poly(2-hydroxyethyl acrylate-*co*-butyl acrylate)	water	2007MUN
Poly(2-hydroxyethyl acrylate-*co*-2-hydroxyethyl methacrylate)	water	2008KHU
Poly(2-hydroxyethyl acrylate-*co*-hydroxypropyl acrylate)	water	1975TAY
Poly(2-hydroxyethyl acrylate-*co*-vinyl butyl ether)	water	2006MUN
Poly(2-hydroxyethyl methacrylate-*g*-ethylene glycol)	water	2005ZH1
Poly(2-hydroxyethyl methacrylate-*b*-*N*-isopropyl-acrylamide)	water	2006CAO
Poly(2-hydroxyisopropyl acrylate-*co*-aminoethyl methacrylate)	water	2009DEN
Poly(2-hydroxypropyl acrylate-*co*-aminoethyl methacrylate)	water	2009DEN
Poly[*N*-(2-hydroxypropyl)methacrylamide-*b*-*N*-isopropylacrylamide]	water	2000KON

Polymer (B)	Solvent (A)	Ref.
Poly[*N*-(2-hydroxypropyl)methacrylamide monolactate-*co*-*N*-(2-hydroxypropyl) methacrylamide dilactate]		
	water	2004SOG
Poly(isobutyl vinyl ether-*co*-2-hydroxyethyl vinyl ether)		
	water	2004SUG
Poly[*N*-isopropylacrylamide-*b*-*N*-(acetylimino)ethylene]		
	water	2003DAV
Poly[*N*-isopropylacrylamide-*g*-*N*-(acetylimino)ethylene]		
	water	2003DAV
Poly(*N*-isopropylacrylamide-*co*-acrylamide)		
	water	2004EEC
	water	2007ERB
Poly(*N*-isopropylacrylamide-*co*-6-acrylaminohexanoic acid)		
	water	2000KUC
Poly(*N*-isopropylacrylamide-*co*-3-acrylaminopropanoic acid)		
	water	2000KUC
Poly(*N*-isopropylacrylamide-*co*-11-acrylaminoundecanoic acid)		
	water	2000KUC
Poly(*N*-isopropylacrylamide-*co*-acrylic acid)		
	water	1993OTA
	water	1997YOO
	water	2000BO1
	water	2001BUL
	water	2000YAM
	water	2004MAE
	water	2006WEN

Polymer (B)	Solvent (A)	Ref.
Poly(*N*-isopropylacrylamide-*co*-acrylic acid-*co*-ethyl methacrylate)	water	2005TIE
Poly(*N*-isopropylacrylamide-*co*-acryloyloxypropylphosphinic acid)	water	2004NON
Poly(*N*-isopropylacrylamide-*b*-*N*-acryloyl-pyrrolidine-*b*-*N,N*-dimethylacrylamide)	water	2007SKR
Poly(*N*-isopropylacrylamide-*co*-*N*-adamantyl-acrylamide)	water	2008WIN
Poly(*N*-isopropylacrylamide-*co*-benzo-15-crown-5-acrylamide)	water	2008MIP
Poly[*N*-isopropylacrylamide-*co*-(*N*-(R,S)-*sec*-butylacrylamide)]	water	2009LIP
Poly[*N*-isopropylacrylamide-*co*-(*N*-(S)-*sec*-butylacrylamide)]	water	2009LIP
Poly(*N*-isopropylacrylamide-*co*-butyl acrylate)	water	2003MAE
Poly(*N*-isopropylacrylamide-*b*-ε-caprolactone)	water	2006CHO
Poly(*N*-isopropylacrylamide-*co*-1-deoxy-1-methacrylamido-D-glucitol)	water	2001GOM

Polymer (B)	**Solvent (A)**	**Ref.**
Poly(*N*-isopropylacrylamide-*co*-*N,N*-diethylacrylamide)	water	2009MA2
Poly(*N*-isopropylacrylamide-*co*-*N,N*-dimethylacrylamide)	water	2005LIU
Poly(*N*-isopropylacrylamide-*b*-*N,N*-dimethyl-acrylamide-*b*-*N*-acryloylpyrrolidine)	water	2007SKR
Poly[(*N*-isopropylacrylamide-*co*-*N,N*-dimethyl-acrylamide)-*b*-(DL-lactide-*co*-glycolide)]	water	2005LIU
Poly[*N*-isopropylacrylamide-*b*-3'-(1',2':5',6'-di-O-isopropylidene-α-D-glucofuranosyl)-6-methacrylamido hexanoate]	water	2007OEZ
Poly[*N*-isopropylacrylamide-*co*-3'-(1',2':5',6'-di-O-isopropylidene-α-D-glucofuranosyl)-6-methacrylamido hexanoate]	water	2007OEZ
Poly[*N*-isopropylacrylamide-*b*-3'-(1',2':5',6'-di-O-isopropylidene-α-D-glucofuranosyl)-6-methacrylamido undecanoate]	water	2007OEZ
Poly[*N*-isopropylacrylamide-*co*-3'-(1',2':5',6'-di-O-isopropylidene-α-D-glucofuranosyl)-6-methacrylamido undecanoate]	water	2007OEZ
Poly[*N*-isopropylacrylamide-*b*-2-(*N,N*-dimethyl-amino)ethyl methacrylate]	water	2008LI3

Polymer (B)	Solvent (A)	Ref.
Poly[*N*-isopropylacrylamide-*co*-2-(*N,N*-dimethyl-amino)ethyl methacrylate-*co*-butyl methacylate]	water	2004TAK
Poly[*N*-isopropylacrylamide-*co*-(*N,N*-dimethyl-amino)propylmethacrylamide]	water	2000BO2
Poly(*N*-isopropylacrylamide-*co*-*N*-dodecyl-acrylamide)	water	2008WIN
Poly(*N*-isopropylacrylamide-*co*-ethyl acrylate)	water	2003MAE
Poly(*N*-isopropylacrylamide-*b*-ethylene glycol)	water	2007VAN
	water	2009TAU
Poly(*N*-isopropylacrylamide-*g*-ethylene glycol)	water	2005BIS
	water	2007VAN
Poly(*N*-isopropylacrylamide-*b*-ethylene oxide)	water	2004NED
Poly(*N*-isopropylacrylamide-*b*-ethylenimine)	water	2002DIN
Poly{*N*-isopropylacrylamide-*b*-[(L-glutamic acid)-*co*-(γ-benzyl-L-glutamate)]}	water	2008HE1
Poly(*N*-isopropylacrylamide-*alt*-2-hydroxyethyl methacrylate)	water	2008FAR
Poly(*N*-isopropylacrylamide-*co*-2-hydroxyethyl methacrylate)	water	2009ZHA

Polymer (B)	Solvent (A)	Ref.
Poly(*N*-isopropylacrylamide-*co*-2-hydroxyethyl methacrylate-*co*-acrylic acid)		
	water	2005LE2
Poly(*N*-isopropylacrylamide-*co*-2-hydroxyethyl methacrylate lactate-*co*-acrylic acid)		
	water	2005LE2
Poly[*N*-isopropylacrylamide-*co*-(2-hydroxyisopropyl)acrylamide]		
	water	2006MA1
	water	2006MA2
Poly(*N*-isopropylacrylamide-*co*-*N*-hydroxymethylacrylamide)		
	water	2006DI2
	water	2008KOT
Poly[*N*-isopropylacrylamide-*b*-*N*-(2-hydroxy-propyl)methacrylamide]		
	water	2000KON
Poly(*N*-isopropylacrylamide-*co*-3H-imidazole-4-carbodithioic acid 4-vinylbenzyl ester)		
	water	2005CAR
Poly[*N*-isopropylacrylamide-*b*-(*N*-isopropyl-acrylamide-*co*-*N*-(hydroxymethyl)acrylamide)]		
	water	2009KOT
Poly(*N*-isopropylacrylamide-*b*-*N*-isopropylacryl-amide(isotactic)-*b*-*N*-isopropylacrylamide)		
	water	2008NU1
	water	2008NU2
Poly[*N*-isopropylacrylamide(isotactic)-*b*-*N*-isopro-pylacrylamide-*b*-*N*-isopropylacrylamide(isotactic)]		
	water	2008NU1
	water	2008NU2

Polymer (B)	**Solvent (A)**	**Ref.**
Poly(*N*-isopropylacrylamide-*co*-*N*-isopropyl-methacrylamide)	water	2005STA
Poly(*N*-isopropylacrylamide-*b*-DL-lactide)	water	2003LI2
Poly(*N*-isopropylacrylamide-*b*-L-lysine)	water	2008ZHA
Poly(*N*-isopropylacrylamide-*co*-maleic anhydride)	water	2007FRA
Poly(*N*-isopropylacrylamide-*co*-maleic acid)	water	2004WEI
Poly(*N*-isopropylacrylamide-*co*-maleimide)	water	2007FRA
Poly[*N*-isopropylacrylamide-*co*-(p-methacryl-amido)acetophenone thiosemicarbazone]	water	2005LIC
Poly{*N*-isopropylacrylamide-*b*-3-[*N*-(3-methacryl-amidopropyl)-dimethylammonio]propane sulfate}	water	2002VIR
Poly{*N*-isopropylacrylamide-*co*-3-[*N*-(3-methacryl-amidopropyl)-dimethylammonio]propane sulfate}	water	2005NED
Poly(*N*-isopropylacrylamide-*co*-methacrylic acid)	water	2006YIN
Poly(*N*-isopropylacrylamide-*co*-*N*-methacryloyl-L-leucine)	water	2000BIG

Polymer (B)	**Solvent (A)**	**Ref.**
Poly{*N*-isopropylacrylamide-*b*-[2-(methacryloyloxy)ethyl phosphorylcholine]-*b*-*N*-isopropylacrylamide}	water	2009CRI
Poly(*N*-isopropylacrylamide-*co*-8-methacryloyloxyoctanoic acid methyl ester)	water	2004SAL
Poly(*N*-isopropylacrylamide-*co*-5-methacryloyloxypentanoic acid methyl ester)	water	2004SAL
Poly(*N*-isopropylacrylamide-*co*-11-methacryloyloxyundecanoic acid methyl ester)	water	2004SAL
Poly(*N*-isopropylacrylamide-*co*-methyl acrylate)	water	2003MAE
Poly[*N*-isopropylacrylamide-*co*-methoxypoly(ethylene glycol) monomethacrylate]	water	2006KI1
	water	2006KI2
Poly(*N*-isopropylacrylamide-*b*-methyl methacrylate)	water	2006WEI
Poly(*N*-isopropylacrylamide-*co*-*N*-methyl-*N*-vinylacetamide)	water	2004EEC
Poly(*N*-isopropylacrylamide-*co*-octadecyl acrylate)	water	2000SHI
Poly[*N*-isopropylacrylamide-*co*-oligo(ethylene glycol) monomethacrylate]	deuterium oxide	2003KOH
	deuterium oxide	2005KOH
	water	2004ALA
	water	2006ALA

Polymer (B)	Solvent (A)	Ref.
Poly[*N*-isopropylacrylamide-*co*-oligo(ethylene glycol) monomethacrylate-*co*-dodecyl methacrylate]	water	2005VIE
Poly(*N*-isopropylacrylamide-*co*-4-pentenoic acid)	water	1999KUN
Poly[*N*-isopropylacrylamide-*co*-4-(2-phenyldiazenyl)benzamido-*N*-(2-aminoethyl)acrylamide]	water	2009JOC
Poly(*N*-isopropylacrylamide-*b*-*N*-propylacrylamide)	water	2009CAO
Poly(*N*-isopropylacrylamide-*co*-*N*-propylacrylamide)	water	2004MAO
Poly(*N*-isopropylacrylamide-*co*-propylacrylic acid)	water	2006YIN
Poly(*N*-isopropylacrylamide-*b*-propylene oxide-*b*-*N*-isopropylacrylamide)	water	2004HAS
Poly(*N*-isopropylacrylamide-*co*-sodium 2-acrylamido-2-methyl-1-propanesulfonate)	water	2003NOW
	water	2008MAS
Poly(*N*-isopropylacrylamide-*co*-sodium acrylate)	water	2006MYL
Poly(*N*-isopropylacrylamide-*co*-sodium styrenesulfonate)	water	2004NOW
Poly{[*N*-isopropylacrylamide-*co*-3-(trimethoxysilyl)propyl methacrylate]-*b*-(2-(*N,N*-diethylamino)ethyl methacrylate)}	water	2009CH1

Polymer (B)	**Solvent (A)**	**Ref.**
Poly(*N*-isopropylacrylamide-*co*-*N*-vinylacetamide)		
	water	2004EEC
Poly(*N*-isopropylacrylamide-*co*-vinyl acetate)		
	water	2003MAE
Poly(*N*-isopropylacrylamide-*co*-1-vinylimidazole)		
	water	2005BIS
Poly(*N*-isopropylacrylamide-*co*-vinyl laurate)		
	water	2005CAO
Poly(*N*-isopropylacrylamide-*co*-p-vinyl-phenylboronic acid)		
	water	2005CIM
Poly(*N*-isopropylacrylamide-*b*-4-vinylpyridine)		
	water	2007XUY
Poly(*N*-isopropylacrylamide-*co*-1-vinyl-2-pyrrolidinone)		
	water	2004EEC
	water	2006DI2
	water	2006GEE
Poly(*N*-isopropylmethacrylamide-*co*-sodium methacrylate)		
	deuterium oxide	2005SPE
Poly(2-isopropyl-2-oxazoline-*co*-2-butyl-2-oxazoline)		
	water	2008HUB
Poly(2-isopropyl-2-oxazoline-*co*-2-nonyl-2-oxazoline)		
	water	2008HUB
Poly(2-isopropyl-2-oxazoline-*co*-2-propyl-2-oxazoline)		
	water	2008HUB

Polymer (B)	**Solvent (A)**	**Ref.**
Poly(L-lysine-*g*-*N*-isopropylacrylamide)	water	2002KON
Poly(maleic anhydride-*alt*-*tert*-butylstyrene)-*g*-poly(ethylene glycol) monomethyl ether	water	2002YIN
Poly(maleic anhydride-*alt*-styrene)-*g*-poly(ethylene glycol) monomethyl ether	water	2002YIN
Poly[methacrylic acid-*co*-butyl methacrylate-*co*-poly(ethylene glycol) monomethyl ether methacrylate]	water	2005JON
Poly(methacrylic acid-*co*-glycidyl methacrylate-*co*-poly(ethylene glycol) monomethyl ether methacrylate)	water	2005JON
Poly(methacrylic acid-*co*-lauryl methacrylate-*co*-poly(ethylene glycol) monomethyl ether methacrylate)	water	2005JON
Poly[2-(2-methoxyethoxy)ethyl methacrylate-*co*-oligo(ethylene glycol) methacrylate]	water	2006LUT
	water	2007LUT
Poly[2-(2-methoxyethoxy)ethyl methacrylate-*co*-oligo(ethylene glycol) methyl ether methacrylate]	water	2007SKR
Poly[methoxypoly(ethylene glycol)-*b*-ε-caprolactone]	water	2006KI2

Polymer (B)	Solvent (A)	Ref.
Poly[methoxytri(ethylene glycol) acrylate-*b*-4-vinylbenzyl methoxytris(oxyethylene) ether]	water	2006HU2
Poly(nonyl acrylate-*co*-2-methyl-5-vinyltetrazole)	nonyl acrylate	2002MIK
Poly[oligo(ethylene glycol) diglycidyl ether-*co*-piperazine-*co*-oligo(propylene glycol) diglycidyl ether)]	water	2009REN
Poly[oligo(ethylene glycol) methyl ether methylacrylate-*co*-ethylene glycol dimethacrylate-*co*-oligo(propylene glycol) methacrylate]	water	2009TAI
Poly[oligo(ethylene glycol) methylacrylate-*co*-oligo(propylene glycol) methacrylate]	water	2009WA2
Poly[oligo(2-ethyl-2-oxazoline)methacrylate-*stat*-methyl methacrylate]	water	2009WEB
Poly[perfluoroalkylacrylate-*co*-poly(ethylene oxide) methacrylate]	water	2006SHA
Poly(*N*-propylacrylamide-*b*-*N*-isopropylacrylamide)	water	2009CAO
Poly(*N*-propylacrylamide-*b*-*N*-isopropylacrylamide-*b*-*N,N*-ethylmethylacrylamide)	water	2009CAO
Poly(propylene oxide-*b*-ethylene oxide-*b*-propylene oxide)	water	2004DER

Polymer (B)	**Solvent (A)**	**Ref.**
Poly(styrene-*co*-acrylonitrile)		
	toluene	2000SCH
	toluene	2002WOL
	toluene	2003LOS
Poly(styrene-*b*-butyl methacrylate)		
	dioctyl phthalate	2006LIC
	n-hexadecane	2006LIC
Poly(styrene-*b*-isoprene)		
	dibutyl phthalate	2002LOD
	diethyl phthalate	2002LOD
	dimethyl phthalate	2002LOD
	propane	2006WIN
Poly(styrene-*b*-*N*-isopropylacrylamide)	water	2008TRO
Poly(styrene-*alt*-maleic anhydride)	water	2006QIU
Poly[(styrene-*alt*-maleic anhydride)-*g*-oligo(oxypropylene)amine]	water	2009LIN
Poly[(styrene-*alt*-maleic anhydride)-*g*-poly(amidoamine) dendrons]	water	2009GAO
Poly(styrenesulfonate-*b*-*N*-isopropylacrylamide)	water	2009TAU
Poly(sulfobetaine methacrylate-*co*-*N*-isopropyl-acrylamide)	water	2009CH2
Poly(*N*-vinylacetamide-*co*-acrylic acid)	water	2006MOR

Polymer (B)	Solvent (A)	Ref.
Poly(*N*-vinylacetamide-*co*-methyl acrylate)		
	water	2004MOR
Poly(*N*-vinylacetamide-*co*-vinyl acetate)		
	water	2003SET
Poly(vinyl alcohol-*co*-sodium acrylate)		
	water	2008PAN
Poly(vinylamine-*co*-vinylamine boronate)		
	water	2009CHE
Poly(*N*-vinylcaprolactam-*co*-acrylic acid)		
	water	2003SHT
Poly(*N*-vinylcaprolactam-*g*-ethylene oxide)		
	water	2003VE1
	water	2004DUR
	water	2005KJO
Poly(*N*-vinylcaprolactam-*co*-methacrylic acid)		
	water	2002MAK
Poly(*N*-vinylcaprolactam-*g*-tetrahydrofuran)		
	water	2003VE1
Poly(*N*-vinylcaprolactam-*co*-1-vinylimidazole)		
	water	2006LOZ
	water	2007SH2
Poly(*N*-vinylcaprolactam-*co*-1-vinyl-2-methylimidazole)		
	water	2007SH2
Poly(*N*-vinylformamide-*co*-vinyl acetate)		
	water	2003SET
Poly(vinylidene fluoride-*co*-hexafluoropropylene)		
	sulfolane	2000CUI

Polymer (B)	**Solvent (A)**	**Ref.**
Poly(*N*-vinylisobutyramide-*co*-*N*-vinylamine)	water	2000KUN
Poly(4-vinylpyridine-*g*-ethylene oxide)	water	2008REN
Poly(5-vinyltetrazole-*co*-2-methyl-5-vinyltetrazole)	water	2009KIZ
Poly(5-vinyltetrazole-*co*-2-nonyl-5-vinyltetrazole)	water	2009KIZ
Poly(5-vinyltetrazole-*co*-2-pentyl-5-vinyltetrazole)	water	2009KIZ
Poly(5-vinyltetrazole-*co*-1,1,7-trihydrododeca-fluoroheptyl methacrylate)	water	2009KIZ
Poly(5-vinyltetrazole-*co*-1-vinylimidazole)	water	2009KIZ
N,N,N-Trimethylchitosan chloride-*g*-poly(*N*-isopropylacrylamide)	water	2007MAO

3.3. Cloud-point and/or coexistence curves of quasiternary and/or quasiquaternary solutions

Polymer (B): **poly(*N,N*-diallylammonioethanoic acid-*co*-sulfur dioxide)** **2002ALM**
Characterization: 50 mol% sulfur dioxide, synthesized in the laboratory
Solvent (A): **water** **H_2O** **7732-18-5**
Polymer (C): **poly(ethylene glycol)**
Characterization: M_n/g.mol^{-1} = 35000, Merck KGaA, Darmstadt, Germany

Type of data: coexistence data (tie lines)

T/K = 296.15

Comments: The concentration of KCl in the feed system is given in mol/l, the concentration of HCl is given in equivalents with respect to the copolymer.

		Total system			Top phase			Bottom phase		
KCl	HCl	w_A	w_B	w_C	w_A	w_B	w_C	w_A	w_B	w_C
1.5	0.70	0.92083	0.05227	0.02690	0.94675	0.00135	0.0519	0.89686	0.0995	0.00364
1.0	0.70	0.92083	0.05227	0.02690	0.94541	0.00409	0.0505	0.89450	0.1040	0.00150
0.5	0.70	0.92095	0.05215	0.02690	0.94635	0.00425	0.0494	0.89242	0.1060	0.00158
0.1	0.70	0.92106	0.05204	0.02690	0.93820	0.01570	0.0461	0.90203	0.0915	0.00647
0.1	0.70	0.89064	0.05468	0.05468	0.91835	0.00175	0.0799	0.83228	0.1660	0.00172
0.1	0.70	0.90740	0.07676	0.01584	0.93700	0.01380	0.0492	0.89647	0.1000	0.00353
0.5	0.00	0.80580	0.11140	0.08280	0.78800	0.00000	0.2120	0.81688	0.1810	0.00212
0.5	0.00	0.83861	0.05934	0.10205	0.83100	0.00000	0.1690	0.84960	0.1480	0.00240
0.5	0.00	0.83440	0.06220	0.10340	0.84200	0.00000	0.1580	0.83920	0.1590	0.00180

Polymer (B): **poly(*N,N*-dimethylacrylamide-*co*-*tert*-butylacrylamide)** **2009FOU**
Characterization: M_n/g.mol^{-1} = 1700, unspecified comonomer content, synthesized in the laboratory
Solvent (A): **water** **H_2O** **7732-18-5**
Salt (C): **monoammonium phosphate** **$(NH_4)H_2PO_4$** **7722-76-1**

Type of data: cloud points (LCST behavior)

T/K = 338.15

w_A	0.8770	0.8750	0.8814	0.8687	0.8513	0.8441	0.8387
w_B	0.0874	0.0930	0.0927	0.1124	0.1297	0.1373	0.1430
w_C	0.0356	0.0320	0.0259	0.0189	0.0190	0.0186	0.0183

Polymer (B):	**poly(*N,N*-dimethylacrylamide-*co*-*tert*-butylacrylamide)**	**2009FOU**
Characterization:	M_n/g.mol^{-1} = 1700, unspecified comonomer content, synthesized in the laboratory	
Solvent (A):	**water** H_2O	**7732-18-5**
Salt (C):	**monopotassium phosphate** KH_2PO_4	**7778-77-0**

Type of data: cloud points (LCST behavior)

T/K = 338.15

w_A	0.8943	0.8877	0.8848	0.8732	0.8662	0.8611	0.8561
w_B	0.0547	0.0840	0.0917	0.1044	0.1119	0.1172	0.1224
w_C	0.0510	0.0283	0.0235	0.0224	0.0219	0.0217	0.0215

Polymer (B):	**poly(*N,N*-dimethylacrylamide-*co*-*tert*-butylacrylamide)**	**2009FOU**
Characterization:	M_n/g.mol^{-1} = 1700, unspecified comonomer content, synthesized in the laboratory	
Solvent (A):	**water** H_2O	**7732-18-5**
Salt (C):	**monosodium carbonate** $NaHCO_3$	**144-55-8**

Type of data: cloud points (LCST behavior)

T/K = 338.15

w_A	0.9242	0.9370	0.9482	0.9400	0.9345
w_B	0.0198	0.0230	0.0268	0.0354	0.0417
w_C	0.0560	0.0400	0.0250	0.0246	0.0238

Polymer (B):	**poly(*N,N*-dimethylacrylamide-*co*-*tert*-butylacrylamide)**	**2009FOU**
Characterization:	M_n/g.mol^{-1} = 1700, unspecified comonomer content, synthesized in the laboratory	
Solvent (A):	**water** H_2O	**7732-18-5**
Salt (C):	**sodium chloride** $NaCl$	**7647-14-5**

Type of data: cloud points (LCST behavior)

T/K = 338.15

w_A	0.9330	0.9268	0.9214	0.9083	0.8956	0.8866	0.8787	0.8731	0.8696
w_B	0.0390	0.0488	0.0588	0.0726	0.0854	0.0953	0.1036	0.1094	0.1129
w_C	0.0280	0.0244	0.0198	0.0191	0.0190	0.0181	0.0177	0.0175	0.0175

Polymer (B):	**poly(ethylene-*co*-vinyl acetate)**	**2003CH2**
Characterization:	M_n/g.mol^{-1} = 19800, 41.0 wt% vinyl acetate	
Solvent (A):	**styrene** C_8H_8	**100-42-5**
Polymer (C):	**polystyrene**	
Characterization:	M_n/g.mol^{-1} = 90000	

Type of data: cloud points

T/K = 358.15

w_A	0.8800	0.9050	0.9150	0.9050	0.8900
w_B	0.0120	0.0285	0.0425	0.0665	0.0990
w_C	0.1080	0.0665	0.0425	0.0285	0.0110

Type of data: coexistence data

T/K = 358.15

Total system			Top phase	Bottom phase
w_A	w_B	w_C	w_A	w_A
0.850	0.075	0.075	0.882	0.796
0.802	0.099	0.099	0.844	0.746
0.733	0.100	0.167	0.780	0.680
0.700	0.150	0.150	0.736	0.653

Polymer (B):	**poly(ethylene oxide-*b*-dimethylsiloxane)**	**2002MAD**
Characterization:	M_n/g.mol^{-1} = 460, M_w/g.mol^{-1} = 600, 75.0 wt% ethylene oxide, Fluka AG, Buchs, Switzerland	
Solvent (A):	**1,2,3,4-tetrahydronaphthalene** $C_{10}H_{12}$	**119-64-2**
Polymer (C):	**poly(ethylene oxide)**	
Characterization:	M_n/g.mol^{-1} = 21000, M_w/g.mol^{-1} = 27000, Fluka AG, Buchs, Switzerland	

Type of data: cloud points

w_A	0.000	0.000	0.000	0.000	0.000	0.000	0.000	0.000	0.000
w_B	0.999	0.990	0.985	0.980	0.950	0.900	0.800	0.700	0.600
w_C	0.001	0.010	0.015	0.020	0.050	0.100	0.200	0.300	0.400
T/K	321.15	330.15	340.15	370.15	388.15	395.65	402.15	405.15	400.15

w_A	0.000	0.000	0.000	0.2506	0.1010	0.0000
w_B	0.500	0.400	0.250	0.5992	0.7490	0.8500
w_C	0.500	0.600	0.750	0.1501	0.1499	0.1500
T/K	390.65	376.15	338.15	335.15	366.15	371.15

Polymer (B):	**poly(ethylene oxide-*b*-dimethylsiloxane)**	**2003JIA**
Characterization:	M_w/g.mol^{-1} = 1800, 77 mol% ethylene oxide, EO_{27}-*b*-DMS_8	
Solvent (A):	**toluene** C_7H_8	**108-88-3**
Polymer (C):	**poly(ethylene oxide)**	
Characterization:	M_w/g.mol^{-1} = 35000, Fluka AG, Buchs, Switzerland	

Type of data: cloud points

w_A	0.000	0.000	0.000	0.000	0.000	0.000	0.000	0.351	0.586
w_B	0.948	0.900	0.799	0.698	0.596	0.496	0.397	0.645	0.375
w_C	0.052	0.100	0.201	0.302	0.404	0.504	0.603	0.004	0.039
T/K	393.0	396.0	403.0	405.0	400.0	391.0	378.0	308.2	308.2
w_A	0.597	0.621	0.607	0.680	0.694	0.619	0.657	0.643	0.641
w_B	0.325	0.292	0.267	0.179	0.161	0.224	0.168	0.164	0.122
w_C	0.078	0.087	0.126	0.141	0.146	0.156	0.175	0.193	0.237
T/K	308.2	308.2	308.2	308.2	308.2	308.2	308.2	308.2	308.2
w_A	0.659	0.624	0.669	0.626	0.569	0.297	0.311	0.366	0.377
w_B	0.095	0.124	0.074	0.108	0.082	0.682	0.648	0.557	0.498
w_C	0.246	0.252	0.258	0.266	0.348	0.021	0.041	0.077	0.125
T/K	308.2	308.2	308.2	308.2	308.2	318.2	318.2	318.2	318.2
w_A	0.387	0.384	0.381	0.342	0.386	0.251	0.245	0.243	0.212
w_B	0.466	0.431	0.370	0.347	0.435	0.556	0.490	0.411	0.381
w_C	0.147	0.185	0.249	0.311	0.179	0.192	0.275	0.346	0.407
T/K	318.2	318.2	318.2	318.2	318.2	328.2	328.2	328.2	328.2
w_A	0.184	0.199	0.233	0.262	0.241	0.245	0.175	0.197	0.209
w_B	0.390	0.774	0.715	0.629	0.603	0.514	0.734	0.659	0.628
w_C	0.426	0.027	0.052	0.108	0.156	0.241	0.091	0.144	0.163
T/K	328.2	328.2	328.2	328.2	328.2	328.2	333.2	333.2	333.2
w_A	0.196	0.156							
w_B	0.597	0.538							
w_C	0.207	0.306							
T/K	333.2	333.2							

Type of data: spinodal points

w_A	0.695	0.695	0.695	0.695	0.695	0.438	0.438	0.438	0.438
w_B	0.169	0.169	0.169	0.169	0.169	0.386	0.386	0.386	0.386
w_C	0.172	0.172	0.172	0.172	0.172	0.176	0.176	0.176	0.176
T/K	309.2	308.2	307.2	306.2	305.2	318.2	317.2	316.2	315.2
P/bar	588	441	342	112	23	617	483	364	159
w_A	0.438	0.255	0.255	0.255	0.255	0.255	0.209	0.209	0.209
w_B	0.386	0.558	0.558	0.558	0.558	0.558	0.626	0.626	0.626
w_C	0.176	0.187	0.187	0.187	0.187	0.187	0.165	0.165	0.165
T/K	314.2	328.2	327.2	326.2	325.2	324.2	333.2	332.2	331.2
P/bar	21	160	223	265	321	383	85	114	163

continued

continued

w_A	0.209	0.209	0.209	0.209	0.209
w_B	0.626	0.626	0.626	0.626	0.626
w_C	0.165	0.165	0.165	0.165	0.165
T/K	330.2	329.2	328.2	327.2	326.2
P/bar	212	258	322	369	408

Polymer (B): **poly(ethylene oxide-*co*-propylene oxide)** **2004PER**
Characterization: M_n/g.mol^{-1} = 3900, 50.0 mol% ethylene oxide, Ucon 50-HB-5100, Union Carbide Corp., NY
Solvent (A): **water** H_2O **7732-18-5**
Salt (C): **ammonium sulfate** $(NH_4)_2SO_4$ **7783-20-2**

Type of data: coexistence data (tie lines)

T/K = 295.15

Total system			Top phase			Bottom phase		
w_A	w_B	w_C	w_A	w_B	w_C	w_A	w_B	w_C
0.8344	0.1001	0.0655	0.7044	0.2652	0.0304	0.8949	0.0206	0.0845
0.7806	0.1614	0.0580	0.6672	0.3072	0.0256	0.8930	0.0134	0.0936
0.7684	0.1696	0.0620	0.6314	0.3472	0.0214	0.8889	0.0091	0.1020
0.7411	0.1891	0.0698	0.5865	0.3967	0.0168	0.8781	0.0038	0.1181

Plait-point composition: w_A = 0.8388 + w_B = 0.1080 + w_C = 0.0532

T/K = 303.15

Total system			Top phase			Bottom phase		
w_A	w_B	w_C	w_A	w_B	w_C	w_A	w_B	w_C
0.8397	0.1203	0.0400	0.8898	0.0608	0.0494	0.7746	0.1966	0.0288
0.8344	0.1206	0.0450	0.9107	0.0289	0.0604	0.7078	0.2700	0.0222
0.8295	0.1205	0.0500	0.9132	0.0196	0.0672	0.6665	0.3151	0.0184
0.8245	0.1203	0.0552	0.9127	0.0130	0.0743	0.6348	0.3496	0.0156

Plait-point composition: w_A = 0.8443 + w_B = 0.1180 + w_C = 0.0377

continued

continued

T/K = 313.15

Total system			Top phase			Bottom phase		
w_A	w_B	w_C	w_A	w_B	w_C	w_A	w_B	w_C
0.8734	0.1009	0.0257	0.9197	0.0494	0.0309	0.7104	0.2772	0.0124
0.8650	0.1000	0.0350	0.9378	0.0192	0.0430	0.6064	0.3841	0.0095
0.8558	0.0995	0.0447	0.9345	0.0105	0.0550	0.5351	0.4573	0.0076
0.8447	0.1002	0.0551	0.9258	0.0060	0.0682	0.4834	0.5103	0.0063

Plait-point composition: w_A = 0.8414 + w_B = 0.1400 + w_C = 0.0186

Polymer (B): **poly(ethylene oxide-*b*-propylene oxide)** **2002MON**
Characterization: M_n/g.mol^{-1} = 1925, 10.0 mol% ethylene oxide
PE61, Oxiteno Brasil S/A, Sao Paulo, Brazil
Solvent (A): **water** **H_2O** **7732-18-5**
Component (C): **D-glucose** **$C_6H_{12}O_6$** **50-99-7**

Type of data: coexistence data (tie lines)

T/K = 298.15

Total system			Top phase			Bottom phase		
w_A	w_B	w_C	w_A	w_B	w_C	w_A	w_B	w_C
0.6213	0.3025	0.0762	0.1613	0.8303	0.0084	0.8553	0.0355	0.1092
0.4989	0.4007	0.1004	0.1229	0.8698	0.0073	0.7912	0.0351	0.1737
0.5353	0.3317	0.1330	0.1070	0.8863	0.0067	0.7654	0.0301	0.2045
0.3984	0.4006	0.2010	0.0782	0.9143	0.0075	0.6306	0.0125	0.3569

Polymer (B): **poly(ethylene oxide-*co*-propylene oxide)** **1995BER**
Characterization: M_n/g.mol^{-1} = 3200, 30 mol% ethylene oxide,
Shearwater Polymers, Huntsville, AL
Solvent (A): **water** **H_2O** **7732-18-5**
Polymer (C): **hydroxypropylstarch**
Characterization: M_w/g.mol^{-1} = 200000, Reppe Glykos AB, Vaxjo, Sweden

Type of data: coexistence data (tie lines)

continued

continued

T/K = 293.15

Total system			Top phase			Bottom phase		
w_A	w_B	w_C	w_A	w_B	w_C	w_A	w_B	w_C
0.850	0.090	0.060	0.865	0.101	0.034	0.775	0.030	0.195
0.830	0.090	0.080	0.859	0.115	0.026	0.760	0.028	0.212
0.790	0.100	0.110	0.832	0.148	0.020	0.722	0.019	0.259
0.780	0.100	0.120	0.829	0.153	0.018	0.714	0.020	0.266

Polymer (B): **poly(ethylene oxide-*co*-propylene oxide)** **1994MOD**
Characterization: M_n/g.mol^{-1} = 4000, 50 mol% ethylene oxide,
Ucon 50-HB-5100, Union Carbide Corp., New York, NY
Solvent (A): **water** **H_2O** **7732-18-5**
Polymer (C): **hydroxypropylstarch** **1995BER**
Characterization: M_w/g.mol^{-1} = 200000, Reppe Glykos AB, Vaxjo, Sweden

Type of data: coexistence data (tie lines)

T/K = 293.15

Total system			Top phase			Bottom phase		
w_A	w_B	w_C	w_A	w_B	w_C	w_A	w_B	w_C
0.856	0.075	0.069	0.873	0.088	0.039	0.797	0.015	0.188
0.845	0.080	0.075	0.866	0.101	0.033	0.786	0.017	0.197
0.820	0.080	0.100	0.857	0.115	0.028	0.750	0.006	0.244
0.808	0.100	0.092	0.840	0.139	0.021	0.734	0.008	0.258
0.759	0.125	0.116	0.803	0.185	0.012	0.676	0.005	0.319

Polymer (B): **poly(ethylene oxide-*co*-propylene oxide)** **2004PER**
Characterization: M_n/g.mol^{-1} = 3900, 50.0 mol% ethylene oxide,
Ucon 50-HB-5100, Union Carbide Corp., NY
Solvent (A): **water** **H_2O** **7732-18-5**
Polymer (C): **hydroxypropylstarch**
Characterization: M_n/g.mol^{-1} = 100000, Reppe Glykos AB, Vaxjo, Sweden

Type of data: coexistence data (tie lines)

continued

continued

T/K = 295.15

Total system			Top phase			Bottom phase		
w_A	w_B	w_C	w_A	w_B	w_C	w_A	w_B	w_C
0.8319	0.0473	0.1208	0.8517	0.0599	0.0884	0.8091	0.0236	0.1673
0.7902	0.0804	0.1294	0.8414	0.1236	0.0350	0.7222	0.0175	0.2603
0.8032	0.0627	0.1341	0.8467	0.0946	0.0587	0.7504	0.0193	0.2303
0.7899	0.0709	0.1392	0.8431	0.1142	0.0427	0.7313	0.0180	0.2507
0.8048	0.0450	0.1502	0.8502	0.0822	0.0676	0.7801	0.0207	0.1992

Plait-point composition: $w_A = 0.8401 + w_B = 0.0369 + w_C = 0.1230$

Polymer (B): **poly(ethylene oxide-*co*-propylene oxide)** **2008SIL**
Characterization: M_n/g.mol^{-1} = 3900, 50.0 mol% ethylene oxide, Ucon 50-HB-5100, Union Carbide Corp., NY
Solvent (A): **water** H_2O **7732-18-5**
Salt (C): **lithium sulfate** Li_2SO_4 **10377-48-7**

Type of data: coexistence data (tie lines)

T/K = 296.15

Total system			Top phase			Bottom phase		
w_A	w_B	w_C	w_A	w_B	w_C	w_A	w_B	w_C
0.8230	0.1200	0.0570	0.7003	0.2721	0.0276	0.9151	0.0058	0.0791
0.8350	0.1100	0.0550	0.7266	0.2427	0.0307	0.9144	0.0129	0.0727
0.8100	0.1300	0.0600	0.6848	0.2894	0.0258	0.9108	0.0017	0.0875

Polymer (B): **poly(ethylene oxide-*co*-propylene oxide)** **2005BOL**
Characterization: M_n/g.mol^{-1} = 1059, M_η/g.mol^{-1} = 1228, 50 mol% ethylene oxide, Dow Chemical Co., San Lorenzo, Argentina
Solvent (A): **water** H_2O **7732-18-5**
Polymer (C): **maltodextrin**
Characterization: M_n/g.mol^{-1} = 838, M_η/g.mol^{-1} = 922, Polimerosa, Kasdorf SA, Buenos Aires, Argentina

Type of data: coexistence data (tie lines)

continued

continued

T/K = 297.15

Total system			Top phase			Bottom phase		
w_A	w_B	w_C	w_A	w_B	w_C	w_A	w_B	w_C
0.6610	0.0950	0.2440	0.7360	0.1640	0.1000	0.5919	0.0306	0.3775
0.6730	0.0930	0.2340	0.7480	0.1550	0.0970	0.6070	0.0312	0.3618
0.6860	0.0890	0.2250	0.7470	0.1376	0.1154	0.6330	0.0366	0.3304
0.7020	0.0860	0.2120	0.7552	0.1400	0.1048	0.6604	0.0460	0.2936
0.7240	0.0790	0.1970	0.7660	0.1145	0.1195	0.6835	0.0386	0.2779
0.7370	0.0750	0.1880	0.7696	0.0967	0.1337	0.6974	0.0404	0.2622

Polymer (B): **poly(ethylene oxide-*b*-propylene oxide)** **2002MON**
Characterization: M_n/g.mol^{-1} = 1925, 10.0 mol% ethylene oxide
PE61, Oxiteno Brasil S/A, Sao Paulo, Brazil
Solvent (A): **water** H_2O **7732-18-5**
Component (C): **maltose** $C_{12}H_{22}O_{11}$ **69-79-4**

Type of data: coexistence data (tie lines)

T/K = 298.15

Total system			Top phase			Bottom phase		
w_A	w_B	w_C	w_A	w_B	w_C	w_A	w_B	w_C
0.5980	0.3503	0.0517	0.1865	0.8081	0.0054	0.8699	0.0504	0.0797
0.5500	0.3504	0.0996	0.1555	0.8373	0.0072	0.8115	0.0295	0.1590
0.4797	0.4002	0.1201	0.1356	0.8578	0.0066	0.7640	0.0221	0.2139
0.3342	0.4968	0.1690	0.0914	0.9045	0.0041	0.6217	0.0039	0.3744

Polymer (B): **poly(ethylene oxide-*b*-propylene oxide)** **1991SA2**
Characterization: M_n/g.mol^{-1} = 3438, 24.8 mol% ethylene oxide,
Polysciences, Inc., Warrington, PA
Solvent (A): **water** H_2O **7732-18-5**
Solvent (C): ***N*-methylacetamide** C_3H_7NO **79-16-3**

Type of data: coexistence data (tie lines)

continued

continued

T/K = 298.15

Top phase			Bottom phase		
w_A	w_B	w_C	w_A	w_B	w_C
0.948	0.000	0.052	0.737	0.199	0.064
0.908	0.000	0.092	0.718	0.199	0.083
0.830	0.000	0.170	0.647	0.200	0.153
0.737	0.000	0.273	0.609	0.144	0.247
0.632	0.000	0.368	0.527	0.115	0.318
0.453	0.000	0.547	0.383	0.112	0.505
0.397	0.000	0.603	0.341	0.096	0.563
0.388	0.000	0.612	0.298	0.127	0.575
0.368	0.000	0.632	0.320	0.098	0.582
0.354	0.000	0.646	0.303	0.113	0.584

Comments: The detection limit for the block copolymer in the top phase was 1 wt%.

Polymer (B): **poly(ethylene oxide-*co*-propylene oxide)** **2009TUB**
Characterization: M_n/g.mol^{-1} = 3900, 50.0 mol% ethylene oxide, Ucon 50-HB-5100, Union Carbide Corp., NY
Solvent (A): **water** **H_2O** **7732-18-5**
Salt (C): **sodium citrate** **$Na_3C_6H_5O_7$** **68-04-2**

Type of data: coexistence data (tie lines)

Total system			Top phase			Bottom phase		
w_A	w_B	w_C	w_A	w_B	w_C	w_A	w_B	w_C
T/K = 278.15, pH = 5.20								
0.8307	0.0896	0.0797	0.7447	0.2231	0.0322	0.8852	0.0051	0.1097
0.8203	0.0898	0.0899	0.7131	0.2588	0.0281	0.8749	0.0039	0.1212
0.8094	0.0900	0.1006	0.6870	0.2882	0.0248	0.8629	0.0034	0.1337
0.8003	0.0898	0.1099	0.6682	0.3096	0.0222	0.8524	0.0032	0.1444
T/K = 293.15, pH = 5.20								
0.8365	0.1065	0.0570	0.7528	0.2140	0.0332	0.8924	0.0348	0.0728
0.8299	0.1102	0.0599	0.7195	0.2521	0.0284	0.8961	0.0252	0.0787
0.7670	0.1708	0.0622	0.6471	0.3340	0.0189	0.8894	0.0042	0.1064
0.7559	0.1787	0.0654	0.6283	0.3553	0.0164	0.8841	0.0021	0.1138

continued

continued

T/K = 313.15, pH = 5.20

0.7662	0.1715	0.0623	0.5506	0.4375	0.0119	0.8990	0.0078	0.0932
0.7571	0.1781	0.0648	0.5260	0.4643	0.0097	0.8994	0.0019	0.0987
0.7342	0.1949	0.0709	0.4700	0.5251	0.0049	0.8876	0.0009	0.1115

T/K = 278.15, pH = 8.20

0.8411	0.0944	0.0645	0.7532	0.2273	0.0195	0.9029	0.0009	0.0962
0.8315	0.0937	0.0748	0.7172	0.2678	0.0150	0.8910	0.0019	0.1071
0.8213	0.0940	0.0847	0.6849	0.3039	0.0112	0.8801	0.0019	0.1180
0.8122	0.0937	0.0941	0.6536	0.3390	0.0074	0.8704	0.0018	0.1278

T/K = 293.15, pH = 8.20

0.8080	0.1488	0.0432	0.7139	0.2645	0.0216	0.9214	0.0093	0.0693
0.7918	0.1633	0.0449	0.6870	0.2937	0.0193	0.9181	0.0061	0.0758
0.7730	0.1770	0.0500	0.6504	0.3333	0.0163	0.9093	0.0033	0.0874
0.7533	0.1918	0.0549	0.6235	0.3624	0.0141	0.8976	0.0022	0.1002

T/K = 313.15, pH = 8.20

0.8216	0.1585	0.0199	0.7218	0.2695	0.0087	0.9434	0.0231	0.0335
0.8141	0.1639	0.0220	0.7038	0.2880	0.0082	0.9442	0.0175	0.0383
0.7715	0.2021	0.0264	0.5789	0.4169	0.0042	0.9493	0.0004	0.0503
0.7351	0.2336	0.0313	0.5353	0.4606	0.0041	0.9309	0.0112	0.0579

Polymer (B): **poly(ethylene oxide-*co*-propylene oxide)** **1997JO1**
Characterization: M_n/g.mol^{-1} = 4000, 50 mol% ethylene oxide,
Ucon 50-HB-5100, Union Carbide Corp., NY
Solvent (A): **water** H_2O **7732-18-5**
Salt (C): **sodium perchlorate** $NaClO_4$ **7601-89-0**

Type of data: cloud points (LCST-behavior)

w_B	0.20
c_C/(mol/l)	0.01
T/K	322.15

Polymer (B): **poly(ethylene oxide-*co*-propylene oxide)** **1997LIM**
Characterization: M_n/g.mol^{-1} = 3000-3500, 33.3 mol% ethylene oxide,
Zhejiang Univ.Chem.Factory, PR China
Solvent (A): **water** H_2O **7732-18-5**
Salt (C): **sodium sulfate** Na_2SO_4 **7757-82-6**

Type of data: cloud points (LCST-behavior)

w_B = 0.10 (was kept constant)

c_C/(mol/l)	0.0	0.1	0.2	0.3
T/K	340.15	324.15	316.15	309.15

Polymer (B): **poly(ethylene oxide-*co*-propylene oxide)** **1997LIM**
Characterization: M_n/g.mol^{-1} = 3000-3500, 50.0 mol% ethylene oxide, Zhejiang Univ.Chem.Factory, PR China
Solvent (A): **water** H_2O **7732-18-5**
Salt (C): **sodium sulfate** Na_2SO_4 **7757-82-6**

Type of data: cloud points (LCST-behavior)

w_B = 0.10 (was kept constant)

c_C/(mol/l)	0.0	0.1	0.2	0.3
T/K	343.15	327.15	321.15	313.15

Polymer (B): **poly(ethylene oxide-*co*-propylene oxide)** **2008SIL**
Characterization: M_n/g.mol^{-1} = 3900, 50.0 mol% ethylene oxide, Ucon 50-HB-5100, Union Carbide Corp., NY
Solvent (A): **water** H_2O **7732-18-5**
Salt (C): **sodium sulfate** Na_2SO_4 **7757-82-6**

Type of data: coexistence data (tie lines)

T/K = 296.15

Total system			Top phase			Bottom phase		
w_A	w_B	w_C	w_A	w_B	w_C	w_A	w_B	w_C
0.7911	0.1676	0.0413	0.6868	0.2970	0.0162	0.9246	0.0021	0.0733
0.7709	0.1751	0.0540	0.6183	0.3712	0.0105	0.9071	0.0001	0.0928
0.7800	0.1700	0.0500	0.6579	0.3285	0.0136	0.9108	0.0001	0.0891

Polymer (B): **poly(ethylene oxide-*co*-propylene oxide)** **1997JO1**
Characterization: M_n/g.mol^{-1} = 4000, 50 mol% ethylene oxide, Ucon 50-HB-5100, Union Carbide Corp., NY
Solvent (A): **water** H_2O **7732-18-5**
Salt (C): **sodium sulfate** Na_2SO_4 **7757-82-6**

Type of data: cloud points (LCST-behavior)

w_B	0.20
c_C/(mol/l)	0.01
T/K	318.15

Polymer (B): **poly(ethylene oxide-*co*-propylene oxide)** **1997LIM**
Characterization: M_n/g.mol^{-1} = 3000-3500, 66.7 mol% ethylene oxide, Zhejiang Univ.Chem.Factory, PR China
Solvent (A): **water** H_2O **7732-18-5**
Salt (C): **sodium sulfate** Na_2SO_4 **7757-82-6**

Type of data: cloud points (LCST-behavior)

w_B = 0.10 (was kept constant)

c_C/(mol/l)	0.0	0.1	0.2	0.3
T/K	359.15	343.15	336.15	329.15

Polymer (B): **poly(ethylene oxide-*co*-propylene oxide)** **2004PER**
Characterization: M_n/g.mol^{-1} = 3900, 50.0 mol% ethylene oxide, Ucon 50-HB-5100, Union Carbide Corp., NY
Solvent (A): **water** H_2O **7732-18-5**
Polymer (C): **poly(vinyl acetate-*co*-vinyl alcohol)**
Characterization: M_n/g.mol^{-1} = 10000, 12.0 mol% vinyl acetate, PVA 10000, 88% hydrolyzed, Scientific Polymer Products, Inc., Ontario, NY

Type of data: coexistence data (tie lines)

T/K = 295.15

Total system			Top phase			Bottom phase		
w_A	w_B	w_C	w_A	w_B	w_C	w_A	w_B	w_C
0.7958	0.1047	0.0995	0.8077	0.1273	0.0650	0.7827	0.0214	0.1959
0.7612	0.1323	0.1065	0.7790	0.1931	0.0279	0.7375	0.0174	0.2451
0.7799	0.1104	0.1097	0.7970	0.1617	0.0413	0.7642	0.0180	0.2178
0.7493	0.1327	0.1180	0.7710	0.2067	0.0223	0.7129	0.0168	0.2703

Plait-point composition: w_A = 0.8100 + w_B = 0.0600 + w_C = 0.1300

Polymer (B): **poly(ethylene oxide-*b*-propylene oxide-*b*-ethylene oxide)** **2004TAD, 2005TAD**
Characterization: M_n/g.mol^{-1} = 4750, 37.0 mol% ethylene oxide, (EO)17-(PO)58-(EO)17, P103, Aldrich Chem. Co., Inc., Milwaukee, WI
Solvent (A): **water** H_2O **7732-18-5**
Polymer (C): **dextran**
Characterization: M_n/g.mol^{-1} = 8200, M_w/g.mol^{-1} = 11600, Dextran 19, Sigma Chemical Co., Inc., St. Louis, MO

Type of data: coexistence data (tie lines)

continued

continued

T/K = 298.15

Total system			Top phase			Bottom phase		
w_A	w_B	w_C	w_A	w_B	w_C	w_A	w_B	w_C
0.8183	0.1015	0.0802	0.8450	0.0460	0.1090	0.7829	0.1634	0.0537
0.8210	0.0904	0.0886	0.8503	0.0276	0.1221	0.7657	0.1916	0.0427
0.8073	0.1129	0.0798	0.8518	0.0230	0.1252	0.7569	0.2036	0.0395

Polymer (B): **poly(ethylene oxide-*b*-propylene oxide-*b*-ethylene oxide)** **2004TAD, 2005TAD**

Characterization: M_n/g.mol^{-1} = 4750, 37.0 mol% ethylene oxide, (EO)17-(PO)58-(EO)17, P103, Aldrich Chem. Co., Inc., Milwaukee, WI

Solvent (A): **water** **H_2O** **7732-18-5**

Polymer (C): **dextran**

Characterization: M_n/g.mol^{-1} = 236000, M_w/g.mol^{-1} = 410000, Dextran 400, Sigma Chemical Co., Inc., St. Louis, MO

Type of data: cloud points

T/K = 298.15

w_A	0.9001	0.9103
w_B	0.0705	0.0343
w_C	0.0294	0.0554

Type of data: coexistence data (tie lines)

T/K = 298.15

Total system			Top phase			Bottom phase		
w_A	w_B	w_C	w_A	w_B	w_C	w_A	w_B	w_C
0.9081	0.0524	0.0395	0.9219	0.0203	0.0578	0.8687	0.1207	0.0106
0.8893	0.0602	0.0505	0.9079	0.0230	0.0691	0.8488	0.1401	0.0111
0.8570	0.0729	0.0701	0.8735	0.0236	0.1029	0.8152	0.1805	0.0043

Polymer (B): **poly(ethylene oxide-*b*-propylene oxide-*b*-ethylene oxide)** **2004TAD, 2005TAD**
Characterization: M_n/g.mol^{-1} = 6500, 54.4 mol% ethylene oxide, (EO)37-(PO)62-(EO)37, F105, ICI Surfactants, Cleveland, UK
Solvent (A): **water** H_2O **7732-18-5**
Polymer (C): **dextran**
Characterization: M_n/g.mol^{-1} = 8200, M_w/g.mol^{-1} = 11600, Dextran 19, Sigma Chemical Co., Inc., St. Louis, MO

Type of data: coexistence data (tie lines)

T/K = 298.15

Total system			Top phase			Bottom phase		
w_A	w_B	w_C	w_A	w_B	w_C	w_A	w_B	w_C
0.7710	0.1169	0.1121	0.7821	0.0022	0.2157	0.7601	0.2239	0.0160
0.8200	0.0800	0.1000	0.8470	0.0083	0.1447	0.7894	0.1835	0.0271
0.7998	0.0901	0.1101	0.8238	0.0025	0.1737	0.7706	0.2085	0.0209

Polymer (B): **poly(ethylene oxide-*b*-propylene oxide-*b*-ethylene oxide)** **2004TAD, 2005TAD**
Characterization: M_n/g.mol^{-1} = 6500, 54.4 mol% ethylene oxide, (EO)37-(PO)62-(EO)37, F105, ICI Surfactants, Cleveland, UK
Solvent (A): **water** H_2O **7732-18-5**
Polymer (C): **dextran**
Characterization: M_n/g.mol^{-1} = 236000, M_w/g.mol^{-1} = 410000, Dextran 400, Sigma Chemical Co., Inc., St. Louis, MO

Type of data: cloud points

T/K = 298.15

w_A	0.8991	0.9095
w_B	0.0720	0.0350
w_C	0.0289	0.0555

Type of data: coexistence data (tie lines)

continued

continued

T/K = 298.15

Total system			Top phase			Bottom phase		
w_A	w_B	w_C	w_A	w_B	w_C	w_A	w_B	w_C
0.8957	0.0442	0.0601	0.9025	0.0142	0.0833	0.8685	0.1236	0.0079
0.8712	0.0601	0.0687	0.8833	0.0124	0.1043	0.8464	0.1485	0.0051
0.8994	0.0511	0.0495	0.9115	0.0173	0.0712	0.8845	0.1027	0.0128
0.8250	0.0875	0.0875	0.8364	0.0076	0.1560	0.8203	0.1767	0.0030

Polymer (B): **poly(ethylene oxide-*b*-propylene oxide-*b*-ethylene oxide)** **1999SVE**
Characterization: M_n/g.mol^{-1} = 6500, 56.9 wt% ethylene oxide, (EO)37-(PO)56-(EO)37, Pluronic P105, BASF, Pasippany, NJ
Solvent (A): **water** H_2O **7732-18-5**
Polymer (C): **dextran**
Characterization: M_w/g.mol^{-1} = 500000, T500, Amersham Pharmacia Biotech, Uppsala, Sweden

Type of data: coexistence data (tie lines)

Total system			Top phase			Bottom phase		
w_A	w_B	w_C	w_A	w_B	w_C	w_A	w_B	w_C
T/K = 278.15								
0.880	0.064	0.056	0.905	0.090	0.005	0.820	0.010	0.170
T/K = 303.15								
0.880	0.064	0.056	0.835	0.160	0.005	0.895	0.030	0.075

Polymer (B): **poly(ethylene oxide-*b*-propylene oxide-*b*-ethylene oxide)** **1999SVE**
Characterization: M_n/g.mol^{-1} = 8400, 84 mol% ethylene oxide, (EO)76-(PO)29-(EO)76, Pluronic F68, BASF, Pasippany, NJ
Solvent (A): **water** H_2O **7732-18-5**
Polymer (C): **dextran**
Characterization: M_w/g.mol^{-1} = 500000, T500, Amersham Pharmacia Biotech, Uppsala, Sweden

Type of data: coexistence data (tie lines)

continued

continued

Total system			Top phase			Bottom phase		
w_A	w_B	w_C	w_A	w_B	w_C	w_A	w_B	w_C
T/K = 278.15								
0.880	0.050	0.070	0.915	0.080	0.005	0.825	0.005	0.170
T/K = 303.15								
0.880	0.050	0.070	0.910	0.085	0.005	0.840	0.005	0.155

Polymer (B): **poly(ethylene oxide-*b*-propylene oxide-*b*-ethylene oxide)** **2004TAD, 2005TAD**
Characterization: M_n/g.mol^{-1} = 8530, 83.5 mol% ethylene oxide, (EO)76-(PO)30-(EO)76, F68, Aldrich Chem. Co., Inc., Milwaukee, WI
Solvent (A): **water** **H_2O** **7732-18-5**
Polymer (C): **dextran**
Characterization: M_n/g.mol^{-1} = 8200, M_w/g.mol^{-1} = 11600, Dextran 19, Sigma Chemical Co., Inc., St. Louis, MO

Type of data: cloud points

T/K = 298.15

w_A	0.8750	0.8649	0.8629	0.8525	0.8401	0.8174	0.8021
w_B	0.1006	0.0826	0.0637	0.0537	0.0405	0.0235	0.0155
w_C	0.0244	0.0525	0.0734	0.0938	0.1194	0.1591	0.1824

Type of data: coexistence data (tie lines)

T/K = 298.15

Total system			Top phase			Bottom phase		
w_A	w_B	w_C	w_A	w_B	w_C	w_A	w_B	w_C
0.8405	0.0700	0.0895	0.8039	0.0236	0.1725	0.8591	0.1035	0.0374
0.8196	0.0799	0.1005	0.7692	0.0101	0.2207	0.8512	0.1307	0.0181
0.8031	0.0899	0.1070	0.7556	0.0070	0.2374	0.8343	0.1481	0.0176

Polymer (B): **poly(ethylene oxide-*b*-propylene oxide-*b*-ethylene oxide)** **2004TAD, 2005TAD**

Characterization: M_n/g.mol^{-1} = 8530, 83.5 mol% ethylene oxide, (EO)76-(PO)30-(EO)76, F68, Aldrich Chem. Co., Inc., Milwaukee, WI

Solvent (A): **water** **H_2O** **7732-18-5**

Polymer (C): **dextran**

Characterization: M_n/g.mol^{-1} = 236000, M_w/g.mol^{-1} = 410000, Dextran 400, Sigma Chemical Co., Inc., St. Louis, MO

Type of data: cloud points

T/K = 298.15

w_A	0.9201	0.9099	0.8892
w_B	0.0404	0.0351	0.0229
w_C	0.0395	0.0550	0.0879

Type of data: coexistence data (tie lines)

T/K = 298.15

Total system			Top phase			Bottom phase		
w_A	w_B	w_C	w_A	w_B	w_C	w_A	w_B	w_C
0.8913	0.0396	0.0691	0.8713	0.0146	0.1141	0.9219	0.0657	0.0124
0.8699	0.0602	0.0699	0.8198	0.0002	0.1800	0.9008	0.0949	0.0043
0.8414	0.0892	0.0694	0.7608	0.0087	0.2305	0.8750	0.1222	0.0028

Polymer (B): **poly(ethylene oxide-*b*-propylene oxide-*b*-ethylene oxide)** **2004TAD, 2005TAD**

Characterization: M_n/g.mol^{-1} = 14000, 83.6 mol% ethylene oxide, (EO)127-(PO)50-(EO)127, P108, ICI Surfactants, Cleveland, UK

Solvent (A): **water** **H_2O** **7732-18-5**

Polymer (C): **dextran**

Characterization: M_n/g.mol^{-1} = 8200, M_w/g.mol^{-1} = 11600, Dextran 19, Sigma Chemical Co., Inc., St. Louis, MO

Type of data: coexistence data (tie lines)

T/K = 298.15

continued

continued

Total system			Top phase			Bottom phase		
w_A	w_B	w_C	w_A	w_B	w_C	w_A	w_B	w_C
0.8499	0.0600	0.0901	0.8433	0.0432	0.1135	0.8595	0.1038	0.0367
0.8389	0.0700	0.0911	0.8283	0.0323	0.1394	0.8463	0.1288	0.0249
0.8210	0.0800	0.0990	0.8017	0.0235	0.1748	0.8330	0.1485	0.0185

Polymer (B): **poly(ethylene oxide-*b*-propylene oxide-*b*-ethylene oxide)** **2004TAD, 2005TAD**

Characterization: M_n/g.mol^{-1} = 14000, 83.6 mol% ethylene oxide, (EO)127-(PO)50-(EO)127, P108, ICI Surfactants, Cleveland, UK

Solvent (A): **water** **H_2O** **7732-18-5**

Polymer (C): **dextran**

Characterization: M_n/g.mol^{-1} = 236000, M_w/g.mol^{-1} = 410000, Dextran 400, Sigma Chemical Co., Inc., St. Louis, MO

Type of data: cloud points

T/K = 298.15

w_A	0.9256	0.9152
w_B	0.0368	0.0157
w_C	0.0376	0.0691

Type of data: coexistence data (tie lines)

T/K = 298.15

Total system			Top phase			Bottom phase		
w_A	w_B	w_C	w_A	w_B	w_C	w_A	w_B	w_C
0.9191	0.0313	0.0496	0.9059	0.0102	0.0839	0.9324	0.0512	0.0164
0.8912	0.0494	0.0594	0.8700	0.0083	0.1217	0.9118	0.0832	0.0050
0.8715	0.0692	0.0593	0.8357	0.0037	0.1606	0.8951	0.1019	0.0030

Polymer (B): **poly(ethylene oxide-*b*-propylene oxide-*b*-ethylene oxide)** **2004TAD, 2005TAD**
Characterization: M_n/g.mol^{-1} = 4800, 83.8 mol% ethylene oxide, (EO)44-(PO)17-(EO)44, F38, ICI Surfactants, Cleveland, UK
Solvent (A): **water** **H_2O** **7732-18-5**
Polymer (C): **dextran**
Characterization: M_n/g.mol^{-1} = 8200, M_w/g.mol^{-1} = 11600, Dextran 19, Sigma Chemical Co., Inc., St. Louis, MO

Type of data: cloud points

T/K = 298.15

w_A	0.8377	0.8508	0.8628	0.8578	0.8631	0.8620	0.8602	0.8506	0.8459
w_B	0.1534	0.1399	0.1187	0.1199	0.1109	0.1101	0.1021	0.0932	0.0863
w_C	0.0089	0.0093	0.0185	0.0223	0.0260	0.0279	0.0377	0.0562	0.0678

w_A	0.8434	0.8365
w_B	0.0817	0.0838
w_C	0.0749	0.0797

Type of data: coexistence data (tie lines)

T/K = 298.15

Total system			Top phase			Bottom phase		
w_A	w_B	w_C	w_A	w_B	w_C	w_A	w_B	w_C
0.8239	0.0853	0.0908	0.7786	0.0000	0.2214	0.8441	0.1087	0.0472
0.7903	0.0998	0.1099	0.7343	0.0000	0.2657	0.8235	0.1548	0.0217
0.7497	0.1303	0.1200	0.6673	0.0000	0.3327	0.7922	0.1950	0.0128
0.7596	0.1202	0.1202	0.6759	0.0000	0.3241	0.8006	0.1849	0.0145

Polymer (B): **poly(ethylene oxide-*b*-propylene oxide-*b*-ethylene oxide)** **2004TAD, 2005TAD**
Characterization: M_n/g.mol^{-1} = 4800, 83.8 mol% ethylene oxide, (EO)44-(PO)17-(EO)44, F38, ICI Surfactants, Cleveland, UK
Solvent (A): **water** **H_2O** **7732-18-5**
Polymer (C): **dextran**
Characterization: M_n/g.mol^{-1} = 236000, M_w/g.mol^{-1} = 410000, Dextran 400, Sigma Chemical Co., Inc., St. Louis, MO

Type of data: cloud points

T/K = 298.15

w_A	0.8973	0.8944	0.8830
w_B	0.0656	0.0456	0.0380
w_C	0.0371	0.0600	0.0790

continued

continued

Type of data: coexistence data (tie lines)

T/K = 298.15

Total system			Top phase			Bottom phase		
w_A	w_B	w_C	w_A	w_B	w_C	w_A	w_B	w_C
0.8805	0.0499	0.0696	0.8527	0.0264	0.1209	0.9092	0.0757	0.0151
0.8600	0.0700	0.0700	0.8016	0.0149	0.1835	0.8997	0.0950	0.0053
0.8407	0.0896	0.0697	0.7693	0.0028	0.2279	0.8712	0.1247	0.0041

Polymer (B): **poly(ethylene oxide-*b*-propylene oxide-*b*-ethylene oxide)** **1997MAT**

Characterization: M_w/g.mol^{-1} = 13000 (2000 PPO block), 80.0 wt% ethylene oxide, Epan 785, Daiichi Kogyo Seiyaku Co., Ltd., Japan

Solvent (A): **water** **H_2O** **7732-18-5**

Salt (C): **dipotassium phosphate** **K_2HPO_4** **7758-11-4**

Type of data: coexistence data (tie lines)

T/K = 298.15

Total system			Top phase			Bottom phase		
w_A	w_B	w_C	w_A	w_B	w_C	w_A	w_B	w_C
0.825	0.091	0.084	0.622	0.359	0.019	0.897	0.0004	0.103
0.783	0.135	0.082	0.610	0.373	0.017	0.881	0.0006	0.118
0.772	0.141	0.087	0.600	0.385	0.015	0.869	0.001	0.130
0.766	0.141	0.093	0.588	0.398	0.014	0.856	0.002	0.142
0.746	0.153	0.101	0.569	0.419	0.012	0.841	0.003	0.156

Polymer (B): **poly(ethylene oxide-*b*-propylene oxide-*b*-ethylene oxide)** **2004HAR**

Characterization: 20.0 wt% ethylene oxide, L62, ICI Surfactants, Cleveland, UK

Solvent (A): **water** **H_2O** **7732-18-5**

Salt (C): **dipotassium phosphate/monopotassium phosphate** **K_2HPO_4/H_2KPO_4** **7758-11-4/7778-77-0**

Type of data: cloud points

continued

continued

T/K = 283.15 pH = 5.0

w_A	0.611	0.626	0.650	0.669	0.691	0.709	0.725	0.753	0.764
w_B	0.319	0.300	0.272	0.246	0.215	0.192	0.170	0.133	0.119
w_C	0.070	0.074	0.078	0.085	0.094	0.099	0.105	0.114	0.117
w_A	0.775								
w_B	0.098								
w_C	0.127								

T/K = 283.15 pH = 7.0

w_A	0.574	0.596	0.620	0.640	0.664	0.684	0.703	0.740	0.756
w_B	0.381	0.357	0.328	0.307	0.277	0.254	0.230	0.182	0.162
w_C	0.045	0.047	0.052	0.053	0.059	0.062	0.067	0.078	0.082
w_A	0.774								
w_B	0.135								
w_C	0.091								

T/K = 298.15 pH = 5.0

w_A	0.727	0.746	0.761	0.777	0.797	0.808	0.822	0.837	0.851
w_B	0.195	0.172	0.158	0.142	0.120	0.104	0.089	0.073	0.057
w_C	0.078	0.082	0.081	0.081	0.083	0.088	0.089	0.090	0.092

T/K = 298.15 pH = 7.0

w_A	0.699	0.739	0.758	0.780	0.816	0.835	0.844	0.870	0.881
w_B	0.240	0.195	0.180	0.154	0.116	0.096	0.082	0.056	0.036
w_C	0.061	0.066	0.062	0.066	0.068	0.069	0.074	0.074	0.083

Type of data: coexistence data (tie lines)

Total system			Top phase			Bottom phase		
w_A	w_B	w_C	w_A	w_B	w_C	w_A	w_B	w_C
T/K = 283.15 pH = 5.0								
0.699	0.182	0.119	0.440	0.522	0.038	0.783	0.065	0.152
0.676	0.194	0.130	0.423	0.541	0.036	0.777	0.048	0.175
T/K = 283.15 pH = 7.0								
0.702	0.210	0.088	0.533	0.433	0.034	0.791	0.094	0.115
0.673	0.225	0.102	0.498	0.473	0.029	0.789	0.062	0.149
0.645	0.237	0.118	0.477	0.497	0.026	0.769	0.048	0.183
0.615	0.254	0.131	0.426	0.549	0.025	0.737	0.068	0.195

continued

continued

T/K = 298.15   pH = 5.0

0.776	0.118	0.106	0.571	0.365	0.064	0.850	0.027	0.123
0.740	0.140	0.120	0.521	0.416	0.063	0.835	0.019	0.146
0.700	0.159	0.141	0.474	0.465	0.061	0.808	0.020	0.172

T/K = 298.15   pH = 7.0

0.780	0.137	0.083	0.554	0.392	0.054	0.880	0.025	0.095
0.760	0.140	0.100	0.519	0.433	0.048	0.862	0.022	0.116
0.724	0.162	0.114	0.494	0.454	0.052	0.841	0.016	0.143
0.694	0.181	0.125	0.460	0.490	0.050	0.821	0.012	0.167

Polymer (B): **poly(ethylene oxide-*b*-propylene oxide-*b*-ethylene oxide)** **2004HAR**

Characterization: 40.0 wt% ethylene oxide, L64, Aldrich Chem. Co., Inc., Milwaukee, WI

Solvent (A): **water** H_2O **7732-18-5**

Salt (C): **dipotassium phosphate/monopotassium phosphate** **K_2HPO_4/H_2KPO_4** **7758-11-4/7778-77-0**

Type of data: cloud points

T/K = 283.15   pH = 5.0

w_A	0.613	0.629	0.656	0.681	0.699	0.708	0.727	0.753	0.764
w_B	0.313	0.293	0.255	0.220	0.199	0.187	0.159	0.117	0.101
w_C	0.074	0.078	0.089	0.099	0.102	0.105	0.114	0.130	0.135

w_A	0.768
w_B	0.085
w_C	0.147

T/K = 283.15   pH = 7.0

w_A	0.576	0.606	0.620	0.642	0.662	0.710	0.704	0.740	0.756
w_B	0.382	0.348	0.328	0.301	0.277	0.213	0.219	0.178	0.155
w_C	0.042	0.046	0.052	0.057	0.061	0.077	0.077	0.082	0.089

w_A	0.771
w_B	0.134
w_C	0.095

T/K = 298.15   pH = 5.0

w_A	0.746	0.771	0.783	0.794	0.803	0.812	0.820	0.828	0.834
w_B	0.147	0.116	0.099	0.083	0.071	0.060	0.052	0.041	0.032
w_C	0.107	0.113	0.118	0.123	0.126	0.128	0.128	0.131	0.134

continued

continued

T/K = 298.15 pH = 7.0

w_A	0.699	0.718	0.736	0.753	0.769	0.786	0.802	0.817	0.833
w_B	0.233	0.211	0.189	0.165	0.144	0.124	0.105	0.085	0.068
w_C	0.068	0.071	0.075	0.082	0.087	0.090	0.093	0.098	0.099

w_A	0.846	0.855
w_B	0.048	0.026
w_C	0.106	0.119

Type of data: coexistence data (tie lines)

Total system			Top phase			Bottom phase		
w_A	w_B	w_C	w_A	w_B	w_C	w_A	w_B	w_C
T/K = 283.15 pH = 5.0								
0.708	0.166	0.126	0.523	0.421	0.056	0.768	0.085	0.147
0.683	0.179	0.138	0.514	0.441	0.045	0.780	0.037	0.183
T/K = 283.15 pH = 7.0								
0.725	0.175	0.100	0.576	0.384	0.040	0.794	0.079	0.127
0.696	0.190	0.114	0.518	0.455	0.027	0.808	0.020	0.172
0.668	0.204	0.128	0.507	0.461	0.032	0.774	0.026	0.200
0.639	0.219	0.142	0.474	0.499	0.027	0.753	0.012	0.235
T/K = 298.15 pH = 5.0								
0.750	0.115	0.135	0.520	0.420	0.060	0.829	0.014	0.157
0.730	0.125	0.145	0.488	0.457	0.055	0.815	0.008	0.177
0.710	0.135	0.155	0.467	0.483	0.050	0.786	0.010	0.204
0.692	0.142	0.166	0.437	0.514	0.049	0.777	0.006	0.217
T/K = 298.15 pH = 7.0								
0.757	0.128	0.115	0.566	0.379	0.055	0.832	0.028	0.140
0.744	0.126	0.130	0.518	0.431	0.051	0.823	0.013	0.164
0.713	0.145	0.142	0.461	0.493	0.046	0.810	0.011	0.179
0.692	0.149	0.159	0.420	0.533	0.047	0.792	0.008	0.200

Polymer (B): **poly(ethylene oxide-*b*-propylene oxide-*b*-ethylene oxide)** **2005SIL**

Characterization: M_n/g.mol^{-1} = 1706, M_w/g.mol^{-1} = 1945, 50.0 wt% ethylene oxide, (EO)11-(PO)16-(EO)11, L35, Aldrich Chem. Co., Inc., Milwaukee, WI

Solvent (A): **water** **H_2O** **7732-18-5**

Salt (C): **dipotassium phosphate/monopotassium phosphate** **K_2HPO_4/H_2KPO_4** **7758-11-4/7778-77-0**

Comments: The mixture of K_2HPO_4/H_2KPO_4 was chosen in a ratio that provides a pH = 7.

Type of data: coexistence data (tie lines)

Total system			Top phase			Bottom phase		
w_A	w_B	w_C	w_A	w_B	w_C	w_A	w_B	w_C
T/K = 283.15								
0.7690	0.1311	0.0999	0.6913	0.3041	0.0046	0.8217	0.0504	0.1279
0.6764	0.1836	0.1400	0.5989	0.4010	0.0001	0.7414	0.0435	0.2151
0.6001	0.2300	0.1699	0.4785	0.5214	0.0001	0.6652	0.0455	0.2893
0.5078	0.2822	0.2100	0.3575	0.6424	0.0001	0.5977	0.0477	0.3546
T/K = 298.15								
0.7839	0.1278	0.0883	0.6500	0.3497	0.0003	0.8563	0.0260	0.1177
0.7087	0.1715	0.1198	0.5144	0.4854	0.0002	0.8160	0.0274	0.1566
0.6501	0.2007	0.1492	0.4334	0.5665	0.0001	0.7531	0.0271	0.2198
0.5903	0.2401	0.1696	0.3699	0.6300	0.0001	0.7286	0.0304	0.2410
T/K = 313.15								
0.7394	0.2107	0.0499	0.6758	0.3225	0.0017	0.8802	0.0312	0.0886
0.6286	0.3015	0.0699	0.4689	0.5310	0.0001	0.8301	0.0164	0.1535
0.5320	0.3803	0.0877	0.3616	0.6382	0.0002	0.7743	0.0210	0.2047
0.4195	0.4706	0.1099	0.2721	0.7277	0.0002	0.6768	0.0271	0.2961

Polymer (B): **poly(ethylene oxide-*b*-propylene oxide-*b*-ethylene oxide)** **2005SIL**

Characterization: M_n/g.mol^{-1} = 5960, M_w/g.mol^{-1} = 8460, 80.0 wt% ethylene oxide, (EO)80-(PO)30-(EO)80, F68, Aldrich Chem. Co., Inc., Milwaukee, WI

Solvent (A): **water** **H_2O** **7732-18-5**

Salt (C): **dipotassium phosphate/monopotassium phosphate** **K_2HPO_4/H_2KPO_4** **7758-11-4/7778-77-0**

continued

continued

Comments: The mixture of K_2HPO_4/H_2KPO_4 was chosen in a ratio that provides a pH = 7.

Type of data: coexistence data (tie lines)

Total system			Top phase			Bottom phase		
w_A	w_B	w_C	w_A	w_B	w_C	w_A	w_B	w_C
T/K = 283.15								
0.7777	0.1425	0.0798	0.7545	0.2278	0.0177	0.8550	0.0164	0.1286
0.7300	0.1702	0.0998	0.6659	0.3269	0.0072	0.8045	0.0257	0.1698
0.6889	0.2012	0.1099	0.6138	0.3782	0.0080	0.7835	0.0304	0.1861
0.6532	0.2189	0.1279	0.5718	0.4176	0.0106	0.7430	0.0410	0.2160
T/K = 298.15								
0.7898	0.1402	0.0700	0.7123	0.2848	0.0029	0.8727	0.0353	0.0920
0.7499	0.1702	0.0799	0.6575	0.3408	0.0017	0.8516	0.0233	0.1251
0.7203	0.1899	0.0898	0.6221	0.3730	0.0049	0.8211	0.0260	0.1529
0.6694	0.2207	0.1099	0.5740	0.4103	0.0157	0.7847	0.0254	0.1899
T/K = 313.15								
0.8101	0.1299	0.0600	0.7435	0.2511	0.0054	0.8854	0.0306	0.0840
0.7602	0.1599	0.0799	0.6443	0.3553	0.0004	0.8590	0.0211	0.1199
0.7201	0.1900	0.0899	0.6090	0.3904	0.0006	0.8464	0.0192	0.1344
0.6910	0.2093	0.0997	0.5712	0.4283	0.0005	0.8124	0.0237	0.1639

Polymer (B): **poly(ethylene oxide-*b*-propylene oxide-*b*-ethylene oxide)** **2004HAR**

Characterization: 80.0 wt% ethylene oxide, F68, Aldrich Chem. Co., Inc., Milwaukee, WI

Solvent (A): **water** H_2O **7732-18-5**

Salt (C): **dipotassium phosphate/monopotassium phosphate** K_2HPO_4/H_2KPO_4 **7758-11-4/7778-77-0**

Type of data: cloud points

T/K = 283.15 pH = 5.0

w_A	0.665	0.685	0.705	0.716	0.750	0.765	0.776
w_B	0.262	0.236	0.205	0.199	0.143	0.116	0.086
w_C	0.073	0.079	0.090	0.085	0.107	0.119	0.138

T/K = 283.15 pH = 7.0

w_A	0.682	0.701	0.742	0.759	0.776
w_B	0.265	0.247	0.191	0.156	0.144
w_C	0.053	0.052	0.067	0.085	0.080

continued

continued

T/K = 298.15 pH = 5.0

w_A	0.746	0.771	0.783	0.794	0.803	0.812	0.820	0.828	0.834
w_B	0.147	0.116	0.099	0.083	0.071	0.060	0.052	0.041	0.032
w_C	0.107	0.113	0.118	0.123	0.126	0.128	0.128	0.131	0.134

T/K = 298.15 pH = 7.0

w_A	0.699	0.734	0.756	0.775	0.790	0.803	0.818	0.821	0.841
w_B	0.242	0.197	0.175	0.149	0.130	0.105	0.086	0.062	0.049
w_C	0.059	0.069	0.069	0.076	0.080	0.092	0.096	0.117	0.110
w_A	0.834								
w_B	0.044								
w_C	0.122								

Type of data: coexistence data (tie lines)

Total system			Top phase			Bottom phase		
w_A	w_B	w_C	w_A	w_B	w_C	w_A	w_B	w_C
T/K = 283.15 pH = 7.0								
0.756	0.149	0.095	0.720	0.231	0.049	0.795	0.074	0.131
0.728	0.164	0.108	0.683	0.270	0.047	0.802	0.009	0.189
0.703	0.177	0.120	0.643	0.324	0.033	0.760	0.017	0.223
0.671	0.193	0.136	0.611	0.358	0.031	0.741	0.012	0.247
T/K = 298.15 pH = 7.0								
0.726	0.178	0.096	0.671	0.280	0.049	0.797	0.024	0.179
0.694	0.196	0.110	0.639	0.315	0.046	0.768	0.022	0.210
0.665	0.210	0.125	0.621	0.337	0.042	0.729	0.017	0.254

Polymer (B): **poly(ethylene oxide-*b*-propylene oxide-*b*-ethylene oxide)** **2005SIL**

Characterization: M_n/g.mol^{-1} = 1706, M_w/g.mol^{-1} = 1945, 50.0 wt% ethylene oxide, (EO)11-(PO)16-(EO)11, L35, Aldrich Chem. Co., Inc., Milwaukee, WI

Solvent (A): **water** **H_2O** **7732-18-5**
Salt (C): **dipotassium phosphate** **K_2HPO_4** **7758-11-4**
Salt (D): **potassium hydroxide** **KOH** **1310-58-3**

Comments: The mixture of K_2HPO_4/KOH was chosen in a ratio that provides a pH = 12.

continued

continued

Type of data: coexistence data (tie lines)

Total system			Top phase			Bottom phase		
w_A	w_B	w_C	w_A	w_B	w_C	w_A	w_B	w_C
T/K = 283.15								
0.4617	0.2869	0.2514	0.4246	0.5699	0.0055	0.5113	0.0596	0.4291
0.3736	0.3402	0.2862	0.3386	0.6601	0.0013	0.4037	0.0365	0.5598
0.2845	0.3895	0.3260	0.2913	0.7060	0.0027	0.3205	0.0286	0.6509
0.2059	0.4265	0.3676	0.2036	0.7930	0.0034	0.1961	0.0188	0.7851
T/K = 298.15								
0.5965	0.2404	0.1634	0.3654	0.6322	0.0024	0.7113	0.0443	0.2444
0.5091	0.2881	0.2028	0.2859	0.7115	0.0026	0.6305	0.0417	0.3278
0.4300	0.3414	0.2286	0.1981	0.8010	0.0009	0.5908	0.0239	0.3853
0.3509	0.3897	0.2594	0.1411	0.8586	0.0003	0.5242	0.0088	0.4670
T/K = 313.15								
0.5708	0.2889	0.1403	0.3103	0.6862	0.0035	0.7423	0.0252	0.2325
0.5103	0.3205	0.1692	0.2706	0.7269	0.0025	0.6897	0.0110	0.2993
0.4383	0.3738	0.1879	0.2123	0.7860	0.0017	0.6192	0.0255	0.3553
0.3643	0.4265	0.2092	0.1324	0.8672	0.0004	0.5846	0.0131	0.4023

Polymer (B): **poly(ethylene oxide-*b*-propylene oxide-*b*-ethylene oxide)** **2009MAR**

Characterization: M_w/g.mol^{-1} = 1900, 50.0 wt% ethylene oxide, (EO)11-(PO)16-(EO)11, L35, Sigma Chemical Co., Inc., St. Louis, MO

Solvent (A): **water** H_2O **7732-18-5**

Salt (C): **disodium (±)tartrate** $C_4H_4Na_2O_6$ **51307-92-7**

Type of data: coexistence data (tie lines)

Total system			Top phase			Bottom phase		
w_A	w_B	w_C	w_A	w_B	w_C	w_A	w_B	w_C
T/K = 283.15								
0.7179	0.2000	0.0821	0.6880	0.2672	0.0448	0.7667	0.0651	0.1682
0.6768	0.2211	0.1021	0.5958	0.3785	0.0257	0.7634	0.0288	0.2078
0.6340	0.2528	0.1132	0.5552	0.4259	0.0189	0.7286	0.0241	0.2473
0.5967	0.2820	0.1213	0.5173	0.4678	0.0149	0.7008	0.0233	0.2759
0.5602	0.3059	0.1339	0.4599	0.5283	0.0118	0.6633	0.0451	0.2916

continued

continued

T/K = 298.15

0.5946	0.3599	0.0455	0.5640	0.4106	0.0254	0.7954	0.0302	0.1744
0.5474	0.3992	0.0534	0.5066	0.4757	0.0177	0.7625	0.0313	0.2044
0.4933	0.4504	0.0563	0.4609	0.5267	0.0124	0.7256	0.0455	0.2289
0.4401	0.4966	0.0633	0.4149	0.5739	0.0112	0.7037	0.0424	0.2539
0.3879	0.5462	0.0659	0.3119	0.6825	0.0056	0.6359	0.0582	0.3059

T/K = 313.15

0.6429	0.2578	0.0993	0.4528	0.5320	0.0152	0.8033	0.0203	0.1764
0.6326	0.2558	0.1116	0.4085	0.5802	0.0113	0.7755	0.0363	0.1882
0.5896	0.2902	0.1202	0.3729	0.6186	0.0085	0.7531	0.0312	0.2157
0.5465	0.3269	0.1266	0.3149	0.6789	0.0062	0.7175	0.0326	0.2499
0.4689	0.3889	0.1422	0.2901	0.7063	0.0036	0.6482	0.0610	0.2908

Polymer (B): **poly(ethylene oxide-*b*-propylene oxide-*b*-ethylene oxide)** **2006SIL**

Characterization: M_n/g.mol^{-1} = 1706, M_w/g.mol^{-1} = 1945, 50.0 wt% ethylene oxide, (EO)11-(PO)16-(EO)11, L35, Aldrich Chem. Co., Inc., Milwaukee, WI

Solvent (A): **water** **H_2O** **7732-18-5**

Salt (C): **lithium sulfate** **Li_2SO_4** **10377-48-7**

Type of data: coexistence data (tie lines)

Total system			Top phase			Bottom phase		
w_A	w_B	w_C	w_A	w_B	w_C	w_A	w_B	w_C
T/K = 298.15								
0.7874	0.0778	0.1348	0.5436	0.4448	0.0116	0.8218	0.0292	0.1490
0.7689	0.0903	0.1408	0.5304	0.4587	0.0109	0.8101	0.0262	0.1637
0.7535	0.0999	0.1466	0.4935	0.4985	0.0080	0.8028	0.0218	0.1754
0.7251	0.1179	0.1570	0.4525	0.5425	0.0050	0.7950	0.0095	0.1955
0.6906	0.1401	0.1693	0.4061	0.5892	0.0047	0.7788	0.0019	0.2193
T/K = 313.15								
0.8024	0.1266	0.0710	0.5473	0.4361	0.0166	0.8890	0.0214	0.0896
0.7879	0.1385	0.0736	0.5212	0.4641	0.0147	0.8857	0.0198	0.0945
0.7684	0.1548	0.0768	0.4984	0.4895	0.0121	0.8792	0.0181	0.1027
0.7166	0.1974	0.0860	0.4827	0.5058	0.0115	0.8692	0.0010	0.1298
0.6850	0.2234	0.0916	0.4695	0.5193	0.0112	0.8520	0.0004	0.1476

Polymer (B): **poly(ethylene oxide-*b*-propylene oxide-*b*-ethylene oxide)** **2009ROD**
Characterization: M_w/g.mol^{-1} = 2900, 40.0 wt% ethylene oxide, (EO)13-(PO)30-(EO)13, L64, Aldrich Chem. Co., Inc., Milwaukee, WI
Solvent (A): **water** **H_2O** **7732-18-5**
Salt (C): **lithium sulfate** **Li_2SO_4** **10377-48-7**

Type of data: coexistence data (tie lines)

Total system			Top phase			Bottom phase		
w_A	w_B	w_C	w_A	w_B	w_C	w_A	w_B	w_C
T/K = 278.15								
0.7108	0.2257	0.0635	0.6216	0.3361	0.0423	0.8623	0.0292	0.1085
0.6884	0.2461	0.0655	0.5632	0.4042	0.0326	0.8678	0.0175	0.1147
0.6661	0.2663	0.0676	0.5482	0.4168	0.0350	0.8638	0.0097	0.1265
0.6435	0.2873	0.0693	0.5195	0.4488	0.0317	0.8598	0.0035	0.1366
T/K = 288.15								
0.7095	0.2273	0.0631	0.5994	0.3591	0.0415	0.8623	0.0420	0.0957
0.6879	0.2463	0.0658	0.5746	0.3872	0.0382	0.8678	0.0259	0.1063
0.6642	0.2684	0.0674	0.5196	0.4497	0.0307	0.8638	0.0246	0.1116
0.6425	0.2880	0.0695	0.4765	0.4990	0.0245	0.8598	0.0194	0.1208
T/K = 298.15								
0.7780	0.1645	0.0575	0.7153	0.2340	0.0507	0.8996	0.0209	0.0795
0.7552	0.1855	0.0593	0.6378	0.3208	0.0414	0.9039	0.0112	0.0848
0.7334	0.2049	0.0616	0.6023	0.3577	0.0401	0.8994	0.0056	0.0950
0.7110	0.2262	0.0628	0.5500	0.4172	0.0327	0.8957	0.0041	0.1001

Polymer (B): **poly(ethylene oxide-*b*-propylene oxide-*b*-ethylene oxide)** **2006SIL**
Characterization: M_n/g.mol^{-1} = 1706, M_w/g.mol^{-1} = 1945, 50.0 wt% ethylene oxide, (EO)11-(PO)16-(EO)11, L35, Aldrich Chem. Co., Inc., Milwaukee, WI
Solvent (A): **water** **H_2O** **7732-18-5**
Salt (C): **magnesium sulfate** **$MgSO_4$** **7487-88-9**

Type of data: coexistence data (tie lines)

continued

continued

Total system			Top phase			Bottom phase		
w_A	w_B	w_C	w_A	w_B	w_C	w_A	w_B	w_C
T/K = 283.15								
0.5769	0.3918	0.0313	0.5382	0.4574	0.0044	0.7618	0.0577	0.1805
0.5215	0.4409	0.0376	0.4860	0.5114	0.0026	0.7396	0.0512	0.2092
0.4865	0.4719	0.0416	0.4341	0.5645	0.0014	0.7265	0.0397	0.2338
0.4365	0.5164	0.0471	0.3728	0.6267	0.0005	0.7086	0.0351	0.2563
0.3931	0.5548	0.0521	0.3266	0.6733	0.0001	0.6967	0.0371	0.2662
T/K = 298.15								
0.6169	0.3500	0.0331	0.5723	0.4195	0.0082	0.8045	0.0483	0.1472
0.5711	0.3908	0.0381	0.5203	0.4739	0.0058	0.7869	0.0476	0.1655
0.5254	0.4315	0.0431	0.4640	0.5318	0.0042	0.7653	0.0401	0.1946
0.4796	0.4723	0.0481	0.4160	0.5815	0.0025	0.7488	0.0312	0.2200
0.4338	0.5131	0.0531	0.3665	0.6323	0.0012	0.7423	0.0196	0.2381
T/K = 313.15								
0.6241	0.3486	0.0273	0.5614	0.4334	0.0052	0.8656	0.0335	0.1009
0.5683	0.3994	0.0323	0.4928	0.5042	0.0030	0.8443	0.0312	0.1245
0.5125	0.4502	0.0373	0.4361	0.5618	0.0021	0.8206	0.0289	0.1505
0.4567	0.5010	0.0423	0.3760	0.6227	0.0013	0.7927	0.0223	0.1850
0.4009	0.5518	0.0473	0.3123	0.6872	0.0005	0.7717	0.0170	0.2113

Polymer (B): **poly(ethylene oxide-*b*-propylene oxide-*b*-ethylene oxide)** **2009ROD**
Characterization: M_w/g.mol^{-1} = 2900, 40.0 wt% ethylene oxide, (EO)13-(PO)30-(EO)13, L64, Aldrich Chem. Co., Inc., Milwaukee, WI
Solvent (A): **water** **H_2O** **7732-18-5**
Salt (C): **magnesium sulfate** **$MgSO_4$** **7487-88-9**

Type of data: coexistence data (tie lines)

Total system			Top phase			Bottom phase		
w_A	w_B	w_C	w_A	w_B	w_C	w_A	w_B	w_C
T/K = 278.15								
0.7005	0.2528	0.0467	0.6340	0.3540	0.0120	0.8495	0.0049	0.1455
0.6702	0.2788	0.0510	0.5962	0.3951	0.0087	0.8397	0.0021	0.1582
0.6415	0.3015	0.0570	0.5504	0.4432	0.0064	0.8282	0.0046	0.1673

continued

continued

0.6202	0.3157	0.0641	0.5080	0.4861	0.0059	0.8164	0.0052	0.1784
0.5889	0.3404	0.0707	0.4664	0.5300	0.0035	0.8043	0.0067	0.1890
T/K = 288.15								
0.7519	0.2030	0.0451	0.6550	0.3256	0.0194	0.8972	0.0088	0.0940
0.6996	0.2532	0.0472	0.5871	0.4005	0.0124	0.8818	0.0045	0.1137
0.6701	0.2787	0.0512	0.5387	0.4515	0.0098	0.8709	0.0002	0.1290
0.6434	0.2997	0.0569	0.5019	0.4910	0.0071	0.8593	0.0016	0.1391
0.6204	0.3158	0.0638	0.4632	0.5313	0.0054	0.8453	0.0033	0.1514
T/K = 298.15								
0.7350	0.2187	0.0463	0.6198	0.3515	0.0287	0.8835	0.0556	0.0609
0.6950	0.2556	0.0494	0.5222	0.4558	0.0220	0.8766	0.0515	0.0718
0.6560	0.2910	0.0530	0.4712	0.5087	0.0201	0.8683	0.0497	0.0820
0.6149	0.3274	0.0577	0.4307	0.5503	0.0191	0.8517	0.0530	0.0953
0.5702	0.3702	0.0596	0.3999	0.5819	0.0182	0.8381	0.0516	0.1103

Polymer (B): **poly(ethylene oxide-*b*-propylene oxide-*b*-ethylene oxide)** **2006SIL**

Characterization: M_n/g.mol^{-1} = 5960, M_w/g.mol^{-1} = 8460, 80.0 wt% ethylene oxide, (EO)80-(PO)30-(EO)80, F68, Aldrich Chem. Co., Inc., Milwaukee, WI

Solvent (A): **water** **H_2O** **7732-18-5**

Salt (C): **magnesium sulfate** **$MgSO_4$** **7487-88-9**

Type of data: coexistence data (tie lines)

Total system			Top phase			Bottom phase		
w_A	w_B	w_C	w_A	w_B	w_C	w_A	w_B	w_C
T/K = 283.15								
0.7822	0.1720	0.0458	0.7220	0.2596	0.0184	0.8585	0.0587	0.0828
0.7531	0.1948	0.0521	0.6809	0.3047	0.0144	0.8449	0.0583	0.0968
0.7299	0.2130	0.0571	0.6599	0.3288	0.0113	0.8242	0.0521	0.1237
0.7067	0.2312	0.0621	0.6407	0.3502	0.0091	0.8069	0.0443	0.1488
0.6835	0.2494	0.0671	0.6203	0.3713	0.0084	0.7925	0.0384	0.1691
T/K = 298.15								
0.8237	0.1322	0.0441	0.7742	0.2071	0.0187	0.8956	0.0267	0.0777
0.8024	0.1498	0.0478	0.7321	0.2519	0.0160	0.8880	0.0188	0.0932
0.7618	0.1832	0.0550	0.6908	0.2959	0.0133	0.8661	0.0156	0.1183
0.7357	0.2047	0.0596	0.6578	0.3322	0.0100	0.8546	0.0112	0.1342
0.7175	0.2197	0.0628	0.6240	0.3676	0.0084	0.8526	0.0042	0.1432

continued

continued

T/K = 313.15

0.7903	0.1694	0.0403	0.7201	0.2633	0.0166	0.8690	0.0638	0.0672
0.7602	0.1945	0.0453	0.6729	0.3137	0.0134	0.8550	0.0636	0.0814
0.7434	0.2085	0.0481	0.6592	0.3290	0.0118	0.8477	0.0602	0.0921
0.7242	0.2245	0.0513	0.6432	0.3461	0.0107	0.8336	0.0589	0.1075
0.6794	0.2618	0.0588	0.5790	0.4109	0.0101	0.8164	0.0586	0.1250

Polymer (B): **poly(ethylene oxide-*b*-propylene oxide-*b*-ethylene oxide)** **2000LAM**
Characterization: M_w/g.mol^{-1} = 4400, 30 mol% ethylene oxide, ρ(298.15 K) = 1.036 g/cm^3
Solvent (A): **water** H_2O **7732-18-5**
Salt (C): **sodium acetate** $C_2H_3NaO_2$ **127-09-3**

Type of data: cloud points (LCST-behavior)

w_B	0.05
c_C/(mol/kg)	1.0
T/K	290.94

Polymer (B): **poly(ethylene oxide-*b*-propylene oxide-*b*-ethylene oxide)** **2000LAM**
Characterization: M_w/g.mol^{-1} = 4400, 30 mol% ethylene oxide, ρ(298.15 K) = 1.036 g/cm^3
Solvent (A): **water** H_2O **7732-18-5**
Salt (C): **sodium bromide** **NaBr** **7647-15-6**

Type of data: cloud points (LCST-behavior)

w_B	0.05
c_C/(mol/kg)	1.0
T/K	283.65

Polymer (B): **poly(ethylene oxide-*b*-propylene oxide-*b*-ethylene oxide)** **2000LAM**
Characterization: M_w/g.mol^{-1} = 4400, 30 mol% ethylene oxide, ρ(298.15 K) = 1.036 g/cm^3
Solvent (A): **water** H_2O **7732-18-5**
Salt (C): **sodium chloride** **NaCl** **7647-14-5**

Type of data: cloud points (LCST-behavior)

w_B	0.05
c_C/(mol/kg)	1.0
T/K	284.65

Polymer (B): **poly(ethylene oxide-*b*-propylene oxide-*b*-ethylene oxide)** **2001DES**
Characterization: M_n/g.mol^{-1} = 12500, 75 mol% ethylene oxide, (EO)99-(PO)65-(EO)99, Pluronic F127, BASF, Parsippany, NJ
Solvent (A): **water** H_2O **7732-18-5**
Salt (C): **sodium chloride** **NaCl** **7647-14-5**

Type of data: cloud points (LCST-behavior)

w_B	0.01	0.01	0.01	0.01
c_C/(mol/l)	0.0	0.5	1.0	2.0
T/K	375.2	368.2	355.2	334.2

Polymer (B): **poly(ethylene oxide-*b*-propylene oxide-*b*-ethylene oxide)** **2009MAR**
Characterization: M_w/g.mol^{-1} = 1900, 50.0 wt% ethylene oxide, (EO)11-(PO)16-(EO)11, L35, Sigma Chemical Co., Inc., St. Louis, MO
Solvent (A): **water** H_2O **7732-18-5**
Salt (C): **sodium citrate** $Na_3C_6H_5O_7$ **68-04-2**

Type of data: coexistence data (tie lines)

Total system			Top phase			Bottom phase		
w_A	w_B	w_C	w_A	w_B	w_C	w_A	w_B	w_C
T/K = 283.15								
0.7221	0.1868	0.0910	0.6205	0.3474	0.0321	0.7829	0.0419	0.1752
0.6834	0.2197	0.0968	0.6011	0.3760	0.0229	0.7580	0.0377	0.2043
0.6453	0.2467	0.1079	0.5568	0.4275	0.0157	0.7159	0.0582	0.2259
0.6052	0.2740	0.1206	0.5011	0.4869	0.0120	0.6775	0.0628	0.2597
0.5661	0.3043	0.1294	0.4554	0.5360	0.0086	0.6518	0.0699	0.2783
T/K = 298.15								
0.6767	0.2732	0.0511	0.6214	0.3494	0.0292	0.8229	0.0243	0.1528
0.6332	0.3106	0.0562	0.5621	0.4168	0.0211	0.8043	0.0233	0.1724
0.5907	0.3482	0.0611	0.5152	0.4693	0.0155	0.7755	0.0243	0.2002
0.5481	0.3857	0.0662	0.4639	0.5251	0.0110	0.7472	0.0225	0.2303
0.5057	0.4231	0.0712	0.4169	0.5756	0.0075	0.7152	0.0332	0.2516
T/K = 313.15								
0.7223	0.1867	0.0910	0.5294	0.4539	0.0167	0.8207	0.0268	0.1525
0.6832	0.2201	0.0967	0.4787	0.5088	0.0125	0.8022	0.0281	0.1697
0.6454	0.2468	0.1078	0.4329	0.5578	0.0093	0.7780	0.0317	0.1903
0.6056	0.2738	0.1206	0.3953	0.5980	0.0067	0.7394	0.0443	0.2163
0.5665	0.3041	0.1294	0.3627	0.6324	0.0049	0.7062	0.0518	0.2420

Polymer (B): **poly(ethylene oxide-*b*-propylene oxide-*b*-ethylene oxide)** **2009MAR**

Characterization: M_w/g.mol^{-1} = 1900, 50.0 wt% ethylene oxide, (EO)11-(PO)16-(EO)11, L35, Sigma Chemical Co., Inc., St. Louis, MO

Solvent (A): **water** **H_2O** **7732-18-5**

Salt (C): **sodium nitrate** **$NaNO_3$** **7631-99-4**

Type of data: coexistence data (tie lines)

Total system			Top phase			Bottom phase		
w_A	w_B	w_C	w_A	w_B	w_C	w_A	w_B	w_C
T/K = 283.15								
0.5376	0.2111	0.2513	0.4565	0.3856	0.1579	0.6451	0.0208	0.3341
0.5135	0.2304	0.2561	0.4166	0.4314	0.1520	0.6366	0.0079	0.3555
0.4873	0.2505	0.2622	0.3739	0.4822	0.1439	0.6093	0.0108	0.3799
0.4638	0.2684	0.2678	0.3483	0.5117	0.1400	0.6000	0.0082	0.3918
0.4399	0.2867	0.2734	0.3284	0.5336	0.1380	0.5764	0.0001	0.4235
T/K = 298.15								
0.5624	0.1916	0.2460	0.4109	0.4488	0.1403	0.6758	0.0171	0.3091
0.5328	0.2128	0.2544	0.3810	0.4871	0.1319	0.6543	0.0171	0.3286
0.5121	0.2312	0.2567	0.3426	0.5311	0.1263	0.6420	0.0130	0.3450
0.4875	0.2514	0.2611	0.3063	0.5735	0.1202	0.6268	0.0101	0.3631
0.4631	0.2714	0.2655	0.2823	0.6006	0.1171	0.5901	0.0320	0.3779
T/K = 313.15								
0.5598	0.1919	0.2483	0.4960	0.3394	0.1728	0.6733	0.0100	0.3167
0.5363	0.2125	0.2512	0.4423	0.3913	0.1664	0.6506	0.0222	0.3272
0.5174	0.2297	0.2529	0.4166	0.4306	0.1528	0.6476	0.0107	0.3417
0.4869	0.2510	0.2621	0.3739	0.4826	0.1435	0.6196	0.0004	0.3800
0.4621	0.2710	0.2669	0.3286	0.5346	0.1368	0.5885	0.0174	0.3941

Polymer (B): **poly(ethylene oxide-*b*-propylene oxide-*b*-ethylene oxide)** **2000LAM**

Characterization: M_w/g.mol^{-1} = 4400, 30 mol% ethylene oxide, ρ(298.15 K) = 1.036 g/cm^3

Solvent (A): **water** **H_2O** **7732-18-5**

Salt (C): **sodium perchlorate** **$NaClO_4$** **7601-89-0**

Type of data: cloud points (LCST-behavior)

w_B	0.05
c_C/(mol/kg)	1.0
T/K	297.65

Polymer (B): **poly(ethylene oxide-*b*-propylene oxide-*b*-ethylene oxide)** **2006SIL**
Characterization: M_n/g.mol^{-1} = 1706, M_w/g.mol^{-1} = 1945, 50.0 wt% ethylene oxide, (EO)11-(PO)16-(EO)11, L35, Aldrich Chem. Co., Inc., Milwaukee, WI
Solvent (A): **water** H_2O **7732-18-5**
Salt (C): **sodium sulfate** Na_2SO_4 **7757-82-6**

Type of data: coexistence data (tie lines)

Total system			Top phase			Bottom phase		
w_A	w_B	w_C	w_A	w_B	w_C	w_A	w_B	w_C
T/K = 298.15								
0.7458	0.1336	0.1206	0.5227	0.4691	0.0082	0.8166	0.0259	0.1575
0.7195	0.1506	0.1299	0.4849	0.5085	0.0066	0.8063	0.0191	0.1746
0.6766	0.1769	0.1465	0.4454	0.5508	0.0038	0.7798	0.0130	0.2072
0.6614	0.1855	0.1531	0.4198	0.5780	0.0022	0.7711	0.0100	0.2189
0.6210	0.2109	0.1681	0.4026	0.5959	0.0015	0.7370	0.0077	0.2553
T/K = 313.15								
0.7749	0.1622	0.0629	0.5228	0.4656	0.0116	0.8910	0.0243	0.0847
0.7525	0.1818	0.0657	0.5136	0.4755	0.0109	0.8894	0.0192	0.0914
0.7428	0.1902	0.0670	0.4701	0.5216	0.0083	0.8855	0.0180	0.0965
0.7084	0.2203	0.0713	0.4044	0.5889	0.0067	0.8774	0.0166	0.1060
0.6609	0.2618	0.0773	0.3933	0.6006	0.0061	0.8693	0.0009	0.1298

Polymer (B): **poly(ethylene oxide-*b*-propylene oxide-*b*-ethylene oxide)** **2009ROD**
Characterization: M_w/g.mol^{-1} = 2900, 40.0 wt% ethylene oxide, (EO)13-(PO)30-(EO)13, L64, Aldrich Chem. Co., Inc., Milwaukee, WI
Solvent (A): **water** H_2O **7732-18-5**
Salt (C): **sodium sulfate** Na_2SO_4 **7757-82-6**

Type of data: coexistence data (tie lines)

Total system			Top phase			Bottom phase		
w_A	w_B	w_C	w_A	w_B	w_C	w_A	w_B	w_C
T/K = 278.15								
0.7489	0.1993	0.0518	0.6820	0.2992	0.0188	0.8759	0.0104	0.1137
0.7134	0.2319	0.0547	0.6340	0.3511	0.0150	0.8617	0.0075	0.1309

continued

continued

0.6776	0.2646	0.0578	0.5811	0.4068	0.0122	0.8509	0.0107	0.1383
0.6425	0.2972	0.0603	0.5311	0.4591	0.0098	0.8409	0.0045	0.1546
0.6052	0.3315	0.0634	0.4919	0.4992	0.0089	0.8229	0.0084	0.1687
0.5744	0.3594	0.0662	0.4715	0.5207	0.0078	0.8057	0.0068	0.1875
T/K = 288.15								
0.7488	0.1995	0.0517	0.6719	0.2950	0.0331	0.9101	0.0002	0.0897
0.7125	0.2329	0.0545	0.5946	0.3832	0.0222	0.8938	0.0001	0.1062
0.6782	0.2642	0.0575	0.5422	0.4407	0.0170	0.8823	0.0001	0.1176
0.6416	0.2978	0.0606	0.4968	0.4899	0.0133	0.8662	0.0002	0.1336
0.6068	0.3299	0.0633	0.4668	0.5226	0.0107	0.8499	0.0001	0.1501
T/K = 298.15								
0.7484	0.1997	0.0519	0.6259	0.3431	0.0310	0.9020	0.0121	0.0858
0.7247	0.2214	0.0539	0.5605	0.4173	0.0223	0.8985	0.0055	0.0960
0.7001	0.2429	0.0570	0.5075	0.4749	0.0176	0.8908	0.0047	0.1046
0.6742	0.2659	0.0598	0.4884	0.4983	0.0133	0.8808	0.0055	0.1137
0.6522	0.2848	0.0630	0.4469	0.5412	0.0119	0.8713	0.0037	0.1250

Polymer (B): **poly(ethylene oxide-*b*-propylene oxide-*b*-ethylene oxide)** **2000LAM**

Characterization: M_w/g.mol^{-1} = 4400, 30 mol% ethylene oxide, ρ(298.15 K) = 1.036 g/cm^3

Solvent (A): **water** H_2O **7732-18-5**

Salt (C): **sodium sulfate** Na_2SO_4 **7757-82-6**

Type of data: cloud points (LCST-behavior)

w_B	0.05
c_C/(mol/kg)	1.0
T/K	274.45

Polymer (B): **poly(ethylene oxide-*b*-propylene oxide-*b*-ethylene oxide)** **2009ROD**

Characterization: M_w/g.mol^{-1} = 2900, 40.0 wt% ethylene oxide, (EO)13-(PO)30-(EO)13, L64, Aldrich Chem. Co., Inc., Milwaukee, WI

Solvent (A): **water** H_2O **7732-18-5**

Salt (C): **zinc sulfate** $ZnSO_4$ **7733-02-0**

Type of data: coexistence data (tie lines)

continued

continued

Total system			Top phase			Bottom phase		
w_A	w_B	w_C	w_A	w_B	w_C	w_A	w_B	w_C
T/K = 278.15								
0.7297	0.1975	0.0728	0.6530	0.3333	0.0137	0.8335	0.0027	0.1638
0.6990	0.2233	0.0777	0.6242	0.3646	0.0112	0.8094	0.0050	0.1855
0.6677	0.2494	0.0829	0.5886	0.4027	0.0087	0.7901	0.0090	0.2009
0.6371	0.2749	0.0880	0.5451	0.4482	0.0067	0.7770	0.0117	0.2113
0.6165	0.2910	0.0925	0.5141	0.4807	0.0051	0.7671	0.0134	0.2195
T/K = 288.15								
0.7602	0.1716	0.0682	0.6433	0.3343	0.0224	0.8731	0.0064	0.1205
0.7292	0.1978	0.0730	0.6033	0.3804	0.0163	0.8634	0.0003	0.1363
0.6982	0.2236	0.0782	0.5542	0.4321	0.0137	0.8497	0.0003	0.1500
0.6673	0.2496	0.0831	0.5298	0.4591	0.0110	0.8343	0.0003	0.1654
0.6382	0.2742	0.0876	0.4969	0.4947	0.0084	0.8213	0.0010	0.1777
T/K = 298.15								
0.7605	0.1712	0.0683	0.6264	0.3414	0.0321	0.8849	0.0029	0.1122
0.7294	0.1975	0.0731	0.5605	0.4195	0.0200	0.8731	0.0036	0.1234
0.6987	0.2228	0.0785	0.5146	0.4706	0.0147	0.8585	0.0079	0.1336
0.6677	0.2494	0.0829	0.4819	0.5054	0.0127	0.8452	0.0031	0.1517
0.6552	0.2581	0.0867	0.4690	0.5198	0.0112	0.8348	0.0078	0.1574

Polymer (B): **poly(styrene-*co*-acrylonitrile)** **2001POS**
Characterization: M_n/g.mol^{-1} = 70000, M_w/g.mol^{-1} = 320000,
35.0 mol% acrylonitrile, ρ = 1.0347 g/cm3 at 303.15 K,
BASF Ludwigshafen, Germany
Solvent (A): ***N,N*-dimethylacetamide** C_4H_9NO **127-19-5**
Polymer (C): **polyacrylonitrile**
Characterization: M_w/g.mol^{-1} = 240000, ρ = 1.175 g/cm3 at 303.15 K,
Acrids Kehlheim GmbH, Germany

Type of data: cloud points

T/K = 298.15

w_A	0.9587	0.9629	0.9686	0.9633	0.9555	0.9674	0.95096
w_B	0.0288	0.0149	0.0094	0.0220	0.0356	0.0076	0.0441
w_C	0.0125	0.0222	0.0220	0.0147	0.0089	0.0250	0.00494

Polymer (B): **poly(styrene-*co*-acrylonitrile)** **2001POS**
Characterization: M_n/g.mol^{-1} = 70000, M_w/g.mol^{-1} = 320000,
35.0 mol% acrylonitrile, ρ = 1.0347 g/cm3 at 303.15 K,
BASF Ludwigshafen, Germany
Solvent (A): ***N,N*-dimethylacetamide** **C_4H_9NO** **127-19-5**
Polymer (C): **polystyrene**
Characterization: M_n/g.mol^{-1} = 29000, M_w/g.mol^{-1} = 100000,
ρ = 1.047 g/cm3 at 303.15 K, Deguss-Hüls, Germany

Type of data: cloud points

T/K = 298.15

w_A	0.9168	0.9155	0.9155	0.91802	0.91651	0.9145	0.9130	0.9106	0.90097
w_B	0.0414	0.0508	0.0590	0.07377	0.06678	0.0370	0.0271	0.01944	0.01493
w_C	0.0418	0.0337	0.0255	0.00821	0.01671	0.0490	0.0599	0.06996	0.0841

w_A	0.91269
w_B	0.08307
w_C	0.00424

Polymer (B): **poly(styrene-*co*-acrylonitrile)** **2002LOS**
Characterization: M_n/g.mol^{-1} = 88600, M_w/g.mol^{-1} = 195000,
48.0 mol% acrylonitrile, BASF Ludwigshafen, Germany
Solvent (A): ***N,N*-dimethylacetamide** **C_4H_9NO** **127-19-5**
Polymer (C): **polystyrene**
Characterization: M_n/g.mol^{-1} = 189000, M_w/g.mol^{-1} = 195000,
Polymer Standards, Mainz, Germany

Type of data: cloud points

T/K = 293.15

w_A	0.8011	0.9503	0.9541	0.9510	0.8534	0.9323	0.9467
w_B	0.0014	0.0037	0.0196	0.0306	0.1427	0.0596	0.0381
w_C	0.1974	0.0460	0.0263	0.0184	0.0040	0.0081	0.0152

Type of data: critical point (determined by phase volume ratio)

$w_{A,crit}$ = 0.953, $w_{B,crit}$ = 0.024, $w_{C,crit}$ = 0.023

Polymer (B): **poly(styrene-*co*-maleic anhydride)** **2009GAR**
Characterization: M_n/g.mol^{-1} = 124000, M_w/g.mol^{-1} = 224000,
7 wt% maleic anhydride, Sigma-Aldrich Inc., St. Louis, MO
Solvent (A): **bisphenol-A diglycidyl ether** **$C_{21}H_{24}O_4$** **1675-54-3**
commercial product Der336, Dow Chemical Company
Polymer (C): **polystyrene**
Characterization: M_n/g.mol^{-1} = 182000, M_w/g.mol^{-1} = 187500

continued

continued

Type of data: cloud points (UCST-behavior)

$w_A/(w_B + w_C)$	w_C/w_B	φ_B	φ_C	T/K
50/50	98/02	0.0104	0.505	418.2
60/40	98/02	0.0084	0.407	418.3
70/30	98/02	0.0063	0.307	406.5
75/25	98/02	0.0053	0.257	401.2
80/20	98/02	0.0042	0.206	394.2
85/15	98/02	0.0032	0.155	377.2
90/10	98/02	0.0021	0.104	374.2
50/50	96/04	0.0208	0.495	414.7
60/40	96/04	0.0167	0.398	415.2
70/30	96/04	0.0126	0.300	400.8
75/25	96/04	0.0106	0.251	399.3
80/20	96/04	0.0084	0.202	388.3
85/15	96/04	0.0064	0.152	374.2
90/10	96/04	0.0043	0.101	369.5
50/50	94/06	0.0312	0.484	410.9
60/40	94/06	0.0251	0.390	411.7
70/30	94/06	0.0190	0.294	398.7
75/25	94/06	0.0159	0.246	397.7
80/20	94/06	0.0127	0.197	383.3
85/15	94/06	0.0096	0.149	372.7
90/10	94/06	0.0064	0.099	367.5

Comments: Calculated tie-lines are given in 2009GAR.

Polymer (B): **poly(styrene-*co*-maleic anhydride)** **2009GAR**
Characterization: M_n/g.mol^{-1} = 110000, M_w/g.mol^{-1} = 200000, 14 wt% maleic anhydride, Sigma-Aldrich Inc., St. Louis, MO
Solvent (A): **bisphenol-A diglycidyl ether** **$C_{21}H_{24}O_4$** **1675-54-3**
commercial product Der336, Dow Chemical Company
Polymer (C): **polystyrene**
Characterization: M_n/g.mol^{-1} = 182000, M_w/g.mol^{-1} = 187500

Type of data: cloud points (UCST-behavior)

$w_A/(w_B + w_C)$	w_C/w_B	φ_B	φ_C	T/K
50/50	99/01	0.0051	0.510	413.3
60/40	99/01	0.0042	0.411	417.2
70/30	99/01	0.0031	0.310	413.3

continued

continued

75/25	99/01	0.0026	0.259	402.9
80/20	99/01	0.0021	0.208	392.2
85/15	99/01	0.0015	0.159	377.0
90/10	99/01	0.0010	0.105	371.3
50/50	98/02	0.0101	0.505	411.5
60/40	98/02	0.0083	0.398	413.3
70/30	98/02	0.0062	0.307	405.3
75/25	98/02	0.0051	0.257	393.6
80/20	98/02	0.0041	0.206	386.2
85/15	98/02	0.0031	0.155	373.4
90/10	98/02	0.0020	0.102	363.2
50/50	97/03	0.0152	0.500	408.0
60/40	97/03	0.0125	0.402	409.2
70/30	97/03	0.0093	0.304	401.4
75/25	97/03	0.0077	0.254	388.7
80/20	97/03	0.0062	0.204	381.2
85/15	97/03	0.0047	0.153	365.5
90/10	97/03	0.0031	0.102	359.0

Comments: Calculated tie-lines are given in 2009GAR.

Polymer (B): **poly(vinyl acetate-*co*-vinyl alcohol)** **2004PER**
Characterization: M_n/g.mol^{-1} = 10000, 12.0 mol% vinyl acetate, PVA 10000, 88% hydrolyzed, Scientific Polymer Products, Inc., Ontario, NY
Solvent (A): **water** **H_2O** **7732-18-5**
Polymer (C): **poly(ethylene glycol)**
Characterization: M_n/g.mol^{-1} = 8000

Type of data: coexistence data (tie lines)

T/K = 298.15

Total system			Top phase			Bottom phase		
w_A	w_B	w_C	w_A	w_B	w_C	w_A	w_B	w_C
0.8640	0.0930	0.0430	0.8780	0.0600	0.0620	0.8540	0.1170	0.0290
0.8530	0.0990	0.0480	0.8810	0.0380	0.0810	0.8300	0.1500	0.0200
0.8420	0.1050	0.0530	0.8770	0.0300	0.0930	0.8100	0.1750	0.0150
0.8210	0.1190	0.0600	0.8650	0.0240	0.1110	0.7790	0.2080	0.0130

Plait-point composition: $w_A = 0.8790 + w_B = 0.0410 + w_C = 0.0800$

Polymer (B):	**poly(vinyl acetate-*co*-vinyl alcohol)**		**1991EAG**
Characterization:	synthesized in the laboratory		
Solvent (A):	**water**	H_2O	**7732-18-5**
Salt (C):	**cesium chloride**	**CsCl**	**7647-17-8**

Type of data: cloud points (LCST-behavior)

w_B 0.02 c_C/(mol/l) 0.15

T/K	329.85	for a copolymer of 18.0 mol% vinyl acetate
T/K	306.45	for a copolymer of 23.7 mol% vinyl acetate
T/K	302.15	for a copolymer of 24.4 mol% vinyl acetate
T/K	311.95	for a copolymer of 30.0 mol% vinyl acetate

Polymer (B):	**poly(vinyl acetate-*co*-vinyl alcohol)**		**1991EAG**
Characterization:	synthesized in the laboratory		
Solvent (A):	**water**	H_2O	**7732-18-5**
Salt (C):	**lithium chloride**	**LiCl**	**7447-41-8**

Type of data: cloud points (LCST-behavior)

w_B 0.02 c_C/(mol/l) 0.15

T/K	336.65	for a copolymer of 18.0 mol% vinyl acetate
T/K	305.85	for a copolymer of 23.7 mol% vinyl acetate
T/K	303.15	for a copolymer of 24.4 mol% vinyl acetate
T/K	312.55	for a copolymer of 30.0 mol% vinyl acetate

Polymer (B):	**poly(vinyl acetate-*co*-vinyl alcohol)**		**1991EAG**
Characterization:	synthesized in the laboratory		
Solvent (A):	**water**	H_2O	**7732-18-5**
Salt (C):	**potassium chloride**	**KCl**	**7447-40-7**

Type of data: cloud points (LCST-behavior)

w_B 0.02 c_C/(mol/l) 0.15

T/K	305.15	for a copolymer of 23.7 mol% vinyl acetate
T/K	301.95	for a copolymer of 24.4 mol% vinyl acetate
T/K	310.35	for a copolymer of 30.0 mol% vinyl acetate

Polymer (B):	**poly(vinyl acetate-*co*-vinyl alcohol)**		**1991EAG**
Characterization:	synthesized in the laboratory		
Solvent (A):	**water**	H_2O	**7732-18-5**
Salt (C):	**sodium bromide**	**NaBr**	**7647-15-6**

Type of data: cloud points (LCST-behavior)

w_B 0.02 c_C/(mol/l) 0.15

continued

continued

T/K	339.55	for a copolymer of 18.0 mol% vinyl acetate
T/K	311.35	for a copolymer of 23.7 mol% vinyl acetate
T/K	306.45	for a copolymer of 24.4 mol% vinyl acetate
T/K	315.15	for a copolymer of 30.0 mol% vinyl acetate

Polymer (B): **poly(vinyl acetate-*co*-vinyl alcohol)** **1991EAG**
Characterization: synthesized in the laboratory
Solvent (A): **water** **H_2O** **7732-18-5**
Salt (C): **sodium chloride** **NaCl** **7647-14-5**

Type of data: cloud points (LCST-behavior)

w_B 0.02 c_C/(mol/l) 0.15

T/K	330.35	for a copolymer of 18.0 mol% vinyl acetate
T/K	305.65	for a copolymer of 23.7 mol% vinyl acetate
T/K	300.65	for a copolymer of 24.4 mol% vinyl acetate
T/K	303.85	for a copolymer of 30.0 mol% vinyl acetate

Polymer (B): **poly(vinyl acetate-*co*-vinyl alcohol)** **1991EAG**
Characterization: synthesized in the laboratory
Solvent (A): **water** **H_2O** **7732-18-5**
Salt (C): **sodium fluoride** **NaF** **7681-49-4**

Type of data: cloud points (LCST-behavior)

w_B 0.02 c_C/(mol/l) 0.15

T/K	353.45	for a copolymer of 13.4 mol% vinyl acetate
T/K	319.85	for a copolymer of 18.0 mol% vinyl acetate
T/K	298.85	for a copolymer of 23.7 mol% vinyl acetate
T/K	295.85	for a copolymer of 24.4 mol% vinyl acetate
T/K	303.95	for a copolymer of 30.0 mol% vinyl acetate

Polymer (B): **poly(vinyl acetate-*co*-vinyl alcohol)** **1991EAG**
Characterization: synthesized in the laboratory
Solvent (A): **water** **H_2O** **7732-18-5**
Salt (C): **sodium iodide** **NaI** **7681-82-5**

Type of data: cloud points (LCST-behavior)

w_B 0.02 c_C/(mol/l) 0.15

T/K	320.15	for a copolymer of 23.7 mol% vinyl acetate
T/K	314.05	for a copolymer of 24.4 mol% vinyl acetate
T/K	323.15	for a copolymer of 30.0 mol% vinyl acetate

Polymer (B): **poly(vinyl acetate-*co*-vinyl alcohol)** **1991EAG**
Characterization: synthesized in the laboratory
Solvent (A): **water** H_2O **7732-18-5**
Salt (C): **sodium sulfate** Na_2SO_4 **7757-82-6**

Type of data: cloud points (LCST-behavior)

w_B 0.02 c_C/(mol/l) 0.15

T/K	310.05	for a copolymer of 13.4 mol% vinyl acetate
T/K	322.65	for a copolymer of 16.0 mol% vinyl acetate
T/K	298.55	for a copolymer of 18.0 mol% vinyl acetate
T/K	283.15	for a copolymer of 23.7 mol% vinyl acetate
T/K	282.05	for a copolymer of 24.4 mol% vinyl acetate
T/K	291.25	for a copolymer of 30.0 mol% vinyl acetate
T/K	334.95	for a copolymer of 65.0 mol% vinyl acetate

Polymer (B): **poly(vinyl acetate-*co*-vinyl alcohol)** **1991EAG**
Characterization: synthesized in the laboratory
Solvent (A): **water** H_2O **7732-18-5**
Salt (C): **sodium thiocyanate** **NaCNS** **540-72-7**

Type of data: cloud points (LCST-behavior)

w_B 0.02 c_C/(mol/l) 0.15

T/K	332.15	for a copolymer of 23.7 mol% vinyl acetate
T/K	325.85	for a copolymer of 24.4 mol% vinyl acetate
T/K	332.15	for a copolymer of 30.0 mol% vinyl acetate

3.4. Table of ternary or quaternary systems where data were published only in graphical form as phase diagrams or related figures

Polymer (B)	Second and third component	Ref.
Poly(acrylamide-*co*-*N*-benzyl-acrylamide)	poly(ethylene glycol) and water	2003ABU
Poly(acrylamide-*co*-*N,N*-dihexyl-acrylamide)	polyacrylamide and water	2000JIM
Poly(acrylamide-*co*-*N*-isopropyl-acrylamide)	1,4-dioxane and water	2006DAL
Poly(acrylamide-*co*-4-methoxy-styrene)	poly(ethylene glycol) and water	2003ABU
Poly(acrylamide-*co*-*N*-phenyl-acrylamide)	poly(ethylene glycol) and water	2003ABU
Poly(acrylamide-*co*-sodium acrylate)	polystyrenesulfonate and water	2003HEL
Poly(acrylic acid-*co*-nonyl acrylate)	acrylic acid and *N,N*-dimethylformamide	2007MIK
Poly(acrylonitrile-*co*-butadiene)	ethyl acetate and poly(butadiene-*co*-α-methylstyrene)	2002VSH
	ethyl acetate and poly(butadiene-*co*-α-methylstyrene)	2004VSH

Polymer (B)	**Second and third component**	**Ref.**
Poly(acrylonitrile-*co*-itaconic acid)		
	N,N-dimethylformamide and water	2008TAN
	dimethylsulfoxide and water	2009TAN
Poly(acrylonitrile-*co*-methyl acrylate-*co*-itaconic acid)		
	dimethylsulfoxide and water	2009DON
Poly(butadiene-*co*-α-methyl-styrene)		
	ethyl acetate and poly(acrylonitrile-*co*-butadiene)	2002VSH
	ethyl acetate and poly(acrylonitrile-*co*-butadiene)	2004VSH
Poly(*N*-*tert*-butylacrylamide-*co*-acrylamide)		
	methanol and water	2009MAH
Poly(butyl acrylate-*b*-acrylic acid)		
	tetrahydrofuran and water	2008CRI
Poly(diacetone acrylamide-*co*-acrylamide)		
	guanidine hydrochloride and water	1975TAY
	potassium hydroxide and water	1975TAY
	sodium benzoate and water	1975TAY
	sodium bromide and water	1975TAY
	sodium chloride and water	1975TAY
	tetrabutylammonium bromide and water	1975TAY
	tetraethylammonium bromide and water	1975TAY
	tetramethylammonium bromide and water	1975TAY
	tetrapropylammonium bromide and water	1975TAY
	urea and water	1975TAY
Poly(diallyaminoethanoate-*co*-dimethylsulfoxide)		
	poly(ethylene glycol) and water	2004WAZ
Poly(*N,N*-diethylacrylamide-*co*-acrylic acid)		
	sodium chloride and water	2008FAN

Polymer (B)	Second and third component	Ref.
Poly(*N,N*-dimethylacrylamide-*co*-*tert*-butylacrylamide)	poly(ethylene glycol)/monopotassium phosphate and water	2009FOU
Poly(*N,N*-dimethylacrylamide-*co*-2-hydroxyethyl methacrylate)	water and β-cyclodextrin	1998GOS
Poly[2-(*N,N*-dimethylamino)ethyl methacrylate-*co*-acrylic acid-*co*-butyl methacrylate]	poly(*N*-isopropylacrylamide-*co*-butyl acrylate-*co*-chlorophyllin sodium copper salt) and water	2009NIN
Poly[2-(*N,N*-dimethylamino)ethyl methacrylate-*b*-*tert*-butyl methacrylate-*b*-methyl methacrylate]	poly(ethylene glycol) and water	2008QIN
Poly(divinyl ether-*alt*-maleic anhydride)	poly(*N*-ethyl-4-vinylpyridinium) bromide and water	2004VOL
	poly(*N*-ethyl-4-vinylpyridinium) bromide and water	2005IZU
	poly(methacrylic acid) and water	2004VOL
	poly(methacrylic acid) and water	2005IZU
	sodium chloride and water	2004VOL
Poly[*N*-(2-ethoxyethyl)acrylamide-*co*-*N*-isopropylacrylamide]	methanol and water	2009MA3
Poly(ethylene-*co*-acrylic acid)-*g*-poly(ethylene glycol) methyl ether	dioctyl phthalate and poly(ethylene glycol) methyl ether	2006ZHO
Poly(ethylene-*co*-propylene-*co*-diene)	n-hexane and propene	2001VLI

Polymer (B)	Second and third component	Ref.
Poly(ethylene-*co*-vinyl acetate)		
	methyl methacrylate and poly(methyl methacrylate)	2003CH1
	styrene and polystyrene	2003CH2
Poly(ethylene-*co*-vinyl alcohol)		
	1,4-butanediol + poly(ethylene glycol)	2007YIN
	dimethylsulfoxide and 2-propanol	2001CHE
	dimethylsulfoxide and water	1999YOU
	dimethylsulfoxide and water	2001CHE
	dimethylsulfoxide and water	2002YOU
	1,3-propanediol and 1,2,3-propanetriol	2005SHA
Poly(ethylene oxide-*b*-dimethylsiloxane)		
	poly(ethylene oxide) and 1,2,3,4-tetrahydronaphthalene	2002MAD
	poly(ethylene oxide) and 1,2,3,4-tetrahydronaphthalene	2003MAD
Poly(ethylene oxide-*b*-isoprene)		
	oligo(ethylene oxide) monododecyl ether and water	2004KUN
Poly(ethylene oxide-*b*-propylene fumarate-*b*-ethylene oxide)		
	sodium chloride and water	2002BEH
Poly(ethylene oxide-*co*-propylene oxide)		
	ammonium acetate and water	1997JO2
	benzoyl dextran and water	1996LUM
	dextran and water	1994ALR
	dextran and water	1997PLA
	dextran and water	1998PLA
	dextran and water	1999BER
	dextran and water	2005OLS
	dextran/polystyrene and water	2005OLS
	dextran/silica and water	2005OLS
	dextran and water	2008MAD
	dipotassium phosphate and water	1997CUN
	ficoll and water	2008MAD
	glycine and water	1997JO2
	hydroxypropyl starch and water	1994MOD

Polymer (B)	Second and third component	Ref.
Poly(ethylene oxide-*co*-propylene oxide)		
	hydroxypropyl starch and water	1997CUN
	hydroxypropyl starch and water	2000PE2
	hydroxypropyl starch and water	2008MAD
	octa(ethylene glycol) dodecyl ether and water	1992ZHA
	phenol and water	1997JO2
	poly(ethylene glycol) and water	2008MAD
	poly(ethylene oxide-*co*-propylene oxide) and water	1999PER
	poly(*N*-isopropylacrylamide-*co*-1-vinylimidazole) and water	2000PE1
	sodium acetate and water	1997JO2
	sodium butyrate and water	1997JO2
	sodium chloride and water	2002HUA
	sodium perchlorate and water	1997JO1
	sodium propionate and water	1997JO2
	sodium sulfate and water	1997JO1
	tetra(ethylene glycol) dodecyl ether and water	1992ZHA
Poly(ethylene oxide-*b*-propylene oxide-*b*-ethylene oxide)		
	acetic acid and water	2008SHA
	ammonium carbamate and water	2007OLI
	ammonium carbamate/bovine serum albumin and water	2007OLI
	ammonium carbamate/γ-globulin and water	2007OLI
	ammonium carbamate/β-lactoglobulin and water	2007OLI
	ammonium carbamate/lysozyme and water	2007OLI
	1-butanol and water	2008SHA
	1-butanol and water	2007BHA
	1-butanol and water	2001KWO
	1-butyl-3-methylimidazolium hexafluorophosphate/1-butanol and water	2009WA1
	1-butyl-3-methylimidazolium hexafluorophosphate/ethanol and water	2009WA1
	1-butyl-3-methylimidazolium hexafluorophosphate/1-propanol and water	2009WA1
	calcium chloride and water	2008SHA
	1,2-dichloroethane and water	2006LAZ
	dimethylurea and water	2008SHA
	ethanol and water	2008SHA
	ethanol and water	2007BHA
	ethanol and water	2001KWO
	formic acid and water	2008SHA

Polymer (B)	Second and third component	Ref.
Poly(ethylene oxide-*b*-propylene oxide-*b*-ethylene oxide)		
	1-hexanol and water	2007BHA
	hydrochloric acid and water	2008SHA
	magnesium chloride and water	2008SHA
	methanol and water	2008SHA
	methanol and water	2007BHA
	nicotinamide and water	2008SHA
	1-pentanol and water	2007BHA
	2-phenylethanol and water	2000FRI
	poly(ethylene oxide-*b*-propylene oxide-*b*-ethylene oxide) and water	2009NAN
	1-propanol and water	2008SHA
	1-propanol and water	2007BHA
	sodium acetate and water	2000LAM
	sodium acetate and water	2008SHA
	sodium benzene sulfonate and water	2004VAR
	sodium bromide and water	2000LAM
	sodium chloride and water	2001DES
	sodium chloride and water	2002DES
	sodium chloride and water	2008SHA
	sodium dodecyl sulfate and water	2008SHA
	sodium formate and water	2008SHA
	sodium hydroxide and water	2008SHA
	sodium oxalate and water	2008SHA
	sodium phosphate and water	2008SHA
	sodium perchlorate and water	2000LAM
	sodium salicylate and water	2008SHA
	sodium sulfate and water	2000LAM
	sodium sulfate and water	2008SHA
	sodium sulfate and water	2008BER
	sodium thiocyanate and water	2002DES
	sodium thiocyanate and water	2008SHA
	sodium toluene sulfonate and water	2004VAR
	sodium xylene sulfonate and water	2004VAR
	surfactants/sodium bromide and water	2006BAK
	tributyl phosphate and water	2006CAU
	Tween surfactant and water	2004MAH
	urea and water	2008SHA
Poly(2-ethylhexylacrylate-*co*-acrylic acid-*co*-vinyl acetate)		
	poly(hexafluoroacetone-*co*-vinylidene fluoride) and tetrahydrofuran	1999KAN

Polymer (B)	Second and third component	Ref.
Poly(2-ethyl-2-oxazoline-*b*-2-nonyl-2-oxazoline)		
	ethanol and water	2009LAM
Poly(2-ethyl-2-oxazoline-*co*-2-nonyl-2-oxazoline)		
	ethanol and water	2009LAM
Poly(2-ethyl-2-oxazoline-*co*-2-phenyl-2-oxazoline)		
	ethanol and water	2008HO2
Poly(hexafluoroacetone-*co*-vinylidene fluoride)		
	poly(2-ethylhexylacrylate-*co*-acrylic acid-*co*-vinyl acetate) and tetrahydrofuran	1999KAN
Poly(3-hydroxybutanoic acid-*co*-3-hydroxypentanoic acid)		
	N,N-dimethylformamide and trichloromethane	1996MAS
	ethanol and trichloromethane	1996MAS
	tetrahydrofuran and trichloromethane	1996MAS
Poly(2-hydroxyethyl acrylate-*co*-butyl acrylate)		
	poly(acrylic acid) and water	2007MUN
Poly(hydroxystearic acid-*b*-ethylene oxide-*b*-hydroxystearic acid)		
	isopropyl myristate and water	2002PLA
	isopropyl myristate/1,2-hexanediol and water	2002PLA
	isopropyl myristate/1,2-octanediol and water	2002PLA
Poly(isoprene-*b*-ethylene oxide)		
	oligo(ethylene oxide) monododecyl ether and water	2004KUN
Poly(*N*-isopropylacrylamide-*co*-acrylic acid)		
	peptides and water	2001BUL

Polymer (B)	**Second and third component**	**Ref.**
Poly(*N*-isopropylacrylamide-*co*-acrylic acid-*co*-ethyl methacrylate)		
	potassium chloride and water	2005TIE
Poly(*N*-isopropylacrylamide-*co*-*N*-(3-aminopropyl)methacrylamide hydrochloride)		
	potassium chloride and water	2009MAI
Poly[*N*-isopropylacrylamide-*co*-*N*-(3-aminopropyl)methacrylamide hydrochloride-*b*-*N*-isopropyl-acrylamide]		
	potassium chloride and water	2009MAI
Poly(*N*-isopropylacrylamide-*co*-benzo-15-crown-5-acrylamide)		
	cesium nitrate and water	2008MIP
	lithium nitrate and water	2008MIP
	potassium nitrate and water	2008MIP
	sodium nitrate and water	2008MIP
Poly(*N*-isopropylacrylamide-*co*-butyl acrylate-*co*-chlorophyllin sodium copper salt)		
	dextran and water	2007KON
	dextran and water	2009SHA
	poly[2-(*N,N*-dimethylamino)ethyl methacrylate-*co*-acrylic acid-*co*-butyl methacrylate] and water	2009NIN
Poly(*N*-isopropylacrylamide-*co*-*N,N*-dimethylacrylamide		
	1,4-dioxane and water	2006PAG
Poly[*N*-isopropylacrylamide-*co*-(2-hydroxyisopropyl)acrylamide]		
	sodium chloride and water	2006MA1
Poly[*N*-isopropylacrylamide-*co*-(p-methacrylamido)acetophenone thiosemicarbazone]		
	monopotassium phosphate/disodium phosphate and water	2005LIC
	sodium chloride and water	2005LIC

Polymer (B)	**Second and third component**	**Ref.**
Poly{*N*-isopropylacrylamide-*b*-3-[*N*-(3-methacrylamidopropyl)-*N,N*-dimethylammonio]propane sulfate}		
	sodium chloride and water	2002VIR
Poly[*N*-isopropylacrylamide-*co*-oligo(ethylene glycol) monomethacrylate]		
	sodium dodecylbenzenesulfonate and water	2004ALA
	sodium dodecylbenzenesulfonate and water	2006ALA
Poly(*N*-isopropylacrylamide-*co*-sodium 2-acrylamido-2-methyl-1-propanesulfonate)		
	dodecyltrimethylammonium chloride and water	2003NOW
Poly(*N*-isopropylacrylamide-*co*-sodium 2-acrylamido-2-methyl-1-propanesulfonate-*co*-*N*-*tert*-butylacrylamide)		
	sodium chloride and water	2007SH1
Poly(*N*-isopropylacrylamide-*co*-sodium styrenesulfonate)		
	dodecyltrimethylammonium chloride and water	2004NOW
Poly(*N*-isopropylacrylamide-*co*-1-vinylimidazole)		
	(ethylene oxide-*co*-propylene oxide) and water	2000PE1
Poly(maleic acid-*co*-acrylic acid)		
	hydroxypropylcellulose and water	2005BUM
Poly(maleic acid-*co*-styrene)		
	hydroxypropylcellulose and water	2005BUM
Poly(maleic acid-*co*-vinyl acetate)		
	hydroxypropylcellulose and water	2005BUM

Polymer (B)	Second and third component	Ref.
Poly[2-(2-methoxyethoxy)ethyl methacrylate-*co*-oligo(ethylene glycol) methacrylate]	sodium chloride and water	2006LUT
Poly(methyl methacrylate-*co*-*N,N*-dimethylacrylamide)	m-xylylene diisocyanate and 4-mercaptomethyl-3,6-dithia-1,8-octanedithiol	2006SOU
Poly(methyl methacrylate-*g*-dimethylsiloxane)	dimethylsulfoxide and tetrachloroethene	2000KAW
Poly(2-methyl-2-oxazoline-*co*-2-phenyl-2-oxazoline)	ethanol and water	2008HO2
Poly[oligo(ethylene glycol) diglycidyl ether-*co*-piperazine-*co*-oligo(propylene glycol) diglycidyl ether)]	sodium chloride and water	2009REN
Poly(styrene-*co*-acrylonitrile)	*N,N*-dimethylacetamide and poly(acrylonitrile)	2001POS
	N,N-dimethylacetamide and poly(acrylonitrile)	2002KUL
	N,N-dimethylacetamide and polystyrene	2001POS
	N,N-dimethylacetamide and polystyrene	2002KUL
	N,N-dimethylacetamide and polystyrene	2003LOS
Poly(styrene-*b*-butyl methacrylate)	dioctyl phthalate and n-hexadecane	2006LIC
Poly[(styrene-*alt*-maleic anhydride)-*g*-poly(amidoamine) dendrons]	sodium chloride and water	2009GAO
Poly(styrene-*co*-methacrylic acid)	trichloromethane and poly(4-vinylpyridine)	2006TO1
	trichloromethane and poly(4-vinylpyridine)	2006TO2
	trichloromethane and poly(4-vinylpyridine)	2007FIG
	trichloromethane and poly(1-vinyl-2-pyrrolidone)	2006TO1

Polymer (B)	**Second and third component**	**Ref.**
Poly(styrene-*alt*-sodium maleate)		
	barium chloride and water	1995JAR
	calcium chloride and water	1995JAR
	magnesium chloride and water	1995JAR
Poly(styrene-*co*-vinylphenol)		
	trichloromethane and poly(4-vinylpyridine)	2006TO1
	trichloromethane and poly(1-vinyl-2-pyrrolidone)	2006TO1
Poly(sulfobetaine methacrylate-*co*-*N*-isopropylacrylamide)		
	methanol and water	2009CH2
Poly[tetrafluoroethylene-*co*-perfluoro(alkyl vinyl ether)]		
	water and xanthan	2003KOE
Poly(*N*-vinylacetamide-*co*-methyl acrylate)		
	1-butanol and water	2004MOR
	ethanol and water	2004MOR
	magnesium sulfate and water	2004MOR
	methanol and water	2004MOR
	potassium sulfate and water	2004MOR
	1-propanol and water	2004MOR
	2-propanol and water	2004MOR
	sodium chloride and water	2004MOR
	sodium nitrate and water	2004MOR
	sodium sulfate and water	2004MOR
	sodium sulfite and water	2004MOR
Poly(vinyl alcohol-*co*-sodium acrylate)		
	sodium chloride and water	2008PAN
Poly(*N*-vinylcaprolactam-*co*-1-vinylimidazole)		
	dextran and water	1997FRA
	potassium bromide and water	2007SH2
	sodium thiocyanate and water	2007SH2
	starch-*g*-polyacrylamide and water	2000PIE

Polymer (B)	**Second and third component**	**Ref.**
Poly(*N*-vinylcaprolactam-*co*-1-vinyl-2-methylimidazole)		
	potassium bromide and water	2007SH2
	sodium thiocyanate and water	2007SH2
Poly(*N*-vinylformamide-*co*-acrylic acid)		
	sodium sulfate and water	2008CHE
Poly(*N*-vinylformamide-*co*-vinyl acetate)		
	potassium chloride and water	2003SET
	potassium iodide and water	2003SET
	potassium sulfate and water	2003SET
Poly(vinyl isobutyl ether-*b*-vinyl methyl ether-*b*-vinyl isobutyl ether)		
	n-decane and water	2003VE2
Poly(vinyl methyl ether-*b*-vinyl isobutyl ether)		
	n-decane and water	2003VE2
Poly(4-vinylpyridine-*g*-ethylene oxide)		
	sodium chloride and water	2008REN
Starch-*g*-polyacrylamide		
	poly(ethylene glycol) and water	2000PIE
	poly(*N*-vinylcaprolactam-*co*-1-vinylimidazole) and water	2000PIE

3.5. Upper critical (UCST) and/or lower critical (LCST) solution temperatures

Polymer (B)	M_n g/mol	M_w g/mol	Solvent (A)	UCST/ K	LCST/ K	Ref.
Poly(acrylamide-*co*-hydroxypropyl acrylate)						
(50 mol% hydroxypropyl acrylate)			water		348.15	1975TAY
(60 mol% hydroxypropyl acrylate)			water		324.15	1975TAY
(70 mol% hydroxypropyl acrylate)			water		305.15	1975TAY
Poly(acrylonitrile-*co*-butadiene)						
(18% acrylonitrile)		840000	ethyl acetate		427	2002VSH
(26% acrylonitrile)		1000000	ethyl acetate		412	2002VSH
Poly(butadiene-*co*-α-methylstyrene)						
(10% α-methylstyrene)		100000	ethyl acetate	387	393	2002VSH
Poly(diacetone acrylamide-*co*-hydroxyethyl acrylate)						
(70 mol% hydroxyethyl acrylate)			water		284.15	1975TAY
(80 mol% hydroxyethyl acrylate)			water		298.15	1975TAY
(90 mol% hydroxyethyl acrylate)			water		326.15	1975TAY
Poly(*N,N*-diethylacrylamide-*co*-acrylic acid)						
(5.98 mol% acrylic acid)		319000	water		305.05	2001CAI
(13.22 mol% acrylic acid)		306000	water		304.15	2001CAI
(20.64 mol% acrylic acid)		308000	water		300.15	2001CAI
Poly(*N,N*-diethylacrylamide-*co*-*N*-ethylacrylamide)						
(40 mol% *N*-ethylacrylamide)			water		305.15	1999LIU
Poly(*N*-ethylacrylamide-*co*-*N*-isopropylacrylamide)						
(50 mol% *N*-isopropylacrylamide)			water		321.15	1975TAY
(60 mol% *N*-isopropylacrylamide)			water		325.15	1975TAY
(70 mol% *N*-isopropylacrylamide)			water		329.15	1975TAY
(80 mol% *N*-isopropylacrylamide)			water		335.15	1975TAY
(90 mol% *N*-isopropylacrylamide)			water		340.15	1975TAY

Polymer (B)	M_n g/mol	M_w g/mol	**Solvent (A)**	**UCST/ K**	**LCST/ K**	**Ref.**
Poly[(ethylene glycol) monomethacrylate-*co*-methyl methacrylate]						
(60 mol% methyl methacrylate)			water		328.95	2004ALI
(70 mol% methyl methacrylate)			water		322.95	2004ALI
(76 mol% methyl methacrylate)			water		315.85	2004ALI
Poly(ethylene oxide-*co*-propylene oxide)						
(20.0 mol% EO)	3400		water		303	2000PE2
(27.0 mol% EO)	3000		water		309	2000PE1
(30.0 mol% EO)	5400		water		313	2000PE2
(33.3 mol% EO)	3000-3500		water		340	1997LIM
(38.5 mol% EO)	5000		water		309	2000PE1
(50.0 mol% EO)	2340	2480	water		338.25	2000LIW
(50.0 mol% EO)	3000-3500		water		343	1997LIM
(50.0 mol% EO)	3640	4040	water		324.85	2000LIW
(50.0 mol% EO)	3900		water		323	2000PE2
(58.8 mol% EO)	3000		water		326.65	2000PE1
(66.7 mol% EO)	3000-3500		water		359	1997LIM
Poly(ethylene oxide-*b*-propylene oxide-*b*-ethylene oxide)						
(30 mol% EO)		4400	water		286.65	2000LAM
Poly(2-ethyl-2-oxazoline-*co*-2-propyl-2-oxazoline)						
(EOz40/POz10)	3500		water		370.45	2008HO1
(EOz35/POz15)	3700		water		355.15	2008HO1
(EOz30/POz20)	3700		water		345.35	2008HO1
(EOz25/POz25)	4000		water		332.95	2008HO1
(EOz20/POz30)	5400		water		324.45	2008HO1
(EOz15/POz35)	3900		water		318.95	2008HO1
(EOz10/POz40)	3800		water		313.15	2008HO1
(EOz05/POz45)	4000		water		307.35	2008HO1
(EOz90/POz10)	15200		water		354.75	2008HO1
(EOz80/POz20)	13600		water		348.65	2008HO1
(EOz70/POz30)	12600		water		337.95	2008HO1
(EOz60/POz40)	13000		water		329.05	2008HO1
(EOz50/POz50)	10700		water		324.25	2008HO1
(EOz40/POz60)	10200		water		317.35	2008HO1
(EOz30/POz70)	9700		water		313.15	2008HO1
(EOz20/POz80)	9600		water		307.95	2008HO1
(EOz10/POz90)	7800		water		304.35	2008HO1
(EOz135/POz15)	17000		water		344.55	2008HO1
(EOz120/POz30)	17700		water		336.25	2008HO1
(EOz105/POz45)	17200		water		326.85	2008HO1
(EOz90/POz60)	17900		water		321.35	2008HO1

Polymer (B)	M_n g/mol	M_w g/mol	**Solvent (A)**	**UCST/ K**	**LCST/ K**	**Ref.**
(EOz75/POz75)	17600		water		315.75	2008HO1
(EOz60/POz90)	18600		water		310.45	2008HO1
(EOz45/POz105)	17900		water		307.25	2008HO1
(EOz30/POz120)	18500		water		302.25	2008HO1
(EOz15/POz135)	18600		water		297.65	2008HO1
Poly(2-hydroxyethyl acrylate-*co*-hydroxypropyl acrylate)						
(20 mol% hydroxypropyl acrylate)			water		333.15	1975TAY
(30 mol% hydroxypropyl acrylate)			water		317.15	1975TAY
(40 mol% hydroxypropyl acrylate)			water		307.15	1975TAY
(50 mol% hydroxypropyl acrylate)			water		305.15	1975TAY
(60 mol% hydroxypropyl acrylate)			water		298.15	1975TAY
(70 mol% hydroxypropyl acrylate)			water		296.15	1975TAY
Poly[*N*-(2-hydroxypropyl)methacrylamide monolactate-*co*-*N*-(2-hydroxypropyl) methacrylamide dilactate]						
(25 mol% dilactate)	7500	17600	water		323.65	2004SOG
(49 mol% dilactate)	8100	16900	water		309.65	2004SOG
(74 mol% dilactate)	6800	14000	water		298.65	2004SOG
Poly(isobutyl vinyl ether-*co*-2-hydroxyethyl vinyl ether)						
(12 mol% isobutyl vinyl ether)			water		338.15	2004SUG
(20 mol% isobutyl vinyl ether)			water		314.15	2004SUG
(33 mol% isobutyl vinyl ether)			water		289.15	2004SUG
Poly(*N*-isopropylacrylamide-*co*-*N,N*-dimethylacrylamide)						
(10.7 mol% *N,N*-dimethylacrylamide)			water		310.05	2006SHE
(13.0 mol% *N,N*-dimethylacrylamide)			water		309.15	2003BAR
(14.7 mol% *N,N*-dimethylacrylamide)			water		311.25	2006SHE
(17.3 mol% *N,N*-dimethylacrylamide)			water		312.95	2006SHE
(21.2 mol% *N,N*-dimethylacrylamide)			water		314.25	2006SHE
(27.0 mol% *N,N*-dimethylacrylamide)			water		312.15	2003BAR
(30.0 mol% *N,N*-dimethylacrylamide)			water		315.15	2003BAR
(31.4 mol% *N,N*-dimethylacrylamide)			water		319.15	2006SHE
(50.0 mol% *N,N*-dimethylacrylamide)			water		323.15	2003BAR
(60.0 mol% *N,N*-dimethylacrylamide)			water		336.15	2003BAR
(66.0 mol% *N,N*-dimethylacrylamide)			water		345.15	2003BAR
Poly(*N*-isopropylacrylamide-*co*-2-hydroxyethyl methacrylate-*co*-acrylic acid)						
(92.0 mol% NIPAM, 5.0 mol% HEMA, 3.0 mol% acrylic acid)						
		160000	water (pH=7.4)		308.15	2005LE2

Polymer (B)	M_n g/mol	M_w g/mol	Solvent (A)	UCST/ K	LCST/ K	Ref.
(92.0 mol% NIPAM, 4.3 mol% HEMA, 3.7 mol% acrylic acid)						
		120000	water (pH=7.4)		309.75	2005LE2
(89.8 mol% NIPAM, 4.2 mol% HEMA, 6.0 mol% acrylic acid)						
		140000	water (pH=7.4)		316.25	2005LE2
Poly(*N*-isopropylacrylamide-*co*-2-hydroxyethyl methacrylate lactate-*co*-acrylic acid)						
(84.4 mol% NIPAM, 9.4 mol% HEMA lactate, 6.2 mol% acrylic acid)						
		230000	water (pH=7.4)		308.05	2005LE2
(81.7 mol% NIPAM, 12.3 mol% HEMA lactate, 6.0 mol% acrylic acid)						
		240000	water (pH=7.4)		301.55	2005LE2
(75.8 mol% NIPAM, 18.5 mol% HEMA lactate, 5.7 mol% acrylic acid)						
		160000	water (pH=7.4)		296.65	2005LE2
Poly[*N*-isopropylacrylamide-*co*-(2-hydroxyisopropyl)acrylamide]						
(10 mol% HIPAAM)			water		309.85	2006MA2
(30 mol% HIPAAM)			water		314.95	2006MA2
(50 mol% HIPAAM)			water		328.15	2006MA2
(80 mol% HIPAAM)			water		353.15	2006MA2
Poly(*N*-isopropylacrylamide-*co*-*N*-isopropylmethacrylamide)						
(10.56 mol% NIPMAM)		177000	water		307.15	2001DJO
(20.30 mol% NIPMAM)		137000	water		308.45	2001DJO
(30.00 mol% NIPMAM)		92000	water		309.75	2001DJO
(39.99 mol% NIPMAM)		74000	water		311.05	2001DJO
(50.22 mol% NIPMAM)		60000	water		312.95	2001DJO
(59.89 mol% NIPMAM)		74000	water		314.65	2001DJO
(69.86 mol% NIPMAM)		56000	water		315.55	2001DJO
(79.81 mol% NIPMAM)		53000	water		317.35	2001DJO
(89.99 mol% NIPMAM)		47000	water		318.75	2001DJO
Poly[*N*-isopropylacrylamide-*co*-oligo(ethylene glycol) monomethacrylate]						
			deuterium oxide		309.15	2003KOH
			water		312.65	2004ALA
Poly(*N*-isopropylacrylamide-*b*-propylene glycol-*b*-*N*-isopropylacrylamide)						
	6600	12700	water		305.15	2005CHE
Poly(*N*-isopropylacrylamide-*co*-1-vinylimidazole)						
(4.76 mol% 1-vinylimidazole)			water		309.65	2000PE1
(33.3 mol% 1-vinylimidazole)			water		318.15	2000PE1
(50.0 mol% 1-vinylimidazole)			water		329.15	2000PE1

Polymer (B)	M_n g/mol	M_w g/mol	Solvent (A)	UCST/ K	LCST/ K	Ref.
Poly(2-isopropyl-2-oxazoline-*co*-2-butyl-2-oxazoline)						
(5.8 mol% 2-butyl)	2750	3470	water		311.15	2008HUB
(8.0 mol% 2-butyl)	3110	3920	water		308.15	2008HUB
(12 mol% 2-butyl)	3220	3960	water		306.15	2008HUB
(20 mol% 2-butyl)	3120	4060	water		294.15	2008HUB
Poly(2-isopropyl-2-oxazoline-*co*-2-nonyl-2-oxazoline)						
(5.9 mol% 2-nonyl)	3510	4140	water		288.15	2008HUB
(7.7 mol% 2-nonyl)	3140	3670	water		284.15	2008HUB
(12 mol% 2-nonyl)	3350	4020	water		282.15	2008HUB
Poly(2-isopropyl-2-oxazoline-*co*-2-propyl-2-oxazoline)						
(11.5 mol% 2-propyl)	2660	3000	water		319.15	2008HUB
(20.8 mol% 2-propyl)	2860	3670	water		310.15	2008HUB
Poly(styrene-*co*-acrylonitrile)						
(25.0 wt% acrylonitrile)		147000	toluene	313.15		2000SCH
Poly(*N*-vinylacetamide-*co*-vinyl acetate)						
(58 mol% VAc)	30000	57000	water		340.15	2003YAM
(63 mol% VAc)	27000	48600	water		323.15	2003YAM
(69 mol% VAc)	26000	49400	water		307.15	2003YAM
(71 mol% VAc)	22000	39600	water		295.15	2003YAM
(78 mol% VAc)	26000	46800	water		282.15	2003YAM
Poly(*N*-vinylcaprolactam-*co*-methacrylic acid)						
(9 mol% methacrylic acid)			water		303.15	2003OKH
(12 mol% methacrylic acid)			water		310.65	2003OKH
(18 mol% methacrylic acid)			water		313.85	2003OKH
(37 mol% methacrylic acid)			water		304.15	2003OKH
Poly(*N*-vinylcaprolactam-*co*-1-vinylimidazole)						
(9.09 mol% vinylimidazole)			water		310.15	2000PE1
(16.66 mol% vinylimidazole)			water		308.15	2000PE1
(33.33 mol% vinylimidazole)			water		310.15	2000PE1
(50.00 mol% vinylimidazole)			water		312.65	2000PE1
Poly(*N*-vinylformamide-*co*-vinyl acetate)						
(60 mol% VAc)	24000	45600	water		310.15	2003YAM
(66 mol% VAc)	25000	47500	water		291.15	2003YAM
(73 mol% VAc)	23000	50600	water		277.15	2003YAM

3.6. References

1975TAY Taylor, L.D., Cerankowski, L.D., Preparation of films exhibiting a balanced temperature dependence to permeation by aqueous solutions. A study of lower consolute behavior, *J. Polym. Sci.: Polym. Chem. Ed.*, 13, 2551, 1975.

1991EAG Eagland, D. and Crowther, N.J., Influence of composition and segment distribution upon lower critical demixing of aqueous poly(vinyl alcohol-*co*-vinyl acetate) solutions, *Eur. Polym. J.*, 27, 299, 1991.

1991SAM Samii, A.A., Karlström, G., and Lindman, B., Phase behavior of nonionic block copolymer in a mixed-solvent system, *J. Phys. Chem.*, 95, 7887, 1991.

1992ZHA Zhang, K.-W., Karlström, G., and Lindman, B., Phase behaviour of systems of a non-ionic surfactant and a non-ionic polymer in aqueous solution, *Colloids Surfaces*, 67, 147, 1992.

1993OTA Otake, K., Karaki, R., Ebina, T., Yokoyama, C., and Takahashi, S., Pressure effects on the aggregation of poly(*N*-isopropylacrylamide) and poly(*N*-isopropylacrylamide-*co*-acrylic acid) in aqueous solutions, *Macromolecules*, 26, 2194, 1993.

1994ALR Alred, P.A., Kozlowski, A., Harris, J.M., and Tjerneld, F., Application of temperature-induced phase partitioning at ambient temperature for enzyme purification, *J. Chromatogr. A*, 659, 289, 1994.

1994MOD Modlin, R.F., Alred, P. A., and Tjerneld, F., Utilization of temperature-induced phase separation for the purification of ecdysone and 20-hydroxyecdysone from spinach, *J. Chromatogr. A*, 668, 229, 1994.

1995BER Berggren, K., Johansson, H.-O., and Tjerneld, F., Effects of salts and the surface hydrophobicity of proteins on partitioning in aqueous two-phase systems containing thermoseparating ethylene oxide-propylene oxide copolymers, *J. Chromatogr. A*, 718, 67, 1995.

1995CHE Chen, G. and Hoffman, A.S., A new temperature- and pH-responsive copolymer for possible use in protein conjugation, *Macromol. Chem. Phys.*, 196, 1251, 1995.

1995JAR Jarm, V., Sertic, S., and Segudovic, N., Stability of aqueous solutions of poly[(maleic acid)-*alt*-styrene] sodium salts in the presence of divalent cations, *J. Appl. Polym. Sci.*, 58, 1973, 1995.

1996LUM Lu, M., Albertson, P.-A., Johansson, G., and Tjerneld, F., Ucon-benzoyl dextran aqueous two-phase systems: Protein purification with phase component recycling, *J. Chromatogr. B*, 680, 65, 1996.

1996MAS Mas, A., Sledz, J., and Schue, F., Membranes de microfiltration en polyhydroxybutyrate et poly(hydroxybutyrate-*co*-hydroxyvalerate): Influence du pourcentage d'unites hydroxyvalerate, *Bull. Soc. Chim. Belg.*, 105, 223, 1996.

1997CUN Cunha, M.T., Cabral, J.M.S., and Aires-Barros, M.R., Quantification of phase composition in aqueous two-phase systems of Breox/phosphate and Breox/Reppal PES 100 by isocratic HPLC, *Biotechnol. Techn.*, 11, 351, 1997.

1997FRA Franco, T.T., Galaev, I.Yu., Hatti-Kaul, R., Holmberg, N., Bülow, L., and Mattiasson, B., Aqueous two-phase system formed by thermoreactive vinyl imidazole/vinyl caprolactam copolymer and dextran for partitioning of a protein with a polyhistidine tail, *Biotechnol. Techn.*, 11, 231, 1997.

1997JO1 Johansson, H.-O., Karlström, G., and Tjerneld, F., Temperature-induced phase partitioning of peptides in water solutions of ethylene oxide and propylene oxide random copolymers, *Biochim. Biophys. Acta*, 1335, 315, 1997.

1997JO2 Johansson, H.-O., Karlström, G., and Tjerneld, F., Effect of solute hydrophobicity on phase behaviour in solutions of thermoseparating polymers, *Colloid Polym. Sci.*, 275, 458, 1997.

1997LIM Li, M., Zhu, Z.-Q., and Mei, L.-H., Partitioning of amino acids by aqueous two-phase systems combined with temperature-induced phase formation, *Biotechnol. Progr.*, 13, 105, 1997.

1997MAT Matsuyama, K., Enjoji, T., Mishima, K., Oka, S., Uchiyama, H., Ide, M., and Nagatani, M., Partition coefficients of amylase, maltose, and starch in aqueous two-phase systems containing polyoxyethylene-polyoxypropylene block copolymer, *Solvent Extraction Res. Developm., Japan*, 4, 80, 1997.

1997PLA Planas, J., Lefebvre, D., Tjerneld, F., and Hahn-Haegerdal, B., Analysis of phase compo-sition in aqueous two-phase systems using a two-column chromatographic method: Application to lactic acid production by extractive fermentation, *Biotechnol. Bioeng.*, 54, 303, 1997.

1997YOO Yoo, M.K., Sung, Y.K., Cho, C.S., and Lee, Y.M., Effect of polymer complex formation on the cloud-point of poly(*N*-isopropylacrylamide) (PNIPAAm) in the poly(NIPAAm-*co*-acrylic acid): Polyelectrolyte complex between poly(acrylic acid) and poly(allylamine), *Polymer*, 38, 2759, 1997.

1998GOS Gosselet, N.M., Borie, C., Amiel, C., and Sebille, B., Aqueous two-phase systems from cyclodextrin polymers and hydrophobically modified acrylic polymers, *J. Dispersion Sci.Technol.*, 19, 805, 1998.

1998PLA Planas, J., Varelas, V., Tjerneld, F., and Hahn-Hägerdal, B., Amine-based aqueous polymers for the simultaneous titration and extraction of lactic acid in aqueous two-phase systems, *J. Chromatogr. B*, 711, 256, 1998.

1999BER Berggren, K., Veide, A., Nygren, P.-A., and Tjerneld, F., Genetic engineering of protein-peptide fusions for control of protein partitioning in thermoseparating aqueous two-phase systems, *Biotechnol. Bioeng.*, 62, 135, 1999.

1999BEY Beyer, C., Oellrich, L.R., and McHugh, M.A., Effect of copolymer composition and solvent polarity on the phase behavior of mixtures of poly(ethylene-*co*-vinyl acetate) with cyclopentane and cyclopentene (exp. data by C. Beyer), *Chem.-Ing. Techn.*, 71, 1306, 1999.

1999JEO Jeong, B., Lee, D.S., Shon, J.-I., Bae, Y.H., and Kim, S.W., Thermoreversible gelation of poly(ethylene oxide) biodegradable polyester block copolymers, *J. Polym. Sci.: Part B: Polym. Phys.*, 37, 751, 1999.

1999JOH Johansson, H.-O., Persson, J., and Tjerneld, F., Thermoseparating water/polymer system: A novel one-polymer aqueous two-phase system for protein purification, *Biotechnol. Bioeng.*, 66, 247, 1999.

1999JON Jones, M.S., Effect of pH on the lower critical solution temperature of random copolymers of *N*-isopropylacrylamide and acrylic acid, *Eur. Polym. J.*, 35, 795, 1999.

1999KAN Kano, Y., Sato, H., Okamoto, M., Kotaka, T., and Akiyama, S., Phase separation process during solution casting of acrylate-copolymer/fluoro-copolymer blends, *J. Adhesion Sci. Technol.*, 13, 1243, 1999.

1999KUN Kunugi, S., Yamazaki, Y., Takano, K., and Tanaka, N., Effects of ionic additives and ionic comonomers on the temperature and pressure responsive behavior of thermoresponsive polymers in aqueous solutions, *Langmuir*, 15, 4056, 1999.

1999LIU Liu, H.Y. and Zhu, X.X., Lower critical solution temperatures of *N*-substituted acrylamide copolymers in aqueous solutions, *Polymer*, 40, 6985, 1999.

1999PER Persson, J., Johansson, H.-O., and Tjerneld, F., Purification of protein and recycling of polymers in a new aqueous two-phase system using two thermoseparating polymers, *J. Chromatogr. A*, 864, 31, 1999.

1999SVE Svensson, M., Berggren, K., Veide, A., and Tjerneld, F., Aqueous two-phase systems containing self-associating block copolymers. Partitioning of hydrophilic and hydrophobic biomolecules, *J. Chromatogr. A*, 839, 71, 1999.

1999YOU Young, T.-H., Cheng, L.-P., You, W.-M., and Chen, L.-Y., Prediction of EVAL membrane morphologies using the phase diagram of water-DMSO-EVAL at different temperatures, *Polymer*, 40, 2189, 1999.

2000AZU Azuma, T., Tyagi, O.S., and Nose, T., Static and dynamic properties of block-copolymer solution in poor solvent I, *Polym. J.*, 32, 151, 2000.

2000BEY Beyer, C., Oellrich, L.R., and McHugh, M.A., Effect of copolymer composition and solvent polarity on the phase behavior of mixtures of poly(ethylene-*co*-vinyl acetate) with cyclopentane and cyclopentene, *Chem. Eng. Technol.*, 23, 592, 2000.

2000BIG Bignotti, F., Penco, M., Sartore, L., Peroni, I. Mendichi, R., and Casolaro, M., Synthesis, characterisation and solution behaviour of thermo- and pH-responsive polymers bearing L-leucine residues in the side chains, *Polymer*, 41, 8247, 2000.

2000BO1 Bokias, G., Staikos, G., and Iliopoulos, I., Solution properties and phase behaviour of copolymers of acrylic acid with *N*-isopropylacrylamide, *Polymer*, 41, 7399, 2000.

2000BO2 Bokias, G., Vasilevskaya, V.V., Iliopoulos, I., Hourdet, D., and Khokhlov, A.R., Influence of migrating ionic groups on the solubility pf polyelectrolytes: Phase behavior of ionic poly(*N*-isopropylacrylamide) copolymers in water, *Macromolecules*, 33, 9757, 2000.

2000FRI Friberg, S.E., Yin, Q., Barber, J.L., and Aikens, P.A., Vapor pressures of phenethyl alcohol in the system water-phenethyl alcohol and the triblock copolymer $EO_{4.5}PO_{59}EO_{4.5}$, *J. Dispersion Sci. Technol.*, 21, 65, 2000.

2000JIM Jimenez-Regalado, E., Selb, J., and Candau, F., Phase behavior and rheological properties of aqueous solutions containing mixtures of associating polymers, *Macromolecules*, 33, 8720, 2000.

2000KAW Kawai, T., Teramachi, S., Tanaka, S., and Maeda, S., Comparison of chemical composition distributions of poly(methyl methacrylate)-*graft*-polydimethylsiloxane by high-performance liquid chromatography and demixing solvent fractionation, *Int. J. Polym. Anal. Char.*, 5, 381, 2000.

2000KIM Kim, C., Lee, S.C., Kang, S.W., Kwon, I.C., and Jeong, S.Y., Phase-transition characteristics of amphiphilic poly(2-ethyl-2-oxazoline)/poly(ε-caprolactone) block copolymers in aqueous solutions *J. Polym. Sci., Part B: Polym. Phys.*, 38, 2400, 2000.

2000KON Konak, C., Oupicky, D., Chytry, V., Ulbrich, K., and Helmstedt, M., Thermally controlled association in aqueous solutions of diblock copolymers of poly[*N*-(2-hydroxypropyl)methacrylamide] and poly(*N*-isopropylacrylamide), *Macromolecules*, 33, 5318, 2000.

2000KUC Kuckling, D., Adler, H.-J.P., Arndt, K.F., Ling, L., and Habicher, W.D., Temperature and pH dependent solubility of novel poly(*N*-isopropylacrylamide) copolymers, *Macromol. Chem. Phys.*, 201, 273, 2000.

2000KUN Kunugi, S., Tada, T., Yamazaki, Y., Yamamoto, K., and Akashi, M., Thermodynamic studies on coil-globule transitions of poly(*N*-vinylisobutyramide-*co*-vinylamine) in aqueous solutions, *Langmuir*, 16, 2042, 2000.

2000LAM LaMesa, C., Phase equilibria in a water-block copolymer system, *J. Therm. Anal. Calorim.*, 61, 493, 2000.

2000LIW Li, W., Zhu, Z.-Q., Li, M., Measurement and calculation of liquid-liquid equilibria of binary aqueous polymer solutions, *Chem. Eng. J.*, 78, 179, 2000.

2000PE1 Persson, J., Johansson, H.-O., Galaev, I., Mattiasson, B., and Tjerneld, F., Aqueous polymer two-phase systems formed by new thermoseparating polymers, *Bioseparation*, 9, 105, 2000.

2000PE2 Persson, J., Kaul, A., and Tjerneld, F., Polymer recycling in aqueous two-phase extractions using thermoseparating ethylene oxide-propylene oxide copolymers, *J. Chromatogr. B*, 743, 115, 2000.

2000PIE Pietruszka, N., Galaev, I.Yu., Kumar, A., Brzozowski, Z.K., and Mattiasson, B., New polymers forming aqueous two-phase polymer systems, *Biotechnol. Progr.*, 16, 408, 2000.

2000PRI Principi, T., Goh, C.C.E., Liu, R.C.W., and Winnik, F.M., Solution properties of hydrophobically modified copolymers of *N*-isopropylacrylamide and *N*-glycineacryl-amide: A study by microcalorimetry and fluorescence spectroscopy, *Macromolecules*, 33, 2958, 2000.

2000SAH Sahakaro, K., Chaibundit, C., Kaligradaki, Z., Mai, S.-M., Heatley, F., Booth, C., Padget, J. C., and Shirley, I.M., Clouding of aqueous solutions of difunctional tapered statistical copolymers of ethylene oxide and 1,2-butylene oxide, *Eur. Polym. J.*, 36, 1835, 2000.

2000SCH Schneider, A. and Wolf, B.A., Specific features of the interfacial tension in the case of phase separated solutions of random copolymers, *Polymer*, 41, 4089, 2000.

2000SHI Shi, X., Li, J., Sun, C., and Wu, S., The aggregation and phase separation behavior of a hydrophobically modified poly(*N*-isopropylacrylamide), *Coll. Surfaces A*, 175, 41, 2000.

2000YAM Yamazaki, Y., Tada, T., and Kunugi, S., Effect of acrylic acid incorporation on the pressure-temperature behavior and the calorimetric properties of poly(*N*-isopropyl-acrylamide) in aqueous solutions, *Colloid Polym. Sci.*, 278, 80, 2000.

2001BUL Dulmus, V., Patir, S., Tuncel, S.A., Piskin, E., Stimuli-responsive properties of conjugates of *N*-isopropylacrylamide-*co*-acrylic acid oligomers with alanine, glycine and serine mono-, di- and tri-peptides, *J. Control. Release*, 76, 265, 2001.

2001CAI Cai, W.S., Gan, L.H., and Tam, K.C., Phase transition of aqueous solutions of poly(*N,N*-diethylacrylamide-*co*-acrylic acid) by differential scanning calorimetric and spectrophotometric methods, *Colloid Polym. Sci.*, 279, 793, 2001.

2001CHE Cheng, L.-P., Young, T.-H., Chuang, W.-Y., Chen, L.-Y., and Chen, L.-W., The formation mechanism of membranes prepared from the nonsolvent-solvent-crystalline polymer systems, *Polymer*, 42, 443, 2001.

2001DES Desai, P.R., Jain, N.J., Sharma, R.K., and Bahadur, P., Effect of additives on the micellization of PEO/PPO/PEO block copolymer F127 in aqueous solution, *Coll. Surfaces A*, 178, 57, 2001.

2001DJO Djokpe, E. and Vogt, W., *N*-Isopropylacrylamide and *N*-isopropylmethacrylamide: Cloud points of mixtures and copolymers, *Macromol. Chem. Phys.*, 202, 750, 2001.

2001GAN Gan, L.-H., Cai, W., and Tam, K.C., Studies of phase transition of aqueous solutions of poly(*N,N*-diethylacrylamide-*co*-acrylic acid) by differential scanning calorimetry and spectrophotometry, *Eur. Polym. J.*, 37, 1773, 2001.

2001GOM Gomes de Azevedo, R., Rebelo, L.P.N., Ramos, A.M., Szydlowski, J., DeSousa, H.C., and Klein, J., Phase behavior of (polyacrylamides + water) solutions: Concentration, pressure and isotope effects, *Fluid Phase Equil.*, 185, 189, 2001.

2001HAM Hamley, I.W., Mai, S.-M., Ryan, A.J., Fairclough, J.P.A., and Booth, C., Aqueous mesophases of block copolymers of ethylene oxide and 1,2-butylene oxide, *Phys. Chem. Chem. Phys.*, 3, 2972, 2001.

2001KWO Kwon, K.W., Park, M.J., Hwang, J., and Char, K., Effects of alcohol addition on gelation in aqueous solution of poly(ethylene oxide)-poly(propylene oxide)-poly(ethylene oxide) triblock copolymer, *Polym. J.*, 33, 404, 2001.

2001LIU Liu, F., Frere, Y., and Francois, J., Association properties of poly(ethylene oxide) modified by pendant aliphatic groups, *Polymer*, 42, 2969, 2001.

2001POS Poshamova, N., Schneider, A., Wünsch, M., Kuleznew, V., Wolf, B.A., Polymer-polymer interaction parameters for homopolymers and copolymers from light scattering and phase separation experiments in a common solvent (experimental data by A. Schneider), *J. Chem. Phys.*, 115, 9536, 2001.

2001VLI Vliet, R.E. van, Tiemersma, T.P., Krooshof, G.J., and Iedema, P.D., The use of liquid-liquid extraction in the EPDM solution polymerization process, *Ind. Eng. Chem. Res.*, 40, 4586, 2001.

2002ALM Al-Muallem, H.A., Wazeer, M.I.M., and Ali, Sk.A., Synthesis and solution properties of a new ionic polymer and its behavior in aqueous two-phase polymer systems, *Polymer*, 43, 1041, 2002.

2002BEH Behravesh, E., Shung, A.K., Jo, S., and Mikos, A.G., Synthesis and characterization of triblock copolymers of methoxy poly(ethylene glycol) and poly(propylene fumarate), *Biomacromolecules*, 3, 153, 2002.

2002CH1 Chaibundit, C., Ricardo, N.M.P.S., Crothers, M., and Booth, C., Micellization of diblock (oxyethylene/oxybutylene) copolymer $E_{11}B_8$ in aqueous solution. Micelle size and shape. Drug solubilization, *Langmuir*, 18, 4277, 2002.

2002CH2 Chang, Y., Bender, J.D., Phelps, M.V.B., and Allcock, H.R., Synthesis and self-association behavior of biodegradable amphiphilic poly[bis(ethyl glycinat-*N*-yl)phosphazene]-poly (ethylene oxide) block copolymers, *Biomacromolecules*, 3, 1364, 2002.

2002DES Desai, P.R., Jain, N.J., and Bahadur, P., Anomalous clouding behavior of an ethylene oxide-propylene oxide block copolymer inaqueous solution, *Coll. Surfaces A*, 197, 19, 2002.

2002DIN Dincer, S., Tuncel, A., and Piskin, E., A potential gene delivery vector: *N*-isopropylacrylamide-ethyleneimine block copolymers, *Macromol. Chem. Phys.*, 203, 1460, 2002.

2002HUA Huang, Y. and Forciniti, D., Ethylene oxide and propylene oxide random copolymer/sodium chloride aqueous two-phase systems: Wetting and adsorption on dodecylagarose and polystyrene, *Biotechnol. Bioeng.*, 77, 786, 2002.

2002KON Konak, C., Reschel, T., Oupicky, D., and Ulbrich, K., Thermally controlled association in aqueous solutions of poly(L-lysine) grafted with poly(*N*-isopropylacrylamide), *Langmuir*, 18, 8217, 2002.

2002KUL Kuleznev, V.N., Wolf, B.A., and Pozharova, N.A., On intermolecular interactions in solutions of polymer blends (Russ.), *Vysokomol. Soedin., Ser. B*, 44, 512, 2002.

2002LOD Lodge, T.P., Pudil, B., and Hanley, K.J., The full phase behavior for block copolymers in solvents of varying selectivity, *Macromolecules*, 35, 4707, 2002.

2002LOS Loske, S., Fraktionierung und Lösungseigenschaften von Polymeren mit komplexer Struktur: Celluloseacetat, PMMA (Sterne, verzweigt) und SAN, *Dissertation*, Johannes-Gutenberg Universität Mainz, 2002.

2002MAD Madbouly, S.A. and Wolf, B.A., Equilibrium phase behavior of poly(ethylene oxide) and of its mixtures with tetrahydronaphthalene or/and poly(ethylene oxide-*block*-dimethylsiloxane) (experimental data by S.A. Madbouly), *J. Chem. Phys.*, 117, 7357, 2002.

2002MAK Makhaeva, E.E., Tenhu, H., and Khokhlov, A.R., Behavior of poly(*N*-vinyl caprolactam-*co*-methacrylic acid) macromolecules in aqueous solution: Interplay between coulombic and hydrophobic interaction, *Macromolecules*, 35, 1870, 2002.

2002MIK Mikhaliov, Yu.M., Ganina, L.V., Kurmaz, S.V., Smirnov, V.S., and Roshchupkin, V.P., Diffusion mobility of reactants, phase equilibrium, and specific features of radical copolymerization kinetics in the nonyl acrylate/2-methyl-5-vinyltetrazole system, *J. Polym. Sci.: Part B: Polym. Phys.*, 40, 1383, 2002.

2002MON Monteiro-Filho, E.S., Coimbra, J.S.R., Minim, L.A., Silva, L.H.M., and Meirelles, A.J.A., Liquid-liquid equilibrium for ternary systems containing a sugar + a synthetic polymer + water, *J. Chem. Eng. Data*, 47, 1346, 2002.

2002PLA Plaza, M., Pons, R., Tadros, Th. F., and Solans, C., Phase behavior and formation of micro-emulsions in water/A-B-A block copolymer (polyhydroxystearic acid-*b*-ethylene oxide-*b*-hydroxystearic acid)/1,2-alkanediol/isopropyl myristate systems, *Langmuir*, 18, 1077, 2002.

2002REB Rebelo, L.P.N., Visak, Z.P., de Sousa, H.C., Szydlowski, J., Gomes de Azevedo, R., Ramos, A.M., Najdanovic-Visak, V., Nunes da Ponte, M., and Klein, J., Double critical phenomena in (water + polyacrylamides) solutions (experimental data by L.P.N. Rebelo), *Macromolecules*, 35, 1887, 2002.

2002SON Soni, S.S., Sastry, N.V., Patra, A.K., Joshi, J.V., and Goyal, P.S., Surface activity, SANS, and viscosity studies in aqueous solutions of oxyethylene and oxybutylene di- and triblock copolymers, *J. Phys. Chem. B*, 106, 13069, 2002.

2002VIR Virtanen, J., Arotcarena, M., Heise, B., Ishaya, S., Laschewsky, A., and Tenhu, H., Dissolution and aggregation of a poly(NIPA-*block*-sulfobetaine) copolymer in water and saline aqueous solutions, Langmuir, 18, 5360, 2002.

2002VSH Vshvikov, S.A., Rusinova, E.V., and Gur'ev, A.A., Phase transitions in solutions of nitrile and methylstyrene rubbers (Russ.), *Vysokomol. Soedin., Ser. B*, 44, 504, 2002.

2002WOL Wolf, B., Kuleznev, V.N., and Pozharnova, N.A., Critical phenomena in solutions of the polystyrene-polyacrylonitrile random copolymer and its blends with polystyrene (Russ.), *Vysokomol. Soedin., Ser. A*, 44, 1212, 2002.

2002YIN Yin, X. and Stoever, H.D.H., Thermosensitive and pH-sensitive polymers based on maleic anhydride copolymers, *Macromolecules*, 35, 100178, 2002.

2002YOU Young, T.-H. and Chuang, W.-Y., Thermodynamic analysis on the cononsolvency of poly(vinyl alcohol) in water-DMSO mixtures through the ternary interaction parameter (exp. data by T.-H. Young), *J. Membrane Sci.*, 210, 349, 2002.

2003ABU Abu-Sharkh, B.F., Hamad, E.Z., and Ali, S.A., Influence of hydrophobe content on phase coexistence curves of aqueous two-phase solutions of associative polyacrylamide copolymers and poly(ethylene glycol), *J. Appl. Polym. Sci.*, 89, 1351, 2003.

2003BAR Barker, I.C., Cowie, J.M.G., Huckeby, T.N., Shaw, D.A., Soutar, I., and Swanson, L., Studies of the 'smart' thermoresponsive behavior of copolymers of *N*-isopropylacrylamide and *N,N*-dimethylacrylamide in dilute aqueous solution, *Macromolecules*, 36, 7765, 2003.

2003CAM Campese, G.M., Rodriguez, E.M.G., Tambourgi, E.B., and Pessoa, Jr, A., Determination of cloud-point temperatures for different copolymers, *Brazil. J. Chem. Eng.*, 20, 335, 2003.

2003CH1 Cheng, S.-K. and Chen, C.-Y., Study on the phase behavior of ethylene-vinyl acetate copolymer and poly(methyl methacrylate) blends by in situ polymerization, *J. Appl. Polym. Sci.*, 90, 1001, 2003.

2003CH2 Cheng, S.-K., Wang, and C.-C., Chen, C.-Y., Study of the phase behavior of EVA/PS blends during in situ polymerization, *Polym. Eng. Sci.*, 43, 1221, 2003.

2003DAV David, G., Alupei, V., Simionescu, B.C., Dincer, S., and Piskin, E., Poly(*N*-isopropylacrylamide)/poly[(*N*-acetylimino)ethylene] thermosensitive block and graft copolymers, *Eur. Polym. J.*, 39, 1209, 2003.

2003HEL Hellebust, S., Nilsson, S., and Blokhus, A.M., Phase behavior of anionic polyelectrolyte mixtures in aqueous solution. Effects of molecular weights, polymer charge density, and ionic strength of solution, *Macromolecules*, 36, 5372, 2003.

2003JIA Jiang, S., An, L., Jiang, B., and Wolf, B.A., Liquid-liquid phase behavior of toluene/polyethylene oxide/poly(ethylene oxide-*b*-dimethylsiloxane) polymer-containing ternary mixtures (exp. data by S. Jiang), *Phys. Chem. Chem. Phys.*, 5, 2066, 2003.

2003KAL Kalaycioglu, E., Patir, S., and Pisukin, E., Poly(*N*-isopropylacrylamide-*co*-2-methacryloamidohistidine) copolymers and their interactions with human immunoglobulin-G, *Langmuir*, 19, 9538, 2003.

2003KOE Könderink, G.H., Aarts, D.G.A.L., Villeneuve, V.W.A.de, Philipse, A.P., Tuinier, R., and Lekkerkerker, H.N.W., Morphology and kinetics of phase separating transparent xanthan-colloid mixtures, *Biomacromolecules*, 4, 129, 2003.

2003KOH Koh, A.Y.C., Heenan, R.K., and Saunders, B.R., A study of temperature-induced aggregation of responsive comb copolymers in aqueous solution, *Phys. Chem. Chem. Phys.*, 5, 2417, 2003.

2003LI1 Liu, S. and Liu, M., Synthesis and characterization of temperature- and pH-sensitive poly(*N*,*N*-diethylacrylamide-*co*-methacrylic acid), *J. Appl. Polym. Sci.*, 90, 3563, 2003.

2003LI2 Liu, S.-Q., Yang, Y.-Y., Liu, X.-M., and Tong, Y.-W., Preparation and characterization of temperature-sensitive poly(*N*-isopropylacrylamide)-*b*-poly(DL-lactide) microspheres for protein delivery, *Biomacromolecules*, 4, 1784, 2003.

2003LOS Loske, S., Schneider, A., and Wolf, B.A., Basis for the preparative fractionation of a statistical copolymer (SAN) with respect to either chain length or chemical composition, *Macromolecules*, 36, 5008, 2003.

2003MAD Madbouly, S.A. and Wolf, B.A., Shear-induced crystallization and shear-induced dissolution of poly(ethylene oxide) in mixtures with tetrahydronaphthalene and oligo(dimethylsiloxane-*b*-ethylene oxide), *Macromol. Chem. Phys.*, 204, 417, 2003.

2003MAE Maeda, Y., Tsubota, M., and Ikeda, I., Fourier transform IR spectroscopic study on phase transitions of copolymers of *N*-isopropylacrylamide and alkyl acrylates in water, *Colloid Polym. Sci.*, 281, 79, 2003.

2003NOW Nowakowska, M., Szczubialka, K., and Grebosz, M., Interactions of temperature-responsive anionic polyelectrolytes with a cationic surfactant, *J. Colloid Interface Sci.*, 265, 214, 2003.

2003OKH Okhapkin, I.M., Nasimova, I.R., Makhaeva, E.E., and Khokhlov, A.R., Effect of complexation of monomer units on pH- and temperature-sensitive properties of poly(*N*-vinylcaprolactam-*co*-methacrylic acid), *Macromolecules*, 36, 8130, 2003.

2003SET Seto, Y., Kameyama, K., Tanaka, N., Kunugi, S., Yamamoto, K., and Akashi, M., High-pressure studies on the coacervation of copoly(*N*-vinylformamide-vinylacetate) and copoly(*N*-vinylacetylamide-vinylacetate), *Colloid Polym. Sci.*, 281, 690, 2003.

2003SH1 Shang, M., Matsuyama, H., Maki, T., Teramoto, M., and Lloyd, D.R., Effect of crystallization and liquid-liquid phase separation on phase-separation kinetics in poly(ethylene-*co*-vinyl alcohol)/glycerol solution, *J. Polym. Sci.: Part B: Polym. Phys.*, 41, 184, 2003.

2003SH2 Shang, M., Matsuyama, H., Maki, T., Teramoto, M., and Lloyd, D.R., Preparation and characterization of poly(ethylene-*co*-vinyl alcohol) membranes via thermally induced liquid-liquid phase separation, *J. Appl. Polym. Sci.*, 87, 853, 2003.

2003SH3 Shang, M., Matsuyama, H., Teramoto, M., Lloyd, D.R., and Kubota, N., Preparation and membrane performance of poly(ethylene-*co*-vinyl alcohol) hollow fiber membrane via thermally induced phase separation, *Polymer*, 44, 7441, 2003.

2003SHT Shtanko, N.I., Lequieu, W., Goethals, E.J., and DuPrez, F.E., pH- and thermo-responsive properties of poly(*N*-vinylcaprolactam-*co*-acrylic acid) copolymers, *Polym. Int.*. 52, 1605, 2003.

2003VE1 Verbrugghe, S., Bernaerts, K., and DuPrez, F.E., Thermo-responsive and emulsifying properties of poly(*N*-vinylcaprolactam) based graft copolymers, *Macromol. Chem. Phys.*, 204, 1217, 2003.

2003VE2 Verdonck, B., Goethals, E.J., and DuPrez, F.E., Block copolymers of methyl vinyl ether and isobutyl vinyl ether with thermo-adjustable amphiphilic properties, *Macromol. Chem. Phys.*, 204, 2090, 2003.

2003YAM Yamamoto, K., Serizawa, T., and Akashi, M., Synthesis and thermosensitive properties of poly[(*N*-vinylamide)-*co*-(vinyl acetate)]s and their hydrogels, *Macromol. Chem. Phys.*, 204, 1027, 2003.

2003YAN Yang, Z., Crothers, M., Attwood, D., Collett, J.H., Ricardo, N.M.P., Martini, L.G.A., and Booth, C., Association properties of ethylene oxide/styrene oxide diblock copolymer E17S8 in aqueous solution, *J. Colloid Interface Sci.*, 263, 312, 2003.

2003YIN Yin, X. and Stoever, H.D.H., Hydrogel microspheres by thermally induced coacervation of poly(*N,N*-dimethylacrylamide-*co*-glycidyl methacrylate) aqueous solutions, *Macromolecules*, 36, 9817, 2003.

2004ABB Abbas, B., Schwahn, D., and Willner, L., Phase behavior of the polybutadiene–polystyrene diblock copolymer with the addition of the nonselective solvent dichlorobenzene in temperature and pressure fields, *J. Polym. Sci.: Part B: Polym. Phys.*, 42, 3179, 2004.

2004ALA Alava, C. and Saunders, B.R., Effect of added surfactant on temperature-induced gelation of emulsions, *Langmuir*, 20, 3107, 2004.

2004ALI Ali, M.M. and Stoever, H.D.H., Well-defined amphiphilic thermosensitive copolymers based on poly(ethylene glycol monomethacrylate) and methyl methacrylate prepared by atom transfer radical polymerization, *Macromolecules*, 37, 5219, 2004.

2004DER D'Errico, G., Paduano, L., and Khan, A., Temperature and concentration effects on supramolecular aggregation and phase behavior for poly(propylene oxide)-*b*-poly(ethylene oxide)-*b*-poly(propylene oxide) copolymers of different composition in aqueous mixtures, *J. Coll. Interface Sci.*, 279, 379, 2004.

2004DUR Durme, K. van, Verbrugghe, S., DuPrez, F.E., and Mele, B. van, Influence of poly(ethylene oxide) grafts on kinetics of LCST behavior in aqueous poly(*N*-vinylcaprolactam) solutions and networks studied by modulated temperature DSC, Macromolecules, 37, 1054, 2004.

2004EEC Eeckman, F., Moës, A.J., and Amighi, K., Synthesis and characterization of thermosensitive copolymers for oral controlled drug delivery, *Eur. Polym. J.*, 40, 873, 2004.

2004HAR Haraguchi, L.H., Mohamed, R.S., Loh, W., and Pessoa Filho, P.A., Phase equilibrium and insulin partitioning in aqueous two-phase systems containing block copolymers and potassium phosphate, *Fluid Phase Equil.*, 215, 1, 2004.

2004HAS Hasan, E., Zhang, M., Müller, A.H.E., and Tsvetanov, Ch.B., Thermoassociative block copolymers of poly(*N*-isopropylacrylamide) and poly(propylene oxide), *J. Macromol. Sci.: Part A: Pure Appl. Chem.*, 41, 467, 2004.

2004KEL Kelarakis, A., Ming, X.-T., Yuan, X.-F., and Booth, C., Aqueous micellar solutions of mixed triblock and diblock copolymers studied using oscillatory shear, *Langmuir*, 20, 2036, 2004.

2004KIT Kitano, H., Hirabayashi, T., Gemmei-Ide, M., and Kyogoku, M., Effect of macrocycles on the temperature-responsiveness of poly[(methoxy diethylene glycol methacrylate)-*graft*-PEG], *Macromol. Chem. Phys.*, 205, 1651, 2004.

2004KUN Kunieda, H., Kaneko, M., Lopez-Quintela, M.A., and Tsukahara, M., Phase behavior of a mixture of poly(isoprene)-poly(oxyethylene) diblock copolymer and poly(oxyethylene) surfactant in water, *Langmuir*, 20, 2164, 2004.

2004MAE Maeda, Y., Yamamoto, H., and Ikeda, I., Effects of ionization on the phase behavior of poly(*N*-isopropylacrylamide-*co*-acrylic acid) and poly(*N, N*-diethylacrylamide-*co*-acrylic acid) in water, *Coll. Polym. Sci.*, 282, 1268, 2004.

2004MAH Mahajan, R.K., Chawla, J., and Bakshi, M.S., Depression in the cloud point of Tween in the presence of glycol additives and triblock polymers, *Colloid Polym. Sci.*, 282, 1165, 2004.

2004MAO Mao, H., Li, C., Zhang, Y. Furyk, S., Cremer, P.S., and Bergbreiter, D.E., High-throughput studies of the effects of polymer structure and solution components on the phase separation of thermoresponsive polymers, *Macromolecules*, 37, 1031, 2004.

2004MIK Mikhailov, Yu.M., Ganina, L.V., Roshchupkin, V.P., Shapaeva, N.V., and Suchkova, L.I., Mutual dissolution of poly(nonyl acrylate-*co*-acrylic acid) in low-molecular-mass solvents and oligomers (Russ.), *Vysokomol. Soedin., Ser. A*, 46, 1583, 2004.

2004MOR Mori, T., Fukuda, Y., Okamura, H., Minagawa, K., Masuda, S., and Tanaka, M., Thermosensitive copolymers having soluble and insoluble monomer units, poly(*N*-vinylacetamide-*co*-methyl acrylate)s: Effect of additives on their lower critical solution temperatures, *J. Polym. Sci.: Part A: Polym. Chem.*, 42, 2651, 2004.

2004NED Nedelcheva, A.N., Vladimirov, N.G., Novakov, C.P., and Berlinova, I.V., Associative block copolymers comprising poly(*N*-isopropylacrylamide) and poly(ethylene oxide) end-functionalized with a fluorophilic or hydrophilic group. Synthesis and aqueous solution properties, *J. Polym. Sci.: Part B: Polym. Phys.*, 42, 5736, 2004.

2004NIE Nie, H., Li, M., Bansil, R., Konak, C., Helmstedt, M., and Lal, J., Structure and dynamics of a pentablock copolymer of polystyrene-polybutadiene in a butadiene-selective solvent, *Polymer*, 45, 8791, 2004.

2004NON Nonaka, T., Hanada, Y., Watanabe, T., Ogata, T., and Kurihara, S., Formation of thermosensitive water-soluble copolymers with phosphinic acid groups and the thermosensitivity of the copolymers and copolymer/metal complexes, *J. Appl. Polym. Sci.*, 92, 116, 2004.

2004NOW Nowakowska, M., Szczubialka, K., and Grebosz, M., Modifying the thermosensitivity of copolymers of sodium styrene sulfonate and *N*-isopropylacrylamide with dodecyltrimethylammonium chloride, *Coll. Polym. Sci.*, 283, 291, 2004.

2004PA1 Park, M.J. and Char, K., Gelation of PEO-PLGA-PEO triblock copolymers induced by macroscopic phase separation, *Langmuir*, 20, 2456, 2004.

2004PER Pereira, M., Wu, Y.T., Madeira, P., Venancio, A., Macedo, E., and Teixeira, J., Liquid-liquid equilibrium phase diagrams of new aqueous two-phase systems: Ucon 50-HB5100 + ammonium sulfate + water, Ucon 50-HB5100 + poly(vinyl alcohol) + water, Ucon 50-HB5100 + hydroxypropyl starch + water, and poly(ethylene glycol) 8000 + poly(vinyl alcohol) + water, *J. Chem. Eng. Data* 49, 43,2004.

2004SAL Salgado-Rodriguez, R., Licea-Claveric, A., and Arndt, K.F., Random copolymers of *N*-isopropylacrylamide and methacrylic acid monomers with hydrophobic spacers. pH-tunable temperature sensitive materials, *Eur. Polym. J.*, 40, 1931, 2004.

2004SOG Soga, O., Nostrum, C.F. van, and Hennik, W.E., Poly[*N*-(2-hydroxypropyl)methacrylamide mono/dilactate]: A new class of biodegradable polymers with tuneable thermosensitivity, *Biomacromolecules*, 5, 818, 2004.

2004SUG Sugihara, S., Kanaoka, S., and Aoshima, S., Thermosensitive random copolymers of hydrophilic and hydrophobic monomers obtained by living cationic copolymerization, *Macromolecules*, 37, 1711, 2004.

2004TAD Tada, E. dos S., Loh, W., and Pessoa-Filho, P. de A., Phase equilibrium in aqueous two-phase systems containing ethylene oxide-propylene oxide block copolymers and dextran, *Fluid Phase Equil.*, 218, 221, 2004.

2004TAK Takeda, N., Nakamura, E., Yokoyama, M., and Okano, T., Temperature-responsive polymeric carriers incorporating hydrophobic monomers for effective transfection in small doses, *J. Control. Release*, 95, 343, 2004.

2004VAR Varade, D., Sharma, R., Aswal, V.K., Goyal, P.S., and Bahadur, P., Effect of hydrotopes on the solution behavior of PEO/PPO/PEO block copolymer L62 in aqueous solutions, *Eur. Polym. J.*, 40, 2457, 2004.

2004VOL Volkova, I.F., Gorshkova, M.Yu., Izumrudov, V.A., and Stotskaya, L.L., Interaction of a polycation with divinyl ether-maleic anhydride copolymer in aqueous solutions (Russ.), *Vysokomol. Soedin., Ser. A*, 46, 1388, 2004.

2004VSH Vshivkov, S.A. and Rusinova, E.V., Thermodynamics of solutions of blends of diene rubbers under deformation (Russ.), *Vysokomol. Soedin., Ser. B*, 46, 912, 2004.

2004WAZ Waziri, S.M., Abu-Sharkh, B.F., and Ali, S.A., Protein partitioning in aqueous two-phase systems composed of a pH-responsive copolymer and poly(ethylene glycol), *Biotechnol. Progr.*, 20, 526, 2004.

2004WEI Weiss-Malik, R.A., Solis, F.J., and Vernon, B.L.. Independent control of lower critical solution temperature and swelling behavior with pH for poly(*N*-isopropylacrylamide-*co*-maleic acid), *J. Appl. Polym. Sci.*, 94, 2110, 2004.

2005AUB Aubrecht, K.B. and Grubbs, R.B., Synthesis and characterization of thermoresponsive amphiphilic block copolymers incorporating a poly(ethylene oxide-*stat*-propylene oxide) block, *J. Polym. Sci.: Part A: Polym. Chem.*, 43, 5156, 2005.

2005BAE Bae, S.J., Suh, J.M., Sohn, Y.S., Bae, Y.H., Kim, S.W., and Jeong, B., Thermogelling of poly(caprolactone-*b*-ethylene glycol-*b*-caprolactone) aqueous solutions, *Macromolecules*, 38, 5260, 2005.

2005BIS Bisht, H.S., Wan, L., Mao, G., and Oupicky, D., pH-Controlled association of PEG-containing terpolymers of *N*-isopropylacrylamide and 1-vinylimidazole, *Polymer*, 46, 7945, 2005.

2005BOL Bolognese, B., Nerli, B., and Pico, G., Application of the aqueous two-phase systems of ethylene and propylene oxide copolymer-maltodextrin for protein purification, *J. Chromatogr. B*, 814, 347, 2005.

2005BUM Bumbu, G.-G., Vasile, C., Chitanu, G.C., and Staikos, G., Interpolymer complexes between hydroxypropylcellulose and copolymers of maleic acid: A comparative study, *Macromol. Chem. Phys.*, 206, 540, 2005.

2005CAO Cao, Z., Liu, W., Gao, P., Yao, K., Li, H., and Wang, G., Toward an understanding of thermoresponsive transition behavior of hydrophobically modified *N*-isopropylacrylamide copolymer solution, *Polymer*, 46, 5268, 2005.

2005CAR Carter, S., Hunt, B., and Rimmer, S., Highly branched poly(*N*-isopropylacrylamide)s with imidazole end groups prepared by radical polymerization in the presence of a styryl monomer containing a dithioester group, *Macromolecules*, 38, 4595, 2005.

2005CHA Chaibundit, C., Sumanatrakool, P., Chinchew, S., Kanatharana, P., Tattershall, C.E., Booth, C., and Yuan, X.-F., Association properties of diblock copolymer of ethylene oxide and 1,2-butylene oxide: $E_{17}B_{12}$ in aqueous solution, *J. Coll. Interface Sci.*, 283, 544, 2005.

2005CHE Chen, X., Ding, X., Zheng, Z., and Peng, Y., Thermosensitive polymeric vesicles self-assembled by PNIPAAm-*b*-PPG-*b*-PNIPAAm triblock copolymers, *Coll. Polym. Sci.*, 283, 452, 2005.

2005CIM Cimen, E.K., Rzaev, Z.M.O., and Piskin, E., Bioengineering functional copolymers: V. Synthesis, LCST,and thermal behavior of poly(*N*-isopropylacrylamide-*co*-p-vinyl-phenylboronic acid), *J. Appl. Polym. Sci.*, 95, 573, 2005.

2005GAO Gaoa, C., Möhwald, H., and Shen, J., Thermosensitive poly(allylamine)-*g*-poly(*N*-isopropylacrylamide): Synthesis, phase separation and particle formation, *Polymer*, 46, 4088, 2005.

2005HEI Heidemann, R.A. , Krenz, R.A., and Laursen, T., Spinodal curves and critical points in mixtures containing polydisperse polymers with many components, *Fluid Phase Equil.*, 228-229, 239, 2005.

2005IZU Izumrudov, V.A., Gorshkova, M.Yu., and Volkova, I.F., Controlled phase separations in solutions of soluble polyelectrolyte complex of DIVEMA (copolymer of divinyl ether and maleic anhydride), *Eur. Polym. J.*, 41, 1251, 2005.

2005JON Jones, J.A., Novo, N., Flagler, K., Pagnucco, C.D., Carew, S., Cheong, C., Kong, Z., Burke, N.A.D., and Stoever, H.D.H., Thermoresponsive copolymers of methacrylic acid and poly(ethylene glycol) methyl ether methacrylate, *J. Polym. Sci.: Part A: Polym. Chem.*, 43, 6095, 2005.

2005KJO Kjøniksen, A.-L., Laukkanen, A., Galant, C., Knudsen, K.D., Tenhu, H., and Nyström, B., Association in aqueous solutions of a thermoresponsive PVCL-*g*-$C_{11}EO_{42}$ copolymer, *Macromolecules*, 38, 948, 2005.

2005KOH Koh, A.Y.C. and Saunders, B.R., Small-angle neutron scattering study of temperature-induced emulsion gelation: The role of sticky microgel particles, *Langmuir*, 21, 6734, 2005.

2005LAU Laukkanen, A., Valtola, L., Winnik, F.M., and Tenhu, H., Thermosensitive graft copolymers of an amphiphilic macromonomer and *N*-vinylcaprolactam: Synthesis and solution properties in dilute aqueous solutions below and above the LCST, *Polymer*, 46, 7055, 2005.

2005LE1 Lee, S.-H., Phase behavior of binary and ternary mixtures of poly(ethylene-*co*-octene)–hydrocarbons (experimental data by S.-H. Lee), *J. Appl. Polym. Sci.*, 95, 161, 2005.

2005LE2 Lee, B.H. and Vernon, B., Copolymers of *N*-isopropylacrylamide, HEMA-lactate and acrylic acid with time-dependent lower critical solution temperature as a bioresorbable carrier, *Polym. Int.*, 54, 418, 2005.

2005LIC Li, C., Meng, L.-Z., Lu, X.-J., Wu, Z.-Q., Zhang, L.-F., and He, Y.-B., Thermo- and pH-sensitivities of thiosemicarbazone-incorporated, fluorescent and amphiphilic poly(*N*-isopropylacrylamide), *Macromol. Chem. Phys.*, 206, 1870, 2005.

2005LIU Liu, S.Q., Tong, Y.W., and Yang, Y.-Y., Incorporation and in vitro release of doxorubicin in thermally sensitive micelles made from poly(*N*-isopropylacrylamide-*co*-*N,N*-dimethylacrylamide)-*b*-poly(DL-lactide-*co*-glycolide) with varying compositions, *Biomaterials*, 26, 5064, 2005.

2005NED Nedelcheva, A.N., Novakov, C.P., Miloshev, S.M., and Berlinova, I.V., Electrostatic self-assembly of thermally responsive zwitterionic poly(*N*-isopropylacrylamide) and poly(ethylene oxide) modified with ionic groups, *Polymer*, 46, 2059, 2005.

2005OLS Olsson, M., Joabsson, F., and Piculell, L., Particle-induced phase separation in mixed polymer solutions, *Langmuir*, 21, 1560, 2005.

2005SHA Shang, M., Matsuyama, H., Teramoto, M., Okuno, J., Lloyd, D.R., and Kubota, N., Effect of diluent on poly(ethylene-*co*-vinyl alcohol) hollow-fiber membrane formation via thermally induced phase separation, *J. Appl. Polym. Sci.*, 95, 219, 2005.

2005SIL Silva, L.H.M. da, Silva, M. do C.H. da, Mesquita, A.F., Nascimento, K.S. do, Coimbra, J.S.R., and Minim, L.A., Equilibrium phase behavior of triblock copolymer + salt + water two-phase systems at different temperatures and pH, *J. Chem. Eng. Data*, 50, 1457, 2005.

2005SPE Spevacek, J., Phase separation in aqueous polymer solutions as studied by NMR methods, *Macromol. Symp.*, 222, 1, 2005.

2005STA Starovoytova, L., Spevacek, J., and Ilavsky, M., ^{1}H-NMR study of temperature-induced phase transitions in D_2O solutions of poly(*N*-isopropylmethacrylamide)/poly(*N*-isopropylacrylamide) mixtures and random copolymers, *Polymer*, 46, 677, 2005.

2005TAD Tada, E. dos S., Loh, W., and Pessoa-Filho, P. de A., Erratum to [*Fluid Phase Equilibr.* 218 (2004) 221–228], *Fluid Phase Equil.*, 231, 250, 2005.

2005TIE Tiera, M.J., Santos, G.R. dos, Oliveira Tiera, V.A. de, Vieira, N.A.B., Frolini, E., Silva, R.C. da, and Loh, W., Aqueous solution behavior of thermosensitive (*N*-isopropylacrylamide-acrylic acid-ethyl methacrylate) terpolymers, *Coll. Polym. Sci.*, 283, 662, 2005.

2005VER Verdonck, B., Gohy, J.-F., Khousakoun, E., Jerome, R., and DuPrez, F., Association behavior of thermo-responsive block copolymers based on poly(vinyl ethers), *Polymer*, 46, 9899, 2005.

2005VIE Vieira, N.A.B., Neto, J.R., and Tiera, M.J., Synthesis, characterization and solution properties of amphiphilic *N*-isopropylacrylamide–poly(ethylene glycol)–dodecyl methacrylate thermosensitive polymers, *Colloids Surfaces A*, 262, 251, 2005.

2005YI1 Yin, X. and Stoever, H.D.H., Probing the influence of polymer architecture on liquid-liquid phase transitions of aqueous poly(*N,N*-dimethylacrylamide) copolymer solutions, *Macromolecules*, 38, 2109, 2005.

2005YI2 Yin, X. and Stoever, H.D.H., Temperature-sensitive hydrogel microspheres formed by liquid-liquid phase transitions of aqueous solutions of poly(*N,N*-dimethylacrylamide-*co*-allyl methacrylate), *J. Polym. Sci.: Part A: Polym. Chem.*, 43, 1641, 2005.

2005ZH1 Zhang, D., Macias, C., and Ortiz, C., Synthesis and solubility of (mono-)end-functionalized poly(2-hydroxyethyl ethacrylate-*g*-ethylene glycol) graft copolymers with varying macromolecular architectures, *Macromolecules*, 38, 2530, 2005.

2005ZH2 Zhang, R., Liu, J., Han, B., Wang, B., Sun, D., and He, J., Effect of PEO–PPO–PEO structure on the compressed ethylene-induced reverse micelle formation and water solubilization, *Polymer*, 46, 3936, 2005.

2005ZH3 Zhang, W., Shi, L., Wu, K., and An, Y., Thermoresponsive micellization of poly(ethylene glycol)-*b*-poly(*N*-isopropylacrylamide) in water, *Macromolecules*, 38, 5743, 2005.

2005ZH4 Zhang, W., Shi, L., Ma, R., An, Y., Xu, Y., and Wu, K., Micellization of thermo- and pH-responsive triblock copolymer of poly(ethylene glycol)-*b*-poly(4-vinylpyridine)-*b*-poly(*N*-isopropylacrylamide), *Macromolecules*, 38, 8850, 2005.

2005ZHO Zhou, J., Yin, J., Lv, R., Du, Q., and Zhong, W., Preparation and properties of MPEG-grafted EAA membranes via thermally induced phase separation, *J. Membrane Sci.*, 267, 90, 2005.

2006ALA Alava, C. and Saunders, B.R., Polymer stabilisers for temperature-induced dispersion gelation: Versatility and control, *J. Colloid Interface Sci.*, 293, 93, 2006.

2006BAK Bakshi, M.S., Kaur, N., Mahajan, R.K., Singh, J., and Sing, N., Estimation of degree of counterion binding and related parameters of monomeric and dimeric cationic surfactants from cloud point measurements by using triblock polymer as probe, *Coll. Polym. Sci.*, 284, 879, 2006.

2006CAO Cao, Z., Liu, W., Ye, G., Zhao, X., Lin, X., Gao, P., and Yao, K., N-Isopropylacrylamide/2-hydroxyethyl methacrylate star diblock copolymers: Synthesis and thermoresponsive behavior, *Macromol. Chem. Phys.*, 207, 2329, 2006.

2006CAU Causse, J., Lagerge, S., Menorval, L.C. de, and Faure, S., Micellar solubilization of tributylphosphate in aqueous solutions of Pluronic block copolymers Part I. Effect of the copolymer structure and temperature on the phase behavior, *J. Colloid Interface Sci.*, 300, 713, 2006.

2006CHA Chalykh, A.E., Gerasimov, V.K., Rusanova, S.N., Stoyanov, O.V., Petukhova, O.G., Kulagina, G.S., and Pisarev, S.A., Phase structure of silanol-modified ethylene-vinyl acetate copolymers, *Polym. Sci., Ser. A*, 48, 1058, 2006.

2006CHO Choi, C., Chae, S.Y., and Nah, J.-W., Thermosensitive poly(*N*-isopropylacrylamide)-*b*-poly(ε-caprolactone) nanoparticles for efficient drug delivery system, *Polymer*, 47, 4571, 2006.

2006DAL Dalkas, G., Pagonis, K., and Bokias, G., Control of the lower critical solution temperature-type cononsolvency properties of poly(*N*-isopropylacrylamide) in water-dioxane mixtures through copolymerisation with acrylamide, *Polymer*, 47, 243, 2006.

2006DI1 Dincer, S., Rzaev, Z.M.O., and Piskin, E., Synthesis and characterization of stimuli-responsive poly(*N*-isopropylacrylamide-*co*-*N*-vinyl-2-pyrrolidone), *J. Polym. Res.*, 13, 121, 2006.

2006DI2 Ding, H., Wu, F., Huang, Y., Zhang, Z., and Nie, Y., Synthesis and characterization of temperature-responsive copolymer of PELGA modified poly(*N*-isopropyl-acrylamide), *Polymer*, 47, 1575, 2006.

2006GAR Garcia Sakai, V., Higgins, J.S., and Trusler, J.P.M., Cloud curves of polystyrene or poly(methyl methacrylate) or poly(styrene-*co*-methyl methacrylate) in cyclohexanol determined with a thermo-optical apparatus, *J. Chem. Eng. Data*, 51, 743, 2006.

2006GEE Geever, L.M., Devine, D.M., Nugent, M.J.D., Kennedy, J.E., Lyons,J.G., Hanley, A., and Higginbotham, C.L., Lower critical solution temperature control and swelling behaviour of physically crosslinked thermosensitive copolymers based on *N*-isopropylacrylamide, *Eur. Polym. J.*, 42, 2540, 2006.

2006HAL Halacheva, S., Rangelov, S., and Tsvetanov, C., Poly(glycidol)-based analogues to pluronic block copolymers. Synthesis and aqueous solution properties, *Macromolecules*, 39, 6845, 2006.

2006HU1 Hua, F., Jiang, X., and Zhao, B., Well-defined thermosensitive, water-soluble polyacrylates and polystyrenics with short pendant oligo(ethylene glycol) groups synthesized by nitroxide-mediated radical polymerization, *J. Polym. Sci.: Part A: Polym. Chem.*, 44, 2454, 2006.

2006HU2 Hua, F., Jiang, X., and Zhao, B., Temperature-induced self-association of doubly thermosensitive diblock copolymers with pendant methoxytris(oxyethylene) groups in dilute aqueous solutions, *Macromolecules*, 39, 3476, 2006.

2006KI1 Kim, Y.-C., Kil, D.-S., and Kim, J.C., Synthesis and phase separation of poly(*N*-isopropylacrylamide-*co*-methoxy polyethyleneglycol monomethacrylate), *J. Appl. Polym. Sci.*, 101, 1833, 2006.

2006KI2 Kim, Y.C., Bang, M.-S., and Kim, J.-C., Synthesis and characterization of poly(*N*-isopropyl acrylamide) copolymer with methoxy polyethyleneglycol monomethacrylate, *J. Ind. Eng. Chem.*, 12, 446, 2006.

2006LAZ Lazzara, G., Milioto, S., and Gradzielski, M., The solubilisation behaviour of some dichloroalkanes in aqueous solutions of PEO–PPO–PEO triblock copolymers: A dynamic light scattering, fluorescence spectroscopy, and SANS study, *Phys. Chem. Chem. Phys.*, 8, 2299, 2006.

2006LIC Li, C., Lee, D.H., Kim, J.K., Ryu, D.Y., and Russel, T.P., Closed-loop phase behavior for weakly interacting block copolymers, *Macromolecules*, 39, 5926, 2006.

2006LOZ Lozinskii, V.I., Simenel, I.A., Semenova, M.G., Belyakova, L.E., Ilin, M.M., Grinberg, V. Ya., Dubovik, A.S., and Khokhlov, A.R., Behavior of protein-like *N*-vinylcaprolactam and *N*-vinylimidazole copolymers in aqueous solutions, *Polym. Sci., Ser. A*, 48, 435, 2006.

2006LUC Lu, C., Guo, S.-R., Zhang, Y., and Yin, M., Synthesis and aggregation behavior of four different shaped PCL-PEG block copolymers, *Polym. Int.*, 55, 694, 2006.

2006MA1 Maeda, T., Kanda, T., Yonekura, Y., Yamamoto, Y., and Aoyagi, T., Hydroxylated poly(*N*-isopropylacrylamide) as functional thermoresponsive materials, *Biomacromolecules*, 7, 545, 2006.

2006MA2 Maeda, T., Yamamoto, K., and Aoyagi, T., Importance of bound water in hydration–dehydration behavior of hydroxylated poly(*N*-isopropylacrylamide), *J. Colloid Interface Sci.*, 302, 467, 2006.

2006MOR Mori, T., Nakashima, M., Fukuda, Y., Minagawa, K., Tanaka, M., and Maeda, Y., Soluble-insoluble-soluble transitions of aqueous poly(*N*-vinylacetamide-*co*-acrylic acid) solutions, *Langmuir*, 22, 4336, 2006.

2006MUN Mun, G.A., Nurkeeva, Z.S., Akhmetkalieva, G.T., Shmakov, S.N., Khutoryanskiy, V.V., Lee, S.C., and Park, K., Novel temperature-responsive water-soluble copolymers based on 2-hydroxyethylacrylate and vinyl butyl ether and their interactions with poly(carboxylic acids), *J. Polym. Sci.: Part B: Polym. Phys.*, 44, 195, 2006.

2006MYL Mylonas, Y., Bokias, G., Iliopoulos, I., and Staikos, G., Interpolymer association between hydrophobically modified poly(sodium acrylate) and poly(*N*-isopropylacrylamide) in water: The role of hydrophobic interactions and polymer structure, *Eur. Polym. J.*, 42, 849, 2006.

2006NAG Nagy, I., Loos, Th.W.de, Krenz, R.A., and Heidemann, R.A., High pressure phase equilibria in the systems linear low density polyethylene + n-hexane and linear low density polyethylene + n-hexane + ethylene: Experimental results and modelling with the Sanchez-Lacombe equation of state, *J. Supercrit. Fluids*, 37, 115, 2006.

2006OSA Osaka, N., Okabe, S., Karino, T., Hirabaru, Y., Aoshima, S., and Shibayama, M., Micro- and macrophase separations of hydrophobically solvated block copolymer aqueous solutions induced by pressure and temperature, *Macromolecules*, 39, 5875, 2006.

2006PAG Pagonis, K. and Bokias, G., Simultaneous lower and upper critical solution temperature-type co-nonsolvency behaviour exhibited in water-dioxane mixtures by linear copolymers and hydrogels containing *N*-isopropylacrylamide and *N,N*-dimethylacrylamide, *Polym. Int.*, 55, 1254, 2006.

2006QIN Qin, S., Geng, Y., Discher, D.E., and Yang, S., Temperature-controlled assembly and release from polymer vesicles of poly(ethylene oxide)-*block*-poly(*N*-isopropyl acrylamide), *Adv. Mater.*, 18, 2905, 2006.

2006QIU Qiu, G.-M., Zhu, B.-K., Xu, Y.-Y., and Geckeler, K.E., Synthesis of ultrahigh molecular weight poly(styrene-*alt*-maleic anhydride) in supercritical carbon dioxide, *Macromolecules*, 39, 3231, 2006.

2006SEE Seetapan, N., Mai-ngam, K., Plucktaveesak, N., and Sirivat, A., Linear viscoelasticity of thermoassociative chitosan-*g*-poly(*N*-isopropylacrylamide) copolymer, *Rheol. Acta*, 45, 1011, 2006.

2006SHA Sharma, S.C., Acharya, D.P., Garcia-Roman, M., Itami, Y., and Kunieda, H., Phase behavior and surface tensions of amphiphilic fluorinated random copolymer aqueous solutions, *Coll. Surfaces A*, 280, 140, 2006.

2006SHE Shen, Z., Terao, K., Maki, Y., Dobashi, T., Ma, G., and Yamamoto, T., Synthesis and phase behavior of aqueous poly(*N*-isopropylacrylamide-*co*-acrylamide), poly(*N*-iso-propylacrylamide-*co*-*N,N*-dimethylacrylamide) and poly(*N*-isopropylacrylamide-*co*-2-hydroxyethyl methacrylate), *Colloid Polym. Sci.*, 284, 1001, 2006.

2006SIL Silva, M. do C.H. da, Silva, L.H.M. da, Amim, J., Guimaraes, R.O., and Martins, J.P., Liquid-liquid equilibrium of aqueous mixture of triblock copolymers L35 and F68 with Na_2SO_4, Li_2SO_4, or $MgSO_4$, *J. Chem. Eng. Data*, 51, 2260, 2006.

2006SOU Soule, E.R., Jaffrennou, B., Mechin, F., Pascault, J.P., Borrajo, J., and Williams, R.J., Thermodynamic analysis of the reaction-induced phase separation of solutions of random copolymers of methyl methacrylate and *N,N*-dimethylacrylamide in the precursors of a polythiourethane network, *J. Polym. Sci.: Part B: Polym. Phys.*, 44, 2821, 2006.

2006TO1 Torrens, F., Soria, V., Codoner, A., Abad, C., and Campos, A., Compatibility between polystyrene copolymers and polymers in solution via hydrogen bonding, *Eur. Polym. J.*, 42, 2807, 2006.

2006TO2 Torrens, F., Soria, V., Monzo, I.S., Abad, C., and Campos, A., Treatment of poly(styrene-*co*-methacrylic acid)/poly(4-vinylpyridine) blends in solution under liquid–liquid phase-separation conditions. A new method for phase-separation data attainment from viscosity measurements, *J. Appl. Polym. Sci.*, 102, 5039, 2006.

2006VER Verezhnikov, V.N., Plaksitskaya, T.V., and Poyarkova, T.N., pH-Thermosensitive behavior of *N,N*-dimethylaminoethyl methacrylate (co)polymers with *N*-vinyl-caprolactam, *Polym. Sci., Ser. A*, 48, 870, 2006.

2006WEN Weng, Y., Ding, Y., Zhang, G., Microcalorimetric investigation on the lower critical solution temperature behavior of *N*-isopropylacrylamide-*co*-acrylic acid copolymer in aqueous solution, *J. Phys. Chem. B*, 110, 11813, 2006.

2006WEI Wei, H., Zhang, X.-Z., Zhou, Y., Cheng, S.-X., and Zhuo, R.-X., Self-assembled thermoresponsive micelles of poly(*N*-isopropylacrylamide-*b*-methyl methacrylate), *Biomaterials*, 27, 2028, 2006.

2006WIN Winoto, W., Adidharma, H., Shen, Y., and Radosz, M., Micellization temperature and pressure for polystyrene-*b*-polyisoprene in subcritical and supercritical propane, *Macromolecules*, 39, 8140, 2006.

2006YIN Yin, X., Hoffman, A.S., and Stayton, P.S., Poly(*N*-isopropylacrylamide-*co*-propylacrylic acid) copolymers that respond sharply to temperature and pH, *Biomacromolecules*, 7, 1381, 2006.

2006ZHO Zhou, J., Lin Y., Du, Q., Zhong, W., and Wang, H., Effect of MPEG on MPEG-grafted EAA membrane formation via thermally induced phase separation, *J. Membrane Sci.*, 283, 310, 2006.

2007BHA Bharatiya, B., Guo, C., Ma, J.H., Hassan, P.A., and Bahadur, P., Aggregation and clouding behavior of aqueous solution of EO–PO block copolymer in presence of n-alkanols, *Eur. Polym. J.*, 43, 1883, 2007.

2007DIM Dimitrov, P., Jamroz-Piegza, M., Trzebicka, B., and Dworak, A., The influence of hydrophobic substitution on self-association of poly(ethylene oxide)-*b*-poly(n-alkyl glycidyl carbamate)s-*b*-poly(ethylene oxide) triblock copolymers in aqueous media, *Polymer*, 48, 1866, 2007.

2007ERB Erbil, C., Gökcören, A.T., and Polat, Y.A., *N*-isopropylacrylamide-acrylamide copolymers initiated by ceric ammonium nitrate in water, *Polym. Int.*, 56, 547, 2007.

2007FIG Figueruelo, J.E., Monzo, I.S., Gomez, C.M., and Soria, V., Ternary polymer solutions with hydrogen bonds 2. Prediction of phase diagrams, *Macromol. Theory Simul.*, 16, 62, 2007.

2007FOU Fournier, D., Hoogenboom, R., Thijs, H.M.L., Paulus, R.M., and Schubert, U.S., Tunable pH- and temperature-sensitive copolymer libraries by reversible addition-fragmentation chain transfer copolymerizations of methacrylates, *Macromolecules*, 40, 915, 2007.

2007FRA Francis, R., Jijil, C.P., Prabhu, C.A., and Suresh, C.H., Synthesis of poly(*N*-isopropylacrylamide) copolymer containing anhydride and imide comonomers: A theoretical study on reversal of LCST, *Polymer*, 48, 6707, 2007.

2007KON Kong, F.Q., Cao, X., Xia, J., and Hur, B.-K., Synthesis and application of a light-sensitive polymer forming aqueous two-phase systems, *J. Ind. Eng. Chem.*, 13, 424, 2007.

2007LUT Lutz, J.-F., Weichenhan, K., Akdemir, O., and Hoth, A., About the phase transitions in aqueous solutions of thermoresponsive copolymers and hydrogels based on 2-(2-methoxyethoxy)ethyl methacrylate and oligo(ethylene glycol) methacrylate, *Macromolecules*, 40, 2503, 2007.

2007LVR Lv, R., Zhou, J., Xu, P., Du, Q., Wang, H., and Zhong, W., Estimation of phase diagrams for copolymer-diluent systems in thermally induced phase separation, *J. Appl. Polym. Sci.*, 105, 3513, 2007.

2007MAN Mansur, C., Spinelli, L., Gonzalez, G., and Lucas, E.F., Evaluation of the physical-chemical properties of poly(ethylene oxide)-*b*-poly(propylene oxide) by different characterization techniques, *Macromol. Symp.*, 258, 5, 2007.

2007MAO Mao, Z., Ma, L., Yan, J., Yan, M., Gao, C., and Shen, J., The gene transfection efficiency of thermoresponsive *N,N,N*-trimethylchitosan chloride-*g*-poly(*N*-isopropyl acrylamide) copolymer, *Biomaterials*, 28, 4488, 2007.

2007MIK Mikhailov, Yu.M., Ganina, L.V., Roshchupkin, V.P., and Shapaeva, N.V., Phase equilibrium and diffusion in binary systems based on nonyl acrylate, acrylic acid, and their homo- and copolymers, *Polym. Sci., Ser. B*, 49, 240, 2007.

2007MUN Mun, G.A., Nurkeeva, Z.S., Beissegul, A.B., Dubolazov, A.V., Urkimbaeva, P.I., Park, K., and Khutoryanskiy, V.V., Temperature-responsive water-soluble copolymers based on 2-hydroxyethyl acrylate and butyl acrylate, *Macromol. Chem. Phys.*, 208, 979, 2007.

2007NAG Nagy, I., Krenz, R.A., Heidemann, R.A., and de Loos, Th.W., High-pressure phase equilibria in the system linear low density polyethylene + isohexane: Experimental results and modelling, *J. Supercrit. Fluids*, 40, 125, 2007.

2007OEZ Oezyuerek, Z., Komber, H., Gramm, S., Schmaljohann, D., Mueller, A.H.E., and Voit, B., Thermoresponsive glycopolymers via controlled radical polymerization, *Macromol. Chem. Phys.*, 208, 1035, 2007.

2007OLI Oliveira, M.C. de, Filho, M.A.N. de A., and Filho, P. de A.P., Phase equilibrium and protein partitioning in aqueous two-phase systems containing ammonium carbamate and block copolymers PEO–PPO–PEO, *Biochem. Eng. J.*, 37, 311, 2007.

2007REN Ren, L. and Agarwal, S., Synthesis, characterization, and properties evaluation of poly[(*N*-isopropylacrylamide)-*co*-ester]s, *Macromol. Chem. Phys.*, 208, 245, 2007.

2007SH1 Shao, D. and Ni, C., Preparations and properties of thermosensitive terpolymers of *N*-isopropylacrylamide, sodium 2-acrylamido-2-methyl-propanesuphonate, and *N*-*tert*-butylacrylamide, *J. Appl. Polym. Sci.*, 105, 2299, 2007.

2007SH2 Shatalov, G.V., Churilina, E.V., Kuznetsov, V.A., and Verezhnikov, V.N., Copolymerization of *N*-vinylcaprolactam with *N*-vinyl(benz)imidazoles and the properties of aqueous solutions of the copolymers, *Polym. Sci., Ser. B*, 49, 57, 2007.

2007SKR Skrabania, K., Kristen, J., Laschewsky, A., Akdemir, O., Hoth, A., and Lutz, J.-F., Design, synthesis, and aqueous aggregation behavior of nonionic single and multiple thermoresponsive polymers, *Langmuir*, 23, 84, 2007.

2007VAN Van Durme, K., Van Assche, G., Aseyev, V., Raula, J., Tenhu, H., and Van Mele, B., Influence of macromolecular architecture on the thermal response rate of amphiphilic copolymers, based on poly(*N*-isopropylacrylamide) and poly(oxyethylene), in water, *Macromolecules*, 40, 3765, 2007.

2007XUY Xu, Y., Shi, L., Ma, R., Zhang, W., An, Y., and Zhu, X.X., Synthesis and micellization of thermo- and pH-responsive block copolymer of poly(*N*-isopropylacrylamide)-*block*-poly(4-vinylpyridine), *Polymer*, 48, 1711, 2007.

2007YIN Yin, J., Lv, R., Zhou, J., Du, Q., and Zhong, W., Preparation of EVOH microporous membranes via thermally induced phase separation using binary solvents, *Chin. J. Polym. Sci.*, 25, 379, 2007.

2008BER Bergbreiter, D.E. and Fu, H., Thermodynamic cloud point assays, *J. Polym. Sci.: Part A: Polym. Chem.*, 46, 186, 2008.

2008CHA Chang, C., Wei, H., Quan, C.-Y., Li, Y.-Y., Liu, J., Wang, Z.-C., Cheng, S.-X., Zhang, X.-Z., and Zhuo, R.-X., Fabrication of thermosensitive PCL-PNIPAAm-PCL triblock copolymeric micelles for drug delivery, *J. Polym. Sci.: Part A: Polym. Chem.*, 46, 3048, 2008.

2008CHE Chen, Q., Liu, X., Xu, K., Song, C., Zhang, W., and Wang, P., Phase behavior and self-assembly of poly[*N*-vinylformamide-*co*-(acrylic acid)] copolymers under highly acidic conditions, *J. Appl. Polym. Sci.*, 109, 2802, 2008.

2008CRI Cristobal, G., Berret, J.-F., Chevallier, C., Talingting-Pabalan, R., Joanicot, M., and Grillo, I., Phase behavior of polyelectrolyte block copolymers in mixed solvents, *Macromolecules*, 41, 1872, 2008.

2008EGG Eggenhuisen, T.M., Becer, C.R., Fijten, M.W.M., Eckardt, R., Hoogenboom, R., and Schubert, U.S., Libraries of statistical hydroxypropyl acrylate containing copolymers with LCST properties prepared by NMP, *Macromolecules*, 41, 5132, 2008.

2008FAN Fang, J., Bian, F., and Shen, W., A study on solution properties of poly(*N,N*-diethylacrylamide-*co*-acrylic acid), *J. Appl. Polym. Sci.*, 110, 3373, 2008.

2008FAR Fares, M.M. and Othman, A.A., Lower critical solution temperature determination of smart, thermosensitive *N*-isopropylacrylamide-*alt*-2-hydroxyethyl methacrylate copolymers: Kinetics and physical properties, *J. Appl. Polym. Sci.*, 110, 2815, 2008.

2008HE1 He, C., Zhao, C., Chen, X., Guo, Z., Zhuang, X., and Jing, X., Novel pH- and temperature-responsive block copolymers with tunable pH-responsive range, *Macromol. Rapid Commun.*, 29, 490, 2008.

2008HE2 He, C., Zhao, C., Guo, X., Guo, Z., Chen, X., Zhuang, X., Liu, S., and Jing, X., Novel temperature- and pH-responsive graft copolymers composed of poly(L-glutamic acid) and poly(*N*-isopropylacrylamide), *J. Polym. Sci.: Part A: Polym. Chem.*, 46, 4140, 2008.

2008HO1 Hoogenboom, R., Thijs, H.M.L., Jochems, M.J.H.C., Lankvelt, B.M. van, Fijten, M.W.M., and Schubert, U.S., Tuning the LCST of poly(2-oxazoline)s by varying composition and molecular weight: Alternatives to poly(*N*-isopropylacrylamide), *Chem. Commun.*, 44, 5758, 2008.

2008HO2 Hoogenboom, R., Thijs, H.M.L., Wouters, D., Hoeppener, S., and Schubert, U.S., Tuning solution polymer properties by binary water–ethanol solvent mixtures, *Soft Matter*, 4, 103, 2008.

2008HUB Huber, S. and Jordan, R., Modulation of the lower critical solution temperature of 2-alkyl-2-oxazoline copolymers, *Coll. Polym. Sci.*, 286, 695, 2008.

2008KHU Khutoryanskaya, O.V., Mayeva, Z.A., Mun, G.A., and Khutoryanskiy, V.V., Designing temperature-responsive biocompatible copolymers and hydrogels based on 2-hydroxyethyl(meth)acrylates, *Biomacromolecules*, 9, 3353, 2008.

2008KOS Kostko, A.F., Lee, S.H., Liu, J., DiNoia, T.P., Kim, Y., and McHugh, M.A., Cloud-point behavior of poly(ethylene-*co*-20.2 mol% 1-butene) (PEB10) in ethane and deuterated ethane and of deuterated PEB10 in pentane isomers, *J. Chem. Eng. Data*, 53, 1626, 2008.

2008KOT Kotsuchinashi, Y., Kuboshima, Y., Yamamoto, K., and Aoyagi, T., Synthesis and characterization of double thermo-responsive block copolymer consisting *N*-isopropylacrylamide by atom transfer radical polymerization, *J. Polym. Sci.: Part A: Polym. Chem.*, 46, 6142, 2008.

2008LI1 Liu, S., Liu, X., Li, F., Fang, Y., Wang, Y., and Yu, J., Phase behavior of temperature- and pH-sensitive poly(acrylic acid-*g*-*N*-isopropylacrylamide) in dilute aqueous solution, *J. Appl. Polym. Sci.*, 109, 4036, 2008.

2008LI2 Liu, R., DeLeonardis, P., Cellesi, F., Tirelli, N., and Saunders, B.R., Cationic temperature-responsive poly(*N*-isopropyl acrylamide) graft copolymers: From triggered association to gelation, *Langmuir*, 24, 7099, 2008.

2008LI3 Liu, L., Wu, C., Zhang, J., Zhang, M., Liu, Y., Wang, X., and Fu, G., Controlled polymerization of 2-(diethylamino)ethyl methacrylate and its block copolymer with *N*-isopropylacrylamide by RAFT polymerization, *J. Polym. Sci.: Part A: Polym. Chem.*, 46, 3294, 2008.

2008LIY Li, Y., Liu, R., Liu, W., Kang, H., Wu, M., and Huang, Y., Synthesis, self-assembly, and thermosensitive properties of ethyl cellulose-*g*-p(PEGMA) amphiphilic copolymers, *J. Polym. Sci.: Part A: Polym. Chem.*, 46, 6907, 2008.

2008LOH Loh, X.J., Wu, Y.-L., Seow, W.T.J., Norimzan, M.N.I., Zhang, Z.-X., Xu, F.-J., Kang, E.-T., Neoh, K.-G., and Li, J., Micellization and phase transition behavior of thermosensitive poly(*N*-isopropylacrylamide)–poly(ε-caprolactone)–poly(*N*-isopropyl acrylamide) triblock copolymers, *Polymer*, 49, 5084, 2008.

2008MAD Madeira, P.P., Teixeira, J.A., Macedo, E.A., Mikheeva, L.M., Zaslavsky, B.Y., $\Delta G(CH_2)$ as solvent descriptor in polymer/polymer aqueous two-phase systems, *J. Chromatogr. A*, 1185, 85, 2008.

2008MAS Masci, G., Diociaiuti, M., and Crescenzi, V., ATRP synthesis and association properties of thermoresponsive anionic block copolymers, *J. Polym. Sci.: Part A: Polym. Chem.*, 46, 4830, 2008.

2008MIP Mi, P., Chu, L.-Y., Ju, X.-J., and Niu, C.H., A smart polymer with ion-induced negative shift of the lower critical solution temperature for phase transition, *Macromol. Rapid Commun.*, 29, 27, 2008.

2008NU1 Nuopponen, M., Kalliomaeki, K., Laukkanen, A., Hietala, S., and Tenhu, H., A-B-A stereoblock copolymers of *N*-isopropylacrylamide, *J. Polym. Sci.: Part A: Polym. Chem.*, 46, 38, 2008.

2008NU2 Nuopponen, M., Kalliomaeki, K., Aseyev, V., and Tenhu, H., Spontaneous and thermally induced self-organization of A-B-A stereoblock polymers of *N*-isopropylacrylamide in aqueous solutions, *Macromolecules*, 41, 4881, 2008.

2008PAN Pan, Y., Xiao, H., Zhao, G., and He, B., Synthesis and characterization of temperature-responsive poly(vinyl alcohol)-based copolymers, *J. Appl. Polym. Sci.*, 110, 2698, 2008.

2008QIN Qin, W. and Cao, X.-J., Synthesis of a novel pH-sensitive methacrylate amphiphilic polymer and its primary application in aqueous two-phase systems, *Appl. Biochem. Biotechnol.*, 150, 171, 2008.

2008REN Ren, Y., Jiang, X., and Yin, J., Copolymer of poly(4-vinylpyridine)-*g*-poly(ethylene oxide) respond sharply to temperature, pH and ionic strength, *Eur. Polym. J.*, 44, 4108, 2008.

2008SHA Shaheen, A., Kaur, N., and Mahajan, R.K., Influence of various series of additives on the clouding behavior of aqueous solutions of triblock copolymers, *Coll. Polym. Sci.*, 286, 319, 2008.

2008SIL Silverio, S.C., Madeira, P.P., Rodriguez, O., Teixeira, J.A., and Macedo, E.A., $\Delta G(CH_2)$ in PEG-salt and Ucon-salt aqueous two-phase systems, *J. Chem. Eng. Data*, 53, 1622, 2008.

2008TAN Tan, L., Pan, D., and Pan, N., Thermodynamic study of a water-dimethylformamide-polyacrylonitrile ternary system, *J. Appl. Polym. Sci.*, 110, 3439, 2008.

2008TRO Troll, K., Kulkarni, A., Wang, W. Darko, C., Koumba, A.M.B., Laschewsky, A., Müller-Buschbaum, P., and Papadakis, C.M., The collapse transition of poly(styrene-*b*-(*N*-isopropylacrylamide)) diblock copolymers in aqueous solution and in thin films, *Colloid Polym. Sci.*, 286, 1079, 2008.

2008WIN Wintgens, V. and Amiel, C., Physical gelation of amphiphilic poly(*N*-isopropylacrylamide): Influence of the hydrophobic groups, *Macromol. Chem. Phys.*, 209, 1553, 2008.

2008YAM Yamamoto, S.-I., Pietrasik, J., and Matyjaszewski, K., The effect of structure on the thermoresponsive nature of well-defined poly(oligo(ethylene oxide) methacrylates) synthesized by ATRP, *J. Polym. Sci.: Part A: Polym. Chem.*, 46, 194, 2008.

2008ZHA Zhao, C., Zhuang, X., He, C., Chen, X., and Jing, X., Synthesis of novel thermo- and pH-responsive poly(L-lysine)-based copolymer and its micellization in water, *Macromol. Rapid Commun.*, 29, 1810, 2008.

2009ALV Alvarez-Ramírez, J.G., Fernández, V.V.A., Macíasa, E.R., Rharbi, Y., Taboada, P., Gámez-Corrales, R., Puig, J.E., and Soltero, J.F.A., Phase behavior of the Pluronic P103/water system in the dilute and semi-dilute regimes, *J. Colloid Interface Sci.*, 333, 655, 2009.

2009CAO Cao, Y., Zhao, N., Wu, K., and Zhu, X.X., Solution properties of a thermosensitive triblock copolymer of *N*-alkyl substituted acrylamides, *Langmuir*, 25, 1699, 2009.

2009CH1 Chang, C., Wei, H., Feng, J., Wang, Z.-C., Wu, X.-J., Wu, D.-Q., Cheng, S.-X., Zhang, X.-Z., and Zhuo, R.-X., Temperature and pH double responsive hybrid cross-linked micelles based on P(NIPAAm-*co*-MPMA)-*b*-P(DEA): RAFT synthesis and "schizophrenic" micellization, *Macromolecules*, 42, 4838, 2009.

2009CH2 Chang, Y., Chen, W.-Y., Yandi, W., Shih, Y.-J., Chu, W.-L., Liu, Y.-L., Chu, C.-W., Ruaan, R.-C., and Higuchi, A., Dual-thermoresponsive phase behavior of blood compatible zwitterionic copolymers containing nonionic poly(*N*-isopropyl-acrylamide), *Biomacromolecules*, 10, 2092, 2009.

2009CHE Chen, W., Pelton, R., and Leung, V., Solution properties of polyvinylamine derivatized with phenylboronic acid, *Macromolecules*, 42, 1300, 2009.

2009CRI Cristiano, C.M.Z., Soldi, V., Li, C., Armes, S.P., Rochas, C., Pignot-Paintrand, I., and Borsali, R., Thermo-responsive copolymers based on poly(*N*-isopropyl acrylamide) and poly[2-(methacryloyloxy)ethyl phosphorylcholine]: Light scattering and microscopy experiments, Macromol. Chem. Phys., 210, 1726, 2009.

2009CUI Cui, Z.-Y., Xu, Y.-X., Zhu, L.-P., Deng, H.-Y., Wang, J.-Y., and Zhu, B.-K., Preparation of PVDF-HFP microporous membranes via thermally induced phase separation process, *J. Macromol. Sci.: Part B: Phys.*, 48, 41, 2009.

2009DEN Deng, K., Tian, H., Zhang, P., Zhong, H., Ren, X., and Wang, H., pH–temperature responsive poly(HPA-*co*-AMHS) hydrogel as a potential drug-release carrier, *J. Appl. Polym. Sci.*, 114, 176, 2009.

2009DON Dong, R., Zhao, J., Zhang, Y., and Pan, D., Morphology control of polyacrylonitrile (PAN) fibers by phase separation technique, *J. Polym. Sci.: Part B: Polym. Phys.*, 47, 261, 2009.

2009FOU Foroutan, M. and Khomami, M.H., Quaternary (liquid + liquid) equilibria of aqueous two-phase poly(ethylene glycol), poly(DMAM-TBAM), and KH_2PO_4: Experimental and generalized Flory-Huggins theory, *J. Chem. Thermodyn.*, 41, 604, 2009.

2009GAO Gao, M., Jia, X., Kuang, G., Li, Y., Liang, D., and Wei, Y., Thermo- and pH-responsive dendronized copolymers of styrene and maleic anhydride pendant with poly(amidoamine) dendrons as side groups, *Macromolecules*, 42, 4273, 2009.

2009GAR Garcia-Lopera, R., Figueruelo, J.E., Monzo, I.S., Abad, C., and Campos, A., Influence of the copolymer content on the miscibility, phase behaviour and morphology of a DGEBA/polystyrene/styrene-*co*-maleic anhydride ternary blend, *Macromol. Chem. Phys.*, 210, 1856, 2009.

2009JOC Jochum, F.D. and Theato, P., Temperature and light sensitive copolymers containing azobenzene moieties prepared via a polymer analogous reaction, *Polymer*, 50, 3079, 2009.

2009KAU Kautharapu, K., Pujari, N.S., Golegaonkar, S.B., Ponrathnam, S., Nene, S.N., and Bhatnagar, D., Vinyl-2-pyrrolidone derivatized guar gum based aqueous two-phase system, *Separation Purification Technol.*, 65, 9, 2009.

2009KIZ Kizhnyaev, V.N., Adamova, L.V., Pokatilov, F.A., Krakhotkina, E.A., Petrova, T.L., and Smirnov, A.I., Thermodynamics of water interaction with 5-vinyltetrazole copolymers with different hydrophilic–hydrophobic balances, *Polym. Sci., Ser. A*, 51, 168, 2009.

2009KOS Kostko, A.F., Harden, J.L., and McHugh, M.A., Dynamic light scattering study of concentrated triblock copolymer micellar solutions under pressure, *Macromolecules*, 42, 5328, 2009.

2009KOT Kotsuchibashi, Y., Yamamoto, K., and Aoyagi, T., Assembly behavior of double thermo-responsive block copolymers with controlled response temperature in aqueous solution, *J. Colloid Interface Sci.*, 336, 67, 2009.

2009LAM Lambermont-Thijs, H.M.L., Hoogenboom, R., Fustin, C.A., Bomal-D'Haese, C., Gohy, J.-F., and Schubert, U.S., Solubility behavior of amphiphilic block and random copolymers based on 2-ethyl-2-oxazoline and 2-nonyl-2-oxazoline in binary water–ethanol mixtures, *J. Polym. Sci.: Part A: Polym. Chem.*, 47, 515, 2009.

2009LIN Lin, J.-J. and Hsu, Y.-C., Temperature and pH-responsive properties of poly(styrene-*co*-maleic anhydride)-grafting-poly(oxypropylene)amines, *J. Colloid Interface Sci.*, 336, 82, 2009.

2009LIP Li, P.-F., Wang, W., Xie, R., Yang, M., Ju, X.-J., and Chu, L.-Y., Lower critical solution temperatures of thermo-responsive poly(*N*-isopropylacrylamide) copolymers with racemate or single enantiomer groups, *Polym. Int.*, 58, 202, 2009.

2009LIX Li, X., Yin, M. Zhang, G., and Zhang, F., Synthesis and characterization of novel temperature and pH responsive hydroxylpropyl cellulose-based graft copolymers, *Chin. J. Chem. Eng.*, 17, 145, 2009.

2009LI1 Liu, C., He, J., Zhao, Q., Zhang, M., and Ni, P., Well-defined poly[(dimethylamimo)ethyl methacrylate]-*b*-poly(fluoroalkyl methacrylate) diblock copolymers: Effects of different fluoroalkyl groups on the solution properties, *J. Polym. Sci.: Part A: Polym. Chem.*, 47, 2702, 2009.

2009LI2 Liu, R., Cellesi, F., Tirelli, N., Saunders, B.R., A study of thermoassociative gelation of aqueous cationic poly(*N*-isopropylacrylamide) graft copolymer solutions, *Polymer*, 50, 1456, 2009.

2009MA2 Maeda, Y. and Yamabe, M., A unique phase behavior of random copolymer of *N*-isopropylacrylamide and *N,N*-diethylacrylamide in water, *Polymer*, 50, 519, 2009.

2009MA3 Maeda, Y., Sakamoto, J., Wang, S.-Y., and Mizuno, Y., Lower critical solution temperature behavior of poly(*N*-(2-ethoxyethyl)acrylamide) as compared with poly(*N*-isopropylacrylamide), *J. Phys. Chem. B*, 113, 12456, 2009.

2009MAH Mahdavi, H., Sadeghzadeh, M., and Qazvini, N.T., Phase behavior study of poly(*N*-*tert*-butylacrylamide-*co*-acrylamide) in the mixture of water–methanol: The role of polymer–nonsolvent second-order interactions, *J. Polym. Sci.: Part B: Polym. Phys.*, 47, 455, 2009.

2009MAI Maingam, K., Boonkitpattarakul, K., Sakulsombat, M., Chumningan, P., and Maingam, B., Synthesis and phase separation of amine-functional temperature responsive copolymers based on poly(*N*-isopropylacrylamide), *Eur. Polym. J.*, 45, 1260, 2009.

2009MAR Martins, J.P., Mageste, A.B., Silva, M.C.H. da, Silva, L.H.M. da, Patricio, P. da R., Coimbra, J.S. dos R., and Minim, L.A., Liquid-liquid equilibria of an aqueous two-phase system formed by a triblock copolymer and sodium salts at different temperatures, *J. Chem. Eng. Data*, 54, 2891, 2009.

2009NAN Nandni, D., Vohra, K.K., and Mahajan, R.K., Study of micellar and phase separation behavior of mixed systems of triblock polymers, *J. Colloid Interface Sci.*, 338, 420, 2009.

2009NIN Ning, B., Wan J., and Cao X., Preparation and recycling of aqueous two-phase systems with pH-sensitive amphiphilic terpolymer PADB, *Biotechnol. Progr.*, 25, 820, 2009.

2009PAT Patrizi, M.L., Piantanida, G., Coluzza, C., and Masci, G., ATRP synthesis and association properties of temperature responsive dextran copolymers grafted with poly(*N*-isopropylacrylamide), *Eur. Polym. J.*, 45, 2779, 2009.

2009REC Recillas, M., Silva, L.L., Peniche, C., Goycoolea, F.M., Rinaudo, M., and Argüelles-Monal, W.M., Thermoresponsive behavior of chitosan-*g*-*N*-isopropylacrylamide copolymer solutions, *Biomacromolecules*, 10, 1633, 2009.

2009REN Ren, Y., Jiang, X., and Yin, J., Poly(ether *tert*-amine): A novel family of multiresponsive polymer, *J. Polym. Sci.: Part A: Polym. Chem.*, 47, 1292, 2009.

2009ROD Rodrigues, G.D., Silva, M. do C.H. da, Silva, L.H.M. da, Teixeira, L. da S., and Andrade, V.M. de, Liquid-liquid phase equilibrium of triblock copolymer L64, poly(ethylene oxide-*b*-propylene oxide-*b*-ethylene oxide), with sulfate salts from (278.15 to 298.15) K, *J. Chem. Eng. Data*, 54, 1894, 2009.

2009SHA Shao, Z., Kong, F., and Cao, X., Phase diagram prediction of recycling aqueous two-phase systems formed by a light-sensitive copolymer and dextran, *Korean J. Chem. Eng.*, 26, 147, 2009.

2009SHI Shinde, V.S. and Pawar, V.U., Synthesis of thermosensitive glycopolymers containing D-glucose residue: Copolymers with *N*-isopropylacrylamide, *J. Appl. Polym. Sci.*, 111, 2607, 2009.

2009TAI Tai, H., Wang, W., Vermonden, T., Heath, F., Hennink, W.E., Alexander, C., Shakesheff, K.M., and Howdle, S.M., Thermoresponsive and photocrosslinkable PEGMEMA-PPGMA-EGDMA copolymers from a one-step ATRP synthesis, *Biomacromolecules*, 10, 822, 2009.

2009TAN Tan, L., Liu, S., and Pan, D., Water effect on the gelation behavior of polyacrylonitrile/dimethyl sulfoxide solution, *Colloids Surfaces A*, 340, 168, 2009.

2009TAU Tauer, K., Gau, D., Schulze, S., Völkel, A., and Dimova, R., Thermal property changes of poly(*N*-isopropylacrylamide) microgel particles and block copolymers, *Colloid Polym. Sci.*, 287, 299, 2009.

2009TRE Trellenkamp, T. and Ritter, H., 3-Ethylated *N*-vinyl-2-pyrrolidone with LCST properties in water, *Macromol. Rapid Commun.*, 30, 1736, 2009.

2009TUB Tubio, G., Nerli, B.B., Pico, G.A., Venancio, A., and Teixeira, J., Liquid-liquid equilibrium of the Ucon 50-HB5100/sodium citrate aqueous two-phase systems, *Separation Purification Technol.*, 65, 3, 2009.

2009WA1 Wang, T., Peng, C., Liu, H., and Hu, Y., Phase behavior and microstructure of the system consisting of 1-butyl-3-methylimidazolium hexafluorophosphate, water, triblock copolymer F127 and short-chain alcohols, *J. Mol. Liq.*, 146, 89, 2009.

2009WA2 Wang, W., Liang, H., Al Ghanami, R.C., Hamilton, L., Fraylich, M., Shakesheff, K.M., Saunders, B., and Alexander, C., Biodegradable thermoresponsive microparticle dispersions for injectable cell delivery prepared using a single-step process, *Adv. Mater*, 21, 1809, 2009.

2009WA3 Wang, Y.-C., Xia, H., Yang, X.-Z., and Wang, J., Synthesis and thermoresponsive behaviors of biodegradable Pluronic analogs, *J. Polym. Sci.: Part A: Polym. Chem.*, 47, 6168, 2009.

2009WEB Weber,C., Becer, C.R., Hoogenboom, R., and Schubert, U.S., Lower critical solution temperature behavior of comb and graft shaped poly[oligo(2-ethyl-2-oxazoline) methacrylate]s, *Macromolecules*, 42, 2965, 2009.

2009ZHA Zhao C., Zhuang X., He P., Xiao C., He C., Sun J., Chen X., and Jing, X., Synthesis of biodegradable thermo- and pH-responsive hydrogels for controlled drug release, *Polymer*, 50, 4308, 2009.

4. HIGH-PRESSURE FLUID PHASE EQUILIBRIUM (HPPE) DATA OF COPOLYMER SOLUTIONS

4.1. Cloud-point and/or coexistence curves of quasibinary solutions

Polymer (B): **poly(cyclohexene oxide-*co*-carbon dioxide)** **2005SCH**
Characterization: M_w/g.mol^{-1} = 12000, M_w/M_n < 1.2, synthesized in the laboratory
Solvent (A): **carbon dioxide** CO_2 **124-38-9**

Type of data: cloud points

w_B 0.0051 was kept constant

T/K	455.06	459.66	465.20	475.95	479.33	479.93
P/MPa	335	309	292	257	254	252

Polymer (B): **poly(ethylene-*co*-acrylic acid)** **2003BUB**
Characterization: M_n/g.mol^{-1} = 21700, M_w/g.mol^{-1} = 86000,
3.5 mol% acrylic acid, synthesized in the laboratory by high-pressure radical polymerization
Solvent (A): **ethene** C_2H_4 **74-85-1**

Type of data: cloud points

w_B 0.03 was kept constant

T/K	433.15	438.15	443.15	448.15	453.15	458.15	463.15	468.15	473.15
P/bar	2815	2645	2535	2425	2270	2135	2040	1950	1830

T/K	483.15	493.15	503.15	513.15	523.15	533.15
P/bar	1725	1610	1520	1435	1370	1299

Polymer (B): **poly(ethylene-*co*-acrylic acid)** **2003BUB**
Characterization: M_n/g.mol^{-1} = 19100, M_w/g.mol^{-1} = 63300,
3.8 mol% acrylic acid, synthesized in the laboratory by high-pressure radical polymerization
Solvent (A): **ethene** C_2H_4 **74-85-1**

Type of data: cloud points

w_B 0.03 was kept constant

T/K	438.15	443.15	448.15	453.15	458.15	463.15	468.15	473.15	478.15
P/bar	2880	2730	2615	2465	2355	2220	2085	1995	1895

continued

continued

T/K	483.15	493.15	503.15	513.15	523.15	533.15
P/bar	1810	1685	1570	1470	1395	1325

Polymer (B): **poly(ethylene-*co*-acrylic acid)** **2003BUB**
Characterization: M_n/g.mol^{-1} = 12600, M_w/g.mol^{-1} = 42100, 5.5 mol% acrylic acid, synthesized in the laboratory by high-pressure radical polymerization
Solvent (A): **ethene** **C_2H_4** **74-85-1**

Type of data: cloud points

w_B 0.03 was kept constant

T/K	473.15	478.15	483.15	488.15	493.15	498.15	503.15	513.15	523.15
P/bar	2820	2700	2540	2390	2255	2110	1990	1785	1630
T/K	533.15								
P/bar	1490								

Polymer (B): **poly(ethylene-*co*-acrylic acid)** **2003BUB**
Characterization: M_n/g.mol^{-1} = 30980, M_w/g.mol^{-1} = 82300, 5.7 mol% acrylic acid, synthesized in the laboratory by high-pressure radical polymerization
Solvent (A): **ethene** **C_2H_4** **74-85-1**

Type of data: cloud points

w_B 0.03 was kept constant

T/K	483.15	488.15	493.15	498.15	503.15	508.15	513.15	518.15	523.15
P/bar	2585	2435	2295	2150	2040	1930	1820	1730	1660
T/K	533.15								
P/bar	1520								

Polymer (B): **poly(ethylene-*co*-acrylic acid)** **2003BUB**
Characterization: M_n/g.mol^{-1} = 9280, M_w/g.mol^{-1} = 31910, 6.7 mol% acrylic acid, synthesized in the laboratory by high-pressure radical polymerization
Solvent (A): **ethene** **C_2H_4** **74-85-1**

Type of data: cloud points

w_B 0.03 was kept constant

T/K	488.15	493.15	498.15	503.15	508.15	513.15	518.15	523.15	529.15
P/bar	2960	2760	2555	2370	2200	2050	1910	1800	1695
T/K	533.15								
P/bar	1610								

Polymer (B): **poly(ethylene-*co*-acrylic acid)** **2002BEY**
Characterization: M_n/g.mol^{-1} = 22700, M_w/g.mol^{-1} = 258000, 6.0 wt% acrylic acid, Exxon Co., Machelen, Belgium
Solvent (A): **ethene** **C_2H_4** **74-85-1**

Type of data: cloud points

w_B	0.05	0.05	0.05	0.05	0.05
T/K	522.65	497.55	472.75	447.05	423.05
P/bar	1302	1394	1539	1779	2153

Polymer (B): **poly(ethylene-*co*-acrylic acid)** **2002BEY**
Characterization: M_n/g.mol^{-1} = 19900, M_w/g.mol^{-1} = 235000, 7.5 wt% acrylic acid, Exxon Co., Machelen, Belgium
Solvent (A): **ethene** **C_2H_4** **74-85-1**

Type of data: cloud points

w_B	0.05	0.05	0.05	0.05	0.05
T/K	523.35	497.85	472.35	447.85	434.25
P/bar	1340	1442	1637	1946	2196

Polymer (B): **poly(ethylene-*co*-acrylic acid)** **2002BEY**
Characterization: 8.0 wt% acrylic acid, Exxon Co., Machelen, Belgium
Solvent (A): **ethene** **C_2H_4** **74-85-1**

Type of data: cloud points

w_B	0.05	0.05	0.05	0.05	0.05
T/K	523.05	497.95	472.65	447.65	433.05
P/bar	1349	1469	1692	2037	2367

Polymer (B): **poly(ethylene-*co*-acrylic acid)** **2002BEY**
Characterization: M_n/g.mol^{-1} = 23400, M_w/g.mol^{-1} = 227000, 9.0 wt% acrylic acid, Exxon Co., Machelen, Belgium
Solvent (A): **ethene** **C_2H_4** **74-85-1**

Type of data: cloud points

w_B	0.05	0.05	0.05	0.05
T/K	522.65	497.75	471.85	447.55
P/bar	1390	1521	1779	2209

Polymer (B): **poly(ethylene-*co*-acrylic acid)** **2002BEY**
Characterization: M_n/g.mol^{-1} = 23700, M_w/g.mol^{-1} = 205000, 11.0 wt% acrylic acid, Exxon Co., Machelen, Belgium
Solvent (A): **ethene** **C_2H_4** **74-85-1**

Type of data: cloud points

continued

continued

w_B	0.05	0.05	0.05
T/K	524.05	497.35	472.35
P/bar	1484	1686	2104

Polymer (B): **poly(ethylene-*co*-acrylic acid)** **2002BEY**
Characterization: 15.0 wt% acrylic acid, Exxon Co., Machelen, Belgium
Solvent (A): **ethene** **C_2H_4** **74-85-1**

Type of data: cloud points

w_B	0.05	0.05	0.05
T/K	522.05	498.05	475.45
P/bar	1732	2066	2647

Polymer (B): **poly(ethylene-*co*-benzyl methacrylate)** **2004LAT**
Characterization: 3.3 mol% benzyl methacrylate, synthesized in the laboratory
Solvent (A): **ethene** **C_2H_4** **74-85-1**

Type of data: cloud points

w_B 0.05 was kept constant

T/K	389.15	411.15	430.15	450.15	470.15	489.15	510.15
P/bar	1697	1581	1501	1426	1366	1311	1268

Polymer (B): **poly(ethylene-*co*-benzyl methacrylate)** **2004LAT**
Characterization: M_n/g.mol^{-1} = 47100, M_w/g.mol^{-1} = 73500,
12.7 mol% benzyl methacrylate, synthesized in the laboratory
Solvent (A): **ethene** **C_2H_4** **74-85-1**

Type of data: cloud points

w_B 0.05 was kept constant

T/K	373.15	393.15	413.15	433.15	453.15	468.15	488.15	508.15	528.15
P/bar	1903	1748	1656	1579	1512	1459	1410	1361	1302

Polymer (B): **poly(ethylene-*co*-benzyl methacrylate)** **2004LAT**
Characterization: M_n/g.mol^{-1} = 36900, M_w/g.mol^{-1} = 60700,
20.0 mol% benzyl methacrylate, synthesized in the laboratory
Solvent (A): **ethene** **C_2H_4** **74-85-1**

Type of data: cloud points

w_B 0.05 was kept constant

T/K	458.15	473.15	483.15	488.15
P/bar	2006	1830	1747	1722

Polymer (B): **poly(ethylene-*co*-benzyl methacrylate)** **2004LAT**
Characterization: M_n/g.mol^{-1} = 36900, M_w/g.mol^{-1} = 61100,
26.1 mol% benzyl methacrylate, synthesized in the laboratory
Solvent (A): **ethene** **C_2H_4** **74-85-1**

Type of data: cloud points

w_B 0.05 was kept constant

T/K	458.15	473.15	483.15	493.15	503.15	513.15	523.15	528.15
P/bar	2380	2299	2230	2123	2055	2014	1949	1881

Polymer (B): **poly(ethylene-*co*-1-butene)** **2004KER**
Characterization: M_n/g.mol^{-1} = 230000, M_w/g.mol^{-1} = 232500,
20.2 mol% 1-butene, 10 ethyl branches per 100 backbone
C-atomes, completely hydrogenated polybutadiene,
synthesized in the laboratory
Solvent (A): **n-butane** **C_4H_{10}** **106-97-8**

Type of data: cloud points

w_B 0.001-0.006 T/K 403.15 P/bar 175

Polymer (B): **poly(ethylene-*co*-1-butene)** **1999CHE**
Characterization: M_n/g.mol^{-1} = 6430, M_w/g.mol^{-1} = 6690, 8.6 mol% 1-butene,
completely hydrogenated polybutadiene, 8.6 mol% 1,2-units,
Polymer Source, Inc., Dorval, Quebec
Solvent (A): **1-butene** **C_4H_8** **106-98-9**

Type of data: cloud points

w_B 0.00075 was kept constant

T/K	428.65	428.15	420.95	415.95	409.15	401.25	395.05	391.95	383.85
P/bar	90.1	88.8	80.5	72.3	63.5	50.3	40.7	34.6	23.5
T/K	381.55	328.85	327.75	326.85					
P/bar	20.8	48.4	89.0	132.3					

Comments: The last three data points are temperature-induced phase transitions.

w_B 0.011 was kept constant

T/K	428.45	428.35	418.85	411.25	402.65	396.75	391.55	381.95	375.35
P/bar	131.6	131.4	119.3	107.3	96.4	85.8	78.8	63.0	52.3
T/K	366.75	362.95	357.65	352.55	350.85	349.35	348.15		
P/bar	37.0	31.0	21.5	12.0	58.6	96.0	162.2		

Comments: The last three data points are temperature-induced phase transitions.

Polymer (B): **poly(ethylene-*co*-1-butene)** **1999CHE**
Characterization: M_n/g.mol^{-1} = 7570, M_w/g.mol^{-1} = 7870, 29.0 mol% 1-butene, completely hydrogenated polybutadiene, 29 mol% 1,2-units, Polymer Source, Inc., Dorval, Quebec
Solvent (A): **1-butene** **C_4H_8** **106-98-9**

Type of data: cloud points

w_B 0.0011 was kept constant

T/K	429.45	426.75	419.55	415.55	409.95	406.35	399.45	396.95	389.95
P/bar	93.1	87.4	79.8	73.6	66.0	59.9	49.3	45.1	33.7
T/K	387.35	381.45	259.65	258.85	258.35				
P/bar	29.0	20.5	31.1	79.6	144.5				

Comments: The last three data points are temperature-induced phase transitions.

w_B 0.0078 was kept constant

T/K	428.95	417.15	412.75	408.95	406.95	398.65	396.55	390.95	386.95
P/bar	134.2	122.8	117.1	107.1	106.4	85.3	83.9	70.6	62.1
T/K	378.75	377.05	372.15	367.45	364.95	277.75	276.85		
P/bar	44.7	39.4	30.1	23.6	15.1	22.2	192.4		

Comments: The last two data points are temperature-induced phase transitions.

Polymer (B): **poly(ethylene-*co*-1-butene)** **1999CHE**
Characterization: M_n/g.mol^{-1} = 115000, M_w/g.mol^{-1} = 120000, 87.5 mol% 1-butene, completely hydrogenated polybutadiene, 87.5 mol% 1,2-units, Polymer Source, Inc., Dorval, Quebec
Solvent (A): **1-butene** **C_4H_8** **106-98-9**

Type of data: cloud points

w_B 0.0012 was kept constant

T/K	431.85	430.25	424.25	419.35	416.15	405.75	400.95	394.15	389.45
P/bar	98.4	96.6	86.7	82.3	74.5	58.4	58.4	39.4	29.2
T/K	386.45	384.45	257.35	255.85					
P/bar	24.9	20.7	52.3	140.3					

Comments: The last two data points are temperature-induced phase transitions.

w_B 0.0091 was kept constant

T/K	429.05	419.75	416.25	415.85	408.05	405.25	396.75	396.25	388.55
P/bar	96.7	81.0	78.2	77.1	64.2	59.2	44.3	43.9	29.2
T/K	386.15	383.15							
P/bar	27.2	20.6							

Polymer (B): **poly(ethylene-*co*-1-butene)** **2005LID**
Characterization: M_n/g.mol^{-1} = 230000, M_w/g.mol^{-1} = 232500, 20.2 mol% 1-butene, 10 ethyl branches per 100 backbone C-atomes, completely hydrogenated polybutadiene, synthesized in the laboratory
Solvent (A): **dimethyl ether** **C_2H_6O** **115-10-6**

Type of data: cloud points

w_B 0.005 was kept constant

T/K	383.15	403.15	423.15	443.15
P/bar	800	586	513	495

Comments: The cloud-point pressures are independent on w_B between w_B = 0.001 and w_B = 0.006. More cloud points at w_B = 0.05 are given in Fig. 2 of 2005LID.

Polymer (B): **poly(ethylene-*co*-1-butene), deuterated** **2008KOS**
Characterization: M_n/g.mol^{-1} = 222700, M_w/g.mol^{-1} = 245000, 20.2 mol% 1-butene, 10 ethyl branches per 100 backbone C-atomes, completely deuterated polybutadiene, synthesized in the laboratory
Solvent (A): **2,2-dimethylpropane** **C_5H_{12}** **463-82-1**

Type of data: cloud points

w_B 0.003 was kept constant

T/K	285.95	289.15	292.75	296.65	299.45	304.15	306.55	310.35	314.75
P/MPa	35.3	32.4	30.3	27.1	26.3	25.0	23.8	22.6	21.7
T/K	317.85	322.65	327.35	329.65	333.15	337.05	339.55	341.35	345.45
P/MPa	21.3	20.8	20.4	20.2	20.0	19.6	19.5	19.4	19.4
T/K	349.45	352.75	358.55	361.65	364.65	366.95	369.35	372.75	375.45
P/MPa	19.4	19.5	19.8	19.9	20.0	20.1	20.2	20.4	20.6
T/K	379.15	381.75	385.35	389.55	391.35	394.85	398.15	401.85	406.45
P/MPa	20.8	20.9	21.1	21.4	21.5	21.7	22.0	22.3	22.6
T/K	409.05	412.95	415.55	418.75	423.05	425.75	428.75	431.25	434.85
P/MPa	22.9	23.1	23.3	23.5	23.9	24.1	24.3	24.5	24.7
T/K	435.85	440.05	441.25	445.65	450.15	454.35	456.05	460.65	461.35
P/MPa	24.7	25.0	25.1	25.3	25.7	26.2	26.4	27.1	27.2

Comments: Solid-liquid equilibrium data are additionally given in 2008KOS.

Polymer (B): **poly(ethylene-*co*-1-butene)** **2008KOS**
Characterization: M_n/g.mol^{-1} = 230200, M_w/g.mol^{-1} = 232500, 20.2 mol% 1-butene, 10 ethyl branches per 100 backbone C-atomes, completely hydrogenated polybutadiene, synthesized in the laboratory
Solvent (A): **ethane-d6** C_2D_6 **1632-99-1**

Type of data: cloud points

w_B 0.048 was kept constant

T/K	368.25	373.65	375.05	384.35	393.85	395.15	404.65	405.05	424.25
P/MPa	117.0	114.3	113.9	112.0	109.5	108.8	107.8	107.2	103.9
T/K	425.05	427.65							
P/MPa	104.3	103.3							

Polymer (B): **poly(ethylene-*co*-1-butene)** **2008KOS**
Characterization: M_n/g.mol^{-1} = 230200, M_w/g.mol^{-1} = 232500, 20.2 mol% 1-butene, 10 ethyl branches per 100 backbone C-atomes, completely hydrogenated polybutadiene, synthesized in the laboratory
Solvent (A): **ethane** C_2H_6 **74-84-0**

Type of data: cloud points

w_B 0.043 was kept constant

T/K	367.65	369.85	376.65	383.35	393.15	403.45	423.65
P/MPa	128.4	126.0	123.2	121.2	118.8	115.9	111.6

Polymer (B): **poly(ethylene-*co*-1-butene)** **2004KER**
Characterization: M_n/g.mol^{-1} = 230000, M_w/g.mol^{-1} = 232500, 20.2 mol% 1-butene, 10 ethyl branches per 100 backbone C-atomes, completely hydrogenated polybutadiene, synthesized in the laboratory
Solvent (A): **ethane** C_2H_6 **74-84-0**

Type of data: cloud points

w_B 0.002-0.007 *T*/K 403.15 *P*/bar 1040

Polymer (B): **poly(ethylene-*co*-1-butene)** **1999CHE**
Characterization: M_n/g.mol^{-1} = 6430, M_w/g.mol^{-1} = 6690, 8.6 mol% 1-butene, completely hydrogenated polybutadiene, 8.6 mol% 1,2-units, Polymer Source, Inc., Dorval, Quebec
Solvent (A): **ethene** C_2H_4 **74-85-1**

Type of data: cloud points

continued

continued

w_B 0.00015 was kept constant

T/K	431.05	430.35	414.65	413.75	398.75	398.55	382.55	367.75	367.55
P/bar	591.1	591.3	596.4	589.1	598.7	596.9	602.2	620.0	621.3
T/K	352.45	352.45	337.65	324.15	317.15	310.35			
P/bar	653.1	652.5	683.5	761.4	998.9	1418			

Comments: The last two data points are temperature-induced phase transitions.

w_B 0.00096 was kept constant

T/K	425.55	424.55	414.45	411.75	404.15	401.05	391.75	383.75	372.35
P/bar	742.8	745.8	751.8	753.4	763.1	765.2	776.8	790.2	812.5
T/K	361.95	355.05	347.95	346.65					
P/bar	855.5	995.2	1280	1471					

Comments: The last three data points are temperature-induced phase transitions.

w_B 0.0065 was kept constant

T/K	426.95	426.85	414.85	413.15	400.05	397.05	389.25	387.55	379.65
P/bar	892.8	896.2	923.8	928.8	970.1	977.6	1010	1018	1055
T/K	377.05	370.45	369.75	368.95	368.25				
P/bar	1063	1163	1290	1370	1469				

Comments: The last three data points are temperature-induced phase transitions.

Polymer (B): **poly(ethylene-*co*-1-butene)** **1999CHE**
Characterization: M_n/g.mol^{-1} = 7570, M_w/g.mol^{-1} = 7870, 29.0 mol% 1-butene, completely hydrogenated polybutadiene, 29 mol% 1,2-units, Polymer Source, Inc., Dorval, Quebec
Solvent (A): **ethene** **C_2H_4** **74-85-1**

Type of data: cloud points

w_B 0.0013 was kept constant

T/K	432.35	419.45	407.15	403.95	388.85	387.85	372.45	372.15	357.55
P/bar	732.5	737.8	739.1	752.0	772.0	770.7	800.0	797.0	833.4
T/K	357.25	345.05	343.35	328.55	327.45	316.65	306.55	295.75	288.35
P/bar	834.2	867.7	870.2	928.9	933.9	985.0	1043	1124	1265

w_B 0.0079 was kept constant

T/K	425.75	412.05	398.55	384.25	383.15	368.65	367.15	353.05	337.75
P/bar	846.8	860.4	884.1	919.3	922.2	965.2	965.4	1024	1092
T/K	323.35	322.65	313.65	307.25					
P/bar	1097	1208	1285	1385					

Polymer (B): **poly(ethylene-*co*-1-butene)** **1999CHE**

Characterization: M_n/g.mol^{-1} = 115000, M_w/g.mol^{-1} = 120000, 87.5 mol% 1-butene, completely hydrogenated polybutadiene, 87.5 mol% 1,2-units, Polymer Source, Inc., Dorval, Quebec

Solvent (A): **ethene** C_2H_4 **74-85-1**

Type of data: cloud points

w_B 0.0012 was kept constant

T/K	424.25	409.25	389.35	372.25	353.25	352.45	333.45	318.25	305.05
P/bar	594.1	595.2	596.4	582.4	577.8	578.8	576.2	583.8	586.8
T/K	288.55	281.35	271.75	266.55	261.15	257.65	253.55	249.75	246.05
P/bar	591.8	593.7	597.9	604.6	629.1	645.8	674.2	693.2	711.1
T/K	235.85	230.25	225.15						
P/bar	768.1	821.4	862.2						

w_B 0.0091 was kept constant

T/K	429.45	407.95	388.05	365.85	350.15	334.95	320.05	300.95	299.35
P/bar	702.9	705.0	709.7	708.3	708.7	712.0	718.5	745.6	754.1
T/K	297.75	292.45	273.75	266.75	260.75	257.45	252.75	245.45	240.15
P/bar	759.6	768.1	818.2	851.6	880.2	901.2	933.7	988.1	1034
T/K	235.25	231.65	229.55						
P/bar	1089	1136	1168						

Polymer (B): **poly(ethylene-*co*-1-butene), deuterated** **2008KOS**

Characterization: M_n/g.mol^{-1} = 222700, M_w/g.mol^{-1} = 245000, 20.2 mol% 1-butene, 10 ethyl branches per 100 backbone C-atomes, completely deuterated polybutadiene, synthesized in the laboratory

Solvent (A): **2-methylbutane** C_5H_{12} **78-78-4**

Type of data: cloud points

w_B 0.005 was kept constant

T/K	371.15	391.75	414.45	437.05
P/MPa	1.5	4.6	8.6	12.0

w_B 0.019 was kept constant

T/K	367.95	378.25	386.25	410.85	426.95	448.15
P/MPa	1.6	3.4	5.1	8.8	11.1	14.3

Comments: Solid-liquid equilibrium data are additionally given in 2008KOS.

Polymer (B): **poly(ethylene-*co*-1-butene)** **2004KER**
Characterization: M_n/g.mol^{-1} = 230000, M_w/g.mol^{-1} = 232500, 20.2 mol% 1-butene, 10 ethyl branches per 100 backbone C-atomes, completely hydrogenated polybutadiene, synthesized in the laboratory
Solvent (A): **n-pentane** **C_5H_{12}** **109-66-0**

Type of data: cloud points

w_B 0.002-0.006 T/K 403.15 P/bar 40

Polymer (B): **poly(ethylene-*co*-1-butene), deuterated** **2008KOS**
Characterization: M_n/g.mol^{-1} = 222700, M_w/g.mol^{-1} = 245000, 20.2 mol% 1-butene, 10 ethyl branches per 100 backbone C-atomes, completely deuterated polybutadiene, synthesized in the laboratory
Solvent (A): **n-pentane** **C_5H_{12}** **109-66-0**

Type of data: cloud points

w_B	0.005	0.005	0.005	0.005	0.018	0.018
T/K	395.65	408.45	420.95	439.55	389.85	408.45
P/MPa	2.6	4.6	6.8	9.5	1.8	5.0

Comments: Solid-liquid equilibrium data are additionally given in 2008KOS.

Polymer (B): **poly(ethylene-*co*-1-butene)** **1998HAN**
Characterization: M_n/g.mol^{-1} = 52450, M_w/g.mol^{-1} = 104900, 5.3 mol% 1-butene, T_m/K = 385.2, synthesized with metallocene catalyst
Solvent (A): **propane** **C_3H_8** **74-98-6**

Type of data: cloud points

w_B	0.05	0.05	0.05	0.05	0.05
T/K	374.55	398.45	423.55	446.95	471.65
P/bar	592.2	580.8	576.9	576.6	577.6

Polymer (B): **poly(ethylene-*co*-1-butene)** **1998HAN**
Characterization: M_n/g.mol^{-1} = 38350, M_w/g.mol^{-1} = 76700, 7.5 mol% 1-butene, T_m/K = 381.2, synthesized with metallocene catalyst
Solvent (A): **propane** **C_3H_8** **74-98-6**

Type of data: cloud points

w_B	0.01	0.01	0.01	0.01	0.01	0.01	0.05	0.05	0.05
T/K	353.85	374.35	397.95	423.35	447.25	471.65	373.15	398.35	423.05
P/bar	587.1	569.4	560.5	561.7	561.5	563.4	571.9	563.6	561.8

w_B	0.05	0.05
T/K	446.55	473.15
P/bar	562.7	565.6

Polymer (B): **poly(ethylene-*co*-1-butene)** **1998HAN**
Characterization: M_n/g.mol^{-1} = 50000, M_w/g.mol^{-1} = 94900, 8.8 mol% 1-butene, T_m/K = 371.2, synthesized with metallocene catalyst
Solvent (A): **propane** **C_3H_8** **74-98-6**

Type of data: cloud points

w_B	0.01	0.01	0.01	0.01	0.01	0.01	0.01	0.05	0.05
T/K	348.35	363.55	373.95	400.35	421.65	446.35	471.05	374.15	397.85
P/bar	580.1	565.8	556.6	556.2	553.4	558.0	561.1	563.5	556.2
w_B	0.05	0.05	0.05						
T/K	423.45	447.15	472.05						
P/bar	556.4	558.4	561.9						

Polymer (B): **poly(ethylene-*co*-1-butene)** **1998HAN**
Characterization: M_n/g.mol^{-1} = 41600, M_w/g.mol^{-1} = 79000, 15.9 mol% 1-butene, T_m/K = 350.2, synthesized with metallocene catalyst
Solvent (A): **propane** **C_3H_8** **74-98-6**

Type of data: cloud points

w_B	0.01	0.01	0.01	0.01	0.01	0.01	0.01	0.01	0.05
T/K	313.55	322.95	348.85	373.15	396.55	423.25	446.55	470.35	373.15
P/bar	616.1	523.2	502.6	506.7	500.3	509.1	515.2	523.5	497.4
w_B	0.05	0.05	0.05	0.05					
T/K	398.65	422.45	447.65	474.35					
P/bar	501.8	508.6	517.0	525.0					

Polymer (B): **poly(ethylene-*co*-1-butene)** **1998HAN**
Characterization: M_n/g.mol^{-1} = 29000, M_w/g.mol^{-1} = 58100, 18.0 mol% 1-butene, T_m/K = 335.2, synthesized with metallocene catalyst
Solvent (A): **propane** **C_3H_8** **74-98-6**

Type of data: cloud points

w_B	0.01	0.01	0.01	0.01	0.01	0.01	0.01	0.01	0.05
T/K	324.25	334.55	348.15	373.55	398.85	422.15	446.65	471.85	348.25
P/bar	448.3	443.7	442.1	448.4	460.4	471.7	482.7	493.7	448.2
w_B	0.05	0.05	0.05	0.05	0.05				
T/K	372.55	395.95	422.95	447.45	469.65				
P/bar	453.2	462.7	475.7	487.7	496.5				

Polymer (B): **poly(ethylene-*co*-1-butene)** **1998HAN**
Characterization: M_n/g.mol^{-1} = 79300, M_w/g.mol^{-1} = 87200, 23.3 mol% 1-butene, T_m/K = 319.2, completely hydrogenated polybutadiene
Solvent (A): **propane** **C_3H_8** **74-98-6**

continued

continued

Type of data: cloud points

w_B	0.01	0.01	0.01	0.01	0.01	0.01	0.01	0.05	0.05
T/K	295.45	324.05	348.75	374.65	398.15	423.05	447.65	295.15	324.15
P/bar	376.5	373.1	384.1	401.2	418.3	437.1	450.6	391.0	385.4

w_B	0.05	0.05	0.05	0.05	0.05
T/K	348.85	373.95	397.95	422.95	447.55
P/bar	396.2	411.0	428.2	445.2	460.4

Polymer (B): **poly(ethylene-*co*-1-butene)** **1998HAN**
Characterization: M_n/g.mol^{-1} = 79800, M_w/g.mol^{-1} = 87800, 60.0 mol% 1-butene, amorphous, completely hydrogenated polybutadiene
Solvent (A): **propane** C_3H_8 **74-98-6**

Type of data: cloud points

w_B	0.01	0.01	0.01	0.01	0.01	0.01	0.01	0.01	0.05
T/K	296.55	323.15	349.05	373.95	397.25	423.55	445.25	472.15	333.85
P/bar	82.4	147.3	200.3	244.4	279.7	313.5	338.0	362.2	289.7

w_B	0.05	0.05	0.05	0.05	0.05	0.05
T/K	348.25	372.65	398.75	422.95	447.65	473.05
P/bar	282.6	317.6	355.5	387.8	412.9	431.3

Polymer (B): **poly(ethylene-*co*-1-butene)** **2004KER**
Characterization: M_n/g.mol^{-1} = 230000, M_w/g.mol^{-1} = 232500, 20.2 mol% 1-butene, 10 ethyl branches per 100 backbone C-atomes, completely hydrogenated polybutadiene, synthesized in the laboratory
Solvent (A): **propane** C_3H_8 **74-98-6**

Type of data: cloud points

w_B 0.002-0.006 T/K 403.15 P/bar 460

Polymer (B): **poly(ethylene-*co*-butyl acrylate)** **1999DIE**
Characterization: M_n/g.mol^{-1} = 16400, M_w/g.mol^{-1} = 294000, 1.0 mol% butyl acrylate, synthesized in the laboratory
Solvent (A): **ethene** C_2H_4 **74-85-1**

Type of data: cloud points

w_B 0.03 was kept constant

T/K	383.15	393.15	403.15	413.15	423.15
P/bar	1900	1803	1725	1670	1600

Polymer (B): **poly(ethylene-*co*-butyl acrylate)** **1999DIE**
Characterization: M_n/g.mol^{-1} = 44300, M_w/g.mol^{-1} = 162000,
5.1 mol% butyl acrylate, synthesized in the laboratory
Solvent (A): **ethene** **C_2H_4** **74-85-1**

Type of data: cloud points

w_B 0.03 was kept constant

T/K	388.15	393.15	393.15	398.15	398.15	403.15	408.15	413.15	418.15
P/bar	1900	1881	1870	1822	1796	1750	1677	1673	1628
T/K	423.15	423.15	433.15	434.15	438.15	443.15	443.15	453.15	463.15
P/bar	1581	1565	1546	1545	1541	1498	1490	1490	1452
T/K	472.15	483.15							
P/bar	1419	1392							

Polymer (B): **poly(ethylene-*co*-butyl acrylate)** **1999DIE**
Characterization: M_n/g.mol^{-1} = 41100, M_w/g.mol^{-1} = 296000,
6.7 mol% butyl acrylate, synthesized in the laboratory
Solvent (A): **ethene** **C_2H_4** **74-85-1**

Type of data: cloud points

w_B 0.03 was kept constant

T/K	363.15	368.15	384.15	403.15	423.15
P/bar	1770	1750	1595	1510	1438

Polymer (B): **poly(ethylene-*co*-butyl acrylate)** **1999DIE**
Characterization: M_n/g.mol^{-1} = 39900, M_w/g.mol^{-1} = 155000,
8.7 mol% butyl acrylate, synthesized in the laboratory
Solvent (A): **ethene** **C_2H_4** **74-85-1**

Type of data: cloud points

w_B 0.03 was kept constant

T/K	333.15	343.15	343.15	353.15	353.15	363.15	363.15	373.15	373.15
P/bar	1655	1582	1562	1503	1472	1460	1432	1415	1402
T/K	383.15	384.15	393.15	403.15	403.15	413.15	423.15	423.15	432.15
P/bar	1355	1380	1320	1320	1318	1289	1270	1259	1255
T/K	443.15	443.15	453.15	463.15	473.15	483.15			
P/bar	1230	1242	1212	1189	1173	1160			

Polymer (B): **poly(ethylene-*co*-butyl acrylate)** **1999DIE**
Characterization: 17.2 mol% butyl acrylate, synthesized in the laboratory
Solvent (A): **ethene** **C_2H_4** **74-85-1**

Type of data: cloud points

continued

continued

w_B 0.03 was kept constant

T/K	328.15	331.15	336.15	341.15	347.15	354.15	364.15	383.15	405.15
P/bar	1400	1400	1370	1340	1324	1300	1290	1220	1190
T/K	421.15	443.15							
P/bar	1260	1130							

Polymer (B): **poly(ethylene-*co*-butyl acrylate)** **1999DIE**
Characterization: 22.2 mol% butyl acrylate, synthesized in the laboratory
Solvent (A): **ethene** C_2H_4 **74-85-1**

Type of data: cloud points

w_B 0.03 was kept constant

T/K	318.15	323.15	333.15	343.15	353.15	363.15	383.15	393.15	403.15
P/bar	1355	1320	1295	1260	1230	1210	1175	1158	1145
T/K	413.15	423.15	433.15	443.15	453.15	463.15	473.15		
P/bar	1130	1127	1125	1120	1092	1090	1080		

Polymer (B): **poly(ethylene-*co*-butyl acrylate)** **1999DIE**
Characterization: M_n/g.mol^{-1} = 29700, M_w/g.mol^{-1} = 110500,
25.2 mol% butyl acrylate, synthesized in the laboratory
Solvent (A): **ethene** C_2H_4 **74-85-1**

Type of data: cloud points

w_B 0.03 was kept constant

T/K	343.15	363.15	383.15	403.15	403.15	423.15	443.15	463.15	473.15
P/bar	1240	1173	1165	1132	1125	1092	1058	1040	1037

Polymer (B): **poly(ethylene-*co*-butyl acrylate)** **2004BEC**
Characterization: M_n/g.mol^{-1} = 49800, M_w/g.mol^{-1} = 159500,
27.0 mol% butyl acrylate, synthesized in the laboratory
Solvent (A): **ethene** C_2H_4 **74-85-1**

Type of data: cloud points

w_B 0.05 was kept constant

T/K	354.15	374.15	393.15	412.15	432.15	452.15	471.15	489.15	510.15
P/bar	1144	1106	1071	1052	1033	1014	993	977	965
T/K	530.15								
P/bar	953								

Polymer (B): **poly(ethylene-*co*-butyl methacrylate)** **2004LAT**
Characterization: 6.7 mol% butyl methacrylate, synthesized in the laboratory
Solvent (A): **ethene** C_2H_4 **74-85-1**

Type of data: cloud points

w_B 0.05 was kept constant

T/K	374.15	393.15	414.15	434.15	453.15	472.15	491.15	513.15	531.15
P/bar	1414	1330	1264	1208	1169	1134	1106	1079	1052

Polymer (B): **poly(ethylene-*co*-butyl methacrylate)** **2004LAT**
Characterization: M_n/g.mol^{-1} = 23400, M_w/g.mol^{-1} = 39800,
11.7 mol% butyl methacrylate, synthesized in the laboratory
Solvent (A): **ethene** C_2H_4 **74-85-1**

Type of data: cloud points

w_B 0.05 was kept constant

T/K	372.15	393.15	411.15	432.15	451.15	470.15	489.15	510.15	531.15
P/bar	1315	1239	1194	1161	1122	1095	1063	1039	1017

Polymer (B): **poly(ethylene-*co*-butyl methacrylate)** **2004BEC**
Characterization: M_n/g.mol^{-1} = 23400, M_w/g.mol^{-1} = 39800,
11.7 mol% butyl methacrylate, synthesized in the laboratory
Solvent (A): **ethene** C_2H_4 **74-85-1**

Type of data: cloud points

w_B 0.05 was kept constant

T/K	373.15	393.15	413.15	433.15	453.15	473.15	488.15	508.15	528.15
P/bar	1250	1200	1151	1114	1089	1062	1041	1021	1000

Polymer (B): **poly(ethylene-*co*-butyl methacrylate)** **2004BEC, 2004LAT**
Characterization: M_n/g.mol^{-1} = 24800, M_w/g.mol^{-1} = 41000,
18.5 mol% butyl methacrylate, synthesized in the laboratory
Solvent (A): **ethene** C_2H_4 **74-85-1**

Type of data: cloud points

w_B 0.05 was kept constant

T/K	353.15	373.15	393.15	413.15	433.15	453.15	473.15	488.15	508.15
P/bar	1082	1045	1016	993	973	958	946	930	915
T/K	528.15								
P/bar	901								

Polymer (B): **poly(ethylene-*co*-butyl methacrylate)** **2004LAT**
Characterization: M_n/g.mol^{-1} = 24500, M_w/g.mol^{-1} = 44400,
22.7 mol% butyl methacrylate, synthesized in the laboratory
Solvent (A): **ethene** **C_2H_4** **74-85-1**

Type of data: cloud points

w_B 0.05 was kept constant

T/K	353.15	374.15	394.15	413.15	432.15	452.15	472.15	491.15	511.15
P/bar	1121	1085	1054	1028	1008	987	970	951	939
T/K	529.15								
P/bar	922								

Polymer (B): **poly(ethylene-*co*-butyl methacrylate)** **2004LAT**
Characterization: M_n/g.mol^{-1} = 55000, M_w/g.mol^{-1} = 96700,
33.8 mol% butyl methacrylate, synthesized in the laboratory
Solvent (A): **ethene** **C_2H_4** **74-85-1**

Type of data: cloud points

w_B 0.05 was kept constant

T/K	350.15	372.15	393.15	413.15	433.15	453.15	471.15	491.15	510.15
P/bar	1002	974	954	940	925	912	901	891	879
T/K	530.15								
P/bar	876								

Polymer (B): **poly(ethylene-*co*-butyl methacrylate)** **2004LAT**
Characterization: M_n/g.mol^{-1} = 58900, M_w/g.mol^{-1} = 101700,
44.0 mol% butyl methacrylate, synthesized in the laboratory
Solvent (A): **ethene** **C_2H_4** **74-85-1**

Type of data: cloud points

w_B 0.05 was kept constant

T/K	362.15	374.15	393.15	413.15	432.15	452.15	471.15	490.15	511.15
P/bar	1064	1038	1023	1005	995	969	957	952	938
T/K	531.15								
P/bar	924								

Polymer (B): **poly(ethylene-*co*-butyl methacrylate)** **2004BEC**
Characterization: M_n/g.mol^{-1} = 58900, M_w/g.mol^{-1} = 101700,
44.0 mol% butyl methacrylate, synthesized in the laboratory
Solvent (A): **ethene** **C_2H_4** **74-85-1**

Type of data: cloud points

continued

continued

w_B 0.05 was kept constant

T/K	383.15	393.15	413.15	433.15	453.15	473.15	493.15	513.15	533.15
P/bar	885	878	866	857	847	840	834	823	813

Polymer (B): **poly(ethylene-*co*-butyl methacrylate-*co*-methacrylic acid)** **2004LAT, 2007TUM**

Characterization: M_n/g.mol^{-1} = 11500, M_w/g.mol^{-1} = 33200, 0.4 mol% butyl methacrylate, 6.6 mol% methacrylic acid, synthesized in the laboratory by partial esterification

Solvent (A): **ethene** **C_2H_4** **74-85-1**

Type of data: cloud points

w_B 0.05 was kept constant

T/K	474	483	490	490	492	493	498	501	503
P/bar	2750	2505	2328	2306	2283	2240	2147	2071	2035
T/K	508	512	518	521					
P/bar	1946	1869	1795	1742					

Polymer (B): **poly(ethylene-*co*-butyl methacrylate-*co*-methacrylic acid)** **2004LAT**

Characterization: 0.6 mol% butyl methacrylate, 6.4 mol% methacrylic acid, synthesized in the laboratory by partial esterification

Solvent (A): **ethene** **C_2H_4** **74-85-1**

Type of data: cloud points

w_B 0.05 was kept constant

T/K	475.85	476.15	477.85	481.65	482.55	489.65	489.95	493.65	493.85
P/bar	2647	2643	2584	2469	2474	2269	2258	2160	2158
T/K	502.15	512.55							
P/bar	1991	1810							

Polymer (B): **poly(ethylene-*co*-butyl methacrylate-*co*-methacrylic acid)** **2004LAT**

Characterization: 0.8 mol% butyl methacrylate, 6.3 mol% methacrylic acid, synthesized in the laboratory by partial esterification

Solvent (A): **ethene** **C_2H_4** **74-85-1**

Type of data: cloud points

w_B 0.05 was kept constant

T/K	467.15	472.15	478.15	483.15	490.15	492.15	498.15	504.15	514.15
P/bar	2707	2565	2429	2327	2222	2175	2062	1972	1845
T/K	523.15	530.15							
P/bar	1762	1681							

Polymer (B): **poly(ethylene-*co*-butyl methacrylate-*co*-methacrylic acid)** **2004LAT, 2007TUM**
Characterization: M_n/g.mol^{-1} = 11500, M_w/g.mol^{-1} = 33200, 0.9 mol% butyl methacrylate, 6.1 mol% methacrylic acid, synthesized in the laboratory by partial esterification
Solvent (A): **ethene** C_2H_4 **74-85-1**

Type of data: cloud points

w_B 0.05 was kept constant

T/K	471	472	478	479	482	487	488	492	497
P/bar	2608	2566	2455	2364	2292	2168	2148	2076	1952
T/K	502	506	508	513	517	518	522	527	528
P/bar	1888	1813	1786	1728	1678	1672	1626	1573	1556

Polymer (B): **poly(ethylene-*co*-butyl methacrylate-*co*-methacrylic acid)** **2004LAT, 2007TUM**
Characterization: M_n/g.mol^{-1} = 11500, M_w/g.mol^{-1} = 33200, 1.7 mol% butyl methacrylate, 5.6 mol% methacrylic acid, synthesized in the laboratory by partial esterification
Solvent (A): **ethene** C_2H_4 **74-85-1**

Type of data: cloud points

w_B 0.05 was kept constant

T/K	479	480	493	500	511	520
P/bar	2070	2076	1875	1735	1680	1605

Polymer (B): **poly(ethylene-*co*-butyl methacrylate-*co*-methacrylic acid)** **2004LAT, 2007TUM**
Characterization: M_n/g.mol^{-1} = 11500, M_w/g.mol^{-1} = 33200, 3.5 mol% butyl methacrylate, 3.8 mol% methacrylic acid, synthesized in the laboratory by partial esterification
Solvent (A): **ethene** C_2H_4 **74-85-1**

Type of data: cloud points

w_B 0.05 was kept constant

T/K	435	451	462	471	481	490	501	521	529
P/bar	2650	2264	2047	1915	1789	1682	1468	1377	1321

Polymer (B): **poly(ethylene-*co*-butyl methacrylate-*co*-methacrylic acid)** **2007TUM**
Characterization: M_n/g.mol^{-1} = 11500, M_w/g.mol^{-1} = 33200, 4.2 mol% butyl methacrylate, 2.9 mol% methacrylic acid, synthesized in the laboratory by partial esterification

continued

continued

Solvent (A): **ethene** C_2H_4 **74-85-1**

Type of data: cloud points

w_B 0.05 was kept constant

T/K	433	446	452	453	465	471	472	482	483
P/bar	2195	2012	1904	1888	1678	1590	1580	1481	1477
T/K	493	501	502	514	522	531	532		
P/bar	1387	1323	1327	1258	1225	1223	1180		

Polymer (B): **poly(ethylene-*co*-butyl methacrylate-*co*-methacrylic acid)** **2004LAT, 2007TUM**

Characterization: M_n/g.mol^{-1} = 11500, M_w/g.mol^{-1} = 33200, 5.5 mol% butyl methacrylate, 1.8 mol% methacrylic acid, synthesized in the laboratory by partial esterification

Solvent (A): **ethene** C_2H_4 **74-85-1**

Type of data: cloud points

w_B 0.05 was kept constant

T/K	373	393	413	433	444	445	452	456	465
P/bar	2130	1860	1645	1460	1366	1374	1315	1301	1247
T/K	466	472	487	491	507	511			
P/bar	1246	1212	1136	1138	1078	1066			

Polymer (B): **poly(ethylene-*co*-butyl methacrylate-*co*-methacrylic acid)** **2004LAT**

Characterization: 5.5 mol% butyl methacrylate, 7.3 mol% methacrylic acid, synthesized in the laboratory by partial esterification

Solvent (A): **ethene** C_2H_4 **74-85-1**

Type of data: cloud points

w_B 0.05 was kept constant

T/K	486.15	494.15	498.15	504.15	508.15	513.15	519.15	528.15	532.15
P/bar	2530	2324	2231	2123	2041	1944	1873	1779	1750

Polymer (B): **poly(ethylene-*co*-butyl methacrylate-*co*-methacrylic acid)** **2004LAT**

Characterization: 6.2 mol% butyl methacrylate, 0.9 mol% methacrylic acid, synthesized in the laboratory by partial esterification

Solvent (A): **ethene** C_2H_4 **74-85-1**

Type of data: cloud points

continued

continued

w_B	0.05	was kept constant							
T/K	373.15	381.85	392.15	403.95	412.15	429.05	432.05	445.15	453.15
P/bar	1717	1625	1529	1448	1405	1340	1319	1272	1236
T/K	461.15	471.15	480.45	490.75	500.65	518.15	520.15		
P/bar	1218	1196	1171	1149	1127	1097	1096		

Polymer (B): **poly(ethylene-*co*-butyl methacrylate-*co*-methacrylic acid)** **2004LAT**
Characterization: 6.8 mol% butyl methacrylate, 0.3 mol% methacrylic acid, synthesized in the laboratory by partial esterification
Solvent (A): **ethene** C_2H_4 **74-85-1**

Type of data: cloud points

w_B	0.05	was kept constant							
T/K	354.15	363.65	372.35	383.05	391.15	394.45	411.65	428.75	430.45
P/bar	1626	1546	1486	1420	1379	1361	1288	1229	1223
T/K	443.15	451.95	464.65	473.15	485.35	492.35	511.25		
P/bar	1190	1170	1144	1126	1104	1092	1064		

Polymer (B): **poly(ethylene-*co*-ethyl acrylate)** **2004BEC, 2004LAT**
Characterization: 3.6 mol% ethyl acrylate, synthesized in the laboratory
Solvent (A): **ethene** C_2H_4 **74-85-1**

Type of data: cloud points

w_B	0.05	was kept constant							
T/K	384.05	395.15	414.85	435.15	454.15	473.15	492.15	514.15	531.15
P/bar	1631	1568	1471	1394	1336	1288	1247	1207	1175

Polymer (B): **poly(ethylene-*co*-ethyl acrylate)** **2004BEC, 2004LAT**
Characterization: 6.3 mol% ethyl acrylate, synthesized in the laboratory
Solvent (A): **ethene** C_2H_4 **74-85-1**

Type of data: cloud points

w_B	0.05	was kept constant							
T/K	372.15	393.15	413.15	433.15	452.15	472.55	492.15	511.15	531.35
P/bar	1510	1420	1351	1295	1252	1212	1181	1147	1122

Polymer (B): **poly(ethylene-*co*-ethyl acrylate)** **2004BEC, 2004LAT**
Characterization: M_n/g.mol^{-1} = 78400, M_w/g.mol^{-1} = 156600, 23.4 mol% ethyl acrylate, synthesized in the laboratory
Solvent (A): **ethene** C_2H_4 **74-85-1**

continued

continued

Type of data: cloud points

w_B 0.05 was kept constant

T/K	353.15	364.15	373.75	393.15	413.15	434.15	454.15	473.15	493.65
P/bar	1280	1256	1236	1190	1155	1123	1095	1072	1047
T/K	513.55	533.15							
P/bar	1025	1000							

Polymer (B): **poly(ethylene-*co*-ethyl acrylate)** **2004BEC, 2004LAT**
Characterization: M_n/g.mol^{-1} = 64400, M_w/g.mol^{-1} = 116700,
28.4 mol% ethyl acrylate, synthesized in the laboratory
Solvent (A): **ethene** C_2H_4 **74-85-1**

Type of data: cloud points

w_B 0.05 was kept constant

T/K	363.15	373.15	383.15	393.15	413.15	433.15	453.15	473.15	493.15
P/bar	1236	1206	1183	1165	1131	1100	1072	1048	1022
T/K	513.15	533.15							
P/bar	1000	982							

Polymer (B): **poly(ethylene-*co*-ethyl acrylate)** **2004LAT**
Characterization: M_n/g.mol^{-1} = 75900, M_w/g.mol^{-1} = 159500,
34.8 mol% ethyl acrylate, synthesized in the laboratory
Solvent (A): **ethene** C_2H_4 **74-85-1**

Type of data: cloud points

w_B 0.05 was kept constant

T/K	366.65	374.65	397.15	412.15	434.15	454.65	474.65	491.15	513.15
P/bar	1296	1272	1219	1191	1151	1120	1090	1066	1039
T/K	532.15								
P/bar	1018								

Polymer (B): **poly(ethylene-*co*-ethyl acrylate)** **2004LAT**
Characterization: M_n/g.mol^{-1} = 67600, M_w/g.mol^{-1} = 142900,
45.7 mol% ethyl acrylate, synthesized in the laboratory
Solvent (A): **ethene** C_2H_4 **74-85-1**

Type of data: cloud points

w_B 0.05 was kept constant

T/K	372.75	391.95	412.15	433.15	453.15	472.15	491.15	511.15	529.15
P/bar	1324	1265	1217	1173	1135	1104	1076	1048	1025

Polymer (B): **poly(ethylene-*co*-2-ethylhexyl acrylate)** **1999DIE**
Characterization: M_n/g.mol^{-1} = 34900, M_w/g.mol^{-1} = 107700,
9.3 mol% 2-ethylhexyl acrylate, synthesized in the laboratory
Solvent (A): **ethene** C_2H_4 **74-85-1**

Type of data: cloud points

w_B 0.03 was kept constant

T/K	363.15	383.15	403.15	423.15	443.15	462.15	483.15
P/bar	1640	1590	1440	1390	1340	1310	1290

Polymer (B): **poly(ethylene-*co*-2-ethylhexyl acrylate)** **1999DIE**
Characterization: M_n/g.mol^{-1} = 21200, M_w/g.mol^{-1} = 49800,
17.0 mol% 2-ethylhexyl acrylate, synthesized in the laboratory
Solvent (A): **ethene** C_2H_4 **74-85-1**

Type of data: cloud points

w_B 0.03 was kept constant

T/K	363.15	383.15	403.15	423.15	443.15	463.15	483.15
P/bar	1240	1220	1145	1125	1085	1060	1010

Polymer (B): **poly(ethylene-*co*-2-ethylhexyl acrylate)** **1999DIE**
Characterization: 22.8 mol% 2-ethylhexyl acrylate, synthesized in the laboratory
Solvent (A): **ethene** C_2H_4 **74-85-1**

Type of data: cloud points

w_B 0.03 was kept constant

T/K	353.15	363.15	383.15	403.15	423.15	443.15	463.15	483.15
P/bar	1070	1055	1025	1005	995	995	965	945

Polymer (B): **poly(ethylene-*co*-1-hexene)** **2001DO1**
Characterization: M_n/g.mol^{-1} = 60000, M_w/g.mol^{-1} = 129000, 16.1 wt% 1-hexene
Solvent (A): **ethene** C_2H_4 **74-85-1**

Type of data: cloud points

w_B	0.150	0.150	0.150	0.150	0.150	0.150
T/K	393.15	413.15	433.15	453.15	473.15	493.15
P/MPa	156.0	144.3	137.3	129.1	123.1	118.6

Polymer (B): **poly(ethylene-*co*-1-hexene)** **2001DO2**
Characterization: M_n/g.mol^{-1} = 60000, M_w/g.mol^{-1} = 129000, 16.1 wt% 1-hexene
Solvent (A): **ethene** C_2H_4 **74-85-1**

Type of data: coexistence data

continued

continued

T/K = 433.15

w_B(total)	0.15	was kept constant			
P/bar	974	1149	1245	1363	1373
w_B(sol phase)	0.009	0.020	0.029	0.039	–
w_B(gel phase)	0.485	0.390	0.320	–	0.150

Polymer (B): **poly(ethylene-*co*-1-hexene)** **1999PAN**
Characterization: M_n/g.mol^{-1} = 46500, M_w/g.mol^{-1} = 102000, 8.5 wt% 1-hexene, MI = 1.2, ρ = 0.912 g/cm^3, T_m/K = 409.95
Solvent (A): **2-methylpropane** C_4H_{10} **75-28-5**

Type of data: cloud points

w_B	0.110	0.110	0.110	0.110	0.110	0.110	0.110	0.110	0.110
T/K	362.15	362.25	376.75	377.55	379.35	393.15	393.35	399.45	405.15
P/bar	353.4	353.4	352.8	352.5	353.0	354.2	354.8	356.0	356.5

w_B	0.110	0.110	0.110	0.110
T/K	407.55	410.45	423.05	423.25
P/bar	357.9	358.5	362.8	364.3

Polymer (B): **poly(ethylene-*co*-1-hexene)** **1999PAN**
Characterization: M_n/g.mol^{-1} = 54800, M_w/g.mol^{-1} = 109500, 13.6 wt% 1-hexene, MI = 1.2, ρ = 0.900 g/cm^3, T_m/K = 406.25
Solvent (A): **2-methylpropane** C_4H_{10} **75-28-5**

Type of data: cloud points

w_B	0.112	0.112	0.112	0.112	0.112	0.112	0.112	0.112	0.112
T/K	355.65	358.35	368.55	368.55	377.85	378.65	392.05	395.25	405.45
P/bar	317.9	318.7	318.0	318.2	319.8	319.7	324.8	324.6	330.4

w_B	0.112	0.112	0.112	0.112
T/K	407.55	410.25	421.75	424.65
P/bar	331.5	331.3	338.0	338.5

Polymer (B): **poly(ethylene-*co*-1-hexene)** **1999PAN**
Characterization: M_n/g.mol^{-1} = 45500, M_w/g.mol^{-1} = 90800, 13.7 wt% 1-hexene, MI = 2.2, ρ = 0.900 g/cm^3, T_m/K = 406.25
Solvent (A): **2-methylpropane** C_4H_{10} **75-28-5**

Type of data: cloud points

w_B	0.111	0.111	0.111	0.111	0.111	0.111	0.111	0.111	0.111
T/K	362.65	362.65	373.15	373.25	383.15	383.25	392.55	393.35	407.95
P/bar	312.5	313.1	314.5	313.3	315.8	315.8	321.1	322.9	327.6

w_B	0.111	0.111	0.111
T/K	408.75	423.25	423.45
P/bar	328.7	335.3	336.2

Polymer (B): **poly(ethylene-*co*-1-hexene)** **1999PAN**
Characterization: M_n/g.mol^{-1} = 42800, M_w/g.mol^{-1} = 83500, 14.2 wt% 1-hexene, MI = 3.5, ρ = 0.900 g/cm^3, T_m/K = 405.95
Solvent (A): **2-methylpropane** C_4H_{10} **75-28-5**

Type of data: cloud points

w_B	0.110	0.110	0.110	0.110	0.110	0.110	0.110	0.110	0.110
T/K	358.25	358.75	368.15	368.95	377.45	380.25	393.25	394.15	406.45
P/bar	305.1	305.7	306.6	307.0	309.4	311.0	316.7	316.4	322.3
w_B	0.110	0.110	0.110						
T/K	408.85	423.45	423.85						
P/bar	323.6	331.2	331.1						

Polymer (B): **poly(ethylene-*co*-1-hexene)** **1999PAN**
Characterization: M_n/g.mol^{-1} = 31300, M_w/g.mol^{-1} = 68800, 14.6wt% 1-hexene, MI = 7.5, ρ = 0.900 g/cm^3, T_m/K = 405.55
Solvent (A): **2-methylpropane** C_4H_{10} **75-28-5**

Type of data: cloud points

w_B	0.112	0.112	0.112	0.112	0.112	0.112	0.112	0.112	0.112
T/K	357.45	360.35	366.35	368.25	376.65	378.45	393.15	394.05	406.15
P/bar	301.3	301.8	302.5	302.5	306.6	306.5	312.9	313.4	320.0
w_B	0.112	0.112	0.112						
T/K	409.15	422.95	423.25						
P/bar	320.9	327.5	327.1						

Polymer (B): **poly(ethylene-*co*-1-hexene)** **1999PAN**
Characterization: M_n/g.mol^{-1} = 47500, M_w/g.mol^{-1} = 94900, 21.6wt% 1-hexene, MI = 2.2, ρ = 0.885 g/cm^3, T_m/K = 400.25
Solvent (A): **2-methylpropane** C_4H_{10} **75-28-5**

Type of data: cloud points

w_B	0.109	0.109	0.109	0.109	0.109	0.109	0.109	0.109	0.109
T/K	358.55	359.75	362.75	367.95	368.95	377.55	377.65	391.95	393.85
P/bar	273.0	273.4	273.8	275.9	276.8	280.8	281.4	289.8	290.4
w_B	0.109	0.109	0.109	0.109	0.109				
T/K	395.35	407.15	409.15	423.55	423.55				
P/bar	292.1	298.9	300.7	309.9	310.2				

Polymer (B): **poly(ethylene-*co*-1-hexene)** **1998HAN**
Characterization: M_n/g.mol^{-1} = 42400, M_w/g.mol^{-1} = 93200, 2.6 mol% 1-hexene, T_m/K = 401.2, synthesized with metallocene catalyst
Solvent (A): **propane** C_3H_8 **74-98-6**

continued

continued

Type of data: cloud points

w_B	0.05	0.05	0.05	0.05
T/K	400.75	423.55	448.35	471.95
P/bar	591.2	586.0	584.2	584.7

Polymer (B): **poly(ethylene-*co*-1-hexene)** **1998HAN**
Characterization: M_n/g.mol^{-1} = 53000, M_w/g.mol^{-1} = 101000, 4.6 mol% 1-hexene, T_m/K = 391.2, synthesized with metallocene catalyst
Solvent (A): **propane** **C_3H_8** **74-98-6**

Type of data: cloud points

w_B	0.05	0.05	0.05	0.05	0.05
T/K	373.65	398.75	427.55	447.55	471.75
P/bar	575.2	565.8	563.9	565.3	567.4

Polymer (B): **poly(ethylene-*co*-1-hexene)** **1998HAN**
Characterization: M_n/g.mol^{-1} = 24100, M_w/g.mol^{-1} = 72300, 13.0 mol% 1-hexene, T_m/K = 349.2, synthesized with metallocene catalyst
Solvent (A): **propane** **C_3H_8** **74-98-6**

Type of data: cloud points

w_B	0.05	0.05	0.05	0.05	0.05	0.05
T/K	353.05	378.45	401.75	426.35	448.95	474.95
P/bar	471.5	472.9	479.3	489.3	495.9	505.5

Polymer (B): **poly(ethylene-*co*-1-hexene)** **1999PAN**
Characterization: M_n/g.mol^{-1} = 46500, M_w/g.mol^{-1} = 102000, 8.5 wt% 1-hexene, MI = 1.2, ρ = 0.912 g/cm^3, T_m/K = 409.95
Solvent (A): **propane** **C_3H_8** **74-98-6**

Type of data: cloud points

w_B	0.111	0.111	0.111	0.111	0.111	0.111	0.111	0.111	0.111
T/K	367.15	367.95	377.55	378.05	390.75	393.35	407.95	408.65	422.75
P/bar	520.8	521.1	520.0	520.4	520.4	521.2	521.8	522.1	525.3

w_B	0.111
T/K	423.45
P/bar	525.0

Polymer (B): **poly(ethylene-*co*-1-hexene)** **1999PAN**
Characterization: M_n/g.mol^{-1} = 54800, M_w/g.mol^{-1} = 109500, 13.6 wt% 1-hexene, MI = 1.2, ρ = 0.900 g/cm^3, T_m/K = 406.25
Solvent (A): **propane** **C_3H_8** **74-98-6**

Type of data: cloud points

continued

continued

w_B	0.109	0.109	0.109	0.109	0.109	0.109	0.109	0.109	0.109
T/K	357.15	358.75	367.35	367.55	379.95	381.65	390.55	394.05	408.65
P/bar	481.8	483.0	484.3	484.3	486.2	486.2	489.1	490.8	495.3

w_B	0.109	0.109	0.109
T/K	412.15	422.45	422.75
P/bar	496.7	500.1	500.1

Polymer (B): **poly(ethylene-*co*-1-hexene)** **1999PAN**
Characterization: M_n/g.mol^{-1} = 45500, M_w/g.mol^{-1} = 90800, 13.7 wt% 1-hexene, MI = 2.2, ρ = 0.900 g/cm^3, T_m/K = 406.25
Solvent (A): **propane** **C_3H_8** **74-98-6**

Type of data: cloud points

w_B	0.109	0.109	0.109	0.109	0.109	0.109	0.109	0.109	0.109
T/K	356.55	359.35	366.65	368.35	376.25	377.85	393.15	395.55	406.15
P/bar	460.9	461.5	461.8	462.1	463.8	464.3	468.8	468.9	473.9

w_B	0.109	0.109	0.109
T/K	408.15	423.15	424.15
P/bar	474.9	481.0	481.3

Polymer (B): **poly(ethylene-*co*-1-hexene)** **1999PAN**
Characterization: M_n/g.mol^{-1} = 42800, M_w/g.mol^{-1} = 85500, 14.2 wt% 1-hexene, MI = 3.5, ρ = 0.900 g/cm^3, T_m/K = 405.95
Solvent (A): **propane** **C_3H_8** **74-98-6**

Type of data: cloud points

w_B	0.109	0.109	0.109	0.109	0.109	0.109	0.109	0.109	0.109
T/K	353.25	355.65	364.05	365.15	375.15	377.75	390.15	393.55	408.45
P/bar	459.0	459.2	459.9	461.3	463.2	464.1	468.3	469.9	474.4

w_B	0.109	0.109	0.109
T/K	408.55	422.75	423.15
P/bar	474.2	478.2	478.4

Polymer (B): **poly(ethylene-*co*-1-hexene)** **1999PAN**
Characterization: M_n/g.mol^{-1} = 47500, M_w/g.mol^{-1} = 94900, 21.6 wt% 1-hexene, MI = 2.2, ρ = 0.885 g/cm^3, T_m/K = 400.25
Solvent (A): **propane** **C_3H_8** **74-98-6**

Type of data: cloud points

w_B	0.110	0.110	0.110	0.110	0.110	0.110	0.110	0.110	0.110
T/K	355.95	358.65	367.45	368.55	376.95	378.35	392.85	395.65	406.25
P/bar	425.1	426.4	429.5	430.5	435.0	435.1	444.4	444.9	451.2

continued

continued

w_B	0.110	0.110	0.110
T/K	409.35	422.85	423.65
P/bar	453.8	459.9	461.5

Polymer (B): **poly(ethylene-*co*-methacrylic acid)** **2003BUB**
Characterization: M_n/g.mol^{-1} = 25700, M_w/g.mol^{-1} = 83200, 1.0 mol% methacrylic acid, synthesized in the laboratory by high-pressure radical polymerization
Solvent (A): **ethene** C_2H_4 **74-85-1**

Type of data: cloud points

w_B 0.03 was kept constant

T/K	383.15	393.15	403.15	418.15	433.15	453.15	464.15	473.15	483.15
P/bar	2730	2390	2190	1940	1760	1595	1520	1475	1425
T/K	493.15	503.15	513.15	523.15	533.15				
P/bar	1375	1345	1320	1285	1260				

Polymer (B): **poly(ethylene-*co*-methacrylic acid)** **2003BUB**
Characterization: 2.2 mol% methacrylic acid, synthesized in the laboratory by high-pressure radical polymerization
Solvent (A): **ethene** C_2H_4 **74-85-1**

Type of data: cloud points

w_B 0.03 was kept constant

T/K	420.15	426.15	431.15	438.15	443.15	453.15	463.15	473.15	482.15
P/bar	2450	2295	2190	2060	1975	1840	1730	1645	1570
T/K	493.15	503.15	512.15	522.15	533.15				
P/bar	1490	1435	1395	1360	1315				

Polymer (B): **poly(ethylene-*co*-methacrylic acid)** **2003BUB**
Characterization: M_n/g.mol^{-1} = 17400, M_w/g.mol^{-1} = 51500, 3.4 mol% methacrylic acid, synthesized in the laboratory by high-pressure radical polymerization
Solvent (A): **ethene** C_2H_4 **74-85-1**

Type of data: cloud points

w_B 0.03 was kept constant

T/K	443.15	453.15	463.15	473.15	483.15	493.15	503.15	513.15	523.15
P/bar	2355	2160	1950	1805	1690	1585	1495	1440	1385
T/K	533.15								
P/bar	1340								

Polymer (B): **poly(ethylene-*co*-methacrylic acid)** **2003BUB**
Characterization: M_n/g.mol^{-1} = 21800, M_w/g.mol^{-1} = 61200,
4.8 mol% methacrylic acid, synthesized in the laboratory by high-pressure radical polymerization
Solvent (A): **ethene** C_2H_4 **74-85-1**

Type of data: cloud points

w_B 0.03 was kept constant

T/K	451.15	453.15	468.15	473.15	478.15	483.15	488.15	493.15	503.15
P/bar	2450	2350	2250	2140	2040	1935	1860	1800	1690

T/K	513.15	523.15	533.15
P/bar	1580	1510	1435

Polymer (B): **poly(ethylene-*co*-methacrylic acid)** **2003BUB**
Characterization: M_n/g.mol^{-1} = 15100, M_w/g.mol^{-1} = 43900,
5.8 mol% methacrylic acid, synthesized in the laboratory by high-pressure radical polymerization
Solvent (A): **ethene** C_2H_4 **74-85-1**

Type of data: cloud points

w_B 0.03 was kept constant

T/K	473.15	483.15	493.15	503.15	513.15	523.15	533.15
P/bar	2390	2120	1940	1780	1660	1575	1440

Polymer (B): **poly(ethylene-*co*-methacrylic acid)** **2003BUB**
Characterization: M_n/g.mol^{-1} = 11500, M_w/g.mol^{-1} = 32200,
7.0 mol% methacrylic acid, synthesized in the laboratory by high-pressure radical polymerization
Solvent (A): **ethene** C_2H_4 **74-85-1**

Type of data: cloud points

w_B 0.03 was kept constant

T/K	494.15	498.15	501.15	512.15	523.15	533.15
P/bar	2320	2190	2055	1805	1625	1480

Polymer (B): **poly(ethylene-*co*-methacrylic acid)** **2004LAT**
Characterization: 7.0 mol% methacrylic acid, synthesized in the laboratory
Solvent (A): **ethene** C_2H_4 **74-85-1**

Type of data: cloud points

w_B 0.05 was kept constant

T/K	488.15	493.65	494.65	501.15	502.15	505.15	506.15	512.15	518.15
P/bar	2620	2408	2393	2203	2198	2109	2089	1980	1872

continued

continued

T/K	522.15	527.15
P/bar	1807	1707

Polymer (B): **poly(ethylene-*co*-methacrylic acid)** **2004LAT**
Characterization: 7.2 mol% methacrylic acid, synthesized in the laboratory
Solvent (A): **ethene** C_2H_4 **74-85-1**

Type of data: cloud points

w_B 0.05 was kept constant

T/K	495.55	496.35	497.35	502.25	502.65	503.35	507.15	507.35	507.85
P/bar	2380	2485	2366	2252	2257	2224	2160	2151	2137
T/K	513.95	514.35	515.45	518.45	519.45	520.15	524.15	529.15	533.15
P/bar	2032	2014	2001	1951	1943	1981	1912	1837	1780

Polymer (B): **poly(ethylene-*co*-methacrylic acid)** **2003BUB**
Characterization: 9.0 mol% methacrylic acid, synthesized in the laboratory by high-pressure radical polymerization
Solvent (A): **ethene** C_2H_4 **74-85-1**

Type of data: cloud points

w_B 0.03 was kept constant

T/K	508.15	513.15	518.15	523.15	528.15	533.15
P/bar	2650	2440	2250	2125	1990	1860

Polymer (B): **poly(ethylene-*co*-methyl acrylate)** **2003BUB**
Characterization: 5.5 mol% mol% methyl acrylate, synthesized in the laboratory
Solvent (A): **ethene** C_2H_4 **74-85-1**

Type of data: cloud points

w_B 0.03 was kept constant

T/K	383.15	393.15	403.15	413.15	423.15	433.15	443.15	453.15	463.15
P/bar	1530	1420	1369	1331	1300	1269	1248	1222	1200
T/K	473.15	483.15							
P/bar	1182	1165							

Polymer (B): **poly(ethylene-*co*-methyl acrylate)** **2004BEC**
Characterization: M_n/g.mol^{-1} = 58400, M_w/g.mol^{-1} = 128480, 6.0 mol% methyl acrylate, synthesized in the laboratory
Solvent (A): **ethene** C_2H_4 **74-85-1**

Type of data: cloud points

continued

continued

w_B 0.05 was kept constant

T/K	373.15	383.15	393.15	403.15	413.15	423.15	433.15	443.15	453.15
P/bar	1642	1583	1533	1492	1448	1412	1381	1354	1324
T/K	463.15	473.15	483.15	493.15					
P/bar	1300	1282	1262	1244					

Polymer (B): **poly(ethylene-*co*-methyl acrylate)** **1999DIE**
Characterization: M_n/g.mol^{-1} = 52500, M_w/g.mol^{-1} = 165000,
8.0 mol% methyl acrylate, synthesized in the laboratory
Solvent (A): **ethene** **C_2H_4** **74-85-1**

Type of data: cloud points

w_B 0.03 was kept constant

T/K	373.15	383.15	393.15	403.15	413.15	423.15	433.15	443.15	453.15
P/bar	1780	1700	1590	1540	1510	1485	1450	1440	1395
T/K	463.15	483.15							
P/bar	1390	1350							

Polymer (B): **poly(ethylene-*co*-methyl acrylate)** **2004BEC**
Characterization: M_n/g.mol^{-1} = 54980, M_w/g.mol^{-1} = 109960,
13.0 mol% methyl acrylate, synthesized in the laboratory
Solvent (A): **ethene** **C_2H_4** **74-85-1**

Type of data: cloud points

w_B 0.05 was kept constant

T/K	493.15	483.15	473.15	463.15	453.15	443.15	433.15	423.15	413.15
P/bar	1185	1215	1223	1239	1260	1276	1288	1310	1351
T/K	403.15	393.15							
P/bar	1388	1408							

Polymer (B): **poly(ethylene-*co*-methyl acrylate)** **1999DIE**
Characterization: M_n/g.mol^{-1} = 55000, M_w/g.mol^{-1} = 150000,
17.9 mol% methyl acrylate, synthesized in the laboratory
Solvent (A): **ethene** **C_2H_4** **74-85-1**

Type of data: cloud points

w_B 0.03 was kept constant

T/K	393.15	403.15	413.15	423.15	433.15	443.15	453.15	463.15	473.15
P/bar	1823	1780	1703	1612	1565	1510	1455	1429	1378
T/K	482.15								
P/bar	1360								

Polymer (B): **poly(ethylene-*co*-methyl acrylate)** **1999DIE**
Characterization: M_n/g.mol^{-1} = 59000, M_w/g.mol^{-1} = 158000,
32.9 mol% methyl acrylate, synthesized in the laboratory
Solvent (A): **ethene** **C_2H_4** **74-85-1**

Type of data: cloud points

w_B 0.03 was kept constant

T/K	391.15	395.15	403.15	413.15	423.15	433.15	443.15	453.15	463.15
P/bar	2040	1980	1849	1768	1685	1605	1545	1490	1450
T/K	473.15	483.15							
P/bar	1410	1382							

Polymer (B): **poly(ethylene-*co*-methyl acrylate)** **2004BEC**
Characterization: M_n/g.mol^{-1} = 56270, M_w/g.mol^{-1} = 112540,
44.0 mol% methyl acrylate, synthesized in the laboratory
Solvent (A): **ethene** **C_2H_4** **74-85-1**

Type of data: cloud points

w_B 0.05 was kept constant

T/K	393.15	403.15	413.15	423.15	433.15	443.15	453.15	463.15	473.15
P/bar	2115	2010	1923	1854	1790	1728	1679	1631	1585
T/K	483.15	493.15							
P/bar	1545	1508							

Polymer (B): **poly(ethylene-*co*-methyl acrylate-*co*-vinyl acetate)** **2007TUM**
Characterization: M_w/g.mol^{-1} = 110000, 23 mol% methyl acrylate,
3.5 mol% vinyl acetate, synthesized in the laboratory
Solvent (A): **ethene** **C_2H_4** **74-85-1**

Type of data: cloud points

w_B 0.03 was kept constant

T/K	383	393	403	413	423	433	443	453	463
P/bar	1880	1810	1710	1650	1600	1560	1490	1460	1410
T/K	473	483	493						
P/bar	1360	1340	1320						

Polymer (B): **poly(ethylene-*co*-methyl acrylate-*co*-vinyl acetate)** **2007TUM**
Characterization: M_w/g.mol^{-1} = 110000, 35 mol% methyl acrylate,
3.5 mol% vinyl acetate, synthesized in the laboratory
Solvent (A): **ethene** **C_2H_4** **74-85-1**

continued

continued

Type of data: cloud points

w_B 0.03 was kept constant

T/K	373	383	393	403	413	423	433	443	453
P/bar	2040	1960	1870	1780	1720	1660	1620	1560	1500
T/K	463	473	483	493	503				
P/bar	1450	1410	1380	1340	1310				

Polymer (B): **poly(ethylene-*co*-methyl acrylate-*co*-vinyl acetate)** **2007TUM**
Characterization: M_w/g.mol^{-1} = 110000, 40 mol% methyl acrylate, 3.5 mol% vinyl acetate, synthesized in the laboratory
Solvent (A): **ethene** **C_2H_4** **74-85-1**

Type of data: cloud points

w_B 0.03 was kept constant

T/K	383	393	403	413	423	433	443	453	463
P/bar	2120	2020	1920	1870	1800	1740	1680	1640	1590
T/K	473	483	493						
P/bar	1550	1510	1490						

Polymer (B): **poly(ethylene-*co*-methyl methacrylate)** **2003BUB**
Characterization: 5.8 mol% methyl methacrylate, synthesized in the laboratory
Solvent (A): **ethene** **C_2H_4** **74-85-1**

Type of data: cloud points

w_B 0.03 was kept constant

T/K	383.15	393.15	403.15	413.15	423.15	433.15	443.15	453.15	473.15
P/bar	1433	1366	1320	1290	1266	1241	1215	1195	1156
T/K	483.15								
P/bar	1140								

Polymer (B): **poly(ethylene-*co*-methyl methacrylate)** **2004LAT**
Characterization: 9.6 mol% methyl methacrylate, synthesized in the laboratory
Solvent (A): **ethene** **C_2H_4** **74-85-1**

Type of data: cloud points

w_B 0.05 was kept constant

T/K	372.45	393.25	402.55	412.05	422.45	432.05	451.65	472.35	491.85
P/bar	1510	1369	1312	1290	1255	1227	1185	1145	1109
T/K	511.55	527.05							
P/bar	1076	1063							

Polymer (B): **poly(ethylene-*co*-methyl methacrylate) 2004BEC, 2004LAT**
Characterization: M_n/g.mol^{-1} = 41400, M_w/g.mol^{-1} = 67100,
16.9 mol% methyl methacrylate, synthesized in the laboratory
Solvent (A): **ethene** **C_2H_4** **74-85-1**

Type of data: cloud points

w_B 0.05 was kept constant

T/K	383.15	393.15	413.15	433.15	453.15	473.15
P/bar	1465	1435	1354	1292	1237	1208

Polymer (B): **poly(ethylene-*co*-methyl methacrylate) 2004BEC, 2004LAT**
Characterization: M_n/g.mol^{-1} = 10900, M_w/g.mol^{-1} = 20000,
18.5 mol% methyl methacrylate, synthesized in the laboratory
Solvent (A): **ethene** **C_2H_4** **74-85-1**

Type of data: cloud points

w_B 0.05 was kept constant

T/K	353.15	373.15	393.15	413.15	433.15	453.15	473.15	488.15	508.15
P/bar	1430	1309	1252	1206	1176	1137	1108	1086	1057
T/K	528.15								
P/bar	1039								

Polymer (B): **poly(ethylene-*co*-methyl methacrylate) 2004LAT**
Characterization: 29.8 mol% methyl methacrylate, synthesized in the laboratory
Solvent (A): **ethene** **C_2H_4** **74-85-1**

Type of data: cloud points

w_B 0.05 was kept constant

T/K	370.25	390.35	411.55	430.85	450.75	470.25	490.45	509.45	529.45
P/bar	1371	1301	1246	1200	1161	1130	1108	1076	1049

Polymer (B): **poly(ethylene-*co*-methyl methacrylate) 2004BEC, 2004LAT**
Characterization: M_n/g.mol^{-1} = 49800, M_w/g.mol^{-1} = 83500,
35.1 mol% methyl methacrylate, synthesized in the laboratory
Solvent (A): **ethene** **C_2H_4** **74-85-1**

Type of data: cloud points

w_B 0.05 was kept constant

T/K	393.15	413.15	433.15	453.15	473.15	488.15	508.15
P/bar	1500	1419	1341	1286	1238	1202	1152

Polymer (B): **poly(ethylene-*co*-methyl methacrylate)** **2004BEC, 2004LAT**
Characterization: M_n/g.mol^{-1} = 28600, M_w/g.mol^{-1} = 52600,
41.5 mol% methyl methacrylate, synthesized in the laboratory
Solvent (A): **ethene** **C_2H_4** **74-85-1**

Type of data: cloud points

w_B 0.05 was kept constant

T/K	393.15	403.15	413.15	423.15	443.15	463.15	473.15	483.15	493.15
P/bar	1673	1585	1533	1478	1378	1306	1271	1242	1210

Polymer (B): **poly(ethylene-*co*-methyl methacrylate)** **2004BEC**
Characterization: M_n/g.mol^{-1} = 31900, M_w/g.mol^{-1} = 62300,
49.6 mol% methyl methacrylate, synthesized in the laboratory
Solvent (A): **ethene** **C_2H_4** **74-85-1**

Type of data: cloud points

w_B 0.05 was kept constant

T/K	393.15	413.15	453.15	473.15	488.15	508.15
P/bar	1855	1716	1507	1436	1366	1328

Polymer (B): **poly(ethylene-*co*-1-octene)** **2000CH4**
Characterization: M_n/g.mol^{-1} = 51200, M_w/g.mol^{-1} = 83000,
13.9 wt% 1-octene, 4 hexyl branches/100 ethyl units
Solvent (A): **ethene** **C_2H_4** **74-85-1**

Type of data: cloud points

w_B	0.099	0.099	0.099	0.099	0.099	0.099	0.099	0.099
T/K	453.15	438.15	423.15	413.15	403.15	393.15	388.15	385.15
P/bar	1362	1410	1477	1510	1556	1614	1644	1668

Polymer (B): **poly(ethylene-*co*-1-octene)** **2000CH4**
Characterization: M_n/g.mol^{-1} = 54800, M_w/g.mol^{-1} = 115000,
25.8 wt% 1-octene, 8 hexyl branches/100 ethyl units
Solvent (A): **ethene** **C_2H_4** **74-85-1**

Type of data: cloud points

w_B	0.105	0.105	0.105	0.105
T/K	453.15	433.15	413.15	393.15
P/bar	1303	1350	1408	1504

Polymer (B): **poly(ethylene-*co*-1-octene)** **2000CH4**
Characterization: M_n/g.mol^{-1} = 11300, M_w/g.mol^{-1} = 26000,
36.5 wt% 1-octene, 12.6 hexyl branches/100 ethyl units
Solvent (A): **ethene** **C_2H_4** **74-85-1**

continued

continued

Type of data: cloud points

w_B	0.099	0.099	0.099	0.099	0.099	0.099	0.099	0.099	0.099
T/K	453.15	443.15	433.15	423.15	413.15	403.15	393.15	383.15	373.15
P/bar	1097	1116	1135	1156	1182	1210	1246	1286	1332

Polymer (B): **poly(ethylene-*co*-1-octene)** **2000CH4**
Characterization: M_n/g.mol^{-1} = 83300, M_w/g.mol^{-1} = 120000,
38.7 wt% 1-octene, 13.6 hexyl branches/100 ethyl units
Solvent (A): **ethene** C_2H_4 **74-85-1**

Type of data: cloud points

w_B	0.047	0.047	0.047	0.047	0.047	0.047	0.047	0.099	0.099
T/K	453.15	443.15	433.15	423.15	413.15	403.15	393.15	453.15	443.15
P/bar	1221	1231	1249	1270	1297	1330	1363	1199	1217
w_B	0.099	0.099	0.099	0.099	0.099	0.124	0.124	0.124	0.124
T/K	433.15	423.15	413.15	403.15	393.15	453.15	443.15	433.15	423.15
P/bar	1264	1254	1281	1311	1345	1168	1182	1194	1215
w_B	0.124	0.124							
T/K	413.15	403.15							
P/bar	1241	1268							

Polymer (B): **poly(ethylene-*co*-1-octene)** **1998HAN**
Characterization: M_n/g.mol^{-1} = 40700, M_w/g.mol^{-1} = 93600, 7.6 mol% 1-octene,
T_m/K = 377.2, synthesized with metallocene catalyst
Solvent (A): **propane** C_3H_8 **74-98-6**

Type of data: cloud points

w_B	0.05	0.05	0.05	0.05	0.05
T/K	374.25	399.25	422.65	446.75	471.55
P/bar	531.2	515.5	520.4	524.0	529.5

Polymer (B): **poly(ethylene-*co*-propylene)** **2000VRI**
Characterization: M_n/g.mol^{-1} = 51000, M_w/g.mol^{-1} = 120000,
M_z/g.mol^{-1} = 210000, 42 mol% propene,
21 methyl groups/100 C-atoms
Solvent (A): **ethene** C_2H_4 **74-85-1**

Type of data: cloud points

w_B	0.0357	0.0357	0.0357	0.0357	0.0357	0.0357	0.0357	0.0498	0.0498
T/K	312.54	317.69	322.51	327.48	332.66	337.28	342.13	312.62	317.58
P/bar	2181	2077	1985	1897	1816	1756	1702	2163	2057

continued

continued

w_B	0.0498	0.0498	0.0498	0.0498	0.0498	0.0756	0.0756	0.0756	0.0756
T/K	322.53	327.10	332.09	337.40	342.30	312.76	317.91	322.27	327.47
P/bar	1961	1885	1809	1741	1684	2115	2006	1926	1843
w_B	0.0756	0.0756	0.0756	0.0989	0.0989	0.0989	0.0989	0.0989	0.0989
T/K	332.22	337.10	341.96	312.54	317.71	322.64	327.50	332.21	337.36
P/bar	1776	1715	1661	2083	1977	1890	1811	1749	1687
w_B	0.0989	0.1230	0.1230	0.1230	0.1230	0.1230	0.1230	0.1230	0.1491
T/K	342.25	312.38	317.41	322.40	327.20	332.31	337.13	342.06	312.62
P/bar	1634	2061	1960	1874	1797	1730	1673	1619	2011
w_B	0.1491	0.1491	0.1491	0.1491	0.1491	0.1491			
T/K	317.54	322.33	327.28	332.16	337.03	341.77			
P/bar	1917	1837	1765	1703	1647	1597			

Polymer (B): **poly(ethylene-*co*-propylene)** **2000VRI**
Characterization: M_n/g.mol^{-1} = 8700, M_w/g.mol^{-1} = 24000, M_z/g.mol^{-1} = 47000, 44 mol% propene, 22 methyl groups/100 C-atoms
Solvent (A): **ethene** C_2H_4 **74-85-1**

Type of data: cloud points

w_B	0.0468	0.0468	0.0468	0.0468	0.0468	0.0468	0.0468	0.1168	0.1168
T/K	312.82	317.56	323.16	327.72	332.74	337.29	342.55	312.66	317.71
P/bar	1827	1748	1666	1610	1555	1512	1465	1759	1682
w_B	0.1668	0.1168	0.1168	0.1168	0.1168	0.1503	0.1503	0.1503	0.1503
T/K	322.51	327.38	332.26	337.28	42.01	312.75	317.60	322.28	327.76
P/bar	1617	1559	1505	1457	1413	1715	1646	1586	1524
w_B	0.1503	0.1503	0.1503	0.1983	0.1983	0.1983	0.1983	0.1983	0.1983
T/K	332.25	337.59	342.52	312.66	317.59	322.98	327.48	332.79	337.26
P/bar	1480	1432	1395	1615	1553	1492	1447	1399	1364
w_B	0.1983								
T/K	342.69								
P/bar	1325								

Polymer (B): **poly(ethylene-*co*-propylene)** **1997HAN**
Characterization: M_n/g.mol^{-1} = 2170, M_w/g.mol^{-1} = 2600, 50 mol% propene, T_g/K = 209.2, alternating ethylene/propylene units from complete hydrogenation (98%) of polyisoprene
Solvent (A): **ethene** C_2H_4 **74-85-1**

Type of data: cloud points

w_B	0.001	0.001	0.001	0.0067	0.0067	0.0067	0.0067	0.0095	0.0095
T/K	323.15	372.95	423.15	294.55	323.35	373.05	423.05	295.05	323.35
P/bar	381.2	396.7	410.5	490.4	484.3	483.1	491.0	539.4	523.0

continued

continued

w_B	0.0095	0.0095	0.026	0.026	0.026	0.026	0.034	0.034	0.034
T/K	373.15	423.15	298.45	323.35	373.05	423.05	323.35	372.95	423.55
P/bar	517.1	522.8	623.3	600.1	576.5	570.4	622.7	596.2	588.9

Type of data: coexistence data

T/K = 423.15

P/ bar	w_B feed phase	gel phase	sol phase
505	0.034	unknown	0.016

Polymer (B): **poly(ethylene-*co*-propylene)** **1998HAN**
Characterization: M_n/g.mol^{-1} = 32300, M_w/g.mol^{-1} = 71000, 8.7 mol% propene, T_m/K = 386.2, synthesized with metallocene catalyst
Solvent (A): **propane** C_3H_8 **74-98-6**

Type of data: cloud points

w_B	0.05	0.05	0.05	0.05	0.05
T/K	372.65	397.65	420.15	445.15	472.15
P/bar	568.7	559.3	559.6	558.9	560.9

Polymer (B): **poly(ethylene-*co*-propylene)** **1998HAN**
Characterization: M_n/g.mol^{-1} = 34500, M_w/g.mol^{-1} = 72500, 10.3 mol% propene, T_m/K = 362.2, synthesized with metallocene catalyst
Solvent (A): **propane** C_3H_8 **74-98-6**

Type of data: cloud points

w_B	0.05	0.05	0.05	0.05	0.05
T/K	376.25	400.35	424.35	446.95	472.15
P/bar	565.4	550.3	551.2	554.0	557.8

Polymer (B): **poly(ethylene-*co*-propylene)** **1998HAN**
Characterization: M_n/g.mol^{-1} = 27300, M_w/g.mol^{-1} = 62800, 28.3 mol% propene, T_m/K = 303.2, synthesized with metallocene catalyst
Solvent (A): **propane** C_3H_8 **74-98-6**

Type of data: cloud points

w_B	0.05	0.05	0.05	0.05	0.05	0.05
T/K	351.55	375.95	400.55	426.05	446.55	470.45
P/bar	471.2	474.4	482.9	493.7	502.3	511.5

Polymer (B): **poly(ethylene-*co*-propylene)** **1997HAN**
Characterization: M_n/g.mol^{-1} = 2170, M_w/g.mol^{-1} = 2600, 50 mol% propene, T_g/K = 209.2, alternating ethylene/propylene units from complete hydrogenation (98%) of polyisoprene
Solvent (A): **propene** C_3H_6 **115-07-1**

Type of data: cloud points

w_B	0.009	0.009	0.009	0.009	0.043	0.043	0.043	0.043	0.132
T/K	324.15	373.55	375.15	422.95	323.65	373.35	375.95	423.25	323.15
P/bar	43.7	139.4	142.5	198.6	53.7	152.0	156.3	224.2	70.4

w_B	0.132	0.132	0.343	0.343	0.343
T/K	372.95	423.45	323.05	373.25	423.35
P/bar	175.6	261.8	32.4	137.2	210.2

Polymer (B): **poly(ethylene-*co*-propylene)** **1997HAN**
Characterization: M_n/g.mol^{-1} = 87300, M_w/g.mol^{-1} = 96000, 50 mol% propene, T_g/K = 220.2, alternating ethylene/propylene units from complete hydrogenation of polyisoprene
Solvent (A): **propene** C_3H_6 **115-07-1**

Type of data: cloud points

w_B	0.045	0.045
T/K	373.15	423.15
P/bar	363.0	413.8

Polymer (B): **poly(ethylene-*co*-propylene)** **1997HAN**
Characterization: M_n/g.mol^{-1} = 53300, M_w/g.mol^{-1} = 96000, 50 mol% propene, T_g/K = 220.2, alternating ethylene/propylene units from complete hydrogenation of polyisoprene
Solvent (A): **propene** C_3H_6 **115-07-1**

Type of data: cloud points

w_B	0.048	0.048	0.048	0.099	0.099	0.099
T/K	323.45	372.95	422.85	323.25	373.15	423.05
P/bar	218.1	288.0	348.5	216.9	286.9	348.9

Polymer (B): **poly(ethylene-*co*-propylene)** **1997HAN**
Characterization: M_n/g.mol^{-1} = 177000, M_w/g.mol^{-1} = 195000, 50 mol% propene, T_g/K = 225.2, alternating ethylene/propylene units from complete hydrogenation (94%) of polyisoprene
Solvent (A): **propene** C_3H_6 **115-07-1**

Type of data: cloud points

w_B	0.008	0.008	0.008	0.050	0.050	0.050	0.100	0.100
T/K	323.35	372.85	423.25	322.95	372.95	423.45	323.35	423.35
P/bar	341.2	387.9	437.0	346.9	393.0	441.9	229.8	438.6

Polymer (B): **poly(ethylene-*co*-propyl methacrylate)** **2004BEC, 2004LAT**
Characterization: 6.9 mol% propyl acrylate, synthesized in the laboratory
Solvent (A): **ethene** **C_2H_4** **74-85-1**

Type of data: cloud points

w_B 0.05 was kept constant

T/K	366.45	376.15	386.65	393.15	413.15	423.15	442.15	452.15	472.15
P/bar	1527	1476	1433	1408	1338	1310	1262	1244	1204
T/K	491.15	511.15	531.15						
P/bar	1173	1141	1115						

Polymer (B): **poly(ethylene-*co*-propyl methacrylate)** **2004BEC, 2004LAT**
Characterization: M_n/g.mol^{-1} = 83000, M_w/g.mol^{-1} = 147300,
14.1 mol% propyl acrylate, synthesized in the laboratory
Solvent (A): **ethene** **C_2H_4** **74-85-1**

Type of data: cloud points

w_B 0.05 was kept constant

T/K	353.15	363.15	373.65	397.85	415.45	434.15	452.15	473.15	492.15
P/bar	1380	1330	1297	1234	1192	1150	1127	1098	1074
T/K	509.15	532.15							
P/bar	1056	1035							

Polymer (B): **poly(ethylene-*co*-propyl methacrylate)** **2004BEC, 2004LAT**
Characterization: M_n/g.mol^{-1} = 54200, M_w/g.mol^{-1} = 112400,
18.9 mol% propyl acrylate, synthesized in the laboratory
Solvent (A): **ethene** **C_2H_4** **74-85-1**

Type of data: cloud points

w_B 0.05 was kept constant

T/K	367.15	373.15	393.15	412.15	431.15	451.15	468.15	491.55	512.15
P/bar	1250	1226	1183	1153	1124	1097	1077	1055	1031
T/K	529.85								
P/bar	1018								

Polymer (B): **poly(ethylene-*co*-propyl methacrylate)** **2004BEC, 2004LAT**
Characterization: M_n/g.mol^{-1} = 58100, M_w/g.mol^{-1} = 127000,
26.2 mol% propyl acrylate, synthesized in the laboratory
Solvent (A): **ethene** **C_2H_4** **74-85-1**

Type of data: cloud points

continued

continued

w_B	0.05	was kept constant							
T/K	354.15	374.15	392.15	413.15	433.15	451.15	471.15	491.15	512.15
P/bar	1208	1182	1147	1112	1083	1062	1038	1017	995
T/K	530.15								
P/bar	978								

Polymer (B): **poly(ethylene-*co*-propyl methacrylate)** **2004LAT**
Characterization: M_n/g.mol^{-1} = 78400, M_w/g.mol^{-1} = 128600,
31.2 mol% propyl acrylate, synthesized in the laboratory
Solvent (A): **ethene** **C_2H_4** **74-85-1**

Type of data: cloud points

w_B	0.05	was kept constant							
T/K	351.55	362.45	373.15	393.15	412.35	432.65	452.15	472.65	492.15
P/bar	1215	1186	1163	1129	1100	1069	1046	1024	1003
T/K	510.15	530.15							
P/bar	985	964							

Polymer (B): **poly(ethylene-*co*-propyl methacrylate)** **2004LAT**
Characterization: M_n/g.mol^{-1} = 49000, M_w/g.mol^{-1} = 126200,
37.4 mol% propyl acrylate, synthesized in the laboratory
Solvent (A): **ethene** **C_2H_4** **74-85-1**

Type of data: cloud points

w_B	0.05	was kept constant							
T/K	351.45	365.15	375.15	391.15	412.15	434.15	453.15	471.65	492.15
P/bar	1218	1193	1172	1139	1106	1072	1050	1029	1008
T/K	512.15	528.15							
P/bar	987	974							

Polymer (B): **poly(ethylene-*co*-vinyl acetate)** **2001DO2**
Characterization: M_n/g.mol^{-1} = 61900, M_w/g.mol^{-1} = 167000,
27.5 wt% vinyl acetate
Solvent (A): **ethene** **C_2H_4** **74-85-1**

Type of data: coexistence data

T/K = 433.15

w_B(total)	0.28	was kept constant						
P/bar	622	633	723	737	760	786	854	872
w_B(gel phase)	–	0.61	–	0.57	–	0.52	0.50	–
w_B(sol phase)	0.06	–	0.06	–	0.07	–	–	0.08

continued

continued

P/bar	920	955	1066					
w_B(gel phase)	–	0.40	0.28					
w_B(sol phase)	0.10	–	–					
w_B(total)	0.20	was kept constant						
P/bar	868	966	1064	1098	1136			
w_B(gel phase)	0.508	0.437	0.350	0.297	0.200			
w_B(sol phase)	0.014	0.028	0.044	0.049	–			
w_B(total)	0.15	was kept constant						
P/bar	820	884	1040	1161	1174			
w_B(gel phase)	0.523	0.477	0.360	0.240	0.150			
w_B(sol phase)	0.009	0.012	0.029	0.052	–			
w_B(total)	0.10	was kept constant						
P/bar	915	940	977	1009	1038	1056	1095	1121
w_B(gel phase)	–	0.48	–	0.42	–	0.37	–	0.30
w_B(sol phase)	0.01	–	0.015	–	0.02	–	0.03	–
P/bar	1130	1170	1192					
w_B(gel phase)	–	0.23	–					
w_B(sol phase)	0.045	–	0.098					

Polymer (B): **poly(ethylene glycol monododecyl ether-*b*-propylene glycol)** **2002LIU**

Characterization: M_n/g.mol^{-1} = 666, 3 EO und 6 PPO units, Ls-36 Surfactant, Henkel AG, Germany

Solvent (A): **carbon dioxide** **CO_2** **124-38-9**

Type of data: polymer solubility

T/K = 308.15

P/MPa	11.10	14.50	16.73	18.71	19.74
c_B/mol.l^{-1}	0.01	0.02	0.03	0.04	0.05

T/K = 318.15

P/MPa	15.31	17.56	19.78	21.33
c_B/mol.l^{-1}	0.01	0.02	0.03	0.04

Polymer (B): **poly(ethylene glycol monododecyl ether-*b*-propylene glycol)** **2002LIU**

Characterization: M_n/g.mol^{-1} = 652, 4 EO und 5 PPO units, Ls-45 Surfactant, Henkel AG, Germany

Solvent (A): **carbon dioxide** **CO_2** **124-38-9**

Type of data: polymer solubility

continued

continued

T/K = 308.15

P/MPa	12.97	16.73	19.32	21.05
c_B/mol.l^{-1}	0.01	0.02	0.03	0.04

T/K = 318.15

P/MPa	16.46	20.11	22.37
c_B/mol.l^{-1}	0.01	0.02	0.03

Polymer (B): **poly(ethylene oxide-*b*-propylene oxide-*b*-ethylene oxide)** **2009STO**

Characterization: M_n/g.mol^{-1} = 2000, 10 wt% PEO, L61, $(EO)_2$-$(PO)_{31}$-$(EO)_2$, BASF SE, Germany

Solvent (A): **carbon dioxide** **CO_2** **124-38-9**

Type of data: cloud points

w_B 0.08 was kept constant

T/K	293.1	299.2	303.7	308.4	313.1	318.2	322.3	328.2	332.5
P/MPa	11.5	13.7	15.5	16.9	18.4	19.8	21.1	22.4	23.9

T/K	337.3
P/MPa	25.1

Polymer (B): **poly(ethylene oxide-*b*-propylene oxide-*b*-ethylene oxide)** **2009STO**

Characterization: M_n/g.mol^{-1} = 2750, 10 wt% PEO, L81, $(EO)_3$-$(PO)_{42}$-$(EO)_3$, BASF SE, Germany

Solvent (A): **carbon dioxide** **CO_2** **124-38-9**

Type of data: cloud points

w_B 0.08 was kept constant

T/K	294.1	298.9	303.3	308.1	312.5	318.1	323.6	328.2	332.8
P/MPa	18.4	19.6	20.7	22.2	23.2	24.4	25.8	26.9	28.2

T/K	337.1
P/MPa	29.0

Polymer (B): **poly(ethylene oxide-*b*-propylene oxide-*b*-ethylene oxide)** **2009STO**

Characterization: M_n/g.mol^{-1} = 2500, 20 wt% PEO, L62, $(EO)_5$-$(PO)_{34}$-$(EO)_5$, BASF SE, Germany

Solvent (A): **carbon dioxide** **CO_2** **124-38-9**

Type of data: cloud points

continued

continued

w_B 0.10 was kept constant

T/K	293.4	298.6	303.4	308.3	313.1	319.2	323.3	328.6	335.5
P/MPa	22.7	23.4	24.6	25.6	26.6	27.7	28.6	29.3	31.3

Polymer (B): **poly(ethylene oxide-*b*-1,1,2,2-tetrahydroperfluorodecyl acrylate)** **2004MA2**

Characterization: M_n/g.mol^{-1} = 22800, 10.3 wt% PEO, molar ratio of PFDA/PEO = 39.5/1 by NMR, synthesized in the laboratory

Solvent (A): **carbon dioxide** **CO_2** **124-38-9**

Type of data: cloud points

w_B 0.01 was kept constant

T/K	338.75	333.25	328.35	323.25	318.15	313.25	308.35	303.65	298.55
P/bar	248.2	232.1	216.8	199.5	182.8	165.5	149.0	130.4	108.0
T/K	293.45								
P/bar	87.1								

Polymer (B): **poly(ethylene oxide-*b*-1,1,2,2-tetrahydroperfluorodecyl acrylate)** **2004MA1**

Characterization: M_n/g.mol^{-1} = 23000, 10.3 wt% PEO, molar ratio of PFDA/PEO = 39.8/1 by NMR, synthesized in the laboratory

Solvent (A): **carbon dioxide** **CO_2** **124-38-9**

Type of data: cloud points

w_B 0.04 was kept constant

T/K	338.15	333.65	328.05	323.15	318.35	313.25	308.35	303.35	298.35
P/bar	241.2	228.8	212.5	196.3	181.3	164.6	148.2	129.9	112.2
T/K	293.15								
P/bar	93.6								

Polymer (B): **poly(ethylene oxide-*b*-1,1,2,2-tetrahydroperfluorodecyl acrylate)** **2004MA2**

Characterization: M_n/g.mol^{-1} = 23000, 10.3 wt% PEO, molar ratio of PFDA/PEO = 39.8/1 by NMR, synthesized in the laboratory

Solvent (A): **carbon dioxide** **CO_2** **124-38-9**

Type of data: cloud points

w_B 0.01 was kept constant

T/K	338.85	333.95	327.75	323.25	318.45	313.35	308.05	303.65	298.45
P/bar	249.8	234.0	213.0	197.4	181.1	164.9	150.3	131.3	111.5
T/K	293.05								
P/bar	93.4								

Polymer (B): **poly(DL-lactide-*co*-glycolide)** **2001CON**
Characterization: M_n/g.mol^{-1} = 61700, M_w/g.mol^{-1} = 95000, 15 mol% glycolide, T_g/K = 320.55, Alkermes, Inc., Cincinnati, OH
Solvent (A): **carbon dioxide** **CO_2** **124-38-9**

Type of data: cloud points

w_B	0.05	0.05	0.05	0.05	0.05
T/K	312.15	317.95	333.95	347.55	364.85
P/bar	1822	1815	1801	1784	1770

Polymer (B): **poly(DL-lactide-*co*-glycolide)** **2001CON**
Characterization: M_n/g.mol^{-1} = 86100, M_w/g.mol^{-1} = 149000, 15 mol% glycolide, T_g/K = 323.85, Alkermes, Inc., Cincinnati, OH
Solvent (A): **carbon dioxide** **CO_2** **124-38-9**

Type of data: cloud points

w_B	0.05	0.05	0.05	0.05	0.05
T/K	309.85	315.75	329.45	345.75	358.65
P/bar	1918	1901	1881	1860	1843

Polymer (B): **poly(DL-lactide-*co*-glycolide)** **2001CON**
Characterization: M_n/g.mol^{-1} = 77700, M_w/g.mol^{-1} = 130000, 25 mol% glycolide, T_g/K = 320.05, Alkermes, Inc., Cincinnati, OH
Solvent (A): **carbon dioxide** **CO_2** **124-38-9**

Type of data: cloud points

w_B	0.05	0.05	0.05	0.05	0.05
T/K	312.05	320.25	335.35	352.85	352.95
P/bar	2394	2373	2322	2249	2198

Polymer (B): **poly(DL-lactide-*co*-glycolide)** **2001CON**
Characterization: M_n/g.mol^{-1} = 57900, M_w/g.mol^{-1} = 83000, 35 mol% glycolide, T_g/K = 317.65, Alkermes, Inc., Cincinnati, OH
Solvent (A): **carbon dioxide** **CO_2** **124-38-9**

Type of data: cloud points

w_B	0.05	0.05	0.05	0.05
T/K	314.65	328.75	343.95	359.85
P/bar	2999	2946	2877	2791

Polymer (B): **poly(DL-lactide-*co*-glycolide)** **2001CON**
Characterization: M_n/g.mol^{-1} = 80100, M_w/g.mol^{-1} = 141000, 35 mol% glycolide, T_g/K = 321.35, Alkermes, Inc., Cincinnati, OH
Solvent (A): **carbon dioxide** **CO_2** **124-38-9**

continued

continued

Type of data: cloud points

w_B	0.05	0.05	0.05	0.05
T/K	316.55	333.55	353.05	369.85
P/bar	3108	3022	2922	2822

Polymer (B): **poly(DL-lactide-*co*-glycolide)** **2000LEE**
Characterization: M_η/g.mol^{-1} = 5000, 10 wt% glycolide,
Polysciences, Inc., Warrington, PA
Solvent (A): **chlorodifluoromethane $CHClF_2$** **75-45-6**

Type of data: cloud points

w_B	0.0370	0.0370	0.0370	0.0370	0.0370	0.0370
T/K	331.85	343.05	352.75	362.85	372.85	383.15
P/MPa	3.73	7.70	11.06	14.05	16.95	19.67

Polymer (B): **poly(DL-lactide-*co*-glycolide)** **2000LEE**
Characterization: M_η/g.mol^{-1} = 5000, 20 wt% glycolide,
Polysciences, Inc., Warrington, PA
Solvent (A): **chlorodifluoromethane $CHClF_2$** **75-45-6**

Type of data: cloud points

w_B	0.0399	0.0399	0.0399	0.0399	0.0399	0.0399	0.0399
T/K	325.05	334.55	344.05	352.95	362.95	374.05	382.55
P/MPa	3.45	7.10	10.65	13.60	16.85	20.15	22.60

Polymer (B): **poly(DL-lactide-*co*-glycolide)** **2001CON**
Characterization: M_n/g.mol^{-1} = 77700, M_w/g.mol^{-1} = 130000, 25 mol% glycolide,
T_g/K = 320.05, Alkermes, Inc., Cincinnati, OH
Solvent (A): **chlorodifluoromethane $CHClF_2$** **75-45-6**

Type of data: cloud points

w_B	0.05	0.05	0.05	0.05
T/K	309.35	315.85	324.95	337.95
P/bar	14.5	25.8	63.8	118.9

Polymer (B): **poly(DL-lactide-*co*-glycolide)** **2000LEE**
Characterization: M_η/g.mol^{-1} = 10000, 30 wt% glycolide,
Polysciences, Inc., Warrington, PA
Solvent (A): **chlorodifluoromethane $CHClF_2$** **75-45-6**

Type of data: cloud points

w_B	0.0314	0.0314	0.0314	0.0314	0.0314	0.0314	0.0314	0.0314
T/K	314.65	324.85	334.75	344.55	352.45	362.85	372.75	382.55
P/MPa	3.80	8.35	12.50	16.50	19.45	23.25	26.62	29.65

Polymer (B): **poly(DL-lactide-*co*-glycolide)** **2000LEE**
Characterization: M_η/g.mol^{-1} = 14500, 45 wt% glycolide, Resomer RG502, Boehringer Ingelheim Chemicals
Solvent (A): **chlorodifluoromethane $CHClF_2$** **75-45-6**

Type of data: cloud points

w_B	0.0324	0.0324	0.0324	0.0324
T/K	313.65	335.25	353.15	373.15
P/MPa	7.55	17.35	24.75	31.45

Polymer (B): **poly(DL-lactide-*co*-glycolide)** **2001CON**
Characterization: M_n/g.mol^{-1} = 50800, M_w/g.mol^{-1} = 69600, 50 mol% glycolide, T_g/K = 320.35, Alkermes, Inc., Cincinnati, OH
Solvent (A): **chlorodifluoromethane $CHClF_2$** **75-45-6**

Type of data: cloud points

w_B	0.05	0.05	0.05
T/K	311.75	325.65	340.05
P/bar	105.1	174.1	241.0

Polymer (B): **poly(DL-lactide-*co*-glycolide)** **2001KUK**
Characterization: M_η/g.mol^{-1} = 5000, 10 wt% glycolide, Polysciences, Inc., Warrington, PA
Solvent (A): **dimethyl ether C_2H_6O** **115-10-6**

Type of data: cloud points

w_B	0.0314	0.0314	0.0314	0.0314	0.0314	0.0314	0.0314	0.0314
T/K	303.65	314.35	323.55	333.15	343.25	352.75	363.05	373.25
P/MPa	8.55	11.37	13.50	15.62	17.65	19.50	21.44	23.29

Polymer (B): **poly(DL-lactide-*co*-glycolide)** **2001KUK**
Characterization: M_η/g.mol^{-1} = 5000, 20 wt% glycolide, Polysciences, Inc., Warrington, PA
Solvent (A): **dimethyl ether C_2H_6O** **115-10-6**

Type of data: cloud points

w_B	0.0293	0.0293	0.0293	0.0293	0.0293	0.0293	0.0293	0.0293
T/K	303.35	312.85	323.75	333.75	343.15	353.45	363.45	373.05
P/MPa	21.65	23.55	25.10	26.50	28.00	29.75	31.25	32.65

Polymer (B): **poly(DL-lactide-*co*-glycolide)** **2001KUK**
Characterization: M_η/g.mol^{-1} = 10000, 30 wt% glycolide, Polysciences, Inc., Warrington, PA
Solvent (A): **dimethyl ether C_2H_6O** **115-10-6**

continued

continued

Type of data: cloud points

w_B	0.0275	0.0275	0.0275	0.0275	0.0275	0.0275	0.0275	0.0275
T/K	305.15	314.15	323.75	333.65	343.25	353.25	362.95	373.45
P/MPa	49.05	48.45	47.95	47.35	46.65	46.05	45.15	44.25

Polymer (B): **poly(DL-lactide-*co*-glycolide)** **2001KUK**
Characterization: M_η/g.mol^{-1} = 14500, 45 wt% glycolide,
Resomer RG502, Boehringer Ingelheim Chemicals
Solvent (A): **dimethyl ether** C_2H_6O **115-10-6**

Type of data: cloud points

w_B	0.0309	0.0309	0.0309	0.0309	0.0309	0.0309
T/K	324.85	333.85	343.35	353.05	363.35	373.35
P/MPa	61.35	58.85	56.65	55.25	53.75	52.35

Polymer (B): **poly(DL-lactide-*co*-glycolide)** **2001CON**
Characterization: M_n/g.mol^{-1} = 77700, M_w/g.mol^{-1} = 130000, 25 mol% glycolide,
T_g/K = 320.05, Alkermes, Inc., Cincinnati, OH
Solvent (A): **trifluoromethane** CHF_3 **75-46-7**

Type of data: cloud points

w_B	0.05	0.05	0.05	0.05	0.05
T/K	301.75	310.55	310.65	339.45	354.25
P/bar	1341	1360	1394	1436	1474

Polymer (B): **poly(L-lactide-*co*-diglycidyl ether of bisphenol A-*co*-4,4'-hexafluoroisopropylidenediphenol)** **2005SHE**
Characterization: M_n/g.mol^{-1} = 16900, M_w/g.mol^{-1} = 32100, 1:1:1 terpolymer,
T_g/K = 358, synthesized in the laboratory
Solvent (A): **dimethyl ether** C_2H_6O **115-10-6**

Type of data: cloud points

w_B 0.028 was kept constant

T/K	299.45	311.85	326.06	343.45	361.15	378.95	395.05
P/bar	279.0	306.9	339.3	375.2	419.7	455.9	489.4
ρ/(g/cm^3)	0.713	0.702	0.690	0.675	0.663	0.650	0.640

Polymer (B): **poly(L-lactide-*co*-diglycidyl ether of bisphenol A-*co*-4,4'-isopropylidenediphenol)** **2005SHE**
Characterization: M_n/g.mol^{-1} = 8900, M_w/g.mol^{-1} = 16000, 1:1:1 terpolymer,
T_g/K = 349, synthesized in the laboratory
Solvent (A): **dimethyl ether** C_2H_6O **115-10-6**

Type of data: cloud points

continued

continued

w_B 0.030 was kept constant

T/K	295.35	314.75	334.55	354.85	377.85	390.65
P/bar	1009.2	931.2	888.9	869.0	856.1	851.6
ρ/(g/cm^3)	0.767	0.749	0.733	0.719	0.703	0.694

Polymer (B): **poly(styrene-*b*-butadiene) (completely deuterated) 2009WI2**
Characterization: M_n/g.mol^{-1} = 39700-*b*-12400, M_w/M_n = 1.02,
synthesized in the laboratory
Solvent (A): **propane** C_3H_8 **74-98-6**

Type of data: cloud points (micellar)

w_B 0.005 was kept constant

T/K	293.15	313.25	313.05	333.25	353.15	373.15	393.25	413.15	433.05
P/bar	1036	927	905	865	808	788	785	774	760

Polymer (B): **poly(styrene-*b*-butadiene) (completely deuterated) 2009WI2**
Characterization: M_n/g.mol^{-1} = 41700-*b*-40900, M_w/M_n = 1.02,
synthesized in the laboratory
Solvent (A): **propane** C_3H_8 **74-98-6**

Type of data: cloud points (micellar)

w_B 0.005 was kept constant

T/K	293.05	313.15	333.05	353.25	373.15	393.15	413.15	433.25	453.15
P/bar	1005	866	783	731	700	680	669	663	686

Polymer (B): **poly(styrene-*b*-butadiene) (completely deuterated) 2009WI2**
Characterization: M_n/g.mol^{-1} = 39700-*b*-117000, M_w/M_n = 1.01,
synthesized in the laboratory
Solvent (A): **propane** C_3H_8 **74-98-6**

Type of data: cloud points (micellar)

w_B 0.005 was kept constant

T/K	293.25	313.15	333.15	353.25	373.15	393.15	413.25	433.15
P/bar	1139	966	863	797	755	728	711	693

Type of data: cloud points

w_B 0.005 was kept constant

T/K	443.25	443.35	433.25	453.25	453.25
P/bar	879	896	911	903	909

Polymer (B): **poly(styrene-*b*-butadiene) (only PS deuterated)** **2009WI2**
Characterization: M_n/g.mol^{-1} = 41700-*b*-40800, M_w/M_n = 1.02, synthesized in the laboratory
Solvent (A): **propane** **C_3H_8** **74-98-6**

Type of data: cloud points (micellar)

w_B 0.005 was kept constant

T/K	293.15	313.15	333.15	353.25	373.15	393.35	413.05	433.25	453.25
P/bar	1163	972	859	791	748	719	702	693	689

Polymer (B): **poly(styrene-*b*-butadiene) (only PB deuterated)** **2009WI2**
Characterization: M_n/g.mol^{-1} = 37200-*b*-36700, M_w/M_n = 1.03, synthesized in the laboratory
Solvent (A): **propane** **C_3H_8** **74-98-6**

Type of data: cloud points (micellar)

w_B 0.005 was kept constant

T/K	373.05	393.15	413.15	433.05	433.05	453.15	453.15
P/bar	1342	1192	1086	1021	1024	945	957

Polymer (B): **poly(styrene-*b*-isoprene) (only PS deuterated)** **2009WI2**
Characterization: M_n/g.mol^{-1} = 41700-*b*-42200, M_w/M_n = 1.02, synthesized in the laboratory
Solvent (A): **propane** **C_3H_8** **74-98-6**

Type of data: cloud points (micellar)

w_B 0.005 was kept constant

T/K	293.15	313.35	333.15	353.15	373.05	393.25	413.15	433.05	435.05
P/bar	483	464	459	462	470	481	492	502	525

Polymer (B): **poly[styrene-*b*-4-(perfluorooctyl(ethyleneoxy) methyl)styrene]** **2004LAC**
Characterization: M_n/g.mol^{-1} = 75700 (by SEC), PS-block: M_n/g.mol^{-1} = 3600, P(PFOEO)MS-block: M_n/g.mol^{-1} = 55700 by NMR or = 72100 by SEC, PS-content is about 6 mol%, synthesized in the laboratory
Solvent (A): **carbon dioxide** **CO_2** **124-38-9**

Type of data: cloud points

w_B 0.040 was kept constant

T/K	298.15	303.15	308.15	313.15	318.15	323.15	328.15	333.15	338.15
P/bar	199.0	212.0	229.0	244.0	260.0	273.0	288.0	307.0	322.0

Polymer (B): **poly[styrene-*b*-4-(perfluorooctyl(ethyleneoxy)methyl)styrene]** **2004LAC**

Characterization: M_n/g.mol^{-1} = 54200 (by SEC), PS-block: M_n/g.mol^{-1} = 3800, P(PFOEO)MS-block: M_n/g.mol^{-1} = 47000 by NMR or = 50400 by SEC, PS-content is about 7 mol%, synthesized in the laboratory

Solvent (A): **carbon dioxide** CO_2 **124-38-9**

Type of data: cloud points

w_B 0.039 was kept constant

T/K	298.35	303.15	308.05	313.15	317.95	322.55	327.75	332.65	337.95
P/bar	121.0	139.0	157.0	176.0	190.0	205.0	221.0	234.0	249.0

Polymer (B): **poly(vinylidene fluoride-*co*-hexafluoropropylene)** **2006AHM**

Characterization: M_n/g.mol^{-1} = 54000, M_w/g.mol^{-1} = 140000, T_g/K = 253, 23.1 mol% hexafluoropropylene, Tecnoflon N215, Solvay-Solexis

Solvent (A): **carbon dioxide** CO_2 **124-38-9**

Type of data: cloud points

T/K = 313.15

w_B	0.0081	0.0095	0.0142	0.0240	0.0295	0.0336	0.0380	0.0449
P/bar	234.3	217.3	211.5	211.2	189.2	185.7	168.7	163.9

4.2. Table of binary systems where data were published only in graphical form as phase diagrams or related figures

Polymer (B)	Solvent (A)	Ref.
Poly[2,2-bis(4-trifluorovinyloxyphenyl)propane-*co*-2,2-bis(4-trifluorovinyloxyphenyl)-1,1,1,3,3,3-hexafluoropropane]		
	carbon dioxide	2007LIU
	propane	2007LIU
Poly(cyclohexene carbonate-*b*-ethylene oxide-*b*-cyclohexene carbonate)		
	carbon dioxide	2000SAR
Poly(1,1-dihydroperfluorooctyl acrylate-*b*-styrene)		
	carbon dioxide	2002TAY
Poly(1,1-dihydroperfluorooctyl acrylate-*b*-vinyl acetate)		
	carbon dioxide	2002TAY
Poly[2-(*N,N*-dimethylamino)ethyl methacrylate-*co*-1H,1H–perfluorooctyl methacrylate]		
	carbon dioxide	2008HWA
Poly[dimethylsiloxane-*co*-(3-acetoxypropyl)-methylsiloxane]		
	carbon dioxide	1999FIN
	carbon dioxide	2003KIL
Poly[dimethylsiloxane-*co*-(4-acetylbutyl)-methylsiloxane]		
	carbon dioxide	2003KIL

Polymer (B)	**Solvent (A)**	**Ref.**
Poly[dimethylsiloxane-*co*-(4,4-dimethylpentyl)-methylsiloxane]	carbon dioxide	2003KIL
Poly[dimethylsiloxane-*co*-(3-ethoxypropyl)-methylsiloxane]	carbon dioxide	2003KIL
Poly[dimethylsiloxane-*co*-glycidyloxypropylsiloxane-*co*-tris(trimethylsiloxy)silylpropylsiloxane]	carbon dioxide	1993HOE
Poly(dimethylsiloxane-*co*-hexylmethylsiloxane)	carbon dioxide	1999FIN
Poly(dimethylsiloxane-*co*-hydromethylsiloxane)	carbon dioxide	1993HOE
	carbon dioxide	1999FIN
	carbon dioxide	2003KIL
Poly[dimethylsiloxane-*co*-(3-methoxycarbonyl-propyl)methylsiloxane]	carbon dioxide	2003KIL
Poly[dimethylsiloxane-*co*-(3-trimethylsilylpropyl)-methylsiloxane]	carbon dioxide	2003KIL
Poly[dodecyl(oligoethylene oxide)-*b*-propylene oxide]	carbon dioxide	2006SUB
Poly(ethylene-*co*-acrylic acid)	ethene	2002LEE
Poly(ethylene-*co*-1-butene copolymer	1-butene	1999CHE
	dimethyl ether	2000BY3
	ethene	1999CHE
	dimethyl ether	2005LID
	propane	1998HAN

Polymer (B)	**Solvent (A)**	**Ref.**
Poly(ethylene-*co*-methacrylic acid)		
	ethene	2006KLE
Poly(ethylene-*co*-1-octene)		
	n-hexane	2005HEI
Poly(ethylene-*co*-propylene)		
	ethene	2001KEM
Poly(ethylene-*co*-propylene-*co*-diene)		
	ethene	2004PFO
Poly(ethylene glycol-*b*-ε-caprolactone)		
	trifluoromethane	2009TYR
Poly(ethylene glycol-*co*-propylene glycol)		
	carbon dioxide	2002DRO
Poly(ethylene oxide-*b*-propylene oxide-*b*-ethylene oxide)		
	carbon dioxide	2002DRO
	water	2009KOS
Poly(ethylene oxide-*b*-1,1,2,2-tetrahydroperfluorodecyl methacrylate)		
	carbon dioxide	2004MA1
	carbon dioxide	2004MA2
Poly(1H,1H,2H,2H-heptadecafluorodecyl methacrylate-*co*-methyl methacrylate)		
	carbon dioxide	2008SH2
Poly(DL-lactic acid-*co*-glycolic acid)		
	carbon dioxide	2006BYU
	chlorodifluoromethane	2006BYU
	dichloromethane	2006BYU
	trichloromethane	2006BYU
	trifluoromethane	2006BYU

Polymer (B)	Solvent (A)	Ref.
Poly(DL-lactic acid-*b*-perfluoropropylene oxide	carbon dioxide	2007EDM
Poly[3-octylthiophene-*co*-2-(3-thienyl)acetyl 3,3,4,4,5,5,6,6,7,7,8,8,8-tridecafluoro-1-octanate]	carbon dioixde	2006GAN
Poly[oligo(ethylene glycol) methacrylate-*b*-1H,1H,2H,2H-perfluorooctyl methacrylate]	carbon dioixde	2004HWA
Poly[oligo(ethylene glycol) methacrylate-*ran*-1H,1H,2H,2H-perfluorooctyl methacrylate]	carbon dioixde	2004HWA
Poly(propylene oxide-*b*-ethylene oxide-*b*-propylene oxide)	carbon dioxide	2002DRO
Poly(propylene oxide-*co*-propylene carbonate)	carbon dioxide	2000SAR
Poly(styrene-*b*-butadiene)	propane	2009WI1
Poly(styrene-*b*-dimethylsiloxane)	carbon dioxide	2000BER
Poly(styrene-*b*-isoprene)	propane	2006WIN
	propane	2009WI1
Poly(styrene-*b*-pentyl methacrylate)	carbon dioxide	2006LAV
Poly{styrene-*b*-4-[perfluorooctyl(ethyleneoxy)-methyl]styrene}	carbon dioxide	2004LAC

Polymer (B)	Solvent (A)	Ref.
Poly(styrene-*b*-1,1,2,2-tetrahydroperfluorodecyl acrylate)	carbon dioxide	2006AND
Poly(styrene-*b*-1,1,2,2-tetrahydroperfluorodecyl methacrylate)	carbon dioxide	2004LAC
Poly(tetrafluoroethylene-*co*-vinyl acetate)	carbon dioxide	2004BAR
Poly(1,1,2,2-tetrahydroperfluorooctyl acrylate-*b*-vinyl acetate)	carbon dioxide	2002COL
Poly(vinyl acetate-*co*-1-methoxyethyl ether)	carbon dioxide	2009WAN
Poly(vinyl acetate-*co*-methoxymethyl ether)	carbon dioxide	2009WAN
Poly(vinylidene fluoride-*co*-chlorotrifluoroethylene)	carbon dioxide	1997DIN
Poly(vinylidene fluoride-*co*-hexafluoropropylene)	carbon dioxide	1997DIN

4.3. Cloud-point and/or coexistence curves of quasiternary and/or quasiquaternary solutions

Polymer (B): **poly(cyclohexene oxide-*co*-carbon dioxide)** **2005SCH**
Characterization: M_w/g.mol^{-1} = 12000, M_w/M_n < 1.2, synthesized in the laboratory
Solvent (A): **carbon dioxide** CO_2 **124-38-9**
Solvent (C): **cyclohexene oxide** $C_6H_{10}O$ **286-20-4**

Type of data: cloud points

w_A	0.8660	0.8660	0.8660	0.8660	0.8660	0.8660	0.8660	0.8660	0.8660
w_B	0.0094	0.0094	0.0094	0.0094	0.0094	0.0094	0.0094	0.0094	0.0094
w_C	0.1246	0.1246	0.1246	0.1246	0.1246	0.1246	0.1246	0.1246	0.1246
T/K	373.08	382.72	393.34	403.44	413.36	422.99	433.27	443.24	453.17
P/MPa	183	176	172	168	165	163	159	157	155
w_A	0.8660								
w_B	0.0094								
w_C	0.1246								
T/K	465.88								
P/MPa	154								

Polymer (B): **poly(cyclohexene oxide-*co*-carbon dioxide)** **2005SCH**
Characterization: M_w/g.mol^{-1} = 25000, M_w/M_n < 1.2, synthesized in the laboratory
Solvent (A): **carbon dioxide** CO_2 **124-38-9**
Solvent (C): **cyclohexene oxide** $C_6H_{10}O$ **286-20-4**

Type of data: cloud points

w_A	0.9173	0.9173	0.9173	0.9173	0.9173	0.8354	0.8354	0.8354	0.8354
w_B	0.0099	0.0099	0.0099	0.0099	0.0099	0.0100	0.0100	0.0100	0.0100
w_C	0.0728	0.0728	0.0728	0.0728	0.0728	0.1546	0.1546	0.1546	0.1546
T/K	432.21	435.60	446.41	456.07	465.54	367.57	380.21	388.37	397.78
P/MPa	351	341	314	297	278	213	204	198	193
w_A	0.8354	0.8354	0.8354	0.8354	0.8354	0.8354	0.8354	0.7474	0.7474
w_B	0.0100	0.0100	0.0100	0.0100	0.0100	0.0100	0.0100	0.0098	0.0098
w_C	0.1546	0.1546	0.1546	0.1546	0.1546	0.1546	0.1546	0.2428	0.2428
T/K	407.09	416.69	426.60	437.84	447.07	455.94	465.93	369.35	380.89
P/MPa	189	184	181	176	174	171	167	115	117
w_A	0.7474	0.7474	0.7474	0.7474	0.7474	0.7474	0.7474	0.7474	0.7474
w_B	0.0098	0.0098	0.0098	0.0098	0.0098	0.0098	0.0098	0.0098	0.0098
w_C	0.2428	0.2428	0.2428	0.2428	0.2428	0.2428	0.2428	0.2428	0.2428
T/K	387.81	395.61	405.75	415.26	427.35	436.35	445.21	454.97	465.99
P/MPa	117	117	118	118	119	119	119	119	119

Polymer (B): **poly(cyclohexene oxide-*co*-carbon dioxide)** **2005SCH**
Characterization: M_w/g.mol^{-1} = 54000, M_w/M_n < 1.2, synthesized in the laboratory
Solvent (A): **carbon dioxide** CO_2 **124-38-9**
Solvent (C): **cyclohexene oxide** $C_6H_{10}O$ **286-20-4**

Type of data: cloud points

w_A	0.8685	0.8685	0.8685	0.8685	0.8685	0.8685	0.8685
w_B	0.0098	0.0098	0.0098	0.0098	0.0098	0.0098	0.0098
w_C	0.1217	0.1217	0.1217	0.1217	0.1217	0.1217	0.1217
T/K	402.69	412.21	421.47	432.80	442.71	451.96	463.80
P/MPa	363	332	310	292	278	265	251

Polymer (B): **poly(ethylene-*co*-acrylic acid)** **2002BEY**
Characterization: M_n/g.mol^{-1} = 22700, M_w/g.mol^{-1} = 258000,
6.0 wt% acrylic acid, Exxon Co., Machelen, Belgium
Solvent (A): **ethene** C_2H_4 **74-85-1**
Solvent (C): **n-decane** $C_{10}H_{22}$ **124-18-5**

Type of data: cloud points

w_A	0.900	0.900	0.900	0.900	0.900	0.900	0.548	0.548	0.548
w_B	0.050	0.050	0.050	0.050	0.050	0.050	0.050	0.050	0.050
w_C	0.050	0.050	0.050	0.050	0.050	0.050	0.402	0.402	0.402
T/K	522.55	497.85	472.85	447.95	423.55	413.65	522.65	498.35	473.15
P/bar	1242	1330	1468	1691	2074	2306	799	844	917

w_A	0.548	0.548	0.548	0.548	0.548
w_B	0.050	0.050	0.050	0.050	0.050
w_C	0.402	0.402	0.402	0.402	0.402
T/K	448.55	423.75	413.55	403.85	393.35
P/bar	1041	1260	1406	1581	1818

Polymer (B): **poly(ethylene-*co*-acrylic acid)** **2002BEY**
Characterization: M_n/g.mol^{-1} = 22700, M_w/g.mol^{-1} = 258000,
6.0 wt% acrylic acid, Exxon Co., Machelen, Belgium
Solvent (A): **ethene** C_2H_4 **74-85-1**
Solvent (C): **ethanol** C_2H_6O **64-17-5**

Type of data: cloud points

w_A	0.900	0.900	0.900	0.900	0.900	0.900	0.900	0.801	0.801
w_B	0.050	0.050	0.050	0.050	0.050	0.050	0.050	0.050	0.050
w_C	0.050	0.050	0.050	0.050	0.050	0.050	0.050	0.149	0.149
T/K	522.85	497.55	472.25	447.55	423.65	413.15	403.65	523.25	497.95
P/bar	1168	1212	1276	1371	1491	1591	1607	1036	1074

continued

continued

w_A	0.801	0.801	0.801	0.801	0.801	0.703	0.703	0.703	0.703
w_B	0.050	0.050	0.050	0.050	0.050	0.050	0.050	0.050	0.050
w_C	0.149	0.149	0.149	0.149	0.149	0.247	0.247	0.247	0.247
T/K	472.95	448.15	424.45	414.15	404.15	523.15	498.05	472.65	447.45
P/bar	1119	1161	1271	1301	1339	966	981	1016	1069
w_A	0.703	0.703	0.703	0.543	0.543	0.543	0.543	0.543	0.543
w_B	0.050	0.050	0.050	0.050	0.050	0.050	0.050	0.050	0.050
w_C	0.247	0.247	0.247	0.407	0.407	0.407	0.407	0.407	0.407
T/K	423.15	413.75	403.65	521.85	497.95	473.05	448.15	423.55	414.05
P/bar	1154	1209	1264	811	834	886	989	1256	1483

Polymer (B): **poly(ethylene-*co*-acrylic acid)** **2002BEY**
Characterization: M_n/g.mol^{-1} = 19900, M_w/g.mol^{-1} = 235000,
7.5 wt% acrylic acid, Exxon Co., Machelen, Belgium
Solvent (A): **ethene** **C_2H_4** **74-85-1**
Solvent (C): **ethanol** **C_2H_6O** **64-17-5**

Type of data: cloud points

w_A	0.893	0.893	0.893	0.893	0.893	0.893	0.893	0.548	0.548
w_B	0.050	0.050	0.050	0.050	0.050	0.050	0.050	0.050	0.050
w_C	0.057	0.057	0.057	0.057	0.057	0.057	0.057	0.402	0.402
T/K	523.35	497.75	472.55	447.95	423.75	414.15	403.85	523.15	497.65
P/bar	1189	1234	1310	1411	1546	1604	1673	766	774
w_A	0.548	0.548	0.548	0.548					
w_B	0.050	0.050	0.050	0.050					
w_C	0.402	0.402	0.402	0.402					
T/K	472.75	448.25	433.25	423.35					
P/bar	809	890	989	1079					

Polymer (B): **poly(ethylene-*co*-acrylic acid)** **2002BEY**
Characterization: M_n/g.mol^{-1} = 23400, M_w/g.mol^{-1} = 227000,
9.0 wt% acrylic acid, Exxon Co., Machelen, Belgium
Solvent (A): **ethene** **C_2H_4** **74-85-1**
Solvent (C): **ethanol** **C_2H_6O** **64-17-5**

Type of data: cloud points

w_A	0.900	0.900	0.900	0.900	0.900	0.900	0.900	0.556	0.556
w_B	0.050	0.050	0.050	0.050	0.050	0.050	0.050	0.050	0.050
w_C	0.050	0.050	0.050	0.050	0.050	0.050	0.050	0.394	0.394
T/K	523.35	497.95	473.05	448.05	423.85	413.85	403.65	524.05	498.15
P/bar	1200	1243	1331	1439	1601	1669	1746	795	804
w_A	0.556	0.556	0.556	0.556	0.556	0.556			
w_B	0.050	0.050	0.050	0.050	0.050	0.050			
w_C	0.394	0.394	0.394	0.394	0.394	0.394			
T/K	472.65	447.95	423.55	413.65	403.65	393.95			
P/bar	826	873	978	1051	1164	1334			

Polymer (B): **poly(ethylene-*co*-acrylic acid)** **2002BEY**
Characterization: M_n/g.mol^{-1} = 22700, M_w/g.mol^{-1} = 258000,
6.0 wt% acrylic acid, Exxon Co., Machelen, Belgium
Solvent (A): **ethene** C_2H_4 **74-85-1**
Solvent (C): **ethyl acetate** $C_4H_8O_2$ **141-78-6**

Type of data: cloud points

w_A	0.850	0.850	0.850	0.850	0.850	0.850	0.552	0.552	0.552
w_B	0.050	0.050	0.050	0.050	0.050	0.050	0.050	0.050	0.050
w_C	0.055	0.055	0.055	0.055	0.055	0.055	0.398	0.398	0.398
T/K	521.95	497.45	472.15	447.25	432.85	423.85	523.25	498.15	472.85
P/bar	1252	1339	1464	1657	1825	1954	936	974	1019
w_A	0.552	0.552	0.552	0.552	0.552	0.258	0.258	0.258	0.258
w_B	0.050	0.050	0.050	0.050	0.050	0.050	0.050	0.050	0.050
w_C	0.398	0.398	0.398	0.398	0.398	0.692	0.692	0.692	0.692
T/K	448.15	433.25	423.65	414.05	403.15	522.95	498.35	473.15	448.65
P/bar	1099	1168	1224	1298	1396	666	677	701	746
w_A	0.258	0.258	0.258	0.258					
w_B	0.050	0.050	0.050	0.050					
w_C	0.692	0.692	0.692	0.692					
T/K	423.75	414.15	403.65	393.75					
P/bar	842	910	1010	1156					

Polymer (B): **poly(ethylene-*co*-acrylic acid)** **2002BEY**
Characterization: M_n/g.mol^{-1} = 22700, M_w/g.mol^{-1} = 258000,
6.0 wt% acrylic acid, Exxon Co., Machelen, Belgium
Solvent (A): **ethene** C_2H_4 **74-85-1**
Solvent (C): **n-heptane** C_7H_{16} **142-82-5**

Type of data: cloud points

w_A	0.900	0.900	0.900	0.900	0.900	0.900	0.552	0.552	0.552
w_B	0.050	0.050	0.050	0.050	0.050	0.050	0.050	0.050	0.050
w_C	0.050	0.050	0.050	0.050	0.050	0.050	0.398	0.398	0.398
T/K	522.45	497.95	472.45	447.75	423.35	413.35	522.85	498.05	473.05
P/bar	1257	1340	1476	1695	2056	2290	863	910	993
w_A	0.552	0.552	0.552	0.552	0.552				
w_B	0.050	0.050	0.050	0.050	0.050				
w_C	0.398	0.398	0.398	0.398	0.398				
T/K	448.35	423.75	413.75	403.45	393.65				
P/bar	1129	1366	1515	1709	1961				

Polymer (B): **poly(ethylene-*co*-acrylic acid)** **2002BEY**
Characterization: M_n/g.mol^{-1} = 22700, M_w/g.mol^{-1} = 258000,
6.0 wt% acrylic acid, Exxon Co., Machelen, Belgium
Solvent (A): **ethene** C_2H_4 **74-85-1**
Solvent (C): **octanoic acid** $C_8H_{16}O_2$ **124-07-2**

Type of data: cloud points

w_A	0.904	0.904	0.904	0.904	0.904	0.904	0.904	0.904	0.904
w_B	0.050	0.050	0.050	0.050	0.050	0.050	0.050	0.050	0.050
w_C	0.046	0.046	0.046	0.046	0.046	0.046	0.046	0.046	0.046
T/K	521.85	497.55	472.55	447.85	423.75	413.75	404.05	393.25	383.75
P/bar	1145	1180	1236	1308	1401	1454	1505	1565	1637
w_A	0.552	0.552	0.552	0.552	0.552	0.552	0.552	0.552	0.552
w_B	0.050	0.050	0.050	0.050	0.050	0.050	0.050	0.050	0.050
w_C	0.398	0.398	0.398	0.398	0.398	0.398	0.398	0.398	0.398
T/K	521.95	498.05	472.75	447.75	423.65	414.15	403.85	393.65	384.05
P/bar	697	691	696	706	719	728	738	749	764
w_A	0.292	0.292	0.292	0.292	0.292	0.292	0.292	0.292	0.292
w_B	0.050	0.050	0.050	0.050	0.050	0.050	0.050	0.050	0.050
w_C	0.658	0.658	0.658	0.658	0.658	0.658	0.658	0.658	0.658
T/K	522.65	497.95	472.65	448.65	424.05	414.25	404.25	393.55	384.15
P/bar	346	325	307	289	269	260	251	243	235

Polymer (B): **poly(ethylene-*co*-acrylic acid)** **2002BEY**
Characterization: M_n/g.mol^{-1} = 22700, M_w/g.mol^{-1} = 258000,
6.0 wt% acrylic acid, Exxon Co., Machelen, Belgium
Solvent (A): **ethene** C_2H_4 **74-85-1**
Solvent (C): **2,2,4-trimethylpentane** C_8H_{18} **540-84-1**

Type of data: cloud points

w_A	0.900	0.900	0.900	0.900	0.900	0.900	0.551	0.551	0.551
w_B	0.050	0.050	0.050	0.050	0.050	0.050	0.050	0.050	0.050
w_C	0.050	0.050	0.050	0.050	0.050	0.050	0.399	0.399	0.399
T/K	522.45	497.95	472.55	447.95	423.65	413.55	522.95	498.05	472.25
P/bar	1233	1314	1445	1648	2005	2211	888	944	1036
w_A	0.551	0.551	0.551	0.551					
w_B	0.050	0.050	0.050	0.050					
w_C	0.399	0.399	0.399	0.399					
T/K	447.95	423.75	413.25	403.55					
P/bar	1181	1430	1586	1784					

Polymer (B): **poly(ethylene-*co*-benzyl methacrylate)** **2004LAT**
Characterization: 13.1 mol% benzyl methacrylate, synthesized in the laboratory
Solvent (A): **ethene** C_2H_4 **74-85-1**
Solvent (C): **benzyl methacrylate** $C_{11}H_{12}O_2$ **2495-37-6**

Type of data: cloud points

w_A	0.9133	0.8908	0.8693	0.849	0.829
w_B	0.050	0.050	0.050	0.050	0.050
w_C	0.0367	0.0592	0.0807	0.101	0.121
T/K	473.15	473.15	473.15	473.15	473.15
P/bar	1339	1308	1277	1245	1227

Polymer (B): **poly(ethylene-*co*-benzyl methacrylate)** **2004LAT**
Characterization: 20.3 mol% benzyl methacrylate, synthesized in the laboratory
Solvent (A): **ethene** C_2H_4 **74-85-1**
Solvent (C): **benzyl methacrylate** $C_{11}H_{12}O_2$ **2495-37-6**

Type of data: cloud points

w_A	0.8868	0.8656	0.845	0.826	0.807
w_B	0.050	0.050	0.050	0.050	0.050
w_C	0.0632	0.0844	0.105	0.124	0.143
T/K	473.15	473.15	473.15	473.15	473.15
P/bar	1401	1368	1330	1299	1275

Polymer (B): **poly(ethylene-*co*-benzyl methacrylate)** **2004LAT**
Characterization: 25.2 mol% benzyl methacrylate, synthesized in the laboratory
Solvent (A): **ethene** C_2H_4 **74-85-1**
Solvent (C): **benzyl methacrylate** $C_{11}H_{12}O_2$ **2495-37-6**

Type of data: cloud points

w_A	0.8629	0.843	0.823	0.805	0.787
w_B	0.050	0.050	0.050	0.050	0.050
w_C	0.0871	0.107	0.127	0.145	0.163
T/K	473.15	473.15	473.15	473.15	473.15
P/bar	1463	1425	1380	1352	1315

Polymer (B): **poly(ethylene-*co*-benzyl methacrylate)** **2004LAT**
Characterization: M_n/g.mol^{-1} = 35600, M_w/g.mol^{-1} = 62200,
30.4 mol% benzyl methacrylate, synthesized in the laboratory
Solvent (A): **ethene** C_2H_4 **74-85-1**
Solvent (C): **benzyl methacrylate** $C_{11}H_{12}O_2$ **2495-37-6**

Type of data: cloud points

w_A	0.833	0.814	0.796	0.778	0.762
w_B	0.050	0.050	0.050	0.050	0.050
w_C	0.117	0.136	0.154	0.172	0.188
T/K	473.15	473.15	473.15	473.15	473.15
P/bar	1508	1460	1417	1375	1340

Polymer (B): **poly(ethylene-*co*-benzyl methacrylate)** **2004LAT**
Characterization: 33.5 mol% benzyl methacrylate, synthesized in the laboratory
Solvent (A): **ethene** C_2H_4 **74-85-1**
Solvent (C): **benzyl methacrylate** $C_{11}H_{12}O_2$ **2495-37-6**

Type of data: cloud points

w_A	0.813	0.795	0.778	0.761	0.745
w_B	0.050	0.050	0.050	0.050	0.050
w_C	0.137	0.155	0.172	0.189	0.205
T/K	473.15	473.15	473.15	473.15	473.15
P/bar	1515	1472	1430	1393	1352

Polymer (B): **poly(ethylene-*co*-1-butene)** **1999CHE**
Characterization: M_n/g.mol^{-1} = 7570, M_w/g.mol^{-1} = 7870, 29.0 mol% 1-butene, completely hydrogenated polybutadiene, 29 mol% 1,2-units, Polymer Source, Inc., Dorval, Quebec
Solvent (A): **ethene** C_2H_4 **74-85-1**
Solvent (C): **1-butene** C_4H_8 **106-98-9**

Type of data: cloud points

w_A 0.49105 w_B 0.00085 w_C 0.50810

T/K	431.85	430.45	418.95	410.95	401.25	399.75	381.75	373.35	368.95
P/bar	400.7	399.3	395.4	391.1	386.6	385.7	376.9	373.0	369.3
T/K	363.75	357.65	353.95	350.05	348.85	347.35			
P/bar	365.9	361.4	353.8	407.4	555.0	983.2			

Comments: The last three data points are temperature-induced phase transitions.

w_A 0.40287 w_B 0.00927 w_C 0.58786

T/K	428.85	428.85	415.05	412.05	398.45	396.95	384.25	381.45	368.65
P/bar	698.7	692.7	676.2	674.2	649.6	645.1	606.2	607.0	544.7
T/K	367.95	353.65	352.75	352.75	352.65	352.55			
P/bar	538.2	498.6	494.0	600.6	995.1	1426			

Comments: The last three data points are temperature-induced phase transitions.

Polymer (B): **poly(ethylene-*co*-1-butene)** **2006NAG**
Characterization: M_n/g.mol^{-1} = 43700, M_w/g.mol^{-1} = 52000, M_z/g.mol^{-1} = 59000, 4.1 mol% 1-butene, 2.05 ethyl branches per 100 backbone C-atomes, hydrogenated polybutadiene PBD 50000, was denoted as LLDPE, DSM, The Netherlands
Solvent (A): **ethene** C_2H_4 **74-85-1**
Solvent (C): **n-hexane** C_6H_{14} **110-54-3**

continued

continued

Type of data: cloud points

w_A	0.0118	0.0118	0.0118	0.0118	0.0118	0.0118	0.0118	0.0118	0.0118
w_B	0.0502	0.0502	0.0502	0.0502	0.0502	0.0502	0.0502	0.0502	0.0502
w_C	0.9380	0.9380	0.9380	0.9380	0.9380	0.9380	0.9380	0.9380	0.9380
T/K	426.28	431.16	436.16	441.26	446.14	451.05	455.97	460.90	465.75
P/MPa	1.846	2.566	3.276	4.096	4.721	5.421	6.096	6.746	7.396
w_A	0.0118	0.0118	0.0118	0.0118	0.0118	0.0118	0.0205	0.0205	0.0205
w_B	0.0502	0.0502	0.0502	0.0502	0.0502	0.0502	0.0501	0.0501	0.0501
w_C	0.9380	0.9380	0.9380	0.9380	0.9380	0.9380	0.9294	0.9294	0.9294
T/K	470.74	475.64	480.51	485.46	490.27	495.18	415.92	420.93	425.96
P/MPa	7.996	8.571	9.176	9.726	10.271	10.841	1.659	2.459	3.224
w_A	0.0205	0.0205	0.0205	0.0205	0.0205	0.0205	0.0205	0.0205	0.0205
w_B	0.0501	0.0501	0.0501	0.0501	0.0501	0.0501	0.0501	0.0501	0.0501
w_C	0.9294	0.9294	0.9294	0.9294	0.9294	0.9294	0.9294	0.9294	0.9294
T/K	430.76	435.65	440.65	445.51	450.52	455.49	460.43	465.36	470.27
P/MPa	3.974	4.734	5.394	6.049	6.769	7.474	8.144	8.724	9.319
w_A	0.0205	0.0205	0.0205	0.0205	0.0205	0.0298	0.0298	0.0298	0.0298
w_B	0.0501	0.0501	0.0501	0.0501	0.0501	0.0499	0.0499	0.0499	0.0499
w_C	0.9294	0.9294	0.9294	0.9294	0.9294	0.9203	0.9203	0.9203	0.9203
T/K	475.28	480.25	485.26	490.18	495.12	411.24	416.19	421.11	426.01
P/MPa	9.869	10.394	10.999	11.549	12.074	2.149	3.021	3.849	4.594
w_A	0.0298	0.0298	0.0298	0.0298	0.0298	0.0298	0.0298	0.0298	0.0298
w_B	0.0499	0.0499	0.0499	0.0499	0.0499	0.0499	0.0499	0.0499	0.0499
w_C	0.9203	0.9203	0.9203	0.9203	0.9203	0.9203	0.9203	0.9203	0.9203
T/K	430.99	435.94	440.90	445.92	450.79	455.84	460.69	465.65	470.62
P/MPa	5.324	6.094	6.799	7.394	8.099	8.749	9.369	9.994	10.699
w_A	0.0298	0.0298	0.0298	0.0298	0.0298	0.0099	0.0099	0.0099	0.0099
w_B	0.0499	0.0499	0.0499	0.0499	0.0499	0.0998	0.0998	0.0998	0.0998
w_C	0.9203	0.9203	0.9203	0.9203	0.9203	0.8903	0.8903	0.8903	0.8903
T/K	475.58	480.51	485.42	490.31	495.24	426.12	430.98	435.95	440.82
P/MPa	11.299	11.799	12.394	12.894	13.369	1.774	2.524	3.194	3.969
w_A	0.0099	0.0099	0.0099	0.0099	0.0099	0.0099	0.0099	0.0099	0.0099
w_B	0.0998	0.0998	0.0998	0.0998	0.0998	0.0998	0.0998	0.0998	0.0998
w_C	0.8903	0.8903	0.8903	0.8903	0.8903	0.8903	0.8903	0.8903	0.8903
T/K	445.76	450.70	455.59	460.54	465.50	470.43	475.42	480.45	485.47
P/MPa	4.739	5.249	5.944	6.609	7.274	7.844	8.574	9.194	9.849
w_A	0.0099	0.0099	0.0196	0.0196	0.0196	0.0196	0.0196	0.0196	0.0196
w_B	0.0998	0.0998	0.1007	0.1007	0.1007	0.1007	0.1007	0.1007	0.1007
w_C	0.8903	0.8903	0.8797	0.8797	0.8797	0.8797	0.8797	0.8797	0.8797
T/K	490.41	495.30	421.26	426.08	431.05	435.99	440.73	445.90	450.91
P/MPa	10.449	10.999	2.449	3.144	3.949	4.719	5.449	6.199	6.844

continued

continued

w_A	0.0196	0.0196	0.0196	0.0196	0.0196	0.0196	0.0196	0.0196	0.0196
w_B	0.1007	0.1007	0.1007	0.1007	0.1007	0.1007	0.1007	0.1007	0.1007
w_C	0.8797	0.8797	0.8797	0.8797	0.8797	0.8797	0.8797	0.8797	0.8797
T/K	455.88	460.64	465.67	470.72	475.58	480.54	485.56	490.41	495.45
P/MPa	7.524	8.224	8.834	9.484	10.094	10.694	11.249	11.819	12.374
w_A	0.0296	0.0296	0.0296	0.0296	0.0296	0.0296	0.0296	0.0296	0.0296
w_B	0.1002	0.1002	0.1002	0.1002	0.1002	0.1002	0.1002	0.1002	0.1002
w_C	0.8702	0.8702	0.8702	0.8702	0.8702	0.8702	0.8702	0.8702	0.8702
T/K	410.90	415.82	420.77	425.74	430.62	435.61	440.43	445.48	450.46
P/MPa	2.199	3.024	3.809	4.629	5.374	6.084	6.824	7.569	8.244
w_A	0.0296	0.0296	0.0296	0.0296	0.0296	0.0296	0.0296	0.0296	0.0098
w_B	0.1002	0.1002	0.1002	0.1002	0.1002	0.1002	0.1002	0.1002	0.1508
w_C	0.8702	0.8702	0.8702	0.8702	0.8702	0.8702	0.8702	0.8702	0.8394
T/K	455.38	460.36	465.45	470.41	475.31	480.19	485.06	489.96	430.69
P/MPa	8.934	9.599	10.057	10.734	11.324	11.949	12.349	12.949	2.074
w_A	0.0098	0.0098	0.0098	0.0098	0.0098	0.0098	0.0098	0.0098	0.0098
w_B	0.1508	0.1508	0.1508	0.1508	0.1508	0.1508	0.1508	0.1508	0.1508
w_C	0.8394	0.8394	0.8394	0.8394	0.8394	0.8394	0.8394	0.8394	0.8394
T/K	435.59	440.64	445.66	450.45	455.37	460.32	465.30	470.23	475.17
P/MPa	2.869	3.619	4.349	5.044	5.694	6.349	6.989	7.614	8.229
w_A	0.0098	0.0098	0.0098	0.0098	0.0210	0.0210	0.0210	0.0210	0.0210
w_B	0.1508	0.1508	0.1508	0.1508	0.1503	0.1503	0.1503	0.1503	0.1503
w_C	0.8394	0.8394	0.8394	0.8394	0.8287	0.8287	0.8287	0.8287	0.8287
T/K	480.07	484.95	489.87	494.80	420.47	425.34	430.32	435.25	440.27
P/MPa	8.824	9.409	9.964	10.514	2.094	2.844	3.644	4.369	5.069
w_A	0.0210	0.0210	0.0210	0.0210	0.0210	0.0210	0.0210	0.0210	0.0210
w_B	0.1503	0.1503	0.1503	0.1503	0.1503	0.1503	0.1503	0.1503	0.1503
w_C	0.8287	0.8287	0.8287	0.8287	0.8287	0.8287	0.8287	0.8287	0.8287
T/K	445.20	450.13	455.10	460.07	464.88	469.80	474.77	479.72	484.58
P/MPa	5.794	6.519	7.319	7.789	8.509	9.024	9.669	10.229	10.769
w_A	0.0210	0.0210	0.0293	0.0293	0.0293	0.0293	0.0293	0.0293	0.0293
w_B	0.1503	0.1503	0.1506	0.1506	0.1506	0.1506	0.1506	0.1506	0.1506
w_C	0.8287	0.8287	0.8201	0.8201	0.8201	0.8201	0.8201	0.8201	0.8201
T/K	489.52	494.48	411.03	415.95	420.84	425.82	431.00	435.98	440.92
P/MPa	11.319	11.884	2.074	2.874	3.749	4.444	5.209	5.969	6.689
w_A	0.0293	0.0293	0.0293	0.0293	0.0293	0.0293	0.0293	0.0293	0.0293
w_B	0.1506	0.1506	0.1506	0.1506	0.1506	0.1506	0.1506	0.1506	0.1506
w_C	0.8201	0.8201	0.8201	0.8201	0.8201	0.8201	0.8201	0.8201	0.8201
T/K	445.91	450.89	455.82	460.76	465.67	470.37	475.46	480.33	485.26
P/MPa	7.339	8.019	8.639	9.304	9.959	10.529	11.124	11.669	12.219

continued

continued

w_A	0.0293	0.0293
w_B	0.1506	0.1506
w_C	0.8201	0.8201
T/K	490.35	495.17
P/MPa	12.759	13.309

Type of data: vapor-liquid equilibrium data

w_A	0.0118	0.0118	0.0118	0.0205	0.0205	0.0298	0.0298	0.0298	0.0099
w_B	0.0502	0.0502	0.0502	0.0501	0.0501	0.0499	0.0499	0.0499	0.0998
w_C	0.9380	0.9380	0.9380	0.9294	0.9294	0.9203	0.9203	0.9203	0.8903
T/K	411.46	416.37	421.20	406.00	410.93	396.49	401.10	406.30	406.11
P/MPa	1.052	1.116	1.181	1.354	1.434	1.529	1.604	1.699	0.969

w_A	0.0099	0.0099	0.0099	0.0196	0.0196	0.0196	0.0196	0.0296	0.0296
w_B	0.0998	0.0998	0.0998	0.1007	0.1007	0.1007	0.1007	0.1002	0.1002
w_C	0.8903	0.8903	0.8903	0.8797	0.8797	0.8797	0.8797	0.8702	0.8702
T/K	411.11	416.06	421.13	401.65	406.43	411.45	416.30	396.26	401.23
P/MPa	1.039	1.109	1.189	1.309	1.374	1.479	1.549	1.580	1.660

w_A	0.0296	0.0098	0.0098	0.0098	0.0210	0.0210	0.0210	0.0210	0.0293
w_B	0.1002	0.1508	0.1508	0.1508	0.1503	0.1503	0.1503	0.1503	0.1506
w_C	0.8702	0.8394	0.8394	0.8394	0.8287	0.8287	0.8287	0.8287	0.8201
T/K	406.21	416.15	421.11	426.07	400.89	405.73	410.70	415.66	396.32
P/MPa	1.750	1.139	1.219	1.299	1.354	1.429	1.504	1.589	1.664

w_A	0.0293	0.0293
w_B	0.1506	0.1506
w_C	0.8201	0.8201
T/K	401.21	406.10
P/MPa	1.744	1.834

Type of data: vapor-liquid-liquid equilibrium data

w_A	0.0118	0.0118	0.0118	0.0118	0.0118	0.0118	0.0118	0.0205	0.0205
w_B	0.0502	0.0502	0.0502	0.0502	0.0502	0.0502	0.0502	0.0501	0.0501
w_C	0.9380	0.9380	0.9380	0.9380	0.9380	0.9380	0.9380	0.9294	0.9294
T/K	431.17	441.13	450.89	460.82	470.58	480.33	490.18	420.93	430.83
P/MPa	1.346	1.561	1.746	1.996	2.251	2.551	2.856	1.589	1.784

w_A	0.0205	0.0205	0.0205	0.0205	0.0205	0.0205	0.0298	0.0298	0.0298
w_B	0.0501	0.0501	0.0501	0.0501	0.0501	0.0501	0.0499	0.0499	0.0499
w_C	0.9294	0.9294	0.9294	0.9294	0.9294	0.9294	0.9203	0.9203	0.9203
T/K	440.72	450.70	460.51	470.39	480.24	490.10	411.32	421.44	431.01
P/MPa	1.974	2.194	2.444	2.699	2.984	3.269	1.774	1.964	2.149

w_A	0.0298	0.0298	0.0298	0.0298	0.0298	0.0298	0.0099	0.0099	0.0099
w_B	0.0499	0.0499	0.0499	0.0499	0.0499	0.0499	0.0998	0.0998	0.0998
w_C	0.9203	0.9203	0.9203	0.9203	0.9203	0.9203	0.8903	0.8903	0.8903
T/K	440.90	450.90	460.75	470.56	480.44	490.32	431.00	440.82	450.72
P/MPa	2.359	2.589	2.834	3.094	3.369	3.649	1.364	1.544	1.754

continued

continued

w_A	0.0099	0.0099	0.0099	0.0099	0.0196	0.0196	0.0196	0.0196	0.0196
w_B	0.0998	0.0998	0.0998	0.0998	0.1007	0.1007	0.1007	0.1007	0.1007
w_C	0.8903	0.8903	0.8903	0.8903	0.8797	0.8797	0.8797	0.8797	0.8797
T/K	460.66	470.51	480.43	490.25	421.34	431.27	440.74	450.63	460.65
P/MPa	1.989	2.254	2.544	2.869	1.629	1.804	2.009	2.234	2.474
w_A	0.0196	0.0196	0.0196	0.0296	0.0296	0.0296	0.0296	0.0296	0.0296
w_B	0.1007	0.1007	0.1007	0.1002	0.1002	0.1002	0.1002	0.1002	0.1002
w_C	0.8797	0.8797	0.8797	0.8702	0.8702	0.8702	0.8702	0.8702	0.8702
T/K	470.39	480.38	490.29	411.27	421.23	431.10	441.04	450.93	460.69
P/MPa	2.734	3.024	3.319	1.829	2.009	2.194	2.429	2.664	2.909
w_A	0.0296	0.0296	0.0296	0.0098	0.0098	0.0098	0.0098	0.0098	0.0098
w_B	0.1002	0.1002	0.1002	0.1508	0.1508	0.1508	0.1508	0.1508	0.1508
w_C	0.8702	0.8702	0.8702	0.8394	0.8394	0.8394	0.8394	0.8394	0.8394
T/K	470.55	480.48	490.31	430.69	440.59	450.49	460.60	470.34	480.41
P/MPa	3.174	3.449	3.709	1.384	1.569	1.774	2.029	2.284	2.579
w_A	0.0098	0.0210	0.0210	0.0210	0.0210	0.0210	0.0210	0.0210	0.0210
w_B	0.1508	0.1503	0.1503	0.1503	0.1503	0.1503	0.1503	0.1503	0.1503
w_C	0.8394	0.8287	0.8287	0.8287	0.8287	0.8287	0.8287	0.8287	0.8287
T/K	490.00	420.58	430.38	440.26	450.09	459.96	469.77	479.68	489.57
P/MPa	2.889	1.674	1.859	2.059	2.269	2.519	2.784	3.044	3.354
w_A	0.0293	0.0293	0.0293	0.0293	0.0293	0.0293	0.0293	0.0293	0.0293
w_B	0.1506	0.1506	0.1506	0.1506	0.1506	0.1506	0.1506	0.1506	0.1506
w_C	0.8201	0.8201	0.8201	0.8201	0.8201	0.8201	0.8201	0.8201	0.8201
T/K	411.03	420.84	430.89	440.85	450.76	460.61	470.42	480.19	490.17
P/MPa	1.899	2.084	2.299	2.519	2.754	3.004	3.264	3.534	3.799

Polymer (B): **poly(ethylene-*co*-butyl acrylate)** **1999DIE**

Characterization: M_n/g.mol^{-1} = 16400, M_w/g.mol^{-1} = 294000,
1.0 mol% butyl acrylate, synthesized in the laboratory

Solvent (A): **ethene** C_2H_4 **74-85-1**

Solvent (C): **butyl acrylate** $C_7H_{12}O_2$ **141-32-2**

Type of data: cloud points

w_B 0.03 was kept constant

x_C	0.0044	0.0044	0.0044	0.0044	0.0044	0.0044	0.0108	0.0108	0.0108
T/K	383.15	388.15	393.15	403.15	413.15	423.15	383.15	388.15	393.15
P/bar	1838	1795	1735	1658	1600	1552	1783	1748	1699
x_C	0.0108	0.0108	0.0108	0.0154	0.0154	0.0154	0.0154	0.0154	0.0154
T/K	403.15	413.15	423.15	383.15	388.15	393.15	403.15	413.15	423.15
P/bar	1628	1570	1507	1770	1730	1684	1607	1545	1485

continued

continued

x_C	0.0268	0.0268	0.0268	0.0268	0.0268	0.0268	0.0396	0.0396	0.0396
T/K	383.15	388.15	393.15	403.15	413.15	423.15	383.15	388.15	393.15
P/bar	1690	1640	1610	1520	1470	1420	1609	1555	1515
x_C	0.0396	0.0396	0.0396						
T/K	403.15	413.15	423.15						
P/bar	1442	1390	1340						

Comments: The mole fraction x_C is given for the monomer mixture, i.e., without the polymer.

Polymer (B): **poly(ethylene-*co*-butyl acrylate)** **1999DIE**
Characterization: M_n/g.mol^{-1} = 39900, M_w/g.mol^{-1} = 155000,
8.7 mol% butyl acrylate, synthesized in the laboratory
Solvent (A): **ethene** C_2H_4 **74-85-1**
Solvent (C): **butyl acrylate** $C_7H_{12}O_2$ **141-32-2**

Type of data: cloud points

w_B 0.03 was kept constant

x_C	0.0075	0.0075	0.0075	0.0075	0.0075	0.0075	0.0075	0.0075	0.0075
T/K	343.15	353.15	363.15	373.15	383.15	393.15	403.15	413.15	423.15
P/bar	1510	1420	1370	1345	1315	1290	1272	1250	1215
x_C	0.0094	0.0094	0.0094	0.0094	0.0094	0.0094	0.0094	0.0210	0.0210
T/K	363.15	373.15	383.15	393.15	403.15	413.15	423.15	343.15	353.15
P/bar	1330	1305	1270	1240	1240	1200	1200	1360	1320
x_C	0.0210	0.0210	0.0210	0.0210	0.0210	0.0210	0.0210	0.0242	0.0242
T/K	363.15	373.15	383.15	393.15	403.15	413.15	423.15	343.15	353.15
P/bar	1290	1240	1210	1190	1165	1160	1140	1320	1270
x_C	0.0242	0.0242	0.0242	0.0242	0.0242	0.0242	0.0242	0.0315	0.0315
T/K	363.15	373.15	383.15	393.15	403.15	413.15	423.15	343.15	353.15
P/bar	1250	1215	1195	1165	1145	1130	1110	1245	1220
x_C	0.0315	0.0315	0.0315	0.0315	0.0315	0.0315	0.0315	0.0382	0.0382
T/K	363.15	373.15	383.15	393.15	403.15	413.15	423.15	343.15	353.15
P/bar	1180	1155	1130	1110	1100	1070	1050	1215	1165
x_C	0.0382	0.0382	0.0382	0.0382	0.0382	0.0382			
T/K	363.15	373.15	383.15	403.15	413.15	423.15			
P/bar	1135	1120	1095	1060	1035	1010			

Comments: The mole fraction x_C is given for the monomer mixture, i.e., without the polymer.

Polymer (B): **poly(ethylene-*co*-butyl acrylate)** **1999DIE**
Characterization: 22.2 mol% butyl acrylate, synthesized in the laboratory
Solvent (A): **ethene** C_2H_4 **74-85-1**
Solvent (C): **butyl acrylate** $C_7H_{12}O_2$ **141-32-2**

Type of data: cloud points

continued

continued

w_B 0.03 was kept constant

x_C	0.0030	0.0030	0.0030	0.0030	0.0030	0.0030	0.0030	0.0030	0.0030
T/K	323.15	333.15	343.15	353.15	363.15	373.15	383.15	393.15	403.15
P/bar	1285	1260	1230	1195	1175	1155	1142	1130	1120
x_C	0.0030	0.0030	0.0130	0.0130	0.0130	0.0130	0.0130	0.0130	0.0130
T/K	413.15	423.15	323.15	333.15	343.15	353.15	363.15	373.15	383.15
P/bar	1115	1095	1200	1160	1145	1120	1110	1090	1080
x_C	0.0130	0.0130	0.0130	0.0130	0.0220	0.0220	0.0220	0.0220	0.0220
T/K	393.15	403.15	413.15	423.15	323.15	333.15	343.15	353.15	363.15
P/bar	1065	1050	1040	1035	1110	1090	1070	1060	1045
x_C	0.0220	0.0220	0.0220	0.0220	0.0220	0.0360	0.0360	0.0360	0.0360
T/K	383.15	393.15	403.15	413.15	423.15	323.15	333.15	343.15	353.15
P/bar	1015	1010	1005	997	990	975	970	955	950
x_C	0.0360	0.0360	0.0360	0.0360	0.0360	0.0360	0.0360		
T/K	363.15	373.15	383.15	393.15	403.15	413.15	423.15		
P/bar	940	930	925	920	915	910	890		

Comments: The mole fraction x_C is given for the monomer mixture, i.e., without the polymer.

Polymer (B): **poly(ethylene-*co*-butyl acrylate)** **1999DIE**
Characterization: M_n/g.mol^{-1} = 39900, M_w/g.mol^{-1} = 155000,
8.7 mol% butyl acrylate, synthesized in the laboratory
Solvent (A): **ethene** C_2H_4 **74-85-1**
Solvent (C): **methyl acrylate** $C_4H_6O_2$ **96-33-3**

Type of data: cloud points

w_B 0.03 was kept constant

x_C	0.0213	0.0213	0.0213	0.0213	0.0213	0.0213	0.0213	0.0431	0.0431
T/K	353.15	363.15	373.15	383.15	393.15	403.15	423.15	353.15	363.15
P/bar	1390	1350	1310	1290	1260	1230	1200	1300	1280
x_C	0.0431	0.0431	0.0431	0.0431	0.0431	0.0431	0.0675	0.0675	0.0675
T/K	373.15	383.15	393.15	403.15	413.15	423.15	353.15	363.15	373.15
P/bar	1250	1220	1200	1170	1160	1130	1260	1200	1160
x_C	0.0675	0.0675	0.0675	0.0675	0.0675				
T/K	383.15	393.15	403.15	413.15	423.15				
P/bar	1150	1125	1110	1095	1065				

Comments: The mole fraction x_C is given for the monomer mixture, i.e., without the polymer.

Polymer (B): **poly(ethylene-*co*-2-ethylhexyl acrylate)** **1999DIE**
Characterization: M_n/g.mol^{-1} = 34900, M_w/g.mol^{-1} = 107700,
9.3 mol% 2-ethylhexyl acrylate, synthesized in the laboratory
Solvent (A): **ethene** C_2H_4 **74-85-1**
Solvent (C): **2-ethylhexyl acrylate** $C_{11}H_{20}O_2$ **103-11-7**

Type of data: cloud points

w_B 0.03 was kept constant

x_C	0.0000	0.0000	0.0000	0.0324	0.0324	0.0324	0.0368	0.0368	0.0368
T/K	403.15	423.15	443.15	403.15	423.15	443.15	403.15	423.15	443.15
P/bar	1440	1390	1335	1240	1200	1100	1170	1140	1080

Comments: The mole fraction x_C is given for the monomer mixture, i.e., without the polymer.

Polymer (B): **poly(ethylene-*co*-ethyl methacrylate)** **2004LAT**
Characterization: M_n/g.mol^{-1} = 25500, M_w/g.mol^{-1} = 47300,
11.4 mol% ethyl methacrylate), synthesized in the laboratory
Solvent (A): **ethene** C_2H_4 **74-85-1**
Solvent (C): **ethyl methacrylate** $C_6H_{10}O_2$ **97-63-2**

Type of data: cloud points

w_A	0.9152	0.8952	0.8760	0.8575	0.840
w_B	0.050	0.050	0.050	0.050	0.050
w_C	0.0348	0.0548	0.0740	0.0925	0.110
T/K	473.15	473.15	473.15	473.15	473.15
P/bar	1034	1004	985	973	955

Polymer (B): **poly(ethylene-*co*-ethyl methacrylate)** **2004LAT**
Characterization: M_n/g.mol^{-1} = 24800, M_w/g.mol^{-1} = 45400,
19.2 mol% ethyl methacrylate), synthesized in the laboratory
Solvent (A): **ethene** C_2H_4 **74-85-1**
Solvent (C): **ethyl methacrylate** $C_6H_{10}O_2$ **97-63-2**

Type of data: cloud points

w_A	0.8918	0.8728	0.8545	0.837	0.820
w_B	0.050	0.050	0.050	0.050	0.050
w_C	0.0582	0.0772	0.0955	0.113	0.130
T/K	473.15	473.15	473.15	473.15	473.15
P/bar	978	960	935	920	890

Polymer (B): **poly(ethylene-*co*-ethyl methacrylate)** **2004LAT**
Characterization: M_n/g.mol^{-1} = 22500, M_w/g.mol^{-1} = 42000,
26.1 mol% ethyl methacrylate), synthesized in the laboratory
Solvent (A): **ethene** C_2H_4 **74-85-1**
Solvent (C): **ethyl methacrylate** $C_6H_{10}O_2$ **97-63-2**

continued

continued

Type of data: cloud points

w_A	0.8706	0.8524	0.835	0.818	0.802
w_B	0.050	0.050	0.050	0.050	0.050
w_C	0.0794	0.0976	0.115	0.132	0.148
T/K	473.15	473.15	473.15	473.15	473.15
P/bar	940	925	906	886	865

Polymer (B): **poly(ethylene-*co*-ethyl methacrylate)** **2004LAT**
Characterization: M_n/g.mol^{-1} = 25100, M_w/g.mol^{-1} = 48700, 29.0 mol% ethyl methacrylate), synthesized in the laboratory
Solvent (A): **ethene** C_2H_4 **74-85-1**
Solvent (C): **ethyl methacrylate** $C_6H_{10}O_2$ **97-63-2**

Type of data: cloud points

w_A	0.837	0.820	0.804	0.789	0.774
w_B	0.050	0.050	0.050	0.050	0.050
w_C	0.113	0.130	0.146	0.161	0.176
T/K	473.15	473.15	473.15	473.15	473.15
P/bar	915	896	879	853	833

Polymer (B): **poly(ethylene-*co*-ethyl methacrylate)** **2004LAT**
Characterization: M_n/g.mol^{-1} = 24000, M_w/g.mol^{-1} = 44300, 38.0 mol% ethyl methacrylate), synthesized in the laboratory
Solvent (A): **ethene** C_2H_4 **74-85-1**
Solvent (C): **ethyl methacrylate** $C_6H_{10}O_2$ **97-63-2**

Type of data: cloud points

w_A	0.808	0.792	0.777	0.763	0.749
w_B	0.050	0.050	0.050	0.050	0.050
w_C	0.142	0.158	0.173	0.187	0.201
T/K	473.15	473.15	473.15	473.15	473.15
P/bar	885	871	853	823	803

Polymer (B): **poly(ethylene-*co*-1-hexene)** **2001DO1**
Characterization: M_n/g.mol^{-1} = 60000, M_w/g.mol^{-1} = 129000, 16.1 wt% 1-hexene
Solvent (A): **ethene** C_2H_4 **74-85-1**
Solvent (C): **n-butane** C_4H_{10} **106-97-8**

Type of data: cloud points

w_A	0.800	0.800	0.800	0.800	0.800	0.800	0.750	0.750	0.750
w_B	0.150	0.150	0.150	0.150	0.150	0.150	0.150	0.150	0.150
w_C	0.050	0.050	0.050	0.050	0.050	0.050	0.100	0.100	0.100
T/K	393.15	413.15	433.15	453.15	473.15	493.15	393.15	413.15	433.15
P/MPa	145.0	135.3	127.3	121.2	116.6	113.0	134.0	125.9	119.1

continued

continued

w_A	0.750	0.750	0.750	0.700	0.700	0.700	0.700	0.700	0.700
w_B	0.150	0.150	0.150	0.150	0.150	0.150	0.150	0.150	0.150
w_C	0.100	0.100	0.100	0.150	0.150	0.150	0.150	0.150	0.150
T/K	453.15	473.15	493.15	393.15	413.15	433.15	453.15	473.15	493.15
P/MPa	113.6	109.6	106.2	125.4	117.8	111.7	107.1	103.7	101.0

Polymer (B): **poly(ethylene-*co*-1-hexene)** **2001DO2**
Characterization: M_n/g.mol^{-1} = 60000, M_w/g.mol^{-1} = 129000, 16.1 wt% 1-hexene
Solvent (A): **ethene** C_2H_4 **74-85-1**
Solvent (C): **n-butane** C_4H_{10} **106-97-8**

Type of data: coexistence data

T/K = 433.15

w_A(total)	0.750	was kept constant
w_B(total)	0.150	was kept constant
w_C(total)	0.100	was kept constant

P/bar	707	875	1017	1172	1191
w_A(gel phase)	0.339	0.431	0.529	0.677	0.750
w_B(gel phase)	0.599	0.496	0.392	0.234	0.150
w_C(gel phase)	0.062	0.073	0.079	0.089	0.100
w_A(sol phase)	0.883	0.870	–	0.832	–
w_B(sol phase)	0.014	0.026	–	0.065	–
w_C(sol phase)	0.103	0.104	–	0.102	–

Polymer (B): **poly(ethylene-*co*-1-hexene)** **2001DO1**
Characterization: M_n/g.mol^{-1} = 60000, M_w/g.mol^{-1} = 129000, 16.1 wt% 1-hexene
Solvent (A): **ethene** C_2H_4 **74-85-1**
Solvent (C): **carbon dioxide** CO_2 **124-38-9**

Type of data: cloud points

w_A	0.800	0.800	0.800	0.800	0.800	0.800	0.750	0.750	0.750
w_B	0.150	0.150	0.150	0.150	0.150	0.150	0.150	0.150	0.150
w_C	0.050	0.050	0.050	0.050	0.050	0.050	0.100	0.100	0.100
T/K	393.15	413.15	433.15	453.15	473.15	493.15	393.15	413.15	433.15
P/MPa	160.0	148.2	138.3	130.8	125.7	122.0	168.3	153.7	142.3

w_A	0.750	0.750	0.750
w_B	0.150	0.150	0.150
w_C	0.100	0.100	0.100
T/K	393.15	413.15	433.15
P/MPa	134.0	128.2	124.0

Polymer (B): **poly(ethylene-*co*-1-hexene)** **2001DO2**
Characterization: M_n/g.mol^{-1} = 60000, M_w/g.mol^{-1} = 129000, 16.1 wt% 1-hexene
Solvent (A): **ethene** C_2H_4 **74-85-1**
Solvent (C): **carbon dioxide** CO_2 **124-38-9**

Type of data: coexistence data

T/K = 433.15

w_A(total) = 0.750 was kept constant w_B(total) = 0.150 was kept constant
w_C(total) = 0.100 was kept constant

P/bar	1151	1302	1403	1432
w_A(gel phase)	0.542	0.602	0.698	0.750
w_B(gel phase)	0.388	0.318	0.216	0.150
w_C(gel phase)	0.070	0.080	0.086	0.100
w_A(sol phase)	0.873	0.862	0.849	–
w_B(sol phase)	0.014	0.031	0.047	–
w_C(sol phase)	0.113	0.107	0.105	–

Polymer (B): **poly(ethylene-*co*-1-hexene)** **2001DO1**
Characterization: M_n/g.mol^{-1} = 60000, M_w/g.mol^{-1} = 129000, 16.1 wt% 1-hexene
Solvent (A): **ethene** C_2H_4 **74-85-1**
Solvent (C): **ethane** C_2H_6 **74-84-0**

Type of data: cloud points

w_A	0.800	0.800	0.800	0.800	0.800	0.800	0.750	0.750	0.750
w_B	0.150	0.150	0.150	0.150	0.150	0.150	0.150	0.150	0.150
w_C	0.050	0.050	0.050	0.050	0.050	0.050	0.100	0.100	0.100
T/K	393.15	413.15	433.15	453.15	473.15	493.15	393.15	413.15	433.15
P/MPa	151.7	140.7	133.4	126.3	120.8	116.5	147.1	136.8	129.1

w_A	0.750	0.750	0.750	0.700	0.700	0.700	0.700	0.700	0.700
w_B	0.150	0.150	0.150	0.150	0.150	0.150	0.150	0.150	0.150
w_C	0.100	0.100	0.100	0.150	0.150	0.150	0.150	0.150	0.150
T/K	453.15	473.15	493.15	393.15	413.15	433.15	453.15	473.15	493.15
P/MPa	123.0	118.1	114.3	143.2	133.5	125.5	120.0	116.0	112.0

Polymer (B): **poly(ethylene-*co*-1-hexene)** **2001DO1**
Characterization: M_n/g.mol^{-1} = 60000, M_w/g.mol^{-1} = 129000, 16.1 wt% 1-hexene
Solvent (A): **ethene** C_2H_4 **74-85-1**
Solvent (C): **helium** He **7440-59-7**

Type of data: cloud points

w_A	0.840	0.840	0.840	0.840	0.840	0.840	0.830	0.830	0.830
w_B	0.150	0.150	0.150	0.150	0.150	0.150	0.150	0.150	0.150
w_C	0.010	0.010	0.010	0.010	0.010	0.010	0.020	0.020	0.020
T/K	393.15	413.15	433.15	453.15	473.15	493.15	433.15	453.15	473.15
P/MPa	195.0	178.6	165.8	155.4	148.2	143.0	214.6	200.7	188.3

continued

continued

w_A	0.830
w_B	0.150
w_C	0.020
T/K	493.15
P/MPa	176.8

Polymer (B): **poly(ethylene-*co*-1-hexene)** **2001DO2**
Characterization: M_n/g.mol^{-1} = 60000, M_w/g.mol^{-1} = 129000, 16.1 wt% 1-hexene
Solvent (A): **ethene** C_2H_4 **74-85-1**
Solvent (C): **helium** **He** **7440-59-7**

Type of data: coexistence data

T/K = 433.15

w_A(total)	0.840	was kept constant
w_B(total)	0.150	was kept constant
w_C(total)	0.010	was kept constant

P/bar	1261	1428	1547	1658
w_A(gel phase)	0.526	0.590	0.655	0.840
w_B(gel phase)	0.470	0.402	0.338	0.150
w_C(gel phase)	0.003	0.005	0.006	0.010
w_A(sol phase)	–	0.978	0.964	–
w_B(sol phase)	–	0.006	0.024	–
w_C(sol phase)	–	0.014	0.012	–

Polymer (B): **poly(ethylene-*co*-1-hexene)** **2001DO1**
Characterization: M_n/g.mol^{-1} = 60000, M_w/g.mol^{-1} = 129000, 16.1 wt% 1-hexene
Solvent (A): **ethene** C_2H_4 **74-85-1**
Solvent (C): **1-hexene** C_6H_{12} **592-41-6**

Type of data: cloud points

w_A	0.6800	0.6800	0.6800	0.6800	0.6800	0.6800	0.4250	0.4250	0.4250
w_B	0.1500	0.1500	0.1500	0.1500	0.1500	0.1500	0.1500	0.1500	0.1500
w_C	0.1700	0.1700	0.1700	0.1700	0.1700	0.1700	0.4250	0.4250	0.4250
T/K	393.15	413.15	433.15	453.15	473.15	493.15	393.15	413.15	433.15
P/MPa	118.5	112.0	106.4	102.5	99.6	96.9	63.7	63.6	63.1

w_A	0.4250	0.4250	0.4250	0.1275	0.1275	0.1275	0.1275	0.1275	0.1725
w_B	0.1500	0.1500	0.1500	0.1500	0.1500	0.1500	0.1500	0.1500	0.1500
w_C	0.4250	0.4250	0.4250	0.7225	0.7225	0.7225	0.7225	0.7225	0.7225
T/K	453.15	473.15	493.15	393.15	413.15	433.15	453.15	473.15	493.15
P/MPa	62.6	62.3	62.1	12.7	15.4	17.8	19.9	21.9	23.5

Polymer (B):	**poly(ethylene-*co*-1-hexene)**		**2001DO2**
Characterization:	M_n/g.mol^{-1} = 60000, M_w/g.mol^{-1} = 129000, 16.1 wt% 1-hexene		
Solvent (A):	**ethene**	C_2H_4	**74-85-1**
Solvent (C):	**1-hexene**	C_6H_{12}	**592-41-6**

Type of data: coexistence data

T/K = 433.15

w_A(total)	0.425	was kept constant
w_B(total)	0.150	was kept constant
w_C(total)	0.425	was kept constant

P/bar	360	442	509	585	591
w_A(gel phase)	0.238	0.278	0.332	0.415	0.425
w_B(gel phase)	0.538	0.460	0.368	0.200	0.150
w_C(gel phase)	0.224	0.262	0.300	0.385	0.425
w_A(sol phase)	0.505	0.498	0.480	0.480	–
w_B(sol phase)	0.010	0.020	0.032	0.053	–
w_C(sol phase)	0.485	0.482	0.488	0.468	–

Polymer (B):	**poly(ethylene-*co*-1-hexene)**		**2001DO1**
Characterization:	M_n/g.mol^{-1} = 60000, M_w/g.mol^{-1} = 129000, 16.1 wt% 1-hexene		
Solvent (A):	**ethene**	C_2H_4	**74-85-1**
Solvent (C):	**1-hexene**	C_6H_{12}	**592-41-6**
Solvent (D):	**nitrogen**	N_2	**7727-37-9**

Type of data: cloud points

w_A	0.6400	0.6400	0.6400	0.6400	0.6400	0.6400	0.6000	0.6000	0.6000
w_B	0.1500	0.1500	0.1500	0.1500	0.1500	0.1500	0.1500	0.1500	0.1500
w_C	0.1600	0.1600	0.1600	0.1600	0.1600	0.1600	0.1500	0.1500	0.1500
w_D	0.0500	0.0500	0.0500	0.0500	0.0500	0.0500	0.1000	0.1000	0.1000
T/K	393.15	413.15	433.15	453.15	473.15	493.15	393.15	413.15	433.15
P/MPa	141.5	131.7	123.8	117.7	112.9	109.5	166.4	151.7	141.4

w_A	0.6000	0.6000	0.6000	0.4000	0.4000	0.4000	0.4000	0.4000	0.4000
w_B	0.1500	0.1500	0.1500	0.1500	0.1500	0.1500	0.1500	0.1500	0.1500
w_C	0.1500	0.1500	0.1500	0.4000	0.4000	0.4000	0.4000	0.4000	0.4000
w_D	0.1000	0.1000	0.1000	0.0500	0.0500	0.0500	0.0500	0.0500	0.0500
T/K	453.15	473.15	493.15	393.15	413.15	433.15	453.15	473.15	493.15
P/MPa	132.8	126.5	122.0	85.3	81.9	79.2	77.0	75.7	74.2

w_A	0.3750	0.3750	0.3750	0.3750	0.3750	0.3750	0.1200	0.1200	0.1200
w_B	0.1500	0.1500	0.1500	0.1500	0.1500	0.1500	0.1500	0.1500	0.1500
w_C	0.3750	0.3750	0.3750	0.3750	0.3750	0.3750	0.6800	0.6800	0.6800
w_D	0.1000	0.1000	0.1000	0.1000	0.1000	0.1000	0.0500	0.0500	0.0500
T/K	393.15	413.15	433.15	453.15	473.15	493.15	393.15	413.15	433.15
P/MPa	114.6	107.7	100.3	96.1	91.6	88.7	28.4	29.9	31.3

continued

continued

w_A	0.1200	0.1200	0.1200	0.1125	0.1125	0.1125	0.1125	0.1125	0.1125
w_B	0.1500	0.1500	0.1500	0.1500	0.1500	0.1500	0.1500	0.1500	0.1500
w_C	0.6800	0.6800	0.6800	0.6375	0.6375	0.6375	0.6375	0.6375	0.6375
w_D	0.0500	0.0500	0.0500	0.1000	0.1000	0.1000	0.1000	0.1000	0.1000
T/K	453.15	473.15	493.15	393.15	413.15	433.15	453.15	473.15	493.15
P/MPa	32.6	33.8	34.9	49.1	48.8	48.6	48.6	48.8	49.0

Polymer (B): **poly(ethylene-*co*-1-hexene)** **2001DO1**
Characterization: M_n/g.mol^{-1} = 60000, M_w/g.mol^{-1} = 129000, 16.1 wt% 1-hexene
Solvent (A): **ethene** C_2H_4 **74-85-1**
Solvent (C): **methane** CH_4 **74-82-8**

Type of data: cloud points

w_A	0.800	0.800	0.800	0.800	0.800	0.800	0.750	0.750	0.750
w_B	0.150	0.150	0.150	0.150	0.150	0.150	0.150	0.150	0.150
w_C	0.050	0.050	0.050	0.050	0.050	0.050	0.100	0.100	0.100
T/K	393.15	413.15	433.15	453.15	473.15	493.15	393.15	413.15	433.15
P/MPa	169.0	156.7	147.1	139.4	132.4	126.4	186.5	170.5	157.8
w_A	0.750	0.750	0.750	0.700	0.700	0.700	0.700	0.700	0.700
w_B	0.150	0.150	0.150	0.150	0.150	0.150	0.150	0.150	0.150
w_C	0.100	0.100	0.100	0.150	0.150	0.150	0.150	0.150	0.150
T/K	453.15	473.15	493.15	393.15	413.15	433.15	453.15	473.15	493.15
P/MPa	149.0	142.0	137.5	202.0	186.9	174.0	162.8	153.7	149.1

Polymer (B): **poly(ethylene-*co*-1-hexene)** **2001DO2**
Characterization: M_n/g.mol^{-1} = 60000, M_w/g.mol^{-1} = 129000, 16.1 wt% 1-hexene
Solvent (A): **ethene** C_2H_4 **74-85-1**
Solvent (C): **methane** CH_4 **74-82-8**

Type of data: coexistence data

T/K = 433.15

w_A(total)	0.750	was kept constant
w_B(total)	0.150	was kept constant
w_C(total)	0.100	was kept constant

P/bar	1069	1158	1313	1440	1556	1578
w_A(gel phase)	0.397	0.436	0.518	0.586	0.669	0.750
w_B(gel phase)	0.550	0.506	0.413	0.335	0.241	0.150
w_C(gel phase)	0.053	0.058	0.069	0.078	0.089	0.100
w_A(sol phase)	0.877	0.877	–	0.862	0.855	–
w_B(sol phase)	0.006	0.007	–	0.023	0.031	–
w_C(sol phase)	0.117	0.116	–	0.115	0.114	–

Polymer (B): **poly(ethylene-*co*-1-hexene)** **2001DO1**
Characterization: M_n/g.mol^{-1} = 60000, M_w/g.mol^{-1} = 129000, 16.1 wt% 1-hexene
Solvent (A): **ethene** C_2H_4 **74-85-1**
Solvent (C): **nitrogen** N_2 **7727-37-9**

Type of data: cloud points

w_A	0.800	0.800	0.800	0.800	0.800	0.800	0.750	0.750	0.750
w_B	0.150	0.150	0.150	0.150	0.150	0.150	0.150	0.150	0.150
w_C	0.050	0.050	0.050	0.050	0.050	0.050	0.100	0.100	0.100
T/K	393.15	413.15	433.15	453.15	473.15	493.15	393.15	413.15	433.15
P/MPa	179.5	166.8	153.3	144.8	138.0	134.4	206.5	188.6	173.2

w_A	0.750	0.750	0.750
w_B	0.150	0.150	0.150
w_C	0.100	0.100	0.100
T/K	453.15	473.15	493.15
P/MPa	161.8	152.8	145.6

Polymer (B): **poly(ethylene-*co*-1-hexene)** **2001DO2**
Characterization: M_n/g.mol^{-1} = 60000, M_w/g.mol^{-1} = 129000, 16.1 wt% 1-hexene
Solvent (A): **ethene** C_2H_4 **74-85-1**
Solvent (C): **nitrogen** N_2 **7727-37-9**

Type of data: coexistence data

T/K = 433.15

w_A(total)	0.750	was kept constant
w_B(total)	0.150	was kept constant
w_C(total)	0.100	was kept constant

P/bar	1409	1599	1697	1732
w_A(gel phase)	0.539	0.630	0.671	0.750
w_B(gel phase)	0.406	0.302	0.251	0.150
w_C(gel phase)	0.065	0.078	0.088	0.100
w_A(sol phase)	0.897	0.887	0.863	–
w_B(sol phase)	0.008	0.018	0.037	–
w_C(sol phase)	0.105	0.105	0.100	–

Polymer (B): **poly(ethylene-*co*-1-hexene)** **2001DO1**
Characterization: M_n/g.mol^{-1} = 60000, M_w/g.mol^{-1} = 129000, 16.1 wt% 1-hexene
Solvent (A): **ethene** C_2H_4 **74-85-1**
Solvent (C): **propane** C_3H_8 **74-98-6**

Type of data: cloud points

w_B	0.150	0.150	0.150	0.150	0.150	0.150	0.150	0.150	0.150
w_C	0.050	0.050	0.050	0.050	0.050	0.050	0.100	0.100	0.100
T/K	393.15	413.15	433.15	453.15	473.15	493.15	393.15	413.15	433.15
P/MPa	145.8	136.0	128.1	122.1	117.4	113.6	138.3	129.3	123.3

continued

continued

w_A	0.750	0.750	0.750	0.700	0.700	0.700	0.700	0.700	0.700
w_B	0.150	0.150	0.150	0.150	0.150	0.150	0.150	0.150	0.150
w_C	0.100	0.100	0.100	0.150	0.150	0.150	0.150	0.150	0.150
T/K	453.15	473.15	493.15	393.15	413.15	433.15	453.15	473.15	493.15
P/MPa	117.1	113.3	110.0	130.3	122.1	116.0	111.8	108.9	106.6

Polymer (B): **poly(ethylene-*co*-1-hexene)** **1999PAN**
Characterization: M_n/g.mol^{-1} = 40900, M_w/g.mol^{-1} = 90000, 13.6 wt% 1-hexene, MI = 1.2, ρ = 0.904 g/cm^3, T_m/K = 406.25
Solvent (A): **2-methylpropane** C_4H_{10} **75-28-5**
Solvent (C): **1-hexene** C_6H_{12} **592-41-6**

Type of data: cloud points

w_A	0.892	0.892	0.892	0.892	0.892	0.892	0.892	0.892	0.892
w_B	0.108	0.108	0.108	0.108	0.108	0.108	0.108	0.108	0.108
w_C	0.000	0.000	0.000	0.000	0.000	0.000	0.000	0.000	0.000
T/K	378.45	386.75	393.05	394.05	400.05	407.75	415.55	421.95	422.75
P/bar	333.5	334.5	336.4	336.8	338.1	340.8	342.8	343.4	344.1

w_A	0.84835	0.84835	0.84835	0.84835	0.84835	0.84835	0.84835	0.84835	0.84835
w_B	0.107	0.107	0.107	0.107	0.107	0.107	0.107	0.107	0.107
w_C	0.04465	0.04465	0.04465	0.04465	0.04465	0.04465	0.04465	0.04465	0.04465
T/K	368.35	373.55	378.85	385.75	393.45	400.05	408.45	415.45	423.75
P/bar	301.55	302.95	304.15	307.55	311.55	313.6	317.95	321.55	324.8

w_A	0.40275	0.40275	0.40275	0.40275	0.40275	0.40275
w_B	0.105	0.105	0.105	0.105	0.105	0.105
w_C	0.49225	0.49225	0.49225	0.49225	0.49225	0.49225
T/K	367.15	373.15	376.35	377.85	383.35	393.15
P/bar	32.5	41.5	46.2	48.9	56.4	71.1

Polymer (B): **poly(ethylene-*co*-methyl acrylate)** **1999DIE**
Characterization: M_n/g.mol^{-1} = 52500, M_w/g.mol^{-1} = 165000, 8.0 mol% methyl acrylate, synthesized in the laboratory
Solvent (A): **ethene** C_2H_4 **74-85-1**
Solvent (C): **butyl acrylate** $C_7H_{12}O_2$ **141-32-2**

Type of data: cloud points

w_B 0.03 was kept constant

x_C	0.0129	0.0129	0.0129	0.0129	0.0129	0.0129	0.0288	0.0288	0.0288
T/K	373.15	383.15	393.15	403.15	413.15	423.15	383.15	393.15	403.15
P/bar	1645	1570	1520	1485	1470	1440	1480	1435	1415

x_C	0.0288	0.0288	0.0305	0.0305	0.0305	0.0305	0.0305	0.0403	0.0403
T/K	413.15	423.15	383.15	393.15	403.15	413.15	423.15	373.15	383.15
P/bar	1390	1360	1450	1410	1390	1370	1370	1440	1420

continued

continued

x_C	0.0403	0.0403	0.0403	0.0403
T/K	393.15	403.15	413.15	423.15
P/bar	1375	1340	1340	1320

Comments: The mole fraction x_C is given for the solvent mixture, i.e., without the polymer.

Polymer (B): **poly(ethylene-*co*-methyl acrylate)** **1999DIE**
Characterization: M_n/g.mol^{-1} = 52500, M_w/g.mol^{-1} = 165000,
8.0 mol% methyl acrylate, synthesized in the laboratory
Solvent (A): **ethene** C_2H_4 **74-85-1**
Solvent (C): **2-ethylhexyl acrylate** $C_{11}H_{20}O_2$ **103-11-7**

Type of data: cloud points

w_B 0.03 was kept constant

x_C	0.0120	0.0120	0.0120	0.0185	0.0185	0.0185	0.0194	0.0194	0.0194
T/K	403.15	423.15	443.15	403.15	423.15	443.15	403.15	423.15	443.15
P/bar	1390	1340	1290	1370	1250	1210	1290	1275	1260

x_C	0.0324	0.0324	0.0324	0.0339	0.0339	0.0339
T/K	403.15	423.15	443.15	403.15	423.15	443.15
P/bar	1240	1200	1140	1250	1170	1190

Comments: The mole fraction x_C is given for the solvent mixture, i.e., without the polymer.

Polymer (B): **poly(ethylene-*co*-methyl acrylate)** **1999DIE**
Characterization: M_n/g.mol^{-1} = 52500, M_w/g.mol^{-1} = 165000,
8.0 mol% methyl acrylate, synthesized in the laboratory
Solvent (A): **ethene** C_2H_4 **74-85-1**
Solvent (C): **n-heptane** C_7H_{16} **142-82-5**

Type of data: cloud points

w_B 0.03 was kept constant

x_C	0.0136	0.0136	0.0136	0.0136	0.0136	0.0261	0.0261	0.0261	0.0261
T/K	383.15	393.15	403.15	413.15	423.15	383.15	393.15	403.15	413.15
P/bar	1500	1460	1405	1380	1360	1420	1400	1370	1340

x_C	0.0261	0.0414	0.0414	0.0414	0.0414	0.0414	0.0599	0.0599	0.0599
T/K	423.15	383.15	393.15	403.15	413.15	423.15	383.15	393.15	403.15
P/bar	1300	1350	1315	1290	1255	1250	1240	1225	1210

x_C	0.0599	0.0599	0.0977	0.0977	0.0977	0.0977
T/K	413.15	423.15	383.15	403.15	413.15	423.15
P/bar	1165	1150	1000	975	970	950

Comments: The mole fraction x_C is given for the solvent mixture, i.e., without the polymer.

Polymer (B):	**poly(ethylene-*co*-methyl acrylate)**		**1999DIE**
Characterization:	M_n/g.mol^{-1} = 52500, M_w/g.mol^{-1} = 165000, 8.0 mol% methyl acrylate, synthesized in the laboratory		
Solvent (A):	**ethene**	C_2H_4	**74-85-1**
Solvent (C):	**methyl acrylate**	$C_4H_6O_2$	**96-33-3**

Type of data: cloud points

w_B 0.03 was kept constant

x_C	0.0144	0.0144	0.0144	0.0144	0.0144	0.0590	0.0590	0.0590	0.0590
T/K	393.15	403.15	423.15	433.15	443.15	383.15	393.15	403.15	413.15
P/bar	1550	1470	1440	1440	1410	1400	1390	1380	1365
x_C	0.0590	0.0855	0.0855	0.0855	0.0855	0.0855	0.0855	0.0855	0.0855
T/K	423.15	373.15	383.15	393.15	403.15	413.15	423.15	433.15	443.15
P/bar	1350	1300	1290	1300	1340	1370	1350	1380	1360

Comments: The mole fraction x_C is given for the monomer mixture, i.e., without the polymer.

Polymer (B):	**poly(ethylene-*co*-methyl acrylate)**		**2007TUM**
Characterization:	M_w/g.mol^{-1} = 110000, 23 mol% methyl acrylate, synthesized in the laboratory		
Solvent (A):	**ethene**	C_2H_4	**74-85-1**
Solvent (C):	**vinyl acetate**	$C_4H_6O_2$	**108-05-4**

Type of data: cloud points

w_A 0.867 was kept constant
w_B 0.030 was kept constant
w_C 0.103 was kept constant

T/K	373	383	393	403	413	423	433	443	453
P/bar	1277	1248	1231	1211	1196	1176	1160	1149	1130
T/K	463	473	483	493					
P/bar	1119	1104	1093	1080					

Polymer (B):	**poly(ethylene-*co*-methyl acrylate-*co*-vinyl acetate)**		**2007TUM**
Characterization:	M_w/g.mol^{-1} = 110000, 35 mol% methyl acrylate, 3.5 mol% vinyl acetate, synthesized in the laboratory		
Solvent (A):	**ethene**	C_2H_4	**74-85-1**
Solvent (C):	**vinyl acetate**	$C_4H_6O_2$	**108-05-4**

Type of data: cloud points

w_A 0.888 was kept constant
w_B 0.030 was kept constant
w_C 0.082 was kept constant

continued

continued

T/K	373	383	393	403	413	423	433	443	453
P/bar	1780	1720	1660	1610	1560	1520	1480	1450	1420
T/K	463	473	483	493					
P/bar	1390	1360	1330	1300					

Polymer (B): **poly(ethylene-*co*-methyl acrylate-*co*-vinyl acetate)** **2007TUM**

Characterization: M_w/g.mol^{-1} = 110000, 40 mol% methyl acrylate, 3.5 mol% vinyl acetate, synthesized in the laboratory

Solvent (A): **ethene** C_2H_4 **74-85-1**

Solvent (C): **vinyl acetate** $C_4H_6O_2$ **108-05-4**

Type of data: cloud points

w_A 0.820 was kept constant
w_B 0.030 was kept constant
w_C 0.150 was kept constant

T/K	373	383	393	403	413	423	433	443	453
P/bar	1480	1440	1410	1370	1330	1310	1280	1250	1230
T/K	463	473	483	493	503				
P/bar	1200	1180	1160	1160	1150				

Polymer (B): **poly(ethylene-*co*-1-octene)** **2005LEE**

Characterization: M_n/g.mol^{-1} = 64810, M_w/g.mol^{-1} = 135500, 15.3 mol% 1-octene, T_m/K = 323.2, T_g/K = 214.2, ρ = 0.86 g/cm^3, DuPont Dow Elastomers Corporation

Solvent (A): **ethene** C_2H_4 **74-85-1**

Solvent (C): **1-octene** C_8H_{16} **111-66-0**

Type of data: cloud points

w_A	0.252	0.252	0.252	0.252	0.252	0.252	0.358	0.358	0.358
w_B	0.048	0.048	0.048	0.048	0.048	0.048	0.051	0.051	0.051
w_C	0.700	0.700	0.700	0.700	0.700	0.700	0.591	0.591	0.591
T/K	326.75	348.55	370.05	401.55	416.05	426.35	323.65	347.85	373.15
P/bar	62.75	98.26	131.01	177.89	195.48	208.58	276.8	291.7	314.8
w_A	0.358	0.358	0.503	0.503	0.503	0.503	0.503	0.503	0.643
w_B	0.051	0.051	0.059	0.059	0.059	0.059	0.059	0.059	0.051
w_C	0.591	0.591	0.438	0.438	0.438	0.438	0.438	0.438	0.306
T/K	396.85	422.85	323.55	327.75	348.85	373.35	397.95	423.25	323.75
P/bar	336.1	357.9	644.7	634.6	601.9	585.1	579.5	579.2	1010.1
w_A	0.643	0.643	0.643	0.643	0.643	0.643			
w_B	0.051	0.051	0.051	0.051	0.051	0.051			
w_C	0.306	0.306	0.306	0.306	0.306	0.306			
T/K	328.15	330.65	348.35	374.35	398.15	428.35			
P/bar	982.2	956.4	889.5	828.8	795.4	766.7			

Polymer (B): **poly(ethylene-*co*-propylene)** **2000VRI**
Characterization: M_n/g.mol^{-1} = 51000, M_w/g.mol^{-1} = 120000, M_z/g.mol^{-1} = 210000, 42 mol% propene, 21 methyl groups/100 C-atoms
Solvent (A): **ethene** C_2H_4 **74-85-1**
Solvent (C): **carbon dioxide** CO_2 **124-38-9**

Type of data: cloud points

w_A	0.9025	0.9025	0.9025	0.9025	0.9025	0.9025	0.9025	0.8550	0.8550
w_B	0.0500	0.0500	0.0500	0.0500	0.0500	0.0500	0.0500	0.0500	0.0500
w_C	0.0475	0.0475	0.0475	0.0475	0.0475	0.0475	0.0475	0.0950	0.0950
T/K	312.59	317.69	322.45	327.31	332.20	336.74	342.05	312.44	317.45
P/bar	2342	2204	2094	1996	1910	1844	1764	2585	2413
w_A	0.8550	0.8550	0.8550	0.8550	0.8550	0.8075	0.8075	0.8075	0.8075
w_B	0.0500	0.0500	0.0500	0.0500	0.0500	0.0500	0.0500	0.0500	0.0500
w_C	0.0950	0.0950	0.0950	0.0950	0.0950	0.1425	0.1425	0.1425	0.1425
T/K	322.59	327.36	332.22	337.07	342.01	312.67	317.61	322.38	327.24
P/bar	2263	2144	2041	1949	1871	2959	2712	2519	2355
w_A	0.8075	0.8075	0.8075	0.7600	0.7600	0.7600	0.7600	0.7600	0.7600
w_B	0.0500	0.0500	0.0500	0.0500	0.0500	0.0500	0.0500	0.0500	0.0500
w_C	0.1425	0.1425	0.1425	0.1900	0.1900	0.1900	0.1900	0.1900	0.1900
T/K	332.15	337.04	341.86	312.43	317.52	322.42	327.33	332.19	337.15
P/bar	2218	2101	2001	3705	3240	2913	2666	2470	2311
w_A	0.7600	0.8550	0.8550	0.8550	0.8550	0.8550	0.8550	0.8550	0.8100
w_B	0.0500	0.1000	0.1000	0.1000	0.1000	0.1000	0.1000	0.1000	0.1000
w_C	0.1900	0.0450	0.0450	0.0450	0.0450	0.0450	0.0450	0.0450	0.0900
T/K	342.00	312.57	317.48	322.32	327.21	332.18	336.83	342.08	312.68
P/bar	2181	2265	2140	2032	1938	1858	1785	1718	2496
w_A	0.8100	0.8100	0.8100	0.8100	0.8100	0.8100	0.7650	0.7650	0.7650
w_B	0.1000	0.1000	0.1000	0.1000	0.1000	0.1000	0.1000	0.1000	0.1000
w_C	0.0900	0.0900	0.0900	0.0900	0.0900	0.0900	0.1350	0.1350	0.1350
T/K	2334	2193	2081	1981	1894	1818	2900	2654	2457
w_A	0.7650	0.7650	0.7650	0.7650	0.7200	0.7200	0.7200	0.7200	0.7200
w_B	0.1000	0.1000	0.1000	0.1000	0.1000	0.1000	0.1000	0.1000	0.1000
w_C	0.1350	0.1350	0.1350	0.1350	0.1800	0.1800	0.1800	0.1800	0.1800
T/K	327.41	332.17	336.99	342.23	312.50	317.48	322.35	327.28	332.16
P/bar	2290	2166	2055	1950	3665	3191	2867	2618	2419
w_A	0.7200	0.7200							
w_B	0.1000	0.1000							
w_C	0.1800	0.1800							
T/K	337.30	342.12							
P/bar	2258	2132							

Polymer (B): **poly(ethylene-*co*-propylene)** **1997HAN**
Characterization: M_n/g.mol^{-1} = 2170, M_w/g.mol^{-1} = 2600, 50 mol% propene, T_g/K = 209.2, alternating ethylene/propylene units from complete hydrogenation (98%) of polyisoprene
Solvent (A): **propene** **C_3H_6** **115-07-1**
Polymer (C): **poly(ethylene-*co*-propylene)**
Characterization: M_n/g.mol^{-1} = 177000, M_w/g.mol^{-1} = 195000, 50 mol% propene, T_g/K = 225.2, alternating ethylene/propylene units from complete hydrogenation (94%) of polyisoprene

Type of data: cloud points

Comments: The weight ratio of w_B/w_C was kept constant at 75/25 = 3/1 and the resulting molar mass averages are M_n/g.mol^{-1} = 3470 and M_w/g.mol^{-1} = 51000.

w_B+w_C	0.012	0.012	0.012	0.050	0.050	0.050	0.100	0.100	0.100
T/K	322.95	372.85	423.15	323.45	373.05	423.05	323.15	372.95	423.35
P/bar	324.5	375.1	424.6	310.3	364.5	417.1	282.0	340.8	396.8

Type of data: coexistence data

T/K = 423.15

P/ bar	w_B feed phase	gel phase	sol phase
356	0.050	unknown	0.041

Polymer (B): **poly(ethylene-*co*-propylene)** **1997HAN**
Characterization: M_n/g.mol^{-1} = 2170, M_w/g.mol^{-1} = 2600, 50 mol% propene, T_g/K = 209.2, alternating ethylene/propylene units from complete hydrogenation (98%) of polyisoprene
Solvent (A): **propene** **C_3H_6** **115-07-1**
Polymer (C): **poly(ethylene-*co*-propylene)**
Characterization: M_n/g.mol^{-1} = 177000, M_w/g.mol^{-1} = 195000, 50 mol% propene, T_g/K = 225.2, alternating ethylene/propylene units from complete hydrogenation (94%) of polyisoprene
Solvent (D): **n-dodecane** **$C_{12}H_{26}$** **112-40-3**

Type of data: cloud points

Comments: The weight ratio of $w_B/w_C/w_D$ was kept constant at 3/1/1. The resulting molar mass averages are M_n/g.mol^{-1} = 3470 and M_w/g.mol^{-1} = 51000.

w_A	0.05	0.05	0.05
T/K	323.15	372.95	423.05
P/bar	282.0	340.8	396.8

Polymer (B):	**poly(ethylene-*co*-vinyl acetate)**		**2001DO2**
Characterization:	M_n/g.mol^{-1} = 61900, M_w/g.mol^{-1} = 167000, 27.5 wt% vinyl acetate		
Solvent (A):	**ethene**	**C_2H_4**	**74-85-1**
Solvent (C):	**n-butane**	**C_4H_{10}**	**106-97-8**

Type of data: coexistence data

T/K = 433.15

w_A(total)	0.750	was kept constant
w_B(total)	0.150	was kept constant
w_C(total)	0.100	was kept constant

P/bar	713	835	946	1032	1060
w_A(gel phase)	0.407	0.506	0.602	0.674	0.750
w_B(gel phase)	0.553	0.449	0.338	0.260	0.150
w_C(gel phase)	0.040	0.045	0.060	0.067	0.100
w_A(sol phase)	0.862	0.846	0.833	0.818	–
w_B(sol phase)	0.019	0.035	0.053	0.072	–
w_C(sol phase)	0.119	0.119	0.115	0.110	–

Polymer (B):	**poly(ethylene-*co*-vinyl acetate)**		**2001DO2**
Characterization:	M_n/g.mol^{-1} = 61900, M_w/g.mol^{-1} = 167000, 27.5 wt% vinyl acetate		
Solvent (A):	**ethene**	**C_2H_4**	**74-85-1**
Solvent (C):	**carbon dioxide**	**CO_2**	**124-38-9**

Type of data: coexistence data

T/K = 433.15

w_A(total)	0.750	was kept constant
w_B(total)	0.150	was kept constant
w_C(total)	0.100	was kept constant

P/bar	742	908	1036	1160	1201
w_A(gel phase)	0.357	0.441	0.530	0.640	0.750
w_B(gel phase)	0.597	0.501	0.400	0.276	0.150
w_C(gel phase)	0.046	0.058	0.070	0.085	0.100
w_A(sol phase)	0.879	0.875	0.865	0.843	–
w_B(sol phase)	0.0047	0.0081	0.020	0.0452	–
w_C(sol phase)	0.117	0.117	0.115	0.113	–

Polymer (B):	**poly(ethylene-*co*-vinyl acetate)**		**2001DO2**
Characterization:	M_n/g.mol^{-1} = 61900, M_w/g.mol^{-1} = 167000, 27.5 wt% vinyl acetate		
Solvent (A):	**ethene**	**C_2H_4**	**74-85-1**
Solvent (C):	**helium**	**He**	**7440-59-7**

Type of data: coexistence data

continued

continued

T/K = 433.15

w_A(total)	0.840	was kept constant
w_B(total)	0.150	was kept constant
w_C(total)	0.010	was kept constant

P/bar	1164	1261	1406	1475	1495
w_A(gel phase)	0.5093	0.565	0.6821	0.7496	0.840
w_B(gel phase)	0.486	0.430	0.313	0.244	0.150
w_C(gel phase)	0.0047	0.005	0.0049	0.0064	0.010
w_A(sol phase)	0.9763	0.9778	0.9589	0.9496	–
w_B(sol phase)	0.0122	0.0112	0.0301	0.040	–
w_C(sol phase)	0.0115	0.0110	0.0110	0.0104	–

Polymer (B): **poly(ethylene-*co*-vinyl acetate)** **2001DO2**
Characterization: M_n/g.mol^{-1} = 61900, M_w/g.mol^{-1} = 167000, 27.5 wt% vinyl acetate
Solvent (A): **ethene** **C_2H_4** **74-85-1**
Solvent (C): **methane** **CH_4** **74-82-8**

Type of data: coexistence data

T/K = 433.15

w_A(total)	0.840	was kept constant
w_B(total)	0.060	was kept constant
w_C(total)	0.100	was kept constant

P/bar	1095	1263	1407	1451
w_A(gel phase)	0.515	0.602	0.705	0.840
w_B(gel phase)	0.425	0.326	0.211	0.060
w_C(gel phase)	0.060	0.072	0.084	0.100
w_A(sol phase)	0.890	0.882	0.875	–
w_B(sol phase)	0.004	0.014	0.021	–
w_C(sol phase)	0.106	0.105	0.104	–

T/K = 433.15

w_A(total)	0.700	was kept constant
w_B(total)	0.150	was kept constant
w_C(total)	0.150	was kept constant

P/bar	1204	1400	1514	1556
w_A(gel phase)	0.422	0.532	0.626	0.700
w_B(gel phase)	0.488	0.354	0.240	0.150
w_C(gel phase)	0.090	0.114	0.134	0.150
w_A(sol phase)	0.812	0.803	0.794	–
w_B(sol phase)	0.014	0.025	0.036	–
w_C(sol phase)	0.174	0.172	0.170	–

continued

continued

T/K = 433.15

w_A(total)	0.750	was kept constant
w_B(total)	0.150	was kept constant
w_C(total)	0.100	was kept constant

P/bar	963	1090	1305	1405	1430
w_A(gel phase)	0.416	0.484	0.626	0.670	0.750
w_B(gel phase)	0.529	0.451	0.291	0.241	0.150
w_C(gel phase)	0.055	0.065	0.083	0.089	0.100
w_A(sol phase)	0.862	0.858	0.837	0.823	–
w_B(sol phase)	0.0227	0.0276	0.0511	0.0676	–
w_C(sol phase)	0.115	0.115	0.112	0.109	–

T/K = 433.15

w_A(total)	0.800	was kept constant
w_B(total)	0.150	was kept constant
w_C(total)	0.050	was kept constant

P/bar	918	1030	1129	1231	1283
w_A(gel phase)	0.400	0.475	0.5545	0.676	0.800
w_B(gel phase)	0.570	0.470	0.390	0.270	0.150
w_C(gel phase)	0.030	0.055	0.0555	0.054	0.050
w_A(sol phase)	–	0.948	0.935	–	–
w_B(sol phase)	–	0.020	0.025	–	–
w_C(sol phase)	–	0.032	0.040	–	–

Polymer (B): **poly(ethylene-*co*-vinyl acetate)** **2001DO2**
Characterization: M_n/g.mol^{-1} = 61900, M_w/g.mol^{-1} = 167000, 27.5 wt% vinyl acetate

Solvent (A): **ethene** C_2H_4 **74-85-1**
Solvent (C): **nitrogen** N_2 **7727-37-9**

Type of data: coexistence data

T/K = 433.15

w_A(total)	0.750	was kept constant
w_B(total)	0.150	was kept constant
w_C(total)	0.100	was kept constant

P/bar	801	1208	1357	1465	1533	1556
w_A(gel phase)	0.378	0.476	0.549	0.620	0.690	0.750
w_B(gel phase)	0.572	0.461	0.377	0.297	0.218	0.150
w_C(gel phase)	0.050	0.063	0.074	0.083	0.092	0.100
w_A(sol phase)	0.881	0.874	0.869	0.852	0.845	–
w_B(sol phase)	0.002	0.009	0.015	0.034	0.042	–
w_C(sol phase)	0.117	0.116	0.116	0.113	0.112	–

Polymer (B):	**poly(ethylene-*co*-vinyl acetate)**		**2001DO2**
Characterization:	M_n/g.mol^{-1} = 61900, M_w/g.mol^{-1} = 167000, 27.5 wt% vinyl acetate		
Solvent (A):	**ethene**	C_2H_4	**74-85-1**
Solvent (C):	**vinyl acetate**	$C_4H_6O_2$	**108-05-4**

Type of data: coexistence data

T/K = 433.15

w_A(total)	0.425	was kept constant
w_B(total)	0.150	was kept constant
w_C(total)	0.425	was kept constant

P/bar	401	451	492	529	570	616
w_A(gel phase)	0.27	0.30	0.33	0.35	0.38	0.425
w_B(gel phase)	0.49	0.44	0.40	0.34	0.27	0.15
w_C(gel phase)	0.24	0.27	0.29	0.31	0.35	0.425
w_A(sol phase)	0.50	0.49	0.49	0.49	0.48	–
w_B(sol phase)	0.02	0.03	0.04	0.04	0.06	–
w_C(sol phase)	0.48	0.48	0.47	0.47	0.46	–

Polymer (B):	**poly(ethylene-*co*-vinyl acetate)**		**2001DO2**
Characterization:	M_n/g.mol^{-1} = 61900, M_w/g.mol^{-1} = 167000, 27.5 wt% vinyl acetate		
Solvent (A):	**ethene**	C_2H_4	**74-85-1**
Solvent (C):	**vinyl acetate**	$C_4H_6O_2$	**108-05-4**
Solvent (D):	**n-butane**	C_4H_{10}	**106-97-8**

Type of data: coexistence data

T/K = 433.15

w_A(total)	0.375	was kept constant
w_B(total)	0.150	was kept constant
w_C(total)	0.375	was kept constant
w_D(total)	0.100	was kept constant

P/bar	323	397	472	552	577
w_A(gel phase)	0.212	0.250	0.265	0.336	0.375
w_B(gel phase)	0.523	0.449	0.380	0.253	0.150
w_C(gel phase)	0.219	0.241	0.290	0.331	0.375
w_D(gel phase)	0.046	0.060	0.065	0.080	0.100
w_A(sol phase)	0.435	0.430	0.423	0.404	–
w_B(sol phase)	0.022	0.027	0.050	0.088	–
w_C(sol phase)	0.432	0.437	0.418	0.402	–
w_D(sol phase)	0.111	0.106	0.109	0.106	–

Polymer (B):	**poly(ethylene-*co*-vinyl acetate)**		**2001DO2**
Characterization:	M_n/g.mol^{-1} = 61900, M_w/g.mol^{-1} = 167000, 27.5 wt% vinyl acetate		
Solvent (A):	**ethene**	C_2H_4	**74-85-1**
Solvent (C):	**vinyl acetate**	$C_4H_6O_2$	**108-05-4**
Solvent (D):	**carbon dioxide**	CO_2	**124-38-9**

Type of data: coexistence data

T/K = 433.15

w_A(total)	0.375	was kept constant
w_B(total)	0.150	was kept constant
w_C(total)	0.375	was kept constant
w_D(total)	0.100	was kept constant

P/bar	366	491	576	674	695
w_A(gel phase)	0.181	0.247	0.300	–	0.375
w_B(gel phase)	0.589	0.455	0.353	–	0.150
w_C(gel phase)	0.180	0.230	0.270	–	0.375
w_D(gel phase)	0.050	0.068	0.083	–	0.100
w_A(sol phase)	0.440	0.430	0.409	0.396	–
w_B(sol phase)	0.000	0.009	0.034	0.071	–
w_C(sol phase)	0.448	0.446	0.451	0.428	–
w_D(sol phase)	0.112	0.115	0.106	0.105	–

Polymer (B):	**poly(ethylene-*co*-vinyl acetate)**		**2001DO2**
Characterization:	M_n/g.mol^{-1} = 61900, M_w/g.mol^{-1} = 167000, 27.5 wt% vinyl acetate		
Solvent (A):	**ethene**	C_2H_4	**74-85-1**
Solvent (C):	**vinyl acetate**	$C_4H_6O_2$	**108-05-4**
Solvent (D):	**helium**	He	**7440-59-7**

Type of data: coexistence data

T/K = 433.15

w_A(total)	0.420	was kept constant
w_B(total)	0.150	was kept constant
w_C(total)	0.420	was kept constant
w_D(total)	0.010	was kept constant

P/bar	711	830	922	950
w_A(gel phase)	0.274	0.330	0.375	0.420
w_B(gel phase)	0.403	0.318	0.230	0.150
w_C(gel phase)	0.320	0.347	0.387	0.420
w_D(gel phase)	0.003	0.005	0.008	0.010
w_A(sol phase)	0.507	0.502	0.478	–
w_B(sol phase)	0.001	0.0135	0.051	–
w_C(sol phase)	0.478	0.471	0.458	–
w_D(sol phase)	0.014	0.0135	0.013	–

Polymer (B):	**poly(ethylene-co-vinyl acetate)**		**2001DO2**
Characterization:	M_n/g.mol^{-1} = 61900, M_w/g.mol^{-1} = 167000, 27.5 wt% vinyl acetate		
Solvent (A):	**ethene**	C_2H_4	**74-85-1**
Solvent (C):	**vinyl acetate**	$C_4H_6O_2$	**108-05-4**
Solvent (D):	**methane**	CH_4	**74-82-8**

Type of data: coexistence data

T/K = 433.15

w_A(total)	0.375	was kept constant
w_B(total)	0.150	was kept constant
w_C(total)	0.375	was kept constant
w_D(total)	0.100	was kept constant

P/bar	477	647	765	850	872
w_A(gel phase)	0.194	0.252	0.291	0.335	0.375
w_B(gel phase)	0.540	0.450	0.360	0.250	0.150
w_C(gel phase)	0.220	0.242	0.280	0.335	0.375
w_D(gel phase)	0.045	0.057	0.069	0.080	0.100
w_A(sol phase)	0.468	0.465	0.451	0.426	–
w_B(sol phase)	0.014	0.027	0.055	0.085	–
w_C(sol phase)	0.392	0.386	0.374	0.378	–
w_D(sol phase)	0.126	0.122	0.120	0.111	–

T/K = 433.15

w_A(total)	0.5625	was kept constant
w_B(total)	0.150	was kept constant
w_C(total)	0.1875	was kept constant
w_D(total)	0.100	was kept constant

P/bar	760	857	974	1106	1139
w_A(gel phase)	0.274	0.297	0.369	0.450	0.5625
w_B(gel phase)	0.557	0.515	0.420	0.308	0.150
w_C(gel phase)	0.120	0.132	0.145	0.162	0.1875
w_D(gel phase)	0.049	0.056	0.066	0.080	0.100
w_A(sol phase)	0.670	0.668	0.660	0.649	–
w_B(sol phase)	0.001	0.005	0.014	0.029	–
w_C(sol phase)	0.210	0.209	0.209	0.208	–
w_D(sol phase)	0.119	0.118	0.117	0.114	–

T/K = 433.15

w_A(total)	0.1875	was kept constant
w_B(total)	0.150	was kept constant
w_C(total)	0.5625	was kept constant
w_D(total)	0.100	was kept constant

continued

continued

P/bar	519	626	640
w_A(gel phase)	0.156	0.183	0.1875
w_B(gel phase)	0.332	0.206	0.150
w_C(gel phase)	0.446	0.533	0.5625
w_D(gel phase)	0.066	0.088	0.100
w_A(sol phase)	0.215	0.204	–
w_B(sol phase)	0.123	0.142	–
w_C(sol phase)	0.575	0.570	–
w_D(sol phase)	0.087	0.084	–

Polymer (B): **poly(ethylene-*co*-vinyl acetate)** **1989WIL**
Characterization: M_n/g.mol^{-1} = 26000, M_w/g.mol^{-1} = 157000, 33.6 wt% vinyl acetate, Bayer AG, Leverkusen, Germany
Solvent (A): **ethene** **C_2H_4** **74-85-1**
Solvent (C): **vinyl acetate** **$C_4H_6O_2$** **108-05-4**
Solvent (D): **2-methyl-2-propanol** **$C_4H_{10}O$** **75-65-0**

Type of data: cloud points

w_A	0.431	0.431	0.431	0.413	0.413	0.413	0.322	0.322	0.322
w_B	0.057	0.057	0.057	0.112	0.112	0.112	0.229	0.229	0.229
w_C	0.173	0.173	0.173	0.148	0.148	0.148	0.113	0.113	0.113
w_D	0.339	0.339	0.339	0.327	0.327	0.327	0.336	0.336	0.336
T/K	333.15	353.15	373.15	333.15	353.15	373.15	333.15	353.15	373.15
P/bar	480	480	480	485	480	480	365	380	390

Polymer (B): **poly(ethylene-*co*-vinyl acetate)** **1989WIL**
Characterization: M_n/g.mol^{-1} = 35000, M_w/g.mol^{-1} = 320000, 45.0 wt% vinyl acetate, Bayer AG, Leverkusen, Germany
Solvent (A): **ethene** **C_2H_4** **74-85-1**
Solvent (C): **vinyl acetate** **$C_4H_6O_2$** **108-05-4**
Solvent (D): **2-methyl-2-propanol** **$C_4H_{10}O$** **75-65-0**

Type of data: cloud points

w_A	0.413	0.413	0.413	0.413	0.368	0.368	0.368	0.368
w_B	0.044	0.044	0.044	0.044	0.093	0.093	0.093	0.093
w_C	0.184	0.184	0.184	0.184	0.168	0.168	0.168	0.168
w_D	0.358	0.358	0.358	0.358	0.372	0.372	0.372	0.372
T/K	313.15	333.15	353.15	373.15	313.15	333.15	353.15	373.15
P/bar	337	345	363	385	300	310	328	355

w_A	0.370	0.370	0.370	0.370	0.316	0.316	0.316	0.316
w_B	0.133	0.133	0.133	0.133	0.227	0.227	0.227	0.227
w_C	0.141	0.141	0.141	0.141	0.099	0.099	0.099	0.099
w_D	0.356	0.356	0.356	0.356	0.357	0.357	0.357	0.357
T/K	313.15	333.15	353.15	373.15	313.15	333.15	353.15	373.15
P/bar	318	326	345	370	295	316	322	344

continued

continued

w_A	0.041	0.041	0.041	0.349	0.349	0.349	0.349	0.403	0.403
w_B	0.077	0.077	0.077	0.024	0.024	0.024	0.024	0.048	0.048
w_C	0.295	0.295	0.295	0.219	0.219	0.219	0.219	0.184	0.184
w_D	0.587	0.587	0.587	0.407	0.407	0.407	0.407	0.365	0.365
T/K	313.15	333.15	353.15	313.15	333.15	353.15	373.15	313.15	333.15
P/bar	21	21	21	248	265	286	309	335	342
w_A	0.403	0.403	0.335	0.335	0.335	0.335	0.295	0.295	0.295
w_B	0.048	0.048	0.097	0.097	0.097	0.097	0.148	0.148	0.148
w_C	0.184	0.184	0.171	0.171	0.171	0.171	0.158	0.158	0.158
w_D	0.365	0.365	0.392	0.392	0.392	0.392	0.399	0.399	0.399
T/K	353.15	373.15	313.15	333.15	353.15	373.15	313.15	333.15	353.15
P/bar	360	382	269	276	294	318	170	195	222
w_A	0.295	0.247	0.247	0.247	0.247	0.205	0.205	0.205	0.205
w_B	0.148	0.203	0.203	0.203	0.203	0.258	0.258	0.258	0.258
w_C	0.158	0.140	0.140	0.140	0.140	0.119	0.119	0.119	0.119
w_D	0.399	0.410	0.410	0.410	0.410	0.419	0.419	0.419	0.419
T/K	373.15	313.15	333.15	353.15	373.15	313.15	333.15	353.15	373.15
P/bar	252	128	156	180	209	75	102	138	175
w_A	0.324	0.324	0.324	0.324	0.304	0.304	0.304	0.304	0.263
w_B	0.023	0.023	0.023	0.023	0.049	0.049	0.049	0.049	0.096
w_C	0.198	0.198	0.198	0.198	0.188	0.188	0.188	0.188	0.173
w_D	0.456	0.456	0.456	0.456	0.459	0.459	0.459	0.459	0.468
T/K	313.15	333.15	353.15	373.15	313.15	333.15	353.15	373.15	313.15
P/bar	187	204	226	256	170	186	210	243	120
w_A	0.263	0.263	0.241	0.241	0.241				
w_B	0.096	0.096	0.089	0.089	0.089				
w_C	0.173	0.173	0.160	0.160	0.160				
w_D	0.468	0.468	0.510	0.510	0.510				
T/K	333.15	353.15	313.15	333.15	353.15				
P/bar	141	170	82	104	130				

Polymer (B): **poly(ethylene-*co*-vinyl acetate)** **1989WIL**
Characterization: M_n/g.mol^{-1} = 51000, M_w/g.mol^{-1} = 360000,
70.0 wt% vinyl acetate, Bayer AG, Leverkusen, Germany
Solvent (A): **ethene** C_2H_4 **74-85-1**
Solvent (C): **vinyl acetate** $C_4H_6O_2$ **108-05-4**
Solvent (D): **2-methyl-2-propanol** $C_4H_{10}O$ **75-65-0**

Type of data: cloud points

w_A	0.145	0.145	0.217	0.217	0.217	0.217	0.371	0.371	0.371
w_B	0.054	0.054	0.062	0.062	0.062	0.062	0.050	0.050	0.050
w_C	0.379	0.379	0.212	0.212	0.212	0.212	0.194	0.194	0.194
w_D	0.422	0.422	0.479	0.479	0.479	0.479	0.385	0.385	0.385
T/K	323.15	353.15	313.15	333.15	353.15	373.15	313.15	333.15	353.15
P/bar	36	48	47	57	67	87	210	230	255

Polymer (B):	**poly(ethylene-*co*-vinyl acetate)**		**2001DO2**
Characterization:	M_n/g.mol^{-1} = 61900, M_w/g.mol^{-1} = 167000, 27.5 wt% vinyl acetate		
Solvent (A):	**ethene**	C_2H_4	**74-85-1**
Solvent (C):	**vinyl acetate**	$C_4H_6O_2$	**108-05-4**
Solvent (D):	**nitrogen**	N_2	**7727-37-9**

Type of data: coexistence data

T/K = 433.15

w_A(total)	0.375	was kept constant
w_B(total)	0.150	was kept constant
w_C(total)	0.375	was kept constant
w_D(total)	0.100	was kept constant

P/bar	676	790	893	992	1011
w_A(gel phase)	0.240	0.256	0.296	0.332	0.375
w_B(gel phase)	0.455	0.410	0.341	0.238	0.150
w_C(gel phase)	0.252	0.270	0.291	0.340	0.375
w_D(gel phase)	0.053	0.064	0.072	0.090	0.100
w_A(sol phase)	0.440	0.435	0.429	–	–
w_B(sol phase)	0.009	0.017	0.030	–	–
w_C(sol phase)	0.427	0.426	0.421	–	–
w_D(sol phase)	0.124	0.122	0.120	–	–

Polymer (B):	**poly(ethylene oxide-*b*-1,1,2,2-tetrahydro-perfluorodecyl acrylate)**		**2004MA2**
Characterization:	M_n/g.mol^{-1} = 22800, 10.3 wt% PEO, molar ratio of PFDA/PEO = 39.5/1 by NMR, synthesized in the laboratory		
Solvent (A):	**carbon dioxide**	CO_2	**124-38-9**
Solvent (C):	**2-hydroxyethyl methacrylate**	$C_6H_{10}O_3$	**868-77-9**

Type of data: cloud points

w_A	0.868	0.868	0.868	0.868	0.868	0.868	0.868	0.868	0.868
w_B	0.012	0.012	0.012	0.012	0.012	0.012	0.012	0.012	0.012
w_C	0.120	0.120	0.120	0.120	0.120	0.120	0.120	0.120	0.120
T/K	338.15	332.85	328.05	322.95	318.45	313.55	308.55	303.85	298.75
P/bar	251.0	230.0	213.0	189.0	171.0	152.0	132.0	113.0	93.0

w_A	0.868
w_B	0.012
w_C	0.120
T/K	294.05
P/bar	76.0

4.4. Table of ternary or quaternary systems where data were published only in graphical form as phase diagrams or related figures

Polymer (B)	Second and third component	Ref.
Poly[2-(*N,N*-dimethylaminoethyl) methacrylate-*co*-1H,1H–perfluoro-octyl methacrylate]		
	carbon dioxide and poly(methyl methacrylate)	2008HWA
Poly(ethylene-*co*-acrylic acid)		
	ethene and methanol	2002LEE
	ethene and 2-propanone	2002LEE
Poly(ethylene-*co*-1-butene)		
	1-butene and ethene	1999CHE
	ethene and cyclohexane/1-butene	2009COS
Poly(ethylene-*co*-1-octene)		
	ethene and cyclohexane/1-octene	2009COS
	ethene and 1-octene	2005LEE
Poly(ethylene-*co*-propylene)		
	ethene and carbon dioxide	2001KEM
Poly(ethylene-*co*-propylene-*co*-diene)		
	n-hexane and propene	2001VLI
Poly(ethylene oxide-*b*-propylene oxide-*b*-ethylene oxide)		
	carbon dioxide and ethanol	2008MUN

Polymer (B)	Second and third component	Ref.
Poly(ethylene oxide-*b*-1,1,2,2-tetrahydroperfluorodecyl methacrylate)	carbon dioxide and 2-hydroxyethyl methacrylate	2004MA2
Poly(DL-lactic acid-*co*-glycolic acid)	carbon dioxide and 2-propanone	2005WAN
Poly(styrene-*b*-dimethylsiloxane)	carbon dioxide and 1-vinyl-2-pyrrolidinone	2000BER

4.5. References

1989WIL Will, B., Löslichkeit von Ethylen-Vinylacetat-Copolymeren in ethylenhaltigen Mischlösungsmitteln, *Dissertation*, Johannes-Gutenberg Universität Mainz, 1989.

1993HOE Hoefling, T.A., Newman, D.A., Enick, R.M., and Beckman, E.J., Effect of structure on the cloud-point curves of silicone-based amphiphiles in supercritical carbon dioxide, *J. Supercrit. Fluids*, 6, 165, 1993.

1997DIN DiNoia, T.P., McHugh, M.A., Cocchiaro, J.E., and Morris, J.B., Solubility and phase behavior of PEP binders in supercritical carbon dioxide, *Waste Managment*, 17, 151, 1997.

1997HAN Han, S.J., Gregg, C.J., and Radosz, M., How the solute polydispersity affects the cloud-point and coexistence pressures in propylene and ethylene solutions of alternating poly(ethylene-*co*-propylene), *Ind. Eng. Chem. Res.*, 36, 5520, 1997.

1998HAN Han, S.J., Lohse, D.J., Radosz, M., and Sperling, L.H., Short chain branching effect on the cloud-point pressures of ethylene copolymers in subcritical and supercritical propane, *Macromolecules*, 31, 2533, 1998.

1999CHE Chen, A.-Q. and Radosz, M., Phase equilibria of dilute poly(ethylene-*co*-1-butene) solutions in ethylene, 1-butene, and 1-butene+ethylene, *J. Chem. Eng. Data*, 44, 854, 1999.

1999DIE Dietzsch, H., Hochdruck-Copolymerisation von Ethen und (Meth)Acrylsäureestern, *Dissertation*, University Göttingen, 1999.

1999FIN Fink, R., Hancu, D., Valentine, R., and Beckman, E.J., Toward the development of "CO_2-philic" hydrocarbons. 1. Use of side-chain functionalization to lower the miscibility pressure of polydimethylsiloxanes in CO_2, *J. Phys. Chem. B*, 103, 6441, 1999.

1999PAN Pan, C. and Radosz, M., Phase behavior of poly(ethylene-*co*-hexene-1) solutions in isobutane and propane, *Ind. Eng. Chem. Res.*, 38, 2842, 1999.

2000BER Berger, B.T., Überkritisches Kohlendioxid als Reaktionsmedium für die Dispersionspolymerisation, *Dissertation*, Johannes-Gutenberg Universität Mainz, 2000.

2000BYU Byun, H.-S., Kim, K., and Lee, H.-S., High-pressure phase behavior and mixture density of binary poly(ethylene-*co*-butene)-dimethyl ether system, *Hwahak Konghak*, 38, 826, 2000.

2000CH4 Chan, A.K.C., Adidharma, H., and Radosz, M., Fluid-liquid transitions of poly(ethylene-*co*-octene-1) in supercritical ethylene solutions, *Ind. Eng. Chem. Res.*, 39, 4370, 2000.

2000LEE Lee, J.M., Lee, B.-C., and Lee, S.-H., Cloud points of biodegradable polymers in compressed liquid and supercritical chlorodifluoromethane, *J. Chem. Eng. Data*, 45, 851, 2000.

2000SAR Sarbu, T., Styranec, T., and Beckman, E.J., Non-fluorous polymers with very high solubility in supercritical CO_2 down to low pressures, *Nature*, 405, 165, 2000.

2000VRI Vries, T.J. de, Somers, P.J.A., Loos, Th.W. de, Vorstman, M.A.G., and Keurentjes, J.T.F., Phase behavior of poly(ethylene-co-propylene) in ethylene and carbon dioxide: Experimental results and modeling with the SAFT equation of state, *Ind. Eng. Chem. Res.*, 39, 4510, 2000.

2001CON Conway, S.E., Byun, H.-S., McHugh, M.A., Wang, J.D., and Mandel, F.S., Poly(lactide-*co*-glycolide) solution behavior in supercritical CO_2, CHF_3, and $CHClF_2$, *J. Appl. Polym. Sci.*, 80, 1155, 2001.

2001DO1 Dörr, H., Kinzl, M., and Luft, G., The influence of inert gases on the high-pressure phase equilibria of EH-copolymer/1-hexene/ethylene-mixtures, *Fluid Phase Equil.*, 178, 191, 2001.

2001DO2 Dörr, H.W., Untersuchungen zum Phasenverhalten von Copolymer-Ethen-Mischungen bei Zusatz der Comonomere und verschiedener Inertgase, *Dissertation*, TU Darmstadt, 2001.

2001KEM Kemmere, M., Vries, T.J. de, Vorstman, M., and Keurentjens, J., A novel process for the catalytic polymerization of olefins in supercritical carbon dioxide, *Chem. Eng. Sci.*, 56, 4197, 2001.

2001KUK Kuk, Y.-M., Lee, B.-C., Lee, Y.W., and Lim, J.S., Phase behavior of biodegradable polymers in dimethyl ether and dimethyl ether + carbon dioxide, *J. Chem. Eng. Data*, 46, 1344, 2001.

2001VLI Vliet, R.E. van, Tiemersma, T.P., Krooshof, G.J., and Iedema, P.D., The use of liquid-liquid extraction in the EPDM solution polymerization process, *Ind. Eng. Chem. Res.*, 40, 4586, 2001.

2002BEY Beyer, C. and Oellrich, L.R., Cosolvent studies with the system ethylene/poly(ethylene-*co*-acrylic acid): Effects of solvent, density, polarity, hydrogen bonding, and copolymer composition, *Helv. Chim. Acta*, 85, 659, 2002.

2002DRO Drohmann, C. and Beckman, E.J., Phase behavior of polymers containing ether groups in carbon dioxide, *J. Supercrit. Fluids*, 22, 103, 2002.

2002COL Colina, C.M., Hall, C.K., and Gubbins, K.E., Phase behavior of PVAC-PTAN block copolymer in supercritical carbon dioxide using SAFT, *Fluid Phase Equil.*, 194-197, 553, 2002.

2002LEE Lee, S.-H. and McHugh, M.A., The effect of hydrogen bonding on the phase behavior of poly(ethylene-*co*-acrylic acid)-ethylene-cosolvent mixtures at high pressures, *Korean J. Chem. Eng.*, 19, 114, 2002.

2002LIU Liu, J., Han, B., Wang, Z., Zhang, J., Li, G., and Yang, G., Solubility of Ls-36 and Ls-45 surfactants in supercritical CO_2 and loading water in the CO_2/water/surfactant system, *Langmuir*, 18, 3086, 2002.

2002TAY Taylor, D.K., Keiper, J.S., and DeSimone, J.M., Polymer self-assembly in carbon dioxide, *Ind. Eng. Chem. Res.*, 41, 4451, 2002.

2003BUB Buback, M. and Latz, H., Cloud-point pressure curves of ethene/poly[ethylene-*co*-((meth)acrylic acid)] mixtures, *Macromol. Chem. Phys.*, 204, 638, 2003.

2003KIL Kilic, S., Michalik, S., Wang, Y., Johnson, J.K., Enick, R.M., and Beckman, E.J., Effect of grafted Lewis base groups on the phase behavior of model poly(dimethyl siloxanes) in CO_2, *Ind. Eng. Chem. Res.*, 42, 6415, 2003.

2004BAR Baradie, B., Shoichet, M.S., Shen, Z., McHugh, M.A., Hong, L., Wang, Y., Johnson, J.K., Beckman, E.J., and Enick, R.M., Synthesis and solubility of linear poly(tetrafluoroethylene-*co*-vinyl acetate) in dense CO_2: Experimental and molecular modeling results, *Macromolecules*, 37, 7799, 2004.

2004BEC Becker, F., Buback, M., Latz, H., Sadowski, G., and Tumakaka, F., Cloud-point curves of ethylene-(meth)acrylate copolymers in fluid ethene up to high pressures and temperatures: Experimental study and PC-SAFT modeling, *Fluid Phase Equil.*, 215, 263, 2004.

2004HWA Hwang, H.S., Kim, H.J., Jeong, Y.T., Gal, Y.-S., and Lim, K.T., Synthesis and properties of semifluorinated copolymers of oligo(ethylene glycol) methacrylate and 1H,1H,2H,2H-perfluorooctyl methacrylate, *Macromolecules*, 37, 9821, 2004.

2004KER Kermis, T.W., Li, D., Guney-Altay, O., Park, I.-H., Zanten, J.H. van, and McHugh, M.A., High-pressure dynamic light scattering of poly(ethylene-*co*-1-butene) in ethane, propane, butane, and pentane at 130°C and kilobar pressures, *Macromolecules*, 37, 9123, 2004.

2004LAC Lacroix-Desmazes, P., Andre, P., Desimone, J.M., Ruzette, A.-V., and Boutevin, B., Macromolecular surfactants for supercritical carbon dioxide applications: Synthesis and characterization of fluorinated block copolymers prepared by nitroxide-mediated radical polymerization (experimental data by P. Lacroix-Desmazes), *J. Polym. Sci.: Part A: Polym. Chem.*, 42, 3537, 2004.

2004LAT Latz, H., Kinetische und thermodynamische Untersuchungen der Hochdruck-Copolymerisation von Ethen mit (Meth)Acrylsäureestern, *Dissertation*, Georg-August-Universität Göttingen, 2004.

2004MA1 Ma, Z. and Lacroix-Desmazes, P., Synthesis of hydrophilic/CO_2-philic poly(ethylene oxide)-*b*-poly(1,1,2,2-tetrahydroperfluorodecyl acrylate) block copolymers via controlled living radical polymerizations and their properties in liquid and supercritical CO_2 (experimental data by P. Lacroix-Desmazes), *J. Polym. Sci.: Part A: Polym. Chem.*, 42, 2405, 2004.

2004MA2 Ma, Z. and Lacroix-Desmazes, P., Dispersion polymerization of 2-hydroxyethyl methacrylate stabilized by a hydrophilic/CO_2-philic poly(ethylene oxide)-*b*-poly(1,1,2,2-tetrahydroperfluorodecyl acrylate) (PEO-*b*-PFDA) diblock copolymer in supercritical carbon dioxide (experimental data by P. Lacroix-Desmazes), *Polymer*, 45, 6789, 2004.

2004PFO Pfohl, O. and Dohrn, R., Provision of thermodynamic properties of polymer systems for industrial applications, *Fluid Phase Equil.*, 217, 189, 2004.

2005HEI Heidemann, R.A. , Krenz, R.A., and Laursen, T., Spinodal curves and critical points in mixtures containing polydisperse polymers with many components, *Fluid Phase Equil.*, 228-229, 239, 2005.

2005LEE Lee, S.-H., Phase behavior of binary and ternary mixtures of poly(ethylene-*co*-octene)–hydrocarbons (experimental data by S.-H. Lee), *J. Appl. Polym. Sci.*, 95, 161, 2005.

2005LID Li, D., McHugh, M.A., and van Zanten, J.H., Density-induced phase separation in poly(ethylene-*co*-1-butene)-dimethyl ether solutions, *Macromolecules*, 38, 2837, 2005.

2005SCH Schilt, M.A. van, Wering, R.M., Meerendonk, W.J. van, Kemmere, M.F., Keurentjes, J.T.F., Kleiner, M., Sadowski, G., and Loos, Th.W.de, High-pressure phase behavior of the system PCHC-CHO-CO_2 for the development of a solvent-free alternative toward polycarbonate production, *Ind. Eng. Chem. Res.*, 44, 3363, 2005.

2005SHE Shen, Z., McHugh, M.A., Smith Jr., D.W., Abayasinghe, N.K., and Jin, J., Impact of hexafluoroisopropylidene on the solubility of aromatic-based polymers in supercritical fluids, *J. Appl. Polym. Sci.*, 97, 1736, 2005.

2005WAN Wang, Y., Pfeffer, R., Dave, R., and Enick, R., Polymer encapsulation of fine particles by a supercritical antisolvent process, *AIChE-J.*, 51, 440, 2005.

2006AHM Ahmed, T.S., DeSimone, J.M., and Roberts, G.W., Copolymerization of vinylidene fluoride with hexafluoropropylene in supercritical carbon dioxide, *Macromolecules*, 39, 15, 2006.

2006AND Andre, P., Lacroix-Desmazes, P., Taylor, D.K., and Boutevin, B., Solubility of fluorinated homopolymer and block copolymer in compressed CO_2, *J. Supercrit. Fluids*, 37, 263, 2006.

2006BYU Byun, H.-S. and Lee, H.-Y., Cloud-point measurement of the biodegradable poly(DL-lactide-*co*-glycolide) solution in supercritical fluid solvents, *Korean J. Chem. Eng.*, 23, 1003, 2006.

2006GAN Ganapathy, H.S., Hwang, H.S., and Lim, K.T., Synthesis and properties of fluorinated ester-functionalized polythiophenes in supercritical carbon dioxide, *Ind. Eng. Chem. Res.*, 45, 3406, 2006.

2006KLE Kleiner, M., Tumakaka, F., Sadowski, G., Latz, H., and Buback, M., Phase equilibria in polydisperse and associating copolymer solutions: Poly(ethene-*co*-(meth)acrylic acid) monomer mixtures, *Fluid Phase Equil.*, 241, 113, 2006.

2006LAV Lavery, K.A., Sievert, J.D., Watkins, J.J., Russell, T.P., Ryu, D.Y., and Kim, J.K., Influence of carbon dioxide swelling on the closed-loop phase behavior of block copolymers, *Macromolecules*, 39, 6580, 2006.

2006NAG Nagy, I., Loos, Th.W.de, Krenz, R.A., and Heidemann, R.A., High pressure phase equilibria in the systems linear low density polyethylene + n-hexane and linear low density polyethylene + n-hexane + ethylene: Experimental results and modelling with the Sanchez-Lacombe equation of state, *J. Supercrit. Fluids*, 37, 115, 2006.

2006SUB Su, B., Lv, X., Yang, Y., and Ren, Q., Solubilities of dodecylpolyoxyethylene polyoxypropylene ether in supercritical carbon dioxide, *J. Chem. Eng. Data*, 51, 542, 2006.

2006WIN Winoto, W., Adidharma, H., Shen, Y., and Radosz, M., Micellization temperature and pressure for polystyrene-*block*-polyisoprene in subcritical and supercritical propane, *Macromolecules*, 39, 8140, 2006.

2007EDM Edmonds, W.F., Hillmyer, M.A., and Lodge, T.P., Block copolymer vesicles in liquid CO_2, *Macromolecules*, 40, 4917, 2007.

2007LIU Liu, J., Spraul, B.K., Topping, C., Smith, Jr., D.W., and McHugh, M.A., Effect of hexafluoroisopropylidene on perfluorocyclobutyl aryl ether copolymer solution behavior in supercritical CO_2 and propane, *Macromolecules*, 40, 5973, 2007.

2007NAG Nagy, I., Krenz, R.A., Heidemann, R.A., and de Loos, Th.W., High-pressure phase equilibria in the system linear low density polyethylene + isohexane: Experimental results and modelling, *J. Supercrit. Fluids*, 40, 125, 2007.

2007TUM Tumakaka, F., Sadowski, G., Latz, H., and Buback, M., Cloud-point pressure curves of ethylene-based terpolymers in fluid ethene and in ethene-comonomer-mixtures. Experimental study and modeling via PC-SAFT, *J. Supercrit. Fluids*, 41, 461, 2007.

2008HAR Haruki, M., Takakura, Y., Sugiura, H., Kihara, S., and Takishima, S., Phase behavior for the supercritical ethylene + hexane + polyethylene systems, *J. Supercrit. Fluids*, 44, 284, 2008.

2008HWA Hwang, H.S., Yuvaraj, H., Kim, W.S., Lee, W.-K., Gal, Y.-S., and Lim, K.T., Dispersion polymerization of MMA in supercritical CO_2 stabilized by random copolymers of 1H,1H–perfluorooctyl methacrylate and 2-(dimethylaminoethyl) methacrylate, *J. Polym. Sci.: Part A: Polym. Chem.*, 46, 1365, 2008.

2008KOS Kostko, A.F., Lee, S.H., Liu, J., DiNoia, T.P., Kim, Y., and McHugh, M.A., Cloud-point behavior of poly(ethylene-*co*-20.2 mol% 1-butene) (PEB10) in ethane and deuterated ethane and of deuterated PEB10 in pentane isomers, *J. Chem. Eng. Data*, 53, 1626, 2008.

2008MUN Munto, M., Ventosa, N., and Veciana, J., Synergistic solubility behaviour of a polyoxyalkylene block co-polymer and its precipitation from liquid CO_2-expanded ethanol as solid microparticles, *J. Supercrit. Fluids*, 47, 290, 2008.

2008SH2 Shin, J., Oh, K.S., Bae, W., Lee, Y.-W., and Kim, H., Dispersion polymerization of methyl methacrylate using poly(HDFDMA-*co*-MMA) as a surfactant in supercritical carbon dioxide, *Ind. Eng. Chem. Res.*, 47, 5680, 2008.

2009COS Costa, G.M.N., Guerrieri, Y., Kislansky, S., Pessoa, F.L.P., Vieira de Melo, S.A.B., and Embirucu, M., Simulation of flash separation in polyethylene industrial processing: Comparison of SRK and SL equations of state, *Ind. Eng. Chem. Res.*, 48, 8613, 2009.

2009KOS Kostko, A.F., Harden, J.L., and McHugh, M.A., Dynamic light scattering study of concentrated triblock copolymer micellar solutions under pressure, *Macromolecules*, 42, 5328, 2009.

2009STO Stoychev, I., Galy, J., Fournel, B., Lacroix-Desmazes, P., Kleiner, M., and Sadowski, G., Modeling the phase behavior of PEO-PPO-PEO surfactants in carbon dioxide using the PC-SAFT equation of state: Application to dry decontamination of solid substrates, *J. Chem. Eng. Data*, 54, 1551, 2009.

2009TYR Tyrrell, Z., Winoto, W., Shen, Y., and Radosz, M., Block copolymer micelles formed in supercritical fluid can become water-dispensable nanoparticles: Poly(ethylene glycol)-*block*-poly(ε-caprolactone) in trifluoromethane, *Ind. Eng. Chem. Res.*, 48, 1928, 2009.

2009WAN Wang, Y., Hong, L., Tapriyal, D., Kim, I.C., Paik, I.-H., Crosthwaite, J.M., Hamilton, A.D., Thies, M.C., Beckman, E.J., Enick, R.M., and Johnson, J.K., Design and evaluation of nonfluorous CO_2-soluble oligomers and polymers, *J. Phys. Chem. B*, 113, 14971, 2009.

2009WI1 Winoto, W., Tan, S.P., Shen, Y., Radosz, M., Hong, K., and Mays, J.W., High-pressure micellar solutions of polystyrene-*block*-polybutadiene and polystyrene-*block*-polyisoprene in propane exhibit cloud-pressure reduction and distinct micellization end points, *Macromolecules*, 42, 3823, 2009.

2009WI2 Winoto, W., Shen, Y., Radosz, M., Hong, K., and Mays, J.W., Deuteration impact on micellization pressure and cloud pressure of polystyrene-*block*-polybutadiene and polystyrene-*block*-polyisoprene in compressible propane, *J. Phys. Chem. B*, 113, 15156, 2009.

5. ENTHALPY CHANGES FOR COPOLYMER SOLUTIONS

5.1. Enthalpies of mixing or intermediary enthalpies of dilution, copolymer partial enthalpies of mixing (at infinite dilution), or copolymer (first) integral enthalpies of solution

Polymer (B):	**poly(ethylene-*co*-vinyl acetate)**	**2002RIG**
Characterization:	85.0 wt% vinyl acetate	
Solvent (A):	**cyclopentanone** C_5H_8O	**120-92-3**

T/K = 298.15

$\Delta_{sol}H_B^{\infty}$ = −0.50 J/(g copolymer)

Polymer (B):	**poly(ethylene-*co*-vinyl acetate)**	**2002RIG**
Characterization:	85.0 wt% vinyl acetate	
Solvent (A):	**cyclopentanone** C_5H_8O	**120-92-3**
Polymer (C):	**poly(vinyl chloride)**	
Characterization:	Fluka AG, Buchs, Switzerland	

T/K = 298.15

w_B/w_C	0/100	05/95	10/90	20/80	50/50	80/20	90/20
$\Delta_{sol}H_{B+C}^{\infty}$/J/(g blend)	−28.0	−19.8	−15.9	−12.2	−4.8	−2.2	−0.7
w_B/w_C	95/05	100/0					
$\Delta_{sol}H_{B+C}^{\infty}$/J/(g blend)	−0.2	−0.5					

Polymer (B):	**poly(styrene-*co*-acrylonitrile)**	**2003CAR**
Characterization:	M_n/g.mol^{-1} = 95000, M_w/g.mol^{-1} = 153800, 4.5 wt% acrylonitrile, T_g/K = 379.75, ENICHEM, Mantova, Italy	
Solvent (A):	**trichloromethane** $CHCl_3$	**67-66-3**

T/K = 298.15

$\Delta_{sol}H_B^{\infty}$ = −50.3 J/(g copolymer) (glassy state polymer)
$\Delta_M H_B^{\infty}$ = −25.0 J/(g copolymer) (liquid state polymer)

Comments: The final concentration was between $0.005 < w_B < 0.020$.

Polymer (B): **poly(styrene-*co*-acrylonitrile)** **2003CAR**
Characterization: M_n/g.mol^{-1} = 97350, M_w/g.mol^{-1} = 161600, 10.5 wt% acrylonitrile, T_g/K = 382.25, ENICHEM, Mantova, Italy
Solvent (A): **trichloromethane** **$CHCl_3$** **67-66-3**

T/K = 298.15

$\Delta_{sol}H_B^{\infty} = -58.7$ J/(g copolymer) (glassy state polymer)
$\Delta_M H_B^{\infty} = -30.1$ J/(g copolymer) (liquid state polymer)

Comments: The final concentration was between $0.005 < w_B < 0.020$.

Polymer (B): **poly(styrene-*co*-acrylonitrile)** **2003CAR**
Characterization: M_n/g.mol^{-1} = 86550, M_w/g.mol^{-1} = 142700, 15.6 wt% acrylonitrile, T_g/K = 382.95, ENICHEM, Mantova, Italy
Solvent (A): **trichloromethane** **$CHCl_3$** **67-66-3**

T/K = 298.15

$\Delta_{sol}H_B^{\infty} = -61.8$ J/(g copolymer) (glassy state polymer)
$\Delta_M H_B^{\infty} = -29.6$ J/(g copolymer) (liquid state polymer)

Comments: The final concentration was between $0.005 < w_B < 0.020$.

Polymer (B): **poly(styrene-*co*-acrylonitrile)** **2003CAR**
Characterization: M_n/g.mol^{-1} = 77950, M_w/g.mol^{-1} = 132400, 19.4 wt% acrylonitrile, T_g/K = 384.85, ENICHEM, Mantova, Italy
Solvent (A): **trichloromethane** **$CHCl_3$** **67-66-3**

T/K = 298.15

$\Delta_{sol}H_B^{\infty} = -67.4$ J/(g copolymer) (glassy state polymer)
$\Delta_M H_B^{\infty} = -32.7$ J/(g copolymer) (liquid state polymer)

Comments: The final concentration was between $0.005 < w_B < 0.020$.

Polymer (B): **poly(styrene-*co*-acrylonitrile)** **2003CAR**
Characterization: M_n/g.mol^{-1} = 48000, M_w/g.mol^{-1} = 78600, 25.0 wt% acrylonitrile, T_g/K = 385.85, ENICHEM, Mantova, Italy
Solvent (A): **trichloromethane** **$CHCl_3$** **67-66-3**

T/K = 298.15

$\Delta_{sol}H_B^{\infty} = -68.1$ J/(g copolymer) (glassy state polymer)
$\Delta_M H_B^{\infty} = -32.1$ J/(g copolymer) (liquid state polymer)

Comments: The final concentration was between $0.005 < w_B < 0.020$.

Polymer (B): **poly(styrene-*co*-acrylonitrile)** **2003CAR**
Characterization: M_n/g.mol^{-1} = 55500, M_w/g.mol^{-1} = 99900, 29.9 wt% acrylonitrile, T_g/K = 386.15, ENICHEM, Mantova, Italy
Solvent (A): **trichloromethane** **$CHCl_3$** **67-66-3**

T/K = 298.15

$\Delta_{sol}H_B^\infty = -66.8$ J/(g copolymer) (glassy state polymer)
$\Delta_M H_B^\infty = -30.7$ J/(g copolymer) (liquid state polymer)

Comments: The final concentration was between $0.005 < w_B < 0.020$.

Polymer (B): **poly(styrene-*co*-acrylonitrile)** **2003CAR**
Characterization: M_n/g.mol^{-1} = 46700, M_w/g.mol^{-1} = 78400, 33.8 wt% acrylonitrile, T_g/K = 386.25, ENICHEM, Mantova, Italy
Solvent (A): **trichloromethane** **$CHCl_3$** **67-66-3**

T/K = 298.15

$\Delta_{sol}H_B^\infty = -73.4$ J/(g copolymer) (glassy state polymer)
$\Delta_M H_B^\infty = -38.2$ J/(g copolymer) (liquid state polymer)

Comments: The final concentration was between $0.005 < w_B < 0.020$.

Polymer (B): **poly(styrene-*co*-acrylonitrile)** **2003CAR**
Characterization: M_n/g.mol^{-1} = 50700, M_w/g.mol^{-1} = 90300, 36.9 wt% acrylonitrile, T_g/K = 386.35, ENICHEM, Mantova, Italy
Solvent (A): **trichloromethane** **$CHCl_3$** **67-66-3**

T/K = 298.15

$\Delta_{sol}H_B^\infty = -71.8$ J/(g copolymer) (glassy state polymer)
$\Delta_M H_B^\infty = -36.5$ J/(g copolymer) (liquid state polymer)

Comments: The final concentration was between $0.005 < w_B < 0.020$.

5.2. Partial molar enthalpies of mixing at infinite dilution and enthalpies of solution of gases/vapors of solvents in molten copolymers from inverse gas-liquid chromatography (IGC)

Polymer (B): **poly(*tert*-butyl acrylate-*b*-methyl methacrylate)** **2008AYD**
Characterization: M_n/g.mol^{-1} = 69000, M_w/g.mol^{-1} = 80700, 5 mol% *tert*-BA

Solvent (A)	T-range/ K	$\Delta_M H_A^\infty$/ kJ/mol	$\Delta_{sol} H_{A(vap)}^\infty$/ kJ/mol
chlorobenzene	423.15-443.15	−1.7	−36.4
n-decane	423.15-443.15	23.4	−18.4
ethyl acetate	423.15-443.15	5.9	−21.8
ethylbenzene	423.15-443.15	10.0	−25.5
isopropylbenzene	423.15-443.15	17.2	−19.7
3-methylbutyl acetate	423.15-443.15	7.95	−31.4
n-nonane	423.15-443.15	26.4	−11.3
n-octane	423.15-443.15	21.3	−12.1
propyl acetate	423.15-443.15	7.1	−24.3
propylbenzene	423.15-443.15	6.7	−31.8
toluene	423.15-443.15	3.8	−27.6

Polymer (B): **poly(*tert*-butyl acrylate-*b*-methyl methacrylate)** **2006SAK**
Characterization: M_n/g.mol^{-1} = 69000 (9000-*b*-60000), 5 mol% *tert*-BA

Solvent (A)	T-range/ K	$\Delta_M H_A^\infty$/ kJ/mol	$\Delta_{sol} H_{A(vap)}^\infty$/ kJ/mol
benzene	413.15-443.15	5.4	−22.2
butyl acetate	413.15-443.15	9.2	−26.0
tert-butyl acetate	413.15-443.15	8.4	−32.7
chlorobenzene	413.15-443.15	−1.7	−36.4
ethyl acetate	413.15-443.15	5.9	−21.8
ethylbenzene	413.15-443.15	10.0	−25.5
isopropylbenzene	413.15-443.15	17.2	−19.7
isobutyl acetate	413.15-443.15	11.3	−21.8
3-methylbutyl acetate	413.15-443.15	7.95	−31.4
propyl acetate	413.15-443.15	7.1	−24.3
2-propyl acetate	413.15-443.15	1.3	−28.1
propylbenzene	413.15-443.15	6.7	−31.8
toluene	413.15-443.15	3.8	−27.6

Polymer (B): **poly(*tert*-butyl acrylate-*b*-methyl methacrylate)** **2008AYD**
Characterization: M_n/g.mol^{-1} = 62000, M_w/g.mol^{-1} = 45000, 18 mol% *tert*-BA

Solvent (A)	T-range/ K	$\Delta_M H_A{}^{\infty}$/ kJ/mol	$\Delta_{sol} H_{A(vap)}{}^{\infty}$/ kJ/mol
chlorobenzene	418.15-443.15	12.1	−22.2
n-decane	418.15-443.15	36.8	−4.6
ethyl acetate	418.15-443.15	9.62	−18.0
ethylbenzene	418.15-443.15	23.0	−12.5
isopropylbenzene	418.15-443.15	34.3	−27.6
3-methylbutyl acetate	418.15-443.15	30.5	−8.8
n-nonane	418.15-443.15	31.8	−5.4
n-octane	418.15-443.15	29.3	−4.2
propyl acetate	418.15-443.15	12.1	−19.2
propylbenzene	418.15-443.15	23.4	−15.1
toluene	418.15-443.15	19.2	−12.6

Polymer (B): **poly(3,4-dichlorobenzyl methacrylate-*co*-ethyl methacrylate)** **2001DEM**
Characterization: 13 mol% ethyl methacrylate, synthesized in the laboratory

Solvent (A)	T-range/ K	$\Delta_M H_A{}^{\infty}$/ kJ/mol
methanol	403.15-423.15	22.86
ethanol	403.15-423.15	54.04
2-propanone	403.15-423.15	20.09
2-butanone	403.15-423.15	18.01
methyl acetate	403.15-423.15	18.71
ethyl acetate	403.15-423.15	20.79
benzene	403.15-423.15	19.40
toluene	403.15-423.15	14.90
1,2-dimethylbenzene	403.15-423.15	13.10
n-octane	403.15-423.15	27.71
n-nonane	403.15-423.15	29.10
n-decane	403.15-423.15	16.63
n-undecane	403.15-423.15	27.02
n-dodecane	403.15-423.15	21.48

Polymer (B): **poly(3,4-dichlorobenzyl methacrylate-*co*-ethyl methacrylate)** **2001KA1**

Characterization: 27 mol% ethyl methacrylate, T_g/K = 336.2, ρ (298 K) = 1.09 g/cm^3, synthesized in the laboratory

Solvent (A)	*T*-range/ K	$\Delta_M H_A^\infty$/ kJ/mol
methanol	403.15-423.15	24.94
ethanol	403.15-423.15	61.66
2-propanone	403.15-423.15	18.71
2-butanone	403.15-423.15	19.40
methyl acetate	403.15-423.15	20.09
ethyl acetate	403.15-423.15	22.86
benzene	403.15-423.15	19.40
toluene	403.15-423.15	22.17
1,2-dimethylbenzene	403.15-423.15	21.48
n-octane	403.15-423.15	31.87
n-nonane	403.15-423.15	28.41
n-decane	403.15-423.15	11.09
n-dodecane	403.15-423.15	21.48

Polymer (B): **poly(3,4-dichlorobenzyl methacrylate-*co*-ethyl methacrylate)** **2001KA1**

Characterization: 60 mol% ethyl methacrylate, T_g/K = 334.2, ρ (298 K) = 1.14 g/cm^3, synthesized in the laboratory

Solvent (A)	*T*-range/ K	$\Delta_M H_A^\infty$/ kJ/mol
methanol	403.15-423.15	22.17
ethanol	403.15-423.15	65.82
2-propanone	403.15-423.15	19.70
2-butanone	403.15-423.15	21.48
methyl acetate	403.15-423.15	19.40
ethyl acetate	403.15-423.15	22.86
benzene	403.15-423.15	22.17
toluene	403.15-423.15	23.56
1,2-dimethylbenzene	403.15-423.15	20.09
n-octane	403.15-423.15	25.64
n-nonane	403.15-423.15	29.79
n-decane	403.15-423.15	31.87
n-dodecane	403.15-423.15	26.33

Polymer (B): **poly(3,4-dichlorobenzyl methacrylate-*co*-ethyl methacrylate)** **2001KA1**

Characterization: 70 mol% ethyl methacrylate, T_g/K = 332.2, ρ(298 K) = 1.14 g/cm^3, synthesized in the laboratory

Solvent (A)	T-range/ K	$\Delta_M H_A^\infty$/ kJ/mol
methanol	403.15-423.15	27.71
ethanol	403.15-423.15	67.21
2-propanone	403.15-423.15	16.63
2-butanone	403.15-423.15	18.71
methyl acetate	403.15-423.15	20.09
ethyl acetate	403.15-423.15	20.09
benzene	403.15-423.15	21.48
toluene	403.15-423.15	23.56
1,2-dimethylbenzene	403.15-423.15	18.71
n-octane	403.15-423.15	27.02
n-nonane	403.15-423.15	28.41
n-decane	403.15-423.15	29.79
n-dodecane	403.15-423.15	24.25

Polymer (B): **poly(3,4-dichlorobenzyl methacrylate-*co*-ethyl methacrylate)** **2001KA1**

Characterization: 82 mol% ethyl methacrylate, T_g/K = 331.2, ρ(298 K) = 1.17 g/cm^3, synthesized in the laboratory

Solvent (A)	T-range/ K	$\Delta_M H_A^\infty$/ kJ/mol
methanol	403.15-423.15	29.79
ethanol	403.15-423.15	65.82
2-propanone	403.15-423.15	35.34
2-butanone	403.15-423.15	17.32
methyl acetate	403.15-423.15	22.86
ethyl acetate	403.15-423.15	18.01
benzene	403.15-423.15	22.17
toluene	403.15-423.15	14.55
1,2-dimethylbenzene	403.15-423.15	13.86
n-octane	403.15-423.15	27.71
n-nonane	403.15-423.15	31.18
n-decane	403.15-423.15	27.02
n-dodecane	403.15-423.15	9.70

Polymer (B): **poly(3,4-dichlorobenzyl methacrylate-*co*-ethyl methacrylate)** **2001DEM**

Characterization: 93 mol% ethyl methacrylate, synthesized in the laboratory

Solvent (A)	T-range/ K	$\Delta_M H_A^\infty$/ kJ/mol
methanol	403.15-423.15	28.41
ethanol	403.15-423.15	73.44
2-propanone	403.15-423.15	27.71
2-butanone	403.15-423.15	23.56
methyl acetate	403.15-423.15	29.79
ethyl acetate	403.15-423.15	21.48
benzene	403.15-423.15	24.94
toluene	403.15-423.15	14.55
1,2-dimethylbenzene	403.15-423.15	18.71
n-octane	403.15-423.15	29.79
n-nonane	403.15-423.15	32.56
n-decane	403.15-423.15	33.26
n-undecane	403.15-423.15	13.86
n-dodecane	403.15-423.15	7.62

Polymer (B): **poly[2-(*N,N*-dimethylamino)ethyl methacrylate]-b-2-(*N*-morpholino)ethyl methacrylate]** **2008YAZ**

Characterization: M_n/g.mol^{-1} = 33500, M_w/g.mol^{-1} = 38200, 50 mol%/50 mol%, T_g/K = 332.5, ρ (298 K) = 1.0892 g/cm^3, synthesized in the laboratory

Solvent (A)	T-range/ K	$\Delta_M H_A^\infty$/ kJ/mol
n-hexane	383.15-413.15	8.55
n-heptane	383.15-413.15	10.9
n-octane	383.15-413.15	7.21
n-nonane	383.15-413.15	3.30
n-decane	383.15-413.15	3.39
cyclopentane	383.15-413.15	7.23
cyclohexane	383.15-413.15	5.31
cycloheptane	383.15-413.15	1.19
benzene	383.15-413.15	−2.32
toluene	383.15-413.15	−1.28

Polymer (B): **poly[dimethylsiloxane-*co*-methyl (4-cyanobiphenoxy)butylsiloxane]** **2002PRI**

Characterization: 50 mol% dimethylsiloxane, 40 repeat units, Merck, Ltd., UK

Solvent (A)	T/ K	$\Delta_{sol}H_A^{\infty}$/ kJ/mol
n-pentane	358.15	−14.9
n-hexane	358.15	−19.9
n-heptane	358.15	−22.5
n-octane	358.15	−25.3
n-nonane	358.15	−29.6
2-methylhexane	358.15	−17.8
3-methylhexane	358.15	−19.4
2,3-dimethylpentane	358.15	−20.2
2,4-dimethylpentane	358.15	−18.3
2,2,3-trimethylbutane	358.15	−17.6
cyclohexane	358.15	−29.0
benzene	358.15	−25.3
toluene	358.15	−28.5
ethylbenzene	358.15	−30.2
1,2-dimethylbenzene	358.15	−32.2
1,3-dimethylbenzene	358.15	−31.8
1,4-dimethylbenzene	358.15	−31.5

Polymer (B): **poly(glycidyl methacrylate-*co*-butyl methacrylate)** **2002KAY**

Characterization: M_n/g.mol^{-1} = 530000, M_w/g.mol^{-1} = 738000, 41 mol% butyl methacrylate, T_g/K = 334, ρ = 1.178 g/cm^3

Solvent (A)	T-range/ K	$\Delta_M H_A^{\infty}$/ kJ/mol
n-pentane	403.15-423.15	3.93
n-hexane	403.15-423.15	5.56
n-heptane	403.15-423.15	6.90
n-octane	403.15-423.15	8.07
n-nonane	403.15-423.15	9.61
n-decane	403.15-423.15	10.7
benzene	403.15-423.15	5.85
toluene	403.15-423.15	6.19
1,4-dimethylbenzene	403.15-423.15	7.73
tetrachloromethane	403.15-423.15	0.63
1-chloropropane	403.15-423.15	3.05
1-chlorobutane	403.15-423.15	3.68

Polymer (B): **poly(glycidyl methacrylate-*co*-ethyl methacrylate)** **2002KAY**

Characterization: M_n/g.mol^{-1} = 441000, M_w/g.mol^{-1} = 700000, 44 mol% ethyl methacrylate, T_g/K = 355, ρ = 1.186 g/cm^3

Solvent (A)	T-range/ K	$\Delta_M H_A^\infty$/ kJ/mol
n-pentane	403.15-423.15	3.47
n-hexane	403.15-423.15	4.89
n-heptane	403.15-423.15	4.89
n-octane	403.15-423.15	7.73
n-nonane	403.15-423.15	9.20
n-decane	403.15-423.15	9.99
benzene	403.15-423.15	5.52
toluene	403.15-423.15	5.48
1,4-dimethylbenzene	403.15-423.15	6.98
tetrachloromethane	403.15-423.15	1.05
1-chloropropane	403.15-423.15	2.84
1-chlorobutane	403.15-423.15	4.39

Polymer (B): **poly(glycidyl methacrylate-*co*-methyl methacrylate)** **2002KAY**

Characterization: M_n/g.mol^{-1} = 555000, M_w/g.mol^{-1} = 710000, 38 mol% methyl methacrylate, T_g/K = 373, ρ= 1.204 g/cm^3

Solvent (A)	T-range/ K	$\Delta_M H_A^\infty$/ kJ/mol
n-pentane	403.15-423.15	3.76
n-hexane	403.15-423.15	5.10
n-heptane	403.15-423.15	6.40
n-octane	403.15-423.15	7.90
n-nonane	403.15-423.15	9.28
n-decane	403.15-423.15	10.4
benzene	403.15-423.15	5.23
toluene	403.15-423.15	5.60
1,4-dimethylbenzene	403.15-423.15	7.23
1,4-dioxane	403.15-423.15	4.10
tetrachloromethane	403.15-423.15	0.59
1-chloropropane	403.15-423.15	2.63
1-chlorobutane	403.15-423.15	2.63

Polymer (B): **poly(3-mesityl-2-hydroxypropyl methacrylate-*co*-1-vinyl-2-pyrrolidinone)** **2005ACI**

Characterization: M_n/g.mol^{-1} = 58000, M_w/g.mol^{-1} = 481000, 45.75 mol% 1-vinyl-2-pyrrolidinone, T_g/K = 380, ρ(298 K) = 1.040 g/cm^3, synthesized in the laboratory

Solvent (A)	T-range/ K	$\Delta_M H_A^\infty$/ kJ/mol	$\Delta_{sol} H_{A(vap)}^\infty$/ kJ/mol
n-hexane	373.15-413.15		−12.35
n-hexane	413.15-453.15	29.18	
n-heptane	373.15-413.15		−12.35
n-heptane	413.15-453.15	30.23	
n-octane	373.15-413.15		−8.42
n-octane	413.15-453.15	33.88	
n-decane	373.15-413.15		−19.91
n-decane	413.15-453.15	43.25	
methanol	373.15-413.15		−9.13
methanol	413.15-453.15	31.86	
ethanol	373.15-413.15		−11.76
ethanol	413.15-453.15	37.26	
1-propanol	373.15-413.15		−10.76
1-propanol	413.15-453.15	40.15	
1-butanol	373.15-413.15		−9.04
1-butanol	413.15-453.15	44.17	
1-pentanol	373.15-413.15		−12.02
1-pentanol	413.15-453.15	44.42	

Polymer (B): **poly(methylhydrosiloxane-*co*-dimethylsiloxane)** **2007COS**

Characterization: HMS-013

Solvent (A)	T-range/ K	$\Delta_M H_A^\infty$/ kJ/mol	$\Delta_{sol} H_{A(vap)}^\infty$/ kJ/mol
methyl acetate	313.15-353.15	6.7	−23.9
ethyl acetate	313.15-353.15	5.0	−28.5
tert-butyl acetate	313.15-353.15	4.1	−33.1
n-pentane	313.15-353.15	1.5	−23.9
n-hexane	313.15-353.15	0.33	−29.7
n-heptane	313.15-353.15	0.46	−33.9

Polymer (B): **poly[(2-phenyl-1,3-dioxolane-4-yl)methyl methacrylate-*co*-butyl methacrylate]** **2003ACI**

Characterization: M_n/g.mol^{-1} = 301100, M_w/g.mol^{-1} = 880200, 55.0 mol% butyl methacrylate, T_g/K = 370, ρ (298 K) = 1.217 g/cm^3, could be a block copolymer, synthesized in the laboratory

Solvent (A)	*T*-range/ K	$\Delta_M H_A^\infty$/ kJ/mol
1-butanol	413.15-453.15	31.06
ethanol	413.15-453.15	29.68
methanol	413.15-453.15	26.79
1-pentanol	413.15-453.15	28.80
1-propanol	413.15-453.15	31.98

Polymer (B): **poly[(2-phenyl-1,3-dioxolane-4-yl)methyl methacrylate-*co*-ethyl methacrylate]** **2009KAR**

Characterization: M_n/g.mol^{-1} = 365400, M_w/g.mol^{-1} = 1215000, 52 mol% ethyl methacrylate, ρ (298 K) = 1.21 g/cm^3, T_g/K = 363, synthesized in the laboratory

Solvent (A)	*T*-range/ K	$\Delta_{sol} H_{A(vap)}^\infty$/ kJ/mol
1-butanol	393.15-433.15	−13.86
ethanol	393.15-433.15	−15.13
n-heptane	393.15-433.15	−11.33
n-hexane	393.15-433.15	−11.05
n-octane	393.15-433.15	−16.61
1-propanol	393.15-433.15	−15.16

Polymer (B): **poly[(2-phenyl-1,3-dioxolane-4-yl)methyl methacrylate-*co*-glycidyl methacrylate]** **2004ILT**

Characterization: M_n/g.mol^{-1} = 244000, M_w/g.mol^{-1} = 623400, 60.0 mol% glycidyl methacrylate, T_g/K = 367, ρ (298 K) = 1.229 g/cm^3, could be a block copolymer, synthesized in the laboratory

Solvent (A)	T-range/ K	$\Delta_M H_A{}^{\infty}$/ kJ/mol
ethanol	413.15-463.15	24.27
1-propanol	413.15-463.15	26.64
1-butanol	413.15-463.15	28.44
1-pentanol	413.15-463.15	25.16
n-hexane	413.15-463.15	15.79
n-heptane	413.15-463.15	20.37
n-octane	413.15-463.15	22.40
n-decane	413.15-463.15	28.10

Polymer (B): **poly[(2-phenyl-1,3-dioxolane-4-yl)methyl methacrylate-*co*-styrene]** **2006KAR**

Characterization: M_n/g.mol^{-1} = 181000, M_w/g.mol^{-1} = 436500, unknown styrene content, T_g/K = 373-383, synthesized in the laboratory

Solvent (A)	T-range/ K	$\Delta_{sol} H_{A(vap)}{}^{\infty}$/ kJ/mol
ethanol	393.15-423.15	−5.9
1-propanol	393.15-423.15	−6.8
1-butanol	393.15-423.15	−13.5
n-hexane	393.15-423.15	−4.2
n-heptane	393.15-423.15	−3.7
n-octane	393.15-423.15	−4.1

Polymer (B): **poly[2-(3-phenyl-3-methylcyclobutyl)-2-hydroxyethyl methacrylate-*co*-methacrylic acid]** **2001KA2**

Characterization: M_n/g.mol^{-1} = 17200, M_w/g.mol^{-1} = 45000, 45 mol% methacrylic acid, synthesized in the laboratory

Solvent (A)	*T*-range/ K	$\Delta_M H_A^\infty$/ kJ/mol
methanol	423.15-453.15	26.63
ethanol	423.15-453.15	29.73
2-propanone	423.15-453.15	24.95
2-butanone	423.15-453.15	27.42
methyl acetate	423.15-453.15	10.47
ethyl acetate	423.15-453.15	15.16
benzene	423.15-453.15	12.64
toluene	423.15-453.15	15.83
1,2-dimethylbenzene	423.15-453.15	14.82
n-dodecane	423.15-453.15	18.30

Polymer (B): **poly[styrene-*b*-(1-butene-*co*-ethylene)-*b*-styrene]** **2009OVE**

Characterization: M_η/g.mol^{-1} = 90000, 32 wt% styren, REPSOL-YPF, Madrid, Spain

Solvent (A)	*T*-range/ K	$\Delta_M H_A^\infty$/ kJ/mol	$\Delta_{sol} H_{A(vap)}^\infty$/ kJ/mol
benzene	303.15-333.15	−0.7	−35.4
cyclohexane	303.15-333.15	−0.5	−33.8
cyclopentane	303.15-333.15	−5.0	−33.0
1,2-dimethylbenzene	303.15-333.15	−2.0	−43.5
ethylbenzene	303.15-333.15	−1.4	−41.7
n-heptane	303.15-333.15	−3.0	−38.4
n-hexane	303.15-333.15	−2.3	−32.8
1-hexene	303.15-333.15	−3.4	−35.2
methylcyclohexane	303.15-333.15	−1.3	−36.0
n-octane	303.15-333.15	−2.6	−42.8
n-pentane	303.15-333.15	−6.8	−32.8
tetrahydrofuran	303.15-333.15	−2.5	−35.6
toluene	303.15-333.15	−0.4	−39.0

Polymer (B): **poly(styrene-*b*-ethylene oxide-*b*-styrene)** **2007ZOU**

Characterization: M_n/g.mol^{-1} = 11900, M_w/g.mol^{-1} = 16700, 42 wt% ethylene oxide, T_m/K = 328.3

Solvent (A)	T-range/ K	$\Delta_M H_A^\infty$/ kJ/mol	$\Delta_{sol} H_{A(vap)}^\infty$/ kJ/mol
n-hexane	343.15-393.15	20.13	−7.72
n-heptane	343.15-393.15	19.31	−13.26
n-octane	343.15-393.15	18.49	−18.96
n-nonane	343.15-393.15	17.28	−24.52
n-decane	343.15-393.15	16.43	−29.77
methyl acetate	343.15-393.15	8.44	−20.63
ethyl acetate	343.15-393.15	7.90	−23.89
propyl acetate	343.15-393.15	6.90	−28.46
butyl acetate	343.15-393.15	6.34	−32.83
pentyl acetate	343.15-393.15	6.80	−37.14
methanol	343.15-393.15	9.13	−26.60
ethanol	343.15-393.15	10.98	−28.08
1-propanol	343.15-393.15	9.91	−32.86
1-butanol	343.15-393.15	8.69	−37.65
1-pentanol	343.15-393.15	7.21	−41.96

Polymer (B): **poly(styrene-*b*-ethylene oxide-*b*-styrene)** **2007ZOU**

Characterization: M_n/g.mol^{-1} = 11200, M_w/g.mol^{-1} = 14300, 60 wt% ethylene oxide, T_m/K = 335

Solvent (A)	T-range/ K	$\Delta_M H_A^\infty$/ kJ/mol	$\Delta_{sol} H_{A(vap)}^\infty$/ kJ/mol
n-hexane	343.15-393.15	17.83	−10.03
n-heptane	343.15-393.15	17.02	−15.56
n-octane	343.15-393.15	15.98	−21.48
n-nonane	343.15-393.15	14.86	−26.94
n-decane	343.15-393.15	14.24	−31.97
methyl acetate	343.15-393.15	7.47	−21.61
ethyl acetate	343.15-393.15	6.76	−25.04
propyl acetate	343.15-393.15	5.91	−29.46
butyl acetate	343.15-393.15	5.24	−33.93
pentyl acetate	343.15-393.15	5.73	−38.21
methanol	343.15-393.15	12.36	−23.37
ethanol	343.15-393.15	11.88	−27.19
1-propanol	343.15-393.15	10.11	−32.66
1-butanol	343.15-393.15	9.20	−37.13
1-pentanol	343.15-393.15	7.95	−41.21

Polymer (B): **poly(styrene-*g*-ethyl methacrylate)** **2006TEM**
Characterization: 68 units EMA/graft, 54% of styrene-units are grafted, T_g/K = 358, synthesized in the laboratory

Solvent (A)	T-range/ K	$\Delta_M H_A^\infty$/ kJ/mol
1-butanol	413.15-463.15	30.73
n-decane	413.15-463.15	30.02
ethanol	413.15-463.15	28.64
n-heptane	413.15-463.15	20.45
n-hexane	413.15-463.15	18.25
methanol	413.15-463.15	24.95
n-nonane	413.15-463.15	27.85
n-octane	413.15-463.15	23.82
1-pentanol	413.15-463.15	30.94
1-propanol	413.15-463.15	28.26
tetrachloromethane	413.15-463.15	15.95

Polymer (B): **poly[tetrafluoroethylene-*co*-perfluoro(methyl vinyl ether)]** **2007FOS**
Characterization: 5 mol% perfluoro(methyl vinyl ether), T_m/K = 549, ρ = 2.121 g/cm^3, 35wt% crystallinity, Hyflon, Solvay Solexis
Comments: From sorption data not from IGC.

Solvent (A)	T-range/ K	$\Delta_M H_A^\infty$/ kJ/mol	$\Delta_{sol} H_{A(vap)}^\infty$/ kJ/mol
ethane	288.15-323.15	+1.9	−6.7
hexafluoroethane	288.15-323.15	−8.2	−19.9
n-hexane	288.15-323.15	+13.0	−19.1

Polymer (B): **poly[tetrafluoroethylene-*co*-perfluoro(methyl vinyl ether)]** **2006DEL**

Characterization: 49.3 mol% perfluoro(methyl vinyl ether), T_g/K = 265.2, ρ = 2.02 g/cm^3, DuPont Performance Elastomers, L.L.C., Wilmington, DE

Solvent (A)	T-range/ K	$\Delta_M H_A{}^\infty$/ kJ/mol	$\Delta_{sol} H_{A(vap)}{}^\infty$/ kJ/mol
n-heptane	318.15-373.15	12.5 ± 1.1	−22.4 ± 1.0
n-octane	318.15-373.15	8.3 ± 0.3	−30.4 ± 0.3
n-nonane	318.15-373.15	9.5 ± 0.2	−33.9 ± 0.2
n-decane	318.15-373.15	11.8 ± 0.2	−36.1 ± 0.3
methylcyclohexane	318.15-373.15	8.8 ± 0.5	−24.9 ± 0.5
benzene	318.15-373.15	8.9 ± 0.4	−23.4 ± 0.4
toluene	318.15-373.15	7.9 ± 0.3	−27.8 ± 0.3
n-perfluorohexane	318.15-373.15	−1.8 ± 0.4	−31.5 ± 0.4
n-perfluorooctane	318.15-373.15	−3.8 ± 0.1	−40.7 ± 0.3
perfluoromethylcyclohexane	318.15-373.15	−0.7 ± 0.2	−31.9 ± 0.2
perfluorobenzene	318.15-373.15	0.7 ± 0.5	−38.7 ± 0.2
perfluorotoluene	318.15-373.15	−7.5 ± 0.2	−32.3 ± 0.3
2,3,4,5,6-pentafluorotoluene	318.15-373.15	2.5 ± 0.2	−36.0 ± 0.2
4-fluorotoluene	318.15-373.15	6.9 ± 0.2	−30.1 ± 0.2

5.3. Table of systems where additional information on enthalpy effects in copolymer solutions can be found

Polymer (B)	Solvent (A)	Enthalpy	T-range	Ref.
Poly(ethylene-*co*-ethyl acrylate)				
	carbon dioxide	$\Delta_{sol}H_{A(vap)}^{\infty}$	423-473 K	2004ARE
Poly(ethylene oxide-*b*-propylene oxide-*b*-ethylene oxide)				
	water	$\Delta_M H$	298 K	2002THU
	water/surfactant	ΔH		2004DEL
	water/surfactant	ΔH		2004JAN
Poly(ethylene oxide-*co*-tetrahydrofuran)				
	2-propanol	$\Delta_M H$	303 K	1998STU
	tetrachloromethane	$\Delta_M H$	303 K	1998STU
	trichloromethane	$\Delta_M H$	303 K	1998STU
Poly(*N*-isopropylacryl-amide-*co*-acrylic acid)				
	water	ΔH		2006WEN
Poly(methyl methacrylate-*co*-methacrylic acid)				
	dibutyl phthalate	$\Delta_M H$	298 K	1991TAG
	didodecyl phthalate	$\Delta_M H$	298 K	1991TAG
	dioctyl phthalate	$\Delta_M H$	298 K	1991TAG
	dioctyl sebacate	$\Delta_M H$	298 K	1991TAG
	tetramethyl pyromellitate	$\Delta_M H$	298 K	1991TAG
	tricresyl phosphate	$\Delta_M H$	298 K	1991TAG
Poly(styrene oxide-*b*-ethylene oxide)				
	water/sodium dodecyl sulfate	ΔH	298 K	2005CAS

5.4. References

1991TAG Tager, A. A., Yushkova, S. M., Adamova, L. V., Kovylin, S. V., Berezov, L. V., Mozzhukhin, V. B., and Guzeev, V. V., Thermodynamics of interaction of methyl methacrylate-methacrylic acid copolymers with plasticizers and their mixtures (Russ.), *Vysokomol. Soedin., Ser. A*, 33, 357, 1991.

1998STU Stumbeck, M., Mischungsthermodynamik von Polyethern in Lösungsmitteln unterschiedlicher Polarität. Theoretische Modelle zur Berechnung von Enthalpie und freier Enthalpie von Mischungen, *Dissertation*, TU München, 1998.

2001DEM Demirelli, K., Kaya, I., and Coskun, M., 3,4-Dichlorobenzyl methacrylate and ethyl methacrylate system. Monomer reactivity ratios and determination of thermodynamic properties using inverse gas chromatography, *Polymer*, 42, 5181, 2001.

2001KA1 Kaya, I. and Demirelli, K., Study of some thermodynamic properties of poly(3,4-dichlorobenzyl methacrylate-*co*-ethyl methacrylate) using inverse gas chromatography, *J. Polym. Eng.*, 21, 1, 2001.

2001KA2 Kaya, I. and Demirelli, K., Determination of the thermodynamic properties of poly[2-(3-phenyl-3-methylcyclobutyl)-2-hydroxyethyl methacrylate-*co*-methacrylic acid] at infinite dilution by inverse gas chromatography, *Turk. J. Chem.*, 25, 11, 2001.

2002KAY Kaya, I., Ilter, Z., and Senol, D., Thermodynamic interactions and characterisation of poly[(glycidyl methacrylate-*co*-methyl, ethyl, butyl) methacrylate] by inverse gas chromatography, *Polymer*, 43, 6455, 2002.

2002PRI Price, G.J. and Shillcock, I.M., Inverse gas chromatography study of the thermodynamic behaviour of thermotropic low molar mass and polymeric liquid crystals, *Phys. Chem. Chem. Phys.*, 4, 5307, 2002.

2002RIG Righetti, M.C., Cardelli, C., Scalari, M., Tombari, E., Conti, G., Thermodynamics of mixing poly(vinyl chloride) and poly(ethylene-*co*-vinyl acetate), *Polymer*, 43, 5035, 2002.

2002THU Thurn, T., Couderc, S., Sidhu, J., Bloor, D.M., Penfold, J., Holzwarth, J.F., and Wyn Jones, E., Study of mixed micelles and interaction parameters for triblock copolymers of the type EO_m-PO_n-EO_m and ionic surfactants: Equilibrium and structure, *Langmuir*, 18, 9267, 2002.

2003ACI Acikses, A., Kaya, I., and Ilter, Z., Study of some thermodynamic properties of poly[(2-phenyl-1,3-dioxolane-4-yl)methyl methacrylate-*co*-butyl methacrylate] by inverse gas chromatography, *Polym.-Plast. Technol. Eng.*, 42, 431, 2003.

2003CAR Cardelli, C., Conti, G., Gianni, P., and Porta, R., Blend formation between homo- and copolymers at 298.15 K. PMMA-SAN blends, *J. Therm. Anal. Calorim.*, 71, 353, 2003.

2004ARE Areerat, S., Funami, E., Hayata, Y., Nakagawa, D., and Ohshima, M., Measurement and prediction of diffusion coefficients of supercritical CO_2 in molten polymers, *Polym. Eng. Sci.*, 44, 1915, 2004.

2004DEL DeLisi, R., Lazzara, G., Milioto, S., and Muratore, N., Thermodynamics of aqueous poly(ethylene oxide)-poly(propylene oxide)-poly(ethylene oxide)/surfactant mixtures. Effect of the copolymer molecular weight and the surfactant alkyl chain length, *J. Phys. Chem. B*, 108, 18214, 2004.

2004ILT Ilter, Z., Kaya, I., and Acikses, A., Determination of poly[(2-phenyl-1,3-dioxolane-4-yl)methyl methacrylate-*co*-glycidyl methacrylate]-probe interactions by inverse gas chromatography, *Polym.-Plast. Technol. Eng.*, 43, 229, 2004.

2004JAN Jansson, J., Schillen, K., Olofsson, G., Cardoso da Silva, R., and Loh, W., The interaction between PEO-PPO-PEO triblock copolymers and ionic surfactants in aqueous solution studied using light scattering and calorimetry, *J. Phys. Chem. B*, 108, 82, 2004.

2005ACI Acikses, A., Kaya, I., Sezek, U., and Kirilmis, C., Synthesis, characterization and thermodynamic properties of poly(3-mesityl-2-hydroxypropyl methacrylate-*co*-*N*-vinyl-2-pyrrolidone), *Polymer*, 46, 11322, 2005.

2005CAS Castro, E., Taboada, P., and Mosquera, V., Behavior of a styrene oxide-ethylene oxide diblock copolymer/surfactant system: A thermodynamic and spectroscopy study, *J. Phys. Chem. B*, 109, 5592, 2005.

2006BEL Belov, N., Yampolskii, Yu., and Coughlin, M.C., Thermodynamics of sorption in an amorphous perfluorinated rubber studied by inverse gas chromatography, *Macromolecules*, 39, 1797, 2006.

2006KAR Karagöz, M.H., Zorer, O.S., and Ilter, Z., Analysis of physical and thermodynamic properties of poly(2-phenyl-1,3-dioxolane-4-yl-methyl-methacrylate-*co*-styrene) with inverse gas chromatography, *Polym.-Plast. Technol. Eng.*, 45, 785, 2006.

2006SAK Sakar, D., Erdogan, T., Cankurtaran, O., Hizal, G., Karaman, F., and Tunca, U., Physicochemical characterization of poly(*tert*-butyl acrylate-*b*-methyl methacrylate) prepared with atom transfer radical polymerization by inverse gas chromatography, *Polymer*, 47, 132, 2006.

2006TEM Temüz, M.M., Coskun, M., and Acikses, A., Determination of some thermodynamic parameters of poly(styrene-*graft*-ethyl methacrylate) using inverse gas chromatography, *J. Macromol. Sci.: Part A: Pure Appl. Chem.*, 43, 609, 2006.

2006WEN Weng, Y., Ding, Y., and Zhang, G., Microcalorimetric investigation on the lower critical solution temperature behavior of *N*-isopropylacrylamide-*co*-acrylic acid copolymer in aqueous solution, *J. Phys. Chem. B*, 110, 11813, 2006.

2007COS Coskun, S., Cankurtaran, O., Eran, B.B., and Sarac, A., A Study of some equation-of-state parameters of poly(methylhydrosiloxane-*co*-dimethylsiloxane) with some solvents by gas chromatography, *J. Appl. Polym. Sci.*, 104, 1627, 2007.

2007FOS Fossati, P., Sanguineti, A, DeAngelis, M.G., Baschetti, M.G., Doghieri, F., and Sarti, G.C., Gas solubility and permeability in MFA, *J. Polym. Sci.: Part B: Polym. Phys.*, 45, 1637, 2007.

2007ZOU Zou, Q.-C. and WU, L.-M., Inverse gas chromatographic characterization of triblock copolymer of polystyrene-*b*-poly(ethylene oxide)-*b*-polystyrene., *J. Polym. Sci.: Part B: Polym. Phys.*, 45, 2015, 2007.

2008AYD Aydin, S., Erdogan, T., Sakar, D., Hizal, G., Cankurtaran, O., Tunca, U., and Karaman, F., Detection of microphase separation in poly(*tert*-butyl acrylate-*b*-methyl methacrylate) synthesized via atom transfer radical polymerization by inverse gas chromatography, *Eur. Polym. J.*, 44, 2115, 2008.

2008YAZ Yazici, D.T., Askin, A., and Bütün, V., Thermodynamic interactions of water-soluble homopolymers and double-hydrophilic diblock copolymer, *J. Chem. Thermodyn.*, 40, 353, 2008.

2009KAR Karagöz, M.H., Erge, H., and Ilter, Z., Physical and thermodynamic properties of poly(2-phenyl-1,3-dioxolane-4-yl-methyl-methacrylate-*co*-ethyl methacrylate) polymer with inverse gas chromatography, *Asian J. Chem.*, 21, 4032, 2009.

2009OVE Ovejero, G., Perez, P., Romero, M.D., Diaz, I., and Diez, E., SEBS triblock copolymer-solvent interaction parameters from inverse gas chromatography measurements, *Eur. Polym. J.*, 45, 590, 2009.

6. PVT DATA OF COPOLYMERS AND SOLUTIONS

6.1. PVT data of copolymers

Polymer (B): **poly(butylene succinate-*co*-butylene adipate)** **2000SA1**

Characterization: M_n/g.mol^{-1} = 53000, M_w/g.mol^{-1} = 180000, 20.0 mol% adipate, T_g/K = 231, T_m/K = 365, 25 wt% crystallinity, linear, random copolymer, Showa Denko K.K., Kawasaki, Japan

P/MPa	*T*/K					
	393.7	413.7	433.5	453.6	473.1	493.1
	V_{spec}/cm^{3}g^{-1}					
0.1	0.8968	0.9092	0.9219	0.9349	0.9489	0.9632
10	0.8918	0.9037	0.9158	0.9282	0.9416	0.9547
20	0.8872	0.8983	0.9100	0.9219	0.9348	0.9467
50	0.8744	0.8848	0.8950	0.9053	0.9171	0.9271
100	0.8572	0.8665	0.8751	0.8841	0.8946	0.9029
150	0.8431	0.8512	0.8592	0.8671	0.8764	0.8839
200	0.8310	0.8381	0.8457	0.8526	0.8612	0.8679

Tait equation parameter functions:

Range of data: *T*/K = 393-493, *P*/MPa = 0.1-200

$$V(P/\text{MPa}, T/\text{K}) = V(0, T/\text{K})\{1 - C^*\ln[1 + (P/\text{MPa})/B(T/\text{K})]\}$$

with $C = 0.0894$

$V(0,T/\text{K})$/cm^{3}g^{-1}	$B(T/\text{K})$/MPa
$0.6775 \exp(7.110\ 10^{-4}\ T)$	$903.5 \exp(-4.441\ 10^{-3}\ T)$

Polymer (B): **poly(ethylene-*co*-1-butene)** **2007SAT**
Characterization: M_n/g.mol^{-1} = 53000, M_w/g.mol^{-1} = 110000, 10 mol% 1-butene, 12 wt% crystallinity, unspecified industrial source

T/K	*P*/MPa 0.1	10	20	50	100	150	200
	V_{spec}/cm^{3}g^{-1}						
373.8	1.236	1.226	1.217	1.194	1.165		
393.9	1.254	1.243	1.233	1.208	1.176	1.152	1.131
414.0	1.271	1.260	1.248	1.222	1.188	1.162	1.140
433.9	1.289	1.276	1.264	1.235	1.199	1.172	1.149
453.7	1.307	1.293	1.280	1.249	1.210	1.181	1.158
473.6	1.326	1.310	1.296	1.262	1.221	1.191	1.166
493.5	1.345	1.328	1.312	1.275	1.232	1.200	1.175

Polymer (B): **poly(ethylene-*co*-1-butene)** **2000CAP**
Characterization: M_n/g.mol^{-1} = 24900, M_w/g.mol^{-1} = 120000, 3.80 mol% 1-butene, ρ = 0.9190 g/cm^3, Nova Chemicals, Calgary, Canada

Tait equation parameter functions:
Range of data: *T*/K = 453-513, *P*/MPa = 0.1-200

$$V(P/\text{MPa}, T/\text{K}) = V(0, T/\text{K})\{1 - C^*\ln[1 + (P/\text{MPa})/B(T/\text{K})]\}$$
with $C = 0.0894$ and $\theta = T/\text{K} - 273.15$

$V(0, \theta/°\text{C})$/cm^{3}g^{-1}	$B(\theta/°\text{C})$/MPa
1.141 exp(7.39 10^{-4} θ)	199.8 exp(−5.01 10^{-3} θ)

Polymer (B): **poly(ethylene-*co*-1-butene)** **2000CAP**
Characterization: M_n/g.mol^{-1} = 24200, M_w/g.mol^{-1} = 98700, 4.03 mol% 1-butene, ρ = 0.9194 g/cm^3, Nova Chemicals, Calgary, Canada

Tait equation parameter functions:
Range of data: *T*/K = 453-513, *P*/MPa = 0.1-200

$$V(P/\text{MPa}, T/\text{K}) = V(0, T/\text{K})\{1 - C^*\ln[1 + (P/\text{MPa})/B(T/\text{K})]\}$$
with $C = 0.0894$ and $\theta = T/\text{K} - 273.15$

$V(0, \theta/°\text{C})$/cm^{3}g^{-1}	$B(\theta/°\text{C})$/MPa
1.137 exp(7.46 10^{-4} θ)	194.4 exp(−4.93 10^{-3} θ)

Polymer (B): **poly(ethylene-*co*-1-hexene)** **2007SAT**
Characterization: M_n/g.mol^{-1} = 28000, M_w/g.mol^{-1} = 63000, 3.5 mol% 1-hexene, 51 wt% crystallinity, unspecified industrial source

T/K	*P*/MPa 0.1	10	20	50	100	150	200
	V_{spec}/cm^{3}g^{-1}						
414.0	1.274	1.263	1.252	1.226			
433.9	1.293	1.280	1.270	1.240	1.204	1.177	
453.8	1.311	1.297	1.284	1.253	1.215	1.186	1.163
473.7	1.329	1.314	1.300	1.266	1.226	1.196	1.171
493.6	1.349	1.331	1.315	1.279	1.236	1.205	1.179

Polymer (B): **poly(ethylene-*co*-1-hexene)** **2000CAP**
Characterization: M_n/g.mol^{-1} = 43000, M_w/g.mol^{-1} = 94000, 2.56 mol% 1-hexene, ρ = 0.9194 g/cm^3, Nova Chemicals, Calgary, Canada

Tait equation parameter functions:
Range of data: *T*/K = 453-513, *P*/MPa = 0.1-200

$$V(P/\text{MPa}, T/\text{K}) = V(0, T/\text{K})\{1 - C{*}\ln[1 + (P/\text{MPa})/B(T/\text{K})]\}$$
with C = 0.0894 and θ = *T*/K – 273.15

$V(0, \theta/°\text{C})$/cm^{3}g^{-1}	$B(\theta/°\text{C})$/MPa
$1.138 \exp(7.44\ 10^{-4}\ \theta)$	$196.3 \exp(-4.99\ 10^{-3}\ \theta)$

Polymer (B): **poly(ethylene-*co*-1-hexene)** **2000CAP**
Characterization: M_n/g.mol^{-1} = 44000, M_w/g.mol^{-1} = 98000, 3.08 mol% 1-hexene, ρ = 0.9192 g/cm^3, Nova Chemicals, Calgary, Canada

Tait equation parameter functions:
Range of data: *T*/K = 453-513, *P*/MPa = 0.1-200

$$V(P/\text{MPa}, T/\text{K}) = V(0, T/\text{K})\{1 - C{*}\ln[1 + (P/\text{MPa})/B(T/\text{K})]\}$$
with C = 0.0894 and θ = *T*/K – 273.15

$V(0, \theta/°\text{C})$/cm^{3}g^{-1}	$B(\theta/°\text{C})$/MPa
$1.138 \exp(7.42\ 10^{-4}\ \theta)$	$199.5 \exp(-5.08\ 10^{-3}\ \theta)$

Polymer (B): **poly(ethylene-*co*-1-hexene)** **2000CAP**
Characterization: M_n/g.mol^{-1} = 36000, M_w/g.mol^{-1} = 111300, 3.77 mol% 1-hexene, ρ = 0.9234 g/cm^3, Nova Chemicals, Calgary, Canada

Tait equation parameter functions:
Range of data: T/K = 453-513, P/MPa = 0.1-200

$$V(P/\text{MPa}, T/\text{K}) = V(0, T/\text{K})\{1 - C^*\ln[1 + (P/\text{MPa})/B(T/\text{K})]\}$$
with $C = 0.0894$ and $\theta = T/\text{K} - 273.15$

$V(0, \theta/°\text{C})/\text{cm}^3\text{g}^{-1}$	$B(\theta/°\text{C})/\text{MPa}$
$1.134 \exp(7.54\ 10^{-4}\ \theta)$	$198.7 \exp(-5.08\ 10^{-3}\ \theta)$

Polymer (B): **poly(ethylene-*co*-1-hexene)** **2000CAP**
Characterization: M_n/g.mol^{-1} = 30000, M_w/g.mol^{-1} = 111000, 3.94 mol% 1-hexene, ρ = 0.9208 g/cm^3, Nova Chemicals, Calgary, Canada

Tait equation parameter functions:
Range of data: T/K = 453-513, P/MPa = 0.1-200

$$V(P/\text{MPa}, T/\text{K}) = V(0, T/\text{K})\{1 - C^*\ln[1 + (P/\text{MPa})/B(T/\text{K})]\}$$
with $C = 0.0894$ and $\theta = T/\text{K} - 273.15$

$V(0, \theta/°\text{C})/\text{cm}^3\text{g}^{-1}$	$B(\theta/°\text{C})/\text{MPa}$
$1.136 \exp(7.52\ 10^{-4}\ \theta)$	$194.5 \exp(-4.99\ 10^{-3}\ \theta)$

Polymer (B): **poly(ethylene-*co*-norbornene)** **2006BLO**
Characterization: M_n/g.mol^{-1} = 56000, M_w/g.mol^{-1} = 101000, T_g/K = 415, 51.8 mol% norbornene, commercial sample

Tait equation parameter functions:
Range of data: T/K = 415-550, P/MPa = 0.1-200

$$V(P/\text{MPa}, T/\text{K}) = V(0, T/\text{K})\{1 - C^*\ln[1 + (P/\text{MPa})/B(T/\text{K})]\}$$
with $C = 0.0894$ and $\theta = T/\text{K} - 273.15$

$V(0, \theta/°\text{C})/\text{cm}^3\text{g}^{-1}$	$B(\theta/°\text{C})/\text{MPa}$
$0.927 + 5.5\ 10^{-4}\ \theta - 3.0\ 10^{-8}\ \theta^2$	$238.0 \exp(-0.3\ 10^{-2}\ \theta)$

Polymer (B): **poly(ethylene-*co*-1-octene)** **2007SAT**

Characterization: M_n/g.mol^{-1} = 106000, M_w/g.mol^{-1} = 196000, 0.9 mol% 1-octene, 17.6 wt% crystallinity, unspecified industrial source

T/K	*P*/MPa 0.1	10	20	50	100	150	200
	V_{spec}/cm^{3}g^{-1}						
373.9	1.238	1.228	1.219				
394.0	1.258	1.247	1.237	1.212	1.180		
414.1	1.275	1.265	1.253	1.226	1.191	1.165	1.143
434.0	1.293	1.281	1.269	1.240	1.203	1.175	1.152
453.8	1.311	1.298	1.285	1.253	1.214	1.185	1.161
473.6	1.330	1.316	1.302	1.267	1.226	1.195	1.170
493.4	1.349	1.333	1.318	1.281	1.237	1.204	1.178

Polymer (B): **poly(ethylene-*co*-1-octene)** **2000CAP**

Characterization: M_n/g.mol^{-1} = 25900, M_w/g.mol^{-1} = 114000, 2.80 mol% 1-octene, ρ = 0.9212 g/cm^3, Nova Chemicals, Calgary, Canada

Tait equation parameter functions:
Range of data: *T*/K = 453-513, *P*/MPa = 0.1-200

$$V(P/\text{MPa}, T/\text{K}) = V(0, T/\text{K})\{1 - C^*\ln[1 + (P/\text{MPa})/B(T/\text{K})]\}$$

with C = 0.0894 and θ = *T*/K – 273.15

$V(0, \theta/°C)$/cm^{3}g^{-1}	$B(\theta/°C)$/MPa
1.140 exp(7.56 10^{-4} θ)	203.7 exp(–5.21 10^{-3} θ)

Polymer (B): **poly(ethylene-*co*-1-octene)** **2000CAP**

Characterization: M_n/g.mol^{-1} = 38000, M_w/g.mol^{-1} = 70000, 3.20 mol% 1-octene, ρ = 0.9180 g/cm^3, Nova Chemicals, Calgary, Canada

Tait equation parameter functions:
Range of data: *T*/K = 453-513, *P*/MPa = 0.1-200

$$V(P/\text{MPa}, T/\text{K}) = V(0, T/\text{K})\{1 - C^*\ln[1 + (P/\text{MPa})/B(T/\text{K})]\}$$

with C = 0.0894 and θ = *T*/K – 273.15

$V(0, \theta/°C)$/cm^{3}g^{-1}	$B(\theta/°C)$/MPa
1.144 exp(7.35 10^{-4} θ)	195.2 exp(–4.91 10^{-3} θ)

Polymer (B): **poly(ethylene-*co*-1-octene)** **2000CAP**
Characterization: M_n/g.mol^{-1} = 17000, M_w/g.mol^{-1} = 106000, 3.20 mol% 1-octene, ρ = 0.9200 g/cm^3, Nova Chemicals, Calgary, Canada

Tait equation parameter functions:
Range of data: T/K = 453-513, P/MPa = 0.1-200

$$V(P/\text{MPa}, T/\text{K}) = V(0, T/\text{K})\{1 - C{*}\ln[1 + (P/\text{MPa})/B(T/\text{K})]\}$$
with C = 0.0894 and θ = T/K – 273.15

$V(0, \theta/°\text{C})/\text{cm}^3\text{g}^{-1}$	$B(\theta/°\text{C})/\text{MPa}$
$1.145 \exp(7.55\ 10^{-4}\ \theta)$	$198.3 \exp(-5.04\ 10^{-3}\ \theta)$

Polymer (B): **poly(ethylene-*co*-1-octene)** **2000CAP**
Characterization: M_n/g.mol^{-1} = 22000, M_w/g.mol^{-1} = 53000, 5.00 mol% 1-octene, ρ = 0.9070 g/cm^3, Nova Chemicals, Calgary, Canada

Tait equation parameter functions:
Range of data: T/K = 453-513, P/MPa = 0.1-200

$$V(P/\text{MPa}, T/\text{K}) = V(0, T/\text{K})\{1 - C{*}\ln[1 + (P/\text{MPa})/B(T/\text{K})]\}$$
with C = 0.0894 and θ = T/K – 273.15

$V(0, \theta/°\text{C})/\text{cm}^3\text{g}^{-1}$	$B(\theta/°\text{C})/\text{MPa}$
$1.148 \exp(7.26\ 10^{-4}\ \theta)$	$194.1 \exp(-4.91\ 10^{-3}\ \theta)$

Polymer (B): **poly(ethylene-*co*-propylene)** **2007SAT**
Characterization: M_n/g.mol^{-1} = 87000, M_w/g.mol^{-1} = 190000, 19 mol% propene, 10 wt% crystallinity, unspecified industrial source

T/K	P/MPa						
	0.1	10	20	50	100	150	200
	V_{spec}/cm^{3}g^{-1}						
333.7	1.200	1.192	1.184	1.165			
353.8	1.217	1.209	1.200	1.179	1.151	1.130	1.111
374.0	1.234	1.225	1.215	1.192	1.163	1.140	1.120
394.0	1.252	1.241	1.230	1.206	1.174	1.150	1.129
414.2	1.269	1.257	1.246	1.219	1.185	1.159	1.138
434.1	1.287	1.274	1.262	1.232	1.196	1.169	1.147
454.0	1.306	1.291	1.278	1.246	1.207	1.179	1.155
473.9	1.324	1.308	1.294	1.259	1.218	1.188	1.164
493.8	1.343	1.326	1.309	1.273	1.229	1.197	1.172

Polymer (B): **poly(ethylene-*co*-propylene)** **2000TOM**

Characterization: M_n/g.mol^{-1} = 71000, M_w/g.mol^{-1} = 141000, 22.0 wt% propene, random copolymer, ρ = 0.88 g/cm^3, Exxelor PE805, Exxon Chemical Company

P/MPa = 0.1		40.0		80.0		120.0		160.0	
T/K	V_{spec}/cm^{3}g^{-1}	*T*/K	V_{spec}/cm^{3}g^{-1}	*T*/K	V_{spec}/cm^{3}g^{-1}	*T*/K	V_{spec}/cm^{3}g^{-1}	*T*/K	V_{spec}/cm^{3}g^{-1}
432.40	1.2876	433.81	1.2441	434.80	1.2143	435.38	1.1906	435.85	1.1705
442.35	1.2972	443.54	1.2512	444.67	1.2204	445.52	1.1964	446.10	1.1760
452.65	1.3076	454.25	1.2587	455.14	1.2264	455.81	1.2012	456.06	1.1804
462.52	1.3170	463.94	1.2657	464.90	1.2324	465.68	1.2069	466.30	1.1857
472.70	1.3269	474.33	1.2733	475.20	1.2387	476.09	1.2125	476.64	1.1908
482.85	1.3368	484.50	1.2812	485.35	1.2448	486.10	1.2175	486.48	1.1952
492.17	1.3460	493.92	1.2882	494.86	1.2509	495.70	1.2229	496.38	1.2002
502.09	1.3564	503.82	1.2953	504.95	1.2569	505.77	1.2282	506.52	1.2052
512.71	1.3673	514.46	1.3033	515.61	1.2633	516.45	1.2339	517.13	1.2099
522.36	1.3776	524.17	1.3101	525.25	1.2695	526.15	1.2389	527.06	1.2146
532.48	1.3884	534.28	1.3173	535.25	1.2752	536.09	1.2438	536.93	1.2189
542.04	1.3996	544.15	1.3250	545.32	1.2816	546.26	1.2495	547.13	1.2239
552.07	1.4109	554.08	1.3321	555.28	1.2875	556.17	1.2550	556.91	1.2284
561.96	1.4227	563.88	1.3396	565.02	1.2930	565.86	1.2596	566.70	1.2325
571.55	1.4340	573.90	1.3475	575.15	1.2997	576.04	1.2654	576.93	1.2378

Polymer (B): **poly[(ethylene-*alt*-propylene)-*co*-carbon monoxide]** **2001PRI**

Characterization: M_w/g.mol^{-1} = 192000, 50 wt% carbon monoxide, synthesized in the laboratory

Tait equation parameter functions:
Range of data: *T*/K = 293-423, *P*/MPa = 30-100

$$V(P/\text{MPa}, \theta/°\text{C})/V(0, \theta/°\text{C}) = 1 - 0.0894 \ln\{1 + (P/\text{MPa})/[92.5 \exp(-3.38\ 10^{-3}\ \theta)]\}$$

Polymer (B): **poly[(ethylene-*alt*-propylene)-*co*-carbon monoxide]** **2001PRI**

Characterization: M_w/g.mol^{-1} = 397000, 40 wt% carbon monoxide, synthesized in the laboratory

Tait equation parameter functions:
Range of data: *T*/K = 293-423, *P*/MPa = 30-100

$$V(P/\text{MPa}, \theta/°\text{C})/V(0, \theta/°\text{C}) = 1 - 0.0894 \ln\{1 + (P/\text{MPa})/[104.2 \exp(-1.93\ 10^{-3}\ \theta)]\}$$

Polymer (B): **poly[(ethylene-*alt*-propylene)-*co*-carbon monoxide] 2001PRI**
Characterization: M_w/g.mol^{-1} = 343000, 30 wt% carbon monoxide, synthesized in the laboratory

Tait equation parameter functions:
Range of data: T/K = 293-423, P/MPa = 30-100

$V(P/\text{MPa}, \theta/°\text{C})/V(0, \theta/°\text{C}) = 1 - 0.0894 \ln\{1 + (P/\text{MPa})/[120.8 \exp(-3.08\ 10^{-3}\ \theta)]\}$

Polymer (B): **poly(ethylene-*co*-propylene-*co*-ethylidene norbornene) 2000TOM**
Characterization: M_n/g.mol^{-1} = 117000, M_w/g.mol^{-1} = 304000, 41.4 wt% propene, 50.2 wt% ethene, 8.7 wt% ethylidene norbornene, random terpolymer, ρ = 0.85 g/cm^3, Royaltuf 490, Uniroyal Chemical

P/MPa = 0.1		40.0		80.0		120.0		160.0	
T/K	V_{spec}/cm^{3}g^{-1}	T/K	V_{spec}/cm^{3}g^{-1}	T/K	V_{spec}/cm^{3}g^{-1}	T/K	V_{spec}/cm^{3}g^{-1}	T/K	V_{spec}/cm^{3}g^{-1}
431.71	1.2543	433.06	1.2140	434.22	1.1860	435.01	1.1639	435.63	1.1436
441.84	1.2631	443.20	1.2210	444.39	1.1918	445.12	1.1690	445.77	1.1499
451.89	1.2724	453.27	1.2278	454.37	1.1975	455.06	1.1740	455.66	1.1545
461.86	1.2814	463.25	1.2346	464.41	1.2035	465.25	1.1794	465.88	1.1593
472.12	1.2913	473.65	1.2420	474.74	1.2098	475.60	1.1848	476.36	1.1642
482.15	1.3018	483.90	1.2494	485.02	1.2160	485.86	1.1905	486.58	1.1693
491.92	1.3110	493.73	1.2564	494.94	1.2217	495.86	1.1955	496.56	1.1739
501.87	1.3204	503.55	1.2628	504.73	1.2268	505.49	1.1999	506.36	1.1779
511.72	1.3299	513.70	1.2703	514.76	1.2325	515.66	1.2048	516.38	1.1822
522.49	1.3412	524.16	1.2778	525.02	1.2382	525.86	1.2095	526.70	1.1865
532.19	1.3513	534.00	1.2850	535.07	1.2444	536.04	1.2150	536.89	1.1917
542.23	1.3629	544.16	1.2931	545.38	1.2511	546.10	1.2208	547.02	1.1964
552.11	1.3753	554.15	1.3017	555.23	1.2589	556.14	1.2272	557.03	1.2022
562.12	1.3896	564.26	1.3111	565.62	1.2667	566.52	1.2340	567.31	1.2084
571.26	1.4048	573.62	1.3214	575.08	1.2753	575.92	1.2414	576.86	1.2151

Polymer (B): **poly(ethylene-*co*-vinyl alcohol)** **2007FU2**
Characterization: 15 mol% ethene, ρ(293.15 K) = 1.2522 g/cm^3, T_g/K = 341,
T_m/K = 484, Kuraray Specialities, Japan

P/MPa	*T*/K 483.55	493.80	503.38	514.05	523.80	534.11
	V_{spec}/cm^{3}g^{-1}					
0.1	0.8783	0.8852	0.8908	0.8964	0.9034	0.9133
10	0.8744	0.8810	0.8863	0.8918	0.8987	0.9083
20	0.8706	0.8769	0.8818	0.8874	0.8940	0.9036
30	0.8671	0.8734	0.8783	0.8837	0.8901	0.8994
40	0.8640	0.8701	0.8749	0.8801	0.8865	0.8958
50	0.8609	0.8670	0.8718	0.8768	0.8831	0.8921
60	0.8581	0.8642	0.8686	0.8736	0.8799	0.8887
70	0.8551	0.8614	0.8659	0.8707	0.8768	0.8854
80	0.8523	0.8587	0.8629	0.8679	0.8738	0.8823
90	0.8492	0.8562	0.8604	0.8651	0.8709	0.8795
100	0.8463	0.8536	0.8577	0.8624	0.8681	0.8765
110	0.8433	0.8512	0.8552	0.8600	0.8655	0.8740
120	0.8404	0.8488	0.8528	0.8574	0.8628	0.8712
130	0.8374	0.8464	0.8505	0.8550	0.8604	0.8687
140	0.8341	0.8443	0.8482	0.8526	0.8581	0.8661
150	0.8313	0.8421	0.8461	0.8505	0.8556	0.8638
160	0.8284	0.8400	0.8439	0.8483	0.8534	0.8615
170	0.8258	0.8379	0.8418	0.8461	0.8513	0.8593
180	0.8234	0.8358	0.8398	0.8440	0.8491	0.8572
190	0.8208	0.8337	0.8379	0.8419	0.8470	0.8550
200	0.8185	0.8316	0.8360	0.8400	0.8451	0.8531

Polymer (B): **poly(ethylene-*co*-vinyl alcohol)** **2007FU2**
Characterization: 27 mol% ethene, ρ(293.15 K) = 1.1959 g/cm^3, T_g/K = 339,
T_m/K = 466, Kuraray Specialities, Japan

P/MPa	*T*/K 466.52	477.39	487.28	497.68	508.00	518.11	528.90	539.09
	V_{spec}/cm^{3}g^{-1}							
0.1	0.9234	0.9290	0.9340	0.9392	0.9453	0.9576	0.9702	0.9821
10	0.9189	0.9243	0.9292	0.9341	0.9402	0.9522	0.9641	0.9754
20	0.9146	0.9196	0.9244	0.9292	0.9352	0.9469	0.9584	0.9691
30	0.9109	0.9158	0.9204	0.9250	0.9309	0.9425	0.9534	0.9638
40	0.9075	0.9121	0.9167	0.9212	0.9271	0.9385	0.9488	0.9589
50	0.9042	0.9088	0.9132	0.9175	0.9235	0.9346	0.9445	0.9543

continued

continued

60	0.9010	0.9055	0.9098	0.9141	0.9200	0.9309	0.9405	0.9500
70	0.8981	0.9024	0.9066	0.9107	0.9167	0.9273	0.9366	0.9458
80	0.8951	0.8994	0.9035	0.9075	0.9136	0.9239	0.9330	0.9419
90	0.8924	0.8965	0.9006	0.9045	0.9105	0.9208	0.9295	0.9382
100	0.8897	0.8938	0.8977	0.9016	0.9076	0.9175	0.9260	0.9346
110	0.8872	0.8910	0.8949	0.8987	0.9048	0.9146	0.9228	0.9311
120	0.8846	0.8885	0.8923	0.8960	0.9020	0.9117	0.9197	0.9278
130	0.8821	0.8860	0.8897	0.8934	0.8994	0.9088	0.9166	0.9246
140	0.8798	0.8835	0.8871	0.8909	0.8968	0.9060	0.9137	0.9215
150	0.8772	0.8812	0.8847	0.8883	0.8944	0.9033	0.9109	0.9186
160	0.8746	0.8789	0.8824	0.8858	0.8919	0.9008	0.9081	0.9156
170	0.8717	0.8766	0.8801	0.8835	0.8895	0.8982	0.9055	0.9129
180	0.8685	0.8745	0.8779	0.8813	0.8873	0.8958	0.9028	0.9101
190	0.8650	0.8724	0.8757	0.8790	0.8850	0.8933	0.9003	0.9075
200	0.8616	0.8703	0.8736	0.8769	0.8828	0.8910	0.8978	0.9049

Polymer (B): **poly(ethylene-*co*-vinyl alcohol)** **2007FU2**
Characterization: 32 mol% ethene, ρ(293.15 K) = 1.1810 g/cm^3, T_g/K = 335, T_m/K = 457, Kuraray Specialities, Japan

P/MPa	T/K 467.46	477.59	488.37	498.75	508.97	519.72	529.57	539.30
	V_{spec}/cm^{3}g^{-1}							
0.1	0.9363	0.9420	0.9479	0.9536	0.9599	0.9673	0.9753	0.9841
10	0.9318	0.9372	0.9428	0.9483	0.9543	0.9613	0.9691	0.9775
20	0.9273	0.9325	0.9378	0.9432	0.9488	0.9555	0.9632	0.9712
30	0.9235	0.9286	0.9335	0.9388	0.9444	0.9507	0.9582	0.9660
40	0.9200	0.9249	0.9297	0.9348	0.9401	0.9464	0.9536	0.9612
50	0.9164	0.9213	0.9260	0.9308	0.9361	0.9423	0.9493	0.9566
60	0.9132	0.9179	0.9225	0.9272	0.9323	0.9383	0.9451	0.9524
70	0.9100	0.9147	0.9190	0.9238	0.9287	0.9347	0.9413	0.9483
80	0.9070	0.9115	0.9159	0.9204	0.9253	0.9311	0.9375	0.9445
90	0.9042	0.9086	0.9128	0.9173	0.9220	0.9277	0.9340	0.9407
100	0.9014	0.9057	0.9098	0.9142	0.9190	0.9245	0.9306	0.9371
110	0.8988	0.9028	0.9069	0.9112	0.9158	0.9213	0.9274	0.9337
120	0.8962	0.9002	0.9042	0.9084	0.9129	0.9183	0.9242	0.9304
130	0.8936	0.8976	0.9015	0.9057	0.9100	0.9153	0.9212	0.9273
140	0.8911	0.8951	0.8989	0.9031	0.9073	0.9125	0.9182	0.9243
150	0.8888	0.8926	0.8964	0.9005	0.9047	0.9097	0.9154	0.9213
160	0.8866	0.8903	0.8940	0.8980	0.9021	0.9071	0.9126	0.9184
170	0.8842	0.8880	0.8916	0.8955	0.8996	0.9045	0.9099	0.9157
180	0.8820	0.8856	0.8893	0.8931	0.8972	0.9020	0.9073	0.9130
190	0.8799	0.8835	0.8871	0.8907	0.8947	0.8996	0.9047	0.9103
200	0.8778	0.8813	0.8849	0.8885	0.8925	0.8972	0.9023	0.9079

Polymer (B): **poly(ethylene-*co*-vinyl alcohol)** **2007FU2**
Characterization: 38 mol% ethene, ρ(293.15 K) = 1.1690 g/cm^3, T_g/K = 332, T_m/K = 450, Kuraray Specialities, Japan

P/MPa	T/K 467.82	477.75	488.37	499.04	509.02	519.75	529.66	539.20
	V_{spec}/cm^{3}g^{-1}							
0.1	0.9530	0.9585	0.9643	0.9705	0.9769	0.9858	0.9963	1.0071
10	0.9482	0.9535	0.9590	0.9649	0.9711	0.9798	0.9899	1.0002
20	0.9434	0.9486	0.9539	0.9595	0.9656	0.9740	0.9837	0.9936
30	0.9394	0.9444	0.9495	0.9549	0.9608	0.9690	0.9784	0.9880
40	0.9355	0.9403	0.9452	0.9505	0.9563	0.9642	0.9733	0.9828
50	0.9318	0.9365	0.9413	0.9464	0.9521	0.9598	0.9686	0.9777
60	0.9283	0.9329	0.9376	0.9425	0.9481	0.9556	0.9642	0.9730
70	0.9250	0.9294	0.9339	0.9388	0.9443	0.9515	0.9600	0.9686
80	0.9218	0.9261	0.9305	0.9353	0.9406	0.9477	0.9560	0.9644
90	0.9187	0.9229	0.9272	0.9318	0.9372	0.9441	0.9521	0.9603
100	0.9156	0.9199	0.9241	0.9286	0.9337	0.9407	0.9485	0.9565
110	0.9128	0.9170	0.9211	0.9253	0.9306	0.9373	0.9449	0.9528
120	0.9100	0.9139	0.9181	0.9224	0.9275	0.9341	0.9416	0.9492
130	0.9073	0.9112	0.9152	0.9194	0.9244	0.9310	0.9383	0.9458
140	0.9047	0.9086	0.9125	0.9167	0.9216	0.9280	0.9353	0.9425
150	0.9021	0.9060	0.9099	0.9138	0.9188	0.9251	0.9322	0.9394
160	0.8998	0.9034	0.9073	0.9112	0.9160	0.9223	0.9292	0.9363
170	0.8974	0.9010	0.9047	0.9086	0.9133	0.9196	0.9263	0.9333
180	0.8951	0.8986	0.9023	0.9062	0.9108	0.9170	0.9236	0.9303
190	0.8927	0.8963	0.8999	0.9037	0.9083	0.9144	0.9209	0.9276
200	0.8906	0.8941	0.8977	0.9013	0.9058	0.9118	0.9183	0.9248

Polymer (B): **poly(ethylene-*co*-vinyl alcohol)** **2007FU2**
Characterization: 44 mol% ethene, ρ(293.15 K) = 1.1359 g/cm^3, T_g/K = 327, T_m/K = 439, Kuraray Specialities, Japan

P/MPa	T/K 450.10	460.43	471.16	481.64	492.41	502.47	512.67	523.20
	V_{spec}/cm^{3}g^{-1}							
0.1	0.9751	0.9813	0.9879	0.9945	1.0012	1.0075	1.0141	1.0214
10	0.9702	0.9762	0.9825	0.9888	0.9952	1.0013	1.0077	1.0145
20	0.9655	0.9713	0.9774	0.9834	0.9895	0.9954	1.0015	1.0081
30	0.9613	0.9669	0.9727	0.9787	0.9845	0.9901	0.9961	1.0024
40	0.9574	0.9627	0.9684	0.9742	0.9798	0.9853	0.9910	0.9971
50	0.9535	0.9587	0.9643	0.9700	0.9752	0.9807	0.9863	0.9921

continued

continued

60	0.9499	0.9550	0.9603	0.9659	0.9711	0.9764	0.9817	0.9875
70	0.9465	0.9515	0.9566	0.9620	0.9671	0.9722	0.9774	0.9830
80	0.9431	0.9479	0.9530	0.9582	0.9632	0.9682	0.9733	0.9788
90	0.9400	0.9446	0.9497	0.9546	0.9596	0.9644	0.9694	0.9748
100	0.9369	0.9415	0.9464	0.9513	0.9560	0.9608	0.9656	0.9709
110	0.9339	0.9385	0.9432	0.9480	0.9526	0.9573	0.9621	0.9671
120	0.9310	0.9354	0.9401	0.9449	0.9493	0.9539	0.9585	0.9636
130	0.9282	0.9327	0.9373	0.9418	0.9462	0.9508	0.9553	0.9601
140	0.9256	0.9300	0.9344	0.9389	0.9433	0.9477	0.9521	0.9569
150	0.9231	0.9272	0.9316	0.9361	0.9403	0.9447	0.9489	0.9537
160	0.9204	0.9247	0.9290	0.9334	0.9375	0.9418	0.9459	0.9507
170	0.9180	0.9221	0.9264	0.9307	0.9347	0.9389	0.9432	0.9476
180	0.9155	0.9197	0.9238	0.9281	0.9321	0.9362	0.9403	0.9449
190	0.9128	0.9173	0.9215	0.9257	0.9295	0.9335	0.9376	0.9420
200	0.9096	0.9150	0.9191	0.9231	0.9269	0.9309	0.9349	0.9392

continued

P/MPa	*T*/K	
	533.89	543.95
	V_{spec}/cm^{3}g^{-1}	
0.1	1.0293	1.0378
10	1.0220	1.0301
20	1.0153	1.0230
30	1.0093	1.0166
40	1.0037	1.0108
50	0.9985	1.0054
60	0.9937	1.0003
70	0.9891	0.9954
80	0.9846	0.9909
90	0.9804	0.9866
100	0.9764	0.9825
110	0.9726	0.9785
120	0.9689	0.9747
130	0.9654	0.9710
140	0.9620	0.9675
150	0.9588	0.9641
160	0.9556	0.9608
170	0.9525	0.9578
180	0.9495	0.9547
190	0.9467	0.9517
200	0.9438	0.9488

Polymer (B): **poly(ethylene-*co*-vinyl alcohol)** **2007FU2**
Characterization: 48 mol% ethene, ρ(293.15 K) = 1.1243 g/cm^3, T_g/K = 321, T_m/K = 436, Kuraray Specialities, Japan

P/MPa	*T*/K 447.09	456.99	467.74	478.12	488.02	493.16	503.16	519.30
	V_{spec}/cm^{3}g^{-1}							
0.1	0.9886	0.9947	1.0010	1.0072	1.0135	1.0199	1.0264	1.0336
10	0.9833	0.9891	0.9951	1.0011	1.0071	1.0132	1.0193	1.0260
20	0.9781	0.9838	0.9895	0.9954	1.0010	1.0067	1.0127	1.0188
30	0.9736	0.9791	0.9845	0.9902	0.9957	1.0013	1.0069	1.0128
40	0.9694	0.9748	0.9800	0.9854	0.9906	0.9961	1.0014	1.0070
50	0.9654	0.9704	0.9757	0.9809	0.9860	0.9912	0.9964	1.0018
60	0.9616	0.9665	0.9715	0.9767	0.9816	0.9866	0.9916	0.9969
70	0.9579	0.9628	0.9676	0.9726	0.9773	0.9824	0.9873	0.9922
80	0.9545	0.9592	0.9638	0.9687	0.9733	0.9781	0.9829	0.9877
90	0.9511	0.9556	0.9602	0.9649	0.9695	0.9742	0.9788	0.9836
100	0.9479	0.9522	0.9568	0.9615	0.9658	0.9705	0.9749	0.9796
110	0.9447	0.9491	0.9535	0.9580	0.9623	0.9667	0.9711	0.9757
120	0.9417	0.9460	0.9504	0.9547	0.9588	0.9632	0.9676	0.9721
130	0.9388	0.9429	0.9473	0.9516	0.9556	0.9599	0.9641	0.9685
140	0.9361	0.9402	0.9443	0.9485	0.9524	0.9567	0.9608	0.9651
150	0.9333	0.9373	0.9414	0.9456	0.9494	0.9535	0.9575	0.9618
160	0.9307	0.9346	0.9386	0.9427	0.9466	0.9505	0.9545	0.9586
170	0.9281	0.9321	0.9359	0.9399	0.9437	0.9476	0.9516	0.9556
180	0.9257	0.9295	0.9334	0.9373	0.9408	0.9448	0.9486	0.9525
190	0.9232	0.9270	0.9308	0.9346	0.9382	0.9421	0.9458	0.9498
200	0.9209	0.9246	0.9282	0.9322	0.9355	0.9394	0.9431	0.9468

continued

P/MPa	*T*/K 528.86	539.38
	V_{spec}/cm^{3}g^{-1}	
0.1	1.0406	1.0491
10	1.0328	1.0408
20	1.0254	1.0330
30	1.0191	1.0264
40	1.0134	1.0202
50	1.0078	1.0146
60	1.0029	1.0093
70	0.9979	1.0042

continued

continued

80	0.9935	0.9996
90	0.9891	0.9951
100	0.9849	0.9909
110	0.9810	0.9867
120	0.9772	0.9827
130	0.9735	0.9790
140	0.9700	0.9754
150	0.9666	0.9719
160	0.9633	0.9685
170	0.9601	0.9653
180	0.9571	0.9621
190	0.9541	0.9590
200	0.9512	0.9562

Polymer (B): **poly[propylene-*b*-(ethylene-*co*-propylene)]** **2000SA2**
Characterization: M_w/g.mol^{-1} = 240000, 90% isotactic, 58.1 wt% crystallinity, 4.9 wt% ethene, Mitsubishi Chemical, Kurashiki, Japan

P/MPa	*T*/K						
	453.8	473.7	493.6	513.5	533.6	553.6	573.4
	V_{spec}/cm^{3}g^{-1}						
0.1	1.3139	1.3326	1.3515	1.3708	1.3943	1.4145	1.4361
10.0	1.2976	1.3145	1.3315	1.3487	1.3691	1.3866	1.4053
20.0	1.2830	1.2983	1.3138	1.3296	1.3480	1.3640	1.3805
50.0	1.2488	1.2617	1.2743	1.2871	1.3009	1.3134	1.3267
100.0	1.2092	1.2191	1.2294	1.2395	1.2498	1.2601	1.2703
150.0	1.1792	1.1883	1.1970	1.2056	1.2142	1.2232	1.2320
200.0	1.1550	1.1636	1.1712	1.1789	1.1864	1.1944	1.2023

Polymer (B): **poly(styrene-*co*-acrylonitrile)** **2005BRI**
Characterization: M_w/g.mol^{-1} = 80000, 33 wt% acrylonitrile

Tait equation parameter functions:
Range of data: *T*/K = 380-460, *P*/MPa = 0.1-200

$$V(P/\text{MPa}, T/\text{K}) = V(0, T/\text{K})\{1 - C^*\ln[1 + (P/\text{MPa})/B(T/\text{K})]\}$$

with $C = 0.0894$ and $\theta = T/\text{K} - 273.15$

$V(0, \theta/°\text{C})/\text{cm}^3\text{g}^{-1}$	$B(\theta/°\text{C})/\text{MPa}$
$0.8892 + 5.23\ 10^{-4}\ \theta - 3.0\ 10^{-8}\ \theta^2$	$294.0\ \exp(-3.86\ 10^{-3}\ \theta)$

Polymer (B): **poly(styrene-*co*-maleic anhydride)** **2005KIL**
Characterization: M_n/g.mol^{-1} = 46000, M_w/g.mol^{-1} = 89000, 11.8 mol% maleic anhydride, T_g/K = 391, Dylark 232, Acro Chemicals

P/MPa	*T*/K							
	429.93	439.99	450.33	460.55	470.95	481.28	491.35	501.55
	V_{spec}/cm^{3}g^{-1}							
0.1	0.9734	0.9789	0.9846	0.9902	0.9959	1.0015	1.0072	1.0128
10	0.9668	0.9720	0.9774	0.9826	0.9879	0.9933	0.9985	1.0038
20	0.9606	0.9656	0.9706	0.9756	0.9806	0.9857	0.9906	0.9955
30	0.9551	0.9598	0.9644	0.9693	0.9741	0.9789	0.9836	0.9882
40	0.9498	0.9543	0.9589	0.9635	0.9680	0.9728	0.9771	0.9815
50	0.9450	0.9493	0.9537	0.9581	0.9625	0.9670	0.9713	0.9754
60	0.9405	0.9447	0.9488	0.9531	0.9573	0.9617	0.9658	0.9697
70	0.9361	0.9402	0.9442	0.9484	0.9525	0.9567	0.9606	0.9645
80	0.9320	0.9360	0.9400	0.9440	0.9479	0.9520	0.9558	0.9595
90	0.9282	0.9320	0.9359	0.9397	0.9436	0.9475	0.9511	0.9548
100	0.9245	0.9283	0.9320	0.9357	0.9395	0.9433	0.9469	0.9503
110	0.9210	0.9246	0.9283	0.9319	0.9356	0.9392	0.9428	0.9461
120	0.9176	0.9211	0.9247	0.9282	0.9318	0.9355	0.9388	0.9421
130	0.9144	0.9178	0.9213	0.9247	0.9283	0.9318	0.9350	0.9382
140	0.9114	0.9145	0.9180	0.9213	0.9248	0.9283	0.9314	0.9345
150	0.9086	0.9114	0.9148	0.9180	0.9215	0.9249	0.9279	0.9310
160	0.9059	0.9085	0.9118	0.9149	0.9183	0.9215	0.9246	0.9275
170	0.9034	0.9057	0.9087	0.9119	0.9152	0.9184	0.9213	0.9242
180	0.9011	0.9029	0.9058	0.9089	0.9121	0.9153	0.9181	0.9209
190	0.8989	0.9003	0.9030	0.9060	0.9092	0.9122	0.9151	0.9179
200	0.8967	0.8978	0.9002	0.9032	0.9063	0.9093	0.9120	0.9147

Polymer (B): **poly(styrene-*co*-maleic anhydride)** **2005KIL**
Characterization: M_n/g.mol^{-1} = 38000, M_w/g.mol^{-1} = 73000, 14.5 mol% maleic anhydride, T_g/K = 403, Dylark 332, Acro Chemicals

P/MPa	*T*/K							
	438.23	453.03	468.23	483.07	497.88	512.98	527.76	543.63
	V_{spec}/cm^{3}g^{-1}							
0.1	0.95312	0.96159	0.97020	0.97900	0.98749	0.99685	1.00636	1.01598
10	0.94661	0.95461	0.96278	0.97117	0.97948	0.98793	0.99676	1.00574
20	0.94047	0.94803	0.95577	0.96382	0.97192	0.97952	0.98791	0.99625
30	0.93511	0.94234	0.94975	0.95748	0.96543	0.97267	0.98049	0.98853
40	0.93010	0.93709	0.94426	0.95172	0.95932	0.96613	0.97361	0.98087
50	0.92552	0.93227	0.93920	0.94633	0.95360	0.96026	0.96731	0.97434

continued

continued

60	0.92108	0.92769	0.93438	0.94137	0.94835	0.95479	0.96157	0.96835
70	0.91696	0.92343	0.92993	0.93663	0.94346	0.94965	0.95619	0.96272
80	0.91317	0.91930	0.92561	0.93221	0.93881	0.94477	0.95114	0.95738
90	0.90939	0.91538	0.92155	0.92801	0.93436	0.94026	0.94640	0.95243
100	0.90588	0.91173	0.91776	0.92398	0.93019	0.93594	0.94183	0.94766
110	0.90270	0.90821	0.91401	0.92017	0.92614	0.93175	0.93749	0.94317
120	0.89964	0.90468	0.91046	0.91646	0.92221	0.92786	0.93336	0.93898
130	0.89684	0.90147	0.90698	0.91297	0.91855	0.92405	0.92940	0.93488
140	0.89423	0.89822	0.90368	0.90952	0.91502	0.92029	0.92574	0.93086
150	0.89170	0.89524	0.90047	0.90623	0.91165	0.91686	0.92204	0.92717
160	0.88941	0.89232	0.89749	0.90306	0.90838	0.91345	0.91864	0.92349
170	0.88715	0.88970	0.89444	0.89995	0.90514	0.91016	0.91521	0.92001
180	0.88503	0.88713	0.89162	0.89714	0.90203	0.90713	0.91202	0.91675
190	0.88301	0.88478	0.88883	0.89424	0.89914	0.90410	0.90891	0.91352
200	0.88108	0.88259	0.88618	0.89154	0.89639	0.90118	0.90599	0.91042

Polymer (B): **poly(styrene-*co*-maleic anhydride)** **2005KIL**

Characterization: M_n/g.mol^{-1} = 27000, M_w/g.mol^{-1} = 52000, 25.3 mol% maleic anhydride, T_g/K = 426, DSM Research, The Netherlands

P/MPa	*T*/K						
	433.38	443.34	453.79	463.81	474.02	484.58	494.50
	V_{spec}/cm^{3}g^{-1}						
0.1	0.90381	0.90882	0.91386	0.91875	0.92377	0.92873	0.93396
10	0.89850	0.90314	0.90789	0.91255	0.91728	0.92207	0.92699
20	0.89344	0.89774	0.90231	0.90672	0.91115	0.91581	0.92040
30	0.88892	0.89294	0.89719	0.90150	0.90574	0.91025	0.91469
40	0.88492	0.88847	0.89250	0.89665	0.90077	0.90501	0.90940
50	0.88133	0.88425	0.88818	0.89210	0.89614	0.90017	0.90438
60	0.87808	0.88035	0.88407	0.88792	0.89178	0.89574	0.89980
70	0.87505	0.87682	0.88013	0.88393	0.88762	0.89144	0.89543
80	0.87238	0.87359	0.87644	0.88011	0.88372	0.88748	0.89127
90	0.86978	0.87061	0.87307	0.87653	0.88004	0.88366	0.88744
100	0.86734	0.86786	0.86982	0.87307	0.87652	0.87999	0.88366
110	0.86506	0.86533	0.86676	0.86981	0.87317	0.87661	0.88014
120	0.86290	0.86291	0.86403	0.86666	0.86995	0.87326	0.87673
130	0.86076	0.86075	0.86139	0.86362	0.86684	0.87000	0.87352
140	0.85863	0.85853	0.85901	0.86086	0.86381	0.86699	0.87036
150	0.85664	0.85644	0.85671	0.85814	0.86093	0.86403	0.86745
160	0.85467	0.85438	0.85454	0.85567	0.85811	0.86115	0.86448
170	0.85279	0.85244	0.85243	0.85324	0.85543	0.85836	0.86164
180	0.85099	0.85063	0.85046	0.85102	0.85281	0.85562	0.85894
190	0.84910	0.84870	0.84853	0.84888	0.85036	0.85299	0.85618
200	0.84736	0.84696	0.84661	0.84688	0.84798	0.85036	0.85346

Polymer (B): **poly(styrene-*co*-maleic anhydride)** **2005KIL**
Characterization: M_n/g.mol^{-1} = 28000, M_w/g.mol^{-1} = 54000, 30.9 mol% maleic anhydride, T_g/K = 439, DSM Research, The Netherlands

P/MPa	*T*/K = 449.64	459.76	470.05	480.33	490.76	501.01
	V_{spec}/cm^{3}g^{-1}					
0.1	0.88563	0.89054	0.89530	0.90032	0.90511	0.91030
10	0.88037	0.88501	0.88956	0.89428	0.89892	0.90383
20	0.87534	0.87973	0.88410	0.88852	0.89309	0.89770
30	0.87093	0.87511	0.87925	0.88354	0.88787	0.89238
40	0.86679	0.87074	0.87472	0.87881	0.88301	0.88742
50	0.86311	0.86668	0.87057	0.87448	0.87853	0.88271
60	0.85976	0.86287	0.86657	0.87037	0.87427	0.87833
70	0.85681	0.85936	0.86286	0.86660	0.87037	0.87430
80	0.85412	0.85597	0.85938	0.86288	0.86661	0.87041
90	0.85160	0.85304	0.85601	0.85944	0.86303	0.86679
100	0.84933	0.85019	0.85276	0.85614	0.85961	0.86332
110	0.84707	0.84771	0.84979	0.85296	0.85639	0.85997
120	0.84500	0.84528	0.84693	0.84992	0.85330	0.85683
130	0.84297	0.84311	0.84432	0.84701	0.85030	0.85373
140	0.84103	0.84099	0.84186	0.84423	0.84742	0.85078
150	0.83916	0.83900	0.83956	0.84157	0.84463	0.84795
160	0.83728	0.83711	0.83743	0.83909	0.84197	0.84515
170	0.83556	0.83521	0.83539	0.83667	0.83930	0.84247
180	0.83384	0.83341	0.83348	0.83440	0.83681	0.83993
190	0.83210	0.83168	0.83156	0.83229	0.83437	0.83736
200	0.83043	0.82993	0.82977	0.83025	0.83197	0.83483

Polymer (B): **poly(styrene-*co*-maleic anhydride)** **2005KIL**
Characterization: M_n/g.mol^{-1} = 30000, M_w/g.mol^{-1} = 58000, 34.7 mol% maleic anhydride, T_g/K = 449, DSM Research, The Netherlands

P/MPa	*T*/K = 460.25	470.41	480.76	491.06	501.28
	V_{spec}/cm^{3}g^{-1}				
0.1	0.87138	0.87623	0.88109	0.88602	0.89124
10	0.86613	0.87072	0.87531	0.87999	0.88490
20	0.86112	0.86542	0.86981	0.87428	0.87890
30	0.85668	0.86087	0.86501	0.86928	0.87372
40	0.85255	0.85653	0.86055	0.86471	0.86895
50	0.84869	0.85252	0.85639	0.86037	0.86454

continued

continued

60	0.84526	0.84872	0.85248	0.85633	0.86023
70	0.84203	0.84511	0.84879	0.85250	0.85636
80	0.83917	0.84175	0.84526	0.84886	0.85264
90	0.83657	0.83862	0.84191	0.84541	0.84913
100	0.83421	0.83568	0.83877	0.84214	0.84573
110	0.83193	0.83304	0.83571	0.83902	0.84257
120	0.82982	0.83053	0.83280	0.83603	0.83948
130	0.82778	0.82827	0.83012	0.83319	0.83653
140	0.82589	0.82609	0.82758	0.83043	0.83372
150	0.82407	0.82411	0.82520	0.82774	0.83096
160	0.82221	0.82212	0.82299	0.82519	0.82829
170	0.82052	0.82026	0.82087	0.82277	0.82567
180	0.81877	0.81848	0.81893	0.82048	0.82319
190	0.81707	0.81679	0.81704	0.81824	0.82074
200	0.81546	0.81506	0.81521	0.81612	0.81838

Polymer (B): **poly[tetrafluoroethylene-*co*-2,2-bis(trifluoromethyl)-4,5-difluoro-1,3-dioxole]** **2008DLU**

Characterization: 13 mol% tetrafluoroethylene, T_g/K = 513, Teflon AF2400, Random Technologies, San Francisco

T/K	*P*/MPa 0.1	10	20	30	40	50	80	120
	V_{spec}/cm^{3}g^{-1}							
515.46	0.60962							
530.86	0.61813	0.60334						
545.85	0.63020	0.60984	0.59384					
561.43	0.64216	0.61663	0.59853	0.58673				
576.08	0.65486	0.62419	0.60382	0.59045	0.58055			
591.38	0.66872	0.63258	0.61011	0.59535	0.58425	0.57576		
606.33	0.68364	0.64084	0.61603	0.59990	0.58794	0.57845	0.56060	
620.76	0.70048	0.64989	0.62279	0.60520	0.59233	0.58215	0.56155	0.54661
635.12	0.71977	0.65951	0.62983	0.61103	0.59729	0.58644	0.56364	0.54727

Tait equation parameter functions:
Range of data: *T*/K = 515-635, *P*/MPa = 0.1-120

$$V(P/\text{MPa}, T/\text{K}) = V(0, T/\text{K})\{1 - C*\ln[1 + (P/\text{MPa})/B(T/\text{K})]\}$$

with $C = 0.0781$ and $\theta = T/\text{K} - 273.15$

$V(0, \theta/°C)$/cm^{3}g^{-1}	$B(\theta/°C)$/MPa
$0.6490 - 8.8\ 10^{-4}\ \theta + 2.97\ 10^{-6}\ \theta^2$	$978.0\ \exp(-1.45\ 10^{-2}\ \theta)$

Polymer (B): **poly[tetrafluoroethylene-*co*-2,2-bis(trifluoromethyl)-4,5-difluoro-1,3-dioxole** **2008DLU**

Characterization: 35 mol% tetrafluoroethylene, T_g/K = 433, Teflon AF1600, Random Technologies, San Francisco

T/K	*P*/MPa 0.1	10	20	30	40	50	80	120
	V_{spec}/cm^{3}g^{-1}							
454.64	0.58246	0.57009						
469.94	0.59302	0.57728	0.56497					
485.28	0.60202	0.58446	0.57119	0.56091				
494.60	0.61134	0.59105	0.57604	0.56521	0.55659			
500.69	0.61170	0.59195	0.57741	0.56647	0.55760			
515.39	0.62201	0.59985	0.58398	0.57223	0.56280	0.55493		
530.90	0.63310	0.60801	0.59068	0.57796	0.56798	0.55979	0.54113	
546.16	0.64493	0.61633	0.59731	0.58374	0.57312	0.56451	0.54493	
561.62	0.65675	0.62455	0.60397	0.58959	0.57826	0.56910	0.54876	0.53038
576.55	0.66967	0.63313	0.61082	0.59535	0.58353	0.57393	0.55259	0.53351
592.00	0.68299	0.64161	0.61745	0.60112	0.58871	0.57854	0.55629	0.53669
607.04	0.69873	0.65091	0.62464	0.60721	0.59390	0.58332	0.56014	0.53994
621.74	0.71654	0.66043	0.63173	0.61320	0.59920	0.58808	0.56414	0.54318
636.12	0.74124	0.67156	0.63966	0.61976	0.60489	0.59319	0.56818	0.54651
520.34	0.62808	0.60361	0.58650	0.57425	0.56463	0.55659	0.53843	0.52454
494.60	0.61134	0.59105	0.57604	0.56521	0.55659	0.54926	0.53510	0.52399

continued

T/K	*P*/MPa 160	200
	V_{spec}/cm^{3}g^{-1}	
607.04	0.52524	0.51358
621.74	0.52817	0.51602
636.12	0.53105	0.51860

Tait equation parameter functions:
Range of data: *T*/K = 435-635, *P*/MPa = 0.1-120

$$V(P/\text{MPa}, T/\text{K}) = V(0, T/\text{K})\{1 - C^*\ln[1 + (P/\text{MPa})/B(T/\text{K})]\}$$
with $C = 0.0760$ and $\theta = T/\text{K} - 273.15$

$V(0, \theta/°\text{C})/\text{cm}^3\text{g}^{-1}$	$B(\theta/°\text{C})/\text{MPa}$
$0.5494 - 1.4\ 10^{-4}\ \theta + 1.80\ 10^{-6}\ \theta^2$	$256.0 \exp(-1.13\ 10^{-2}\ \theta)$

Polymer (B): **poly[tetrafluoroethylene-*co*-perfluoro(methyl vinyl ether)]** **2004DLU**

Characterization: M_w/g.mol^{-1} = 100000, 15-55 mol% perfluoro(methyl vinyl ether), T_g/K = 271, Dyneon Perfluoroelastomer PFE, Dyneon GmbH & Co. KG, Burgkirchen, Germany

P/MPa	*T*/K							
	298.94	299.465	300.77	305.746	310.59	315.415	320.21	324.945
	V_{spec}/cm^{3}g^{-1}							
0.1	0.48071	0.48126	0.48153	0.48376	0.48600	0.48813	0.49042	0.49279
10	0.47712	0.47765	0.47790	0.47996	0.48204	0.48410	0.48621	0.48841
20	0.47372	0.47426	0.47449	0.47640	0.47832	0.48041	0.48235	0.48437
30	0.47071	0.47124	0.47141	0.47321	0.47505	0.47704	0.47886	0.48080
40	0.46793	0.46837	0.46866	0.47039	0.47208	0.47395	0.47571	0.47748
50	0.46540	0.46582	0.46600	0.46777	0.46943	0.47121	0.47282	0.47452
60	0.46304	0.46345	0.46370	0.46524	0.46684	0.46854	0.47016	0.47184
70	0.46084	0.46125	0.46146	0.46303	0.46453	0.46612	0.46772	0.46928
80	0.45884	0.45916	0.45941	0.46087	0.46232	0.46395	0.46540	0.46697
90	0.45695	0.45726	0.45744	0.45890	0.46032	0.46189	0.46318	0.46477
100	0.45511	0.45545	0.45566	0.45706	0.45840	0.45987	0.46128	0.46266
110	0.45351	0.45376	0.45398	0.45519	0.45661	0.45801	0.45936	0.46071
120	0.45193	0.45224	0.45239	0.45361	0.45484	0.45628	0.45757	0.45887
130	0.45055	0.45078	0.45096	0.45204	0.45321	0.45460	0.45581	0.45721
140	0.44923	0.44943	0.44957	0.45058	0.45170	0.45298	0.45426	0.45552
150	0.44791	0.44820	0.44826	0.44918	0.45019	0.45142	0.45266	0.45394
160	0.44676	0.44695	0.44702	0.44788	0.44882	0.45002	0.45110	0.45241
170	0.44563	0.44582	0.44585	0.44662	0.44742	0.44861	0.44976	0.45084
180	0.44452	0.44470	0.44477	0.44541	0.44627	0.44723	0.44832	0.44954
190	0.44343	0.44361	0.44369	0.44436	0.44506	0.44599	0.44701	0.44816
200	0.44242	0.44258	0.44266	0.44322	0.44392	0.44481	0.44567	0.44676

continued

P/MPa	*T*/K							
	329.77	334.65	364.545	379.32	394.37	409.58	424.67	439.995
	V_{spec}/cm^{3}g^{-1}							
0.1	0.49520	0.49745	0.51195	0.51929	0.52691	0.53482	0.54252	0.55074
10	0.49061	0.49271	0.50598	0.51266	0.51952	0.52655	0.53330	0.54041
20	0.48639	0.48837	0.50055	0.50673	0.51302	0.51934	0.52541	0.53170
30	0.48264	0.48451	0.49599	0.50175	0.50756	0.51336	0.51890	0.52470
40	0.47945	0.48103	0.49190	0.49731	0.50279	0.50820	0.51337	0.51865
50	0.47626	0.47791	0.48826	0.49338	0.49852	0.50358	0.50850	0.51349
60	0.47346	0.47505	0.48494	0.48979	0.49466	0.49952	0.50412	0.50880

continued

continued

70	0.47096	0.47246	0.48189	0.48655	0.49125	0.49588	0.50018	0.50469
80	0.46853	0.46997	0.47916	0.48354	0.48807	0.49252	0.49670	0.50088
90	0.46624	0.46773	0.47648	0.48076	0.48512	0.48939	0.49340	0.49743
100	0.46419	0.46555	0.47409	0.47822	0.48244	0.48652	0.49034	0.49431
110	0.46222	0.46357	0.47181	0.47581	0.47989	0.48386	0.48750	0.49127
120	0.46034	0.46153	0.46968	0.47350	0.47752	0.48124	0.48489	0.48853
130	0.45847	0.45977	0.46768	0.47141	0.47524	0.47902	0.48239	0.48589
140	0.45672	0.45802	0.46567	0.46939	0.47313	0.47665	0.48003	0.48346
150	0.45513	0.45636	0.46387	0.46750	0.47107	0.47458	0.47781	0.48110
160	0.45359	0.45478	0.46212	0.46567	0.46918	0.47257	0.47571	0.47888
170	0.45209	0.45318	0.46039	0.46382	0.46733	0.47060	0.47365	0.47682
180	0.45064	0.45179	0.45881	0.46219	0.46550	0.46875	0.47174	0.47473
190	0.44922	0.45039	0.45715	0.46048	0.46379	0.46693	0.46986	0.47291
200	0.44789	0.44897	0.45558	0.45885	0.46211	0.46517	0.46805	0.47099

continued

P/MPa	T/K							
	455.57	470.84	486.094	501.76	516.876	532.605	547.58	563.503
	V_{spec}/cm^{3}g^{-1}							
0.1	0.55918	0.56803	0.57704	0.58677	0.59692	0.60765	0.61919	0.63239
10	0.54763	0.55503	0.56253	0.57050	0.57862	0.58717	0.59594	0.60571
20	0.53807	0.54446	0.55103	0.55784	0.56481	0.57220	0.57940	0.58748
30	0.53042	0.53616	0.54196	0.54819	0.55429	0.56079	0.56720	0.57425
40	0.52403	0.52922	0.53460	0.54012	0.54576	0.55170	0.55735	0.56373
50	0.51842	0.52327	0.52822	0.53327	0.53858	0.54396	0.54926	0.55514
60	0.51345	0.51811	0.52263	0.52743	0.53235	0.53729	0.54220	0.54772
70	0.50910	0.51340	0.51768	0.52218	0.52679	0.53149	0.53613	0.54126
80	0.50509	0.50925	0.51334	0.51764	0.52191	0.52631	0.53066	0.53556
90	0.50142	0.50530	0.50926	0.51334	0.51748	0.52170	0.52591	0.53049
100	0.49808	0.50178	0.50558	0.50952	0.51339	0.51749	0.52140	0.52594
110	0.49496	0.49864	0.50221	0.50600	0.50975	0.51358	0.51737	0.52174
120	0.49211	0.49556	0.49911	0.50260	0.50625	0.51003	0.51358	0.51789
130	0.48939	0.49276	0.49614	0.49960	0.50307	0.50661	0.51018	0.51428
140	0.48683	0.49005	0.49329	0.49668	0.50015	0.50351	0.50701	0.51089
150	0.48440	0.48752	0.49069	0.49400	0.49720	0.50054	0.50394	0.50768
160	0.48209	0.48515	0.48820	0.49137	0.49448	0.49781	0.50102	0.50473
170	0.47982	0.48284	0.48579	0.48898	0.49200	0.49513	0.49831	0.50202
180	0.47771	0.48076	0.48353	0.48662	0.48957	0.49270	0.49574	0.49930
190	0.47574	0.47852	0.48133	0.48431	0.48729	0.49021	0.49331	0.49668
200	0.47374	0.47647	0.47923	0.48212	0.48504	0.48780	0.49075	0.49425

continued

continued

P/MPa	*T*/K 578.20	593.788
	V_{spec}/cm³g⁻¹	
0.1	0.64774	0.66445
10	0.61667	0.62784
20	0.59638	0.60530
30	0.58195	0.58955
40	0.57068	0.57742
50	0.56140	0.56751
60	0.55359	0.55915
70	0.54677	0.55204
80	0.54072	0.54569
90	0.53548	0.54008
100	0.53064	0.53501
110	0.52626	0.53044
120	0.52224	0.52617
130	0.51841	0.52227
140	0.51490	0.51863
150	0.51161	0.51522
160	0.50851	0.51203
170	0.50566	0.50896
180	0.50287	0.50608
190	0.50014	0.50333
200	0.49764	0.50070

Polymer (B): **poly(vinylidene fluoride-*co*-hexafluoropropylene) 2001MEK**
Characterization: M_n/g.mol^{-1} = 145000, M_w/g.mol^{-1} = 480000, 3.1 mol% hexafluoropropylene, ATOFINA Chemicals, Philadelphia, PA

P/MPa	*T*/K 433.15	438.15	443.15	448.15	453.15	458.15	463.15	468.15
	V_{spec}/cm³g⁻¹							
0.1	0.6500	0.6520	0.6546	0.6569	0.6589	0.6615	0.6639	0.6660
20.0	0.6402	0.6422	0.6444	0.6466	0.6485	0.6508	0.6530	0.6549
40.0	0.6304	0.6323	0.6342	0.6362	0.6381	0.6402	0.6420	0.6439
80.0	0.6170	0.6184	0.6200	0.6216	0.6232	0.6247	0.6263	0.6280
120.0	0.6064	0.6079	0.6092	0.6107	0.6120	0.6135	0.6148	0.6163

continued

continued

P/MPa	*T*/K				
	473.15	478.15	483.15	488.15	493.15
	V_{spec}/cm^{3}g^{-1}				
0.1	0.6682	0.6707	0.6728	0.6747	0.6771
20.0	0.6567	0.6587	0.6605	0.6625	0.6642
40.0	0.6459	0.6479	0.6498	0.6515	0.6525
80.0	0.6294	0.6316	0.6330	0.6346	0.6361
120.0	0.6176	0.6189	0.6202	0.6216	0.6224

Tait equation parameter functions:
Range of data: *T*/K = 433-493, *P*/MPa = 0.1-120

$$V(P/\text{MPa}, T/\text{K}) = V(0, T/\text{K})\{1 - C*\ln[1 + (P/\text{MPa})/B(T/\text{K})]\}$$

with $C = 0.0894$ and $\theta = T/\text{K} - 273.15$

$V(0, \theta/°\text{C})/\text{cm}^3\text{g}^{-1}$	$B(\theta/°\text{C})/\text{MPa}$
$0.587 + 4.138\ 10^{-4}\ \theta$	$157.7\ \exp(-2.83\ 10^{-3}\ \theta)$

Polymer (B): **poly(vinylidene fluoride-*co*-hexafluoropropylene) 2001MEK**
Characterization: M_n/g.mol^{-1} = 111000, M_w/g.mol^{-1} = 321000, 10.5 mol% hexafluoropropylene, ATOFINA Chemicals, Philadelphia, PA

P/MPa	*T*/K							
	433.15	438.15	443.15	448.15	453.15	458.15	463.15	468.15
	V_{spec}/cm^{3}g^{-1}							
0.1	0.6468	0.6492	0.6512	0.6535	0.6558	0.6577	0.6601	0.6624
20.0	0.6377	0.6399	0.6419	0.6438	0.6459	0.6477	0.6501	0.6520
40.0	0.6282	0.6302	0.6322	0.6338	0.6355	0.6375	0.6396	0.6416
80.0	0.6146	0.6159	0.6175	0.6196	0.6210	0.6221	0.6242	0.6259
120.0	0.6043	0.6057	0.6070	0.6083	0.6098	0.6112	0.6125	0.6138

continued

continued

P/MPa	T/K 473.15	478.15	483.15	488.15	493.15
	V_{spec}/cm^{3}g^{-1}				
0.1	0.6639	0.6661	0.6681	0.6698	0.6723
20.0	0.6534	0.6558	0.6577	0.6594	0.6612
40.0	0.6435	0.6455	0.6473	0.6491	0.6503
80.0	0.6270	0.6288	0.6304	0.6315	0.6328
120.0	0.6151	0.6164	0.6177	0.6187	0.6196

Tait equation parameter functions:
Range of data: T/K = 433-493, P/MPa = 0.1-120

$$V(P/\text{MPa}, T/\text{K}) = V(0, T/\text{K})\{1 - C{*}\ln[1 + (P/\text{MPa})/B(T/\text{K})]\}$$
with $C = 0.0894$ and $\theta = T/\text{K} - 273.15$

$V(0, \theta/°\text{C})$/cm^{3}g^{-1}	$B(\theta/°\text{C})$/MPa
$0.577 + 4.543\ 10^{-4}\ \theta$	$207.1\ \exp(-4.15\ 10^{-3}\ \theta)$

Polymer (B): **poly(vinylidene fluoride-*co*-hexafluoropropylene) 2001MEK**
Characterization: M_n/g.mol^{-1} = 143000, M_w/g.mol^{-1} = 471000, 11.1 mol% hexafluoropropylene, ATOFINA Chemicals, Philadelphia, PA

P/MPa	T/K 448.15	453.15	458.15	463.15	468.15	473.15	478.15	483.15
	V_{spec}/cm^{3}g^{-1}							
0.1	0.6589	0.6613	0.6639	0.6657	0.6681	0.6700	0.6718	0.6739
20.0	0.6488	0.6506	0.6532	0.6549	0.6570	0.6587	0.6608	0.6629
40.0	0.6380	0.6401	0.6420	0.6440	0.6461	0.6481	0.6500	0.6520
80.0	0.6233	0.6250	0.6266	0.6282	0.6296	0.6313	0.6329	0.6343
120.0	0.6122	0.6136	0.6151	0.6165	0.6179	0.6192	0.6207	0.6220

continued

continued

P/MPa	*T*/K 488.15	493.15
	V_{spec}/cm^{3}g^{-1}	
0.1	0.6760	0.6776
20.0	0.6648	0.6666
40.0	0.6536	0.6554
80.0	0.6358	0.6370
120.0	0.6230	0.6234

Tait equation parameter functions:
Range of data: *T*/K = 448-493, *P*/MPa = 0.1-120

$$V(P/\text{MPa}, T/\text{K}) = V(0, T/\text{K})\{1 - C^*\ln[1 + (P/\text{MPa})/B(T/\text{K})]\}$$
with $C = 0.0894$ and $\theta = T/\text{K} - 273.15$

$V(0, \theta/°\text{C})/\text{cm}^3\text{g}^{-1}$	$B(\theta/°\text{C})/\text{MPa}$
$0.580 + 4.201\ 10^{-4}\ \theta$	$183.15\ \exp(-3.35\ 10^{-3}\ \theta)$

Polymer (B): **poly(vinylidene fluoride-*co*-hexafluoropropylene) 2004DLU**
Characterization: M_w/g.mol^{-1} = 85000, 22.0 mol% hexafluoropropylene, T_g/K = 255, Fluorel FC-2175, Dyneon GmbH & Co. KG, Burgkirchen, Germany

P/MPa	*T*/K 302.885	305.956	310.925	315.605	320.35	325.19	329.995	334.935
	V_{spec}/cm^{3}g^{-1}							
0.1	0.55660	0.55805	0.56011	0.56223	0.56447	0.56675	0.56887	0.57099
10	0.55358	0.55495	0.55696	0.55897	0.56106	0.56321	0.56523	0.56728
20	0.55065	0.55193	0.55394	0.55582	0.55787	0.55982	0.56176	0.56377
30	0.54808	0.54935	0.55120	0.55309	0.55497	0.55686	0.55872	0.56064
40	0.54569	0.54693	0.54870	0.55053	0.55234	0.55417	0.55590	0.55782
50	0.54343	0.54463	0.54633	0.54810	0.54981	0.55163	0.55342	0.55518
60	0.54133	0.54249	0.54418	0.54591	0.54761	0.54932	0.55098	0.55277
70	0.53940	0.54049	0.54214	0.54378	0.54537	0.54712	0.54876	0.55041
80	0.53749	0.53853	0.54017	0.54178	0.54333	0.54501	0.54663	0.54827
90	0.53565	0.53677	0.53823	0.53991	0.54140	0.54304	0.54450	0.54615
100	0.53396	0.53501	0.53643	0.53804	0.53958	0.54112	0.54264	0.54422
110	0.53221	0.53333	0.53480	0.53631	0.53782	0.53930	0.54073	0.54233

continued

continued

120	0.53066	0.53170	0.53316	0.53464	0.53609	0.53752	0.53900	0.54051
130	0.52918	0.53019	0.53149	0.53301	0.53444	0.53591	0.53729	0.53880
140	0.52763	0.52869	0.53001	0.53144	0.53282	0.53421	0.53564	0.53712
150	0.52615	0.52720	0.52850	0.52993	0.53128	0.53276	0.53405	0.53545
160	0.52477	0.52572	0.52714	0.52850	0.52984	0.53118	0.53256	0.53389
170	0.52349	0.52441	0.52572	0.52703	0.52845	0.52971	0.53105	0.53239
180	0.52215	0.52302	0.52438	0.52558	0.52699	0.52831	0.52959	0.53093
190	0.52081	0.52176	0.52298	0.52428	0.52562	0.52697	0.52812	0.52944
200	0.51953	0.52041	0.52171	0.52299	0.52421	0.52559	0.52680	0.52805

continued

P/MPa	*T*/K							
	349.80	364.62	379.445	394.216	409.46	424.545	439.605	454.74
	V_{spec}/cm^{3}g^{-1}							
0.1	0.57769	0.58435	0.59123	0.59827	0.60540	0.61294	0.62032	0.62744
10	0.57356	0.57983	0.58628	0.59284	0.59946	0.60644	0.61322	0.61975
20	0.56971	0.57563	0.58170	0.58779	0.59404	0.60059	0.60689	0.61288
30	0.56623	0.57190	0.57767	0.58355	0.58938	0.59549	0.60147	0.60710
40	0.56319	0.56863	0.57411	0.57969	0.58527	0.59106	0.59665	0.60202
50	0.56033	0.56552	0.57083	0.57611	0.58144	0.58707	0.59231	0.59740
60	0.55765	0.56272	0.56776	0.57286	0.57806	0.58335	0.58840	0.59319
70	0.55520	0.56007	0.56495	0.56984	0.57481	0.57995	0.58475	0.58938
80	0.55282	0.55756	0.56230	0.56702	0.57188	0.57678	0.58137	0.58584
90	0.55069	0.55527	0.55981	0.56448	0.56910	0.57384	0.57824	0.58255
100	0.54855	0.55302	0.55745	0.56186	0.56643	0.57104	0.57530	0.57949
110	0.54655	0.55089	0.55522	0.55956	0.56398	0.56849	0.57259	0.57661
120	0.54460	0.54892	0.55313	0.55738	0.56164	0.56598	0.56997	0.57386
130	0.54282	0.54703	0.55108	0.55524	0.55938	0.56359	0.56745	0.57123
140	0.54103	0.54515	0.54915	0.55318	0.55722	0.56137	0.56509	0.56879
150	0.53936	0.54337	0.54727	0.55125	0.55516	0.55922	0.56282	0.56635
160	0.53775	0.54162	0.54547	0.54937	0.55325	0.55710	0.56065	0.56413
170	0.53621	0.53996	0.54374	0.54756	0.55129	0.55517	0.55859	0.56194
180	0.53461	0.53835	0.54207	0.54574	0.54949	0.55313	0.55655	0.55985
190	0.53311	0.53679	0.54046	0.54401	0.54762	0.55131	0.55461	0.55780
200	0.53163	0.53527	0.53886	0.54236	0.54592	0.54949	0.55268	0.55579

continued

continued

P/MPa	T/K 470.07	485.945	501.34	516.63	531.755	547.27	562.18	576.84
	V_{spec}/cm^{3}g^{-1}							
0.1	0.63500	0.64298	0.65111	0.65955	0.66830	0.67810	0.68808	0.69928
10	0.62653	0.63368	0.64087	0.64826	0.65593	0.66419	0.67261	0.68196
20	0.61908	0.62559	0.63210	0.63869	0.64560	0.65275	0.66017	0.66837
30	0.61287	0.61886	0.62481	0.63088	0.63724	0.64374	0.65032	0.65774
40	0.60735	0.61294	0.61853	0.62416	0.63006	0.63601	0.64214	0.64887
50	0.60244	0.60776	0.61295	0.61818	0.62378	0.62926	0.63504	0.64125
60	0.59799	0.60304	0.60795	0.61296	0.61811	0.62328	0.62876	0.63461
70	0.59400	0.59874	0.60344	0.60814	0.61309	0.61796	0.62306	0.62870
80	0.59021	0.59479	0.59932	0.60382	0.60845	0.61315	0.61801	0.62331
90	0.58676	0.59116	0.59549	0.59978	0.60422	0.60870	0.61329	0.61844
100	0.58355	0.58779	0.59187	0.59607	0.60036	0.60465	0.60907	0.61389
110	0.58053	0.58454	0.58854	0.59262	0.59671	0.60086	0.60510	0.60977
120	0.57767	0.58160	0.58544	0.58931	0.59331	0.59728	0.60138	0.60594
130	0.57499	0.57878	0.58251	0.58619	0.59014	0.59393	0.59791	0.60227
140	0.57236	0.57607	0.57975	0.58332	0.58714	0.59076	0.59469	0.59890
150	0.56987	0.57353	0.57705	0.58056	0.58423	0.58786	0.59154	0.59570
160	0.56753	0.57107	0.57451	0.57795	0.58148	0.58494	0.58855	0.59262
170	0.56529	0.56868	0.57212	0.57539	0.57888	0.58221	0.58578	0.58972
180	0.56315	0.56646	0.56974	0.57293	0.57632	0.57966	0.58308	0.58694
190	0.56106	0.56421	0.56752	0.57058	0.57389	0.57713	0.58046	0.58424
200	0.55896	0.56207	0.56531	0.56830	0.57162	0.57463	0.57796	0.58170

6.2. Excess volumes and/or densities of copolymer solutions

Polymer (B): **poly(ethylene-*co*-1-butene)** **2000BY2**
Characterization: M_n/g.mol^{-1} = 175000, M_w/g.mol^{-1} = 177000, 94 mol% 1-butene, Exxon Research and Engineering Co.
Solvent (A): **dimethyl ether** **C_2H_6O** **115-10-6**

T/K = 383.15 w_B 0.05

P/bar	346.6	415.5	519.0	691.4	1036.2	1381.0	1725.9	2070.7	2415.5
ρ/(g/cm^3)	0.669	0.683	0.701	0.725	0.763	0.793	0.818	0.840	0.860

T/K = 393.15 w_B 0.05

P/bar	346.6	415.5	519.0	691.4	1036.2	1381.0	1725.9	2070.7	2415.5
ρ/(g/cm^3)	0.660	0.675	0.694	0.719	0.759	0.790	0.816	0.839	0.859

T/K = 403.15 w_B 0.05

P/bar	346.6	415.5	519.0	691.4	1036.2	1381.0	1725.9	2070.7	2415.5
ρ/(g/cm^3)	0.643	0.659	0.678	0.705	0.746	0.777	0.803	0.826	0.846

T/K = 423.15 w_B 0.05

P/bar	346.6	415.5	519.0	691.4	1036.2	1381.0	1725.9	2070.7	2415.5
ρ/(g/cm^3)	0.612	0.632	0.655	0.684	0.727	0.760	0.787	0.811	0.831

Polymer (B): **poly(ethylene-*co*-1-butene)** **2000BY2**
Characterization: M_n/g.mol^{-1} = 230000, M_w/g.mol^{-1} = 232000, 20.2 mol% 1-butene, Exxon Research and Engineering Co.
Solvent (A): **dimethyl ether-d6** **C_2D_6O** **17222-37-6**

T/K = 383.15 w_B 0.05

P/bar	1001.7	1036.2	1105.2	1174.1	1280.6	1381.0	1484.5	1725.9	1932.8
ρ/(g/cm^3)	0.850	0.857	0.860	0.868	0.872	0.889	0.898	0.921	0.938
P/bar	2070.7	2312.1	2401.7						
ρ/(g/cm^3)	0.949	0.969	0.980						

T/K = 393.15 w_B 0.05

P/bar	760.3	829.3	1036.2	1381.0	1725.9	2070.7	2415.5
ρ/(g/cm^3)	0.813	0.823	0.848	0.883	0.915	0.945	0.959

T/K = 403.15 w_B 0.05

P/bar	650.0	670.7	691.4	829.3	1036.2	1381.0	1725.9	2139.7	2415.5
ρ/(g/cm^3)	0.778	0.781	0.785	0.804	0.830	0.866	0.896	0.929	0.959

continued

continued

T/K = 423.15 w_B 0.05

P/bar	622.4	691.4	829.3	1036.2	1381.0	1725.9	2070.7	2415.5
ρ/(g/cm^3)	0.748	0.760	0.782	0.809	0.846	0.878	0.906	0.933

Polymer (B): **poly(ethylene-*co*-1-butene)** **2000BY2**
Characterization: M_n/g.mol^{-1} = 175000, M_w/g.mol^{-1} = 177000, 94 mol% 1-butene, Exxon Research and Engineering Co.
Solvent (A): **dimethyl ether-d6** **C_2D_6O** **17222-37-6**

T/K = 383.15 w_B 0.05

P/bar	346.6	415.5	519.0	691.4	1036.2	1381.0	1725.9	2070.7	2415.5
ρ/(g/cm^3)	0.685	0.698	0.715	0.739	0.776	0.804	0.828	0.849	0.867

T/K = 393.15 w_B 0.05

P/bar	346.6	415.5	519.0	691.4	1036.2	1381.0	1725.9	2070.7	2415.5
ρ/(g/cm^3)	0.676	0.690	0.708	0.733	0.772	0.801	0.826	0.847	0.866

T/K = 403.15 w_B 0.05

P/bar	346.6	415.5	519.0	691.4	1036.2	1381.0	1725.9	2070.7	2415.5
ρ/(g/cm^3)	0.659	0.675	0.694	0.720	0.759	0.789	0.814	0.835	0.854

T/K = 423.15 w_B 0.05

P/bar	346.6	415.5	519.0	691.4	1036.2	1381.0	1725.9	2070.7	2415.5
ρ/(g/cm^3)	0.630	0.649	0.671	0.699	0.740	0.772	0.798	0.821	0.840

Polymer (B): **poly(ethylene-*co*-1-butene) (deuterated)** **2000BY1**
Characterization: M_n/g.mol^{-1} = 155000, M_w/g.mol^{-1} = 169000, 3 mol% 1-butene, synthesized from polymerization of deuterated 1,3-butadiene by controlled 1,2-addition, Polymer Source, Inc., Dorval, Quebec
Solvent (A): **ethane** **C_2H_6** **74-84-0**

T/K = 393.15 w_B 0.024

P/bar	1242.8	1311.7	1380.7	1518.5	1725.4	1897.8	2070.1	2242.5	2414.9
ρ/(g/cm^3)	0.511	0.518	0.523	0.532	0.546	0.554	0.563	0.571	0.580

T/K = 403.15 w_B 0.024

P/bar	1242.8	1311.7	1380.7	1518.5	1725.4	1897.8	2070.1	2242.5	2414.9
ρ/(g/cm^3)	0.506	0.513	0.518	0.528	0.541	0.551	0.561	0.570	0.578

T/K = 423.15 w_B 0.024

P/bar	1242.8	1311.7	1380.7	1518.5	1725.4	1897.8	2070.1	2242.5	2414.9
ρ/(g/cm^3)	0.497	0.503	0.508	0.519	0.533	0.543	0.553	0.562	0.571

Polymer (B):	**poly(ethylene-*co*-1-butene)**		**2000BY1**
Characterization:	M_n/g.mol^{-1} = 230000, M_w/g.mol^{-1} = 232000, 20.2 mol% 1-butene, Exxon Research and Engineering Co.		
Solvent (A):	**ethane**	C_2H_6	**74-84-0**
Solvent (C):	**n-pentane-d12**	C_5D_{12}	**2031-90-5**

T/K = 383.15

w_A	0.231	0.231	0.231	0.231	0.231	0.231	0.231	0.231	0.231
w_B	0.049	0.049	0.049	0.049	0.049	0.049	0.049	0.049	0.049
P/bar	277.5	312.0	346.5	380.9	415.4	484.3	553.3	691.2	1035.9
ρ/(g/cm^3)	0.643	0.650	0.656	0.662	0.667	0.677	0.687	0.702	0.736

w_A	0.231	0.231	0.231	0.231
w_B	0.049	0.049	0.049	0.049
P/bar	1380.7	1725.4	2070.1	2414.9
ρ/(g/cm^3)	0.762	0.785	0.805	0.822

T/K = 403.15

w_A	0.231	0.231	0.231	0.231	0.231	0.231	0.231	0.231	0.231
w_B	0.049	0.049	0.049	0.049	0.049	0.049	0.049	0.049	0.049
P/bar	277.5	312.0	346.5	380.9	415.4	484.3	553.3	691.2	1035.9
ρ/(g/cm^3)	0.622	0.630	0.634	0.645	0.651	0.662	0.672	0.689	0.724

w_A	0.231	0.231	0.231	0.231
w_B	0.049	0.049	0.049	0.049
P/bar	1380.7	1725.4	2070.1	2414.9
ρ/(g/cm^3)	0.752	0.775	0.796	0.814

T/K = 423.15

w_A	0.231	0.231	0.231	0.231	0.231	0.231	0.231	0.231	0.231
w_B	0.049	0.049	0.049	0.049	0.049	0.049	0.049	0.049	0.049
P/bar	346.5	380.9	415.4	484.3	553.3	691.2	1035.9	1380.7	1725.4
ρ/(g/cm^3)	0.619	0.627	0.634	0.646	0.657	0.676	0.712	0.741	0.765

w_A	0.231	0.231
w_B	0.049	0.049
P/bar	2070.1	2414.9
ρ/(g/cm^3)	0.786	0.805

Polymer (B):	**poly(ethylene-*co*-1-butene)**		**2000BY1**
Characterization:	M_n/g.mol^{-1} = 230000, M_w/g.mol^{-1} = 232000, 20.2 mol% 1-butene, Exxon Research and Engineering Co.		
Solvent (A):	**n-pentane**	C_5H_{12}	**109-66-0**

T/K = 383.15 w_B 0.053

P/bar	70.7	139.6	208.6	277.5	346.5	518.8	691.2	1035.9	1380.7
ρ/(g/cm^3)	0.560	0.575	0.586	0.596	0.605	0.623	0.638	0.664	0.685

continued

continued

P/bar	2070.1	2414.9
ρ/(g/cm^3)	0.719	0.733

T/K = 403.15 w_B 0.053

P/bar	45.2	70.7	139.6	208.6	277.5	346.5	415.4	484.4	553.3
ρ/(g/cm^3)	0.528	0.536	0.555	0.569	0.581	0.591	0.599	0.607	0.614
P/bar	622.2	691.2	760.1	829.1	898.0	967.0	1035.9	1104.9	1173.8
ρ/(g/cm^3)	0.621	0.627	0.633	0.639	0.644	0.649	0.654	0.659	0.663
P/bar	1242.8	1311.7	1380.7	1449.6	1518.5	1587.5	1656.4	1725.4	1794.3
ρ/(g/cm^3)	0.668	0.672	0.676	0.680	0.684	0.688	0.691	0.695	0.698
P/bar	1863.3	1932.2	2001.2	2070.1					
ρ/(g/cm^3)	0.702	0.705	0.708	0.712					

T/K = 423.15 w_B 0.053

P/bar	105.1	139.6	208.6	277.5	346.5	518.8	691.2	1035.9	1380.7
ρ/(g/cm^3)	0.525	0.535	0.552	0.566	0.577	0.599	0.617	0.645	0.668
P/bar	2070.1	2414.9							
ρ/(g/cm^3)	0.704	0.720							

Polymer (B): **poly(ethylene-*co*-1-butene)** **2000BY1**
Characterization: M_n/g.mol^{-1} = 230000, M_w/g.mol^{-1} = 232000, 20.2 mol% 1-butene, Exxon Research and Engineering Co.
Solvent (A): **n-pentane-d12** C_5D_{12} **2031-90-5**

T/K = 383.15 w_B 0.051

P/bar	70.7	139.6	208.6	277.5	346.5	518.8	691.2	1035.9	1380.7
ρ/(g/cm^3)	0.644	0.662	0.676	0.688	0.699	0.722	0.740	0.771	0.796
P/bar	2070.1	2414.9							
ρ/(g/cm^3)	0.837	0.855							

T/K = 403.15 w_B 0.051

P/bar	70.7	77.6	84.5	91.3	105.1	125.8	139.6	174.1	208.6
ρ/(g/cm^3)	0.617	0.619	0.621	0.624	0.628	0.634	0.638	0.647	0.655
P/bar	277.5	346.5	518.8	689.5	1034.2	1380.7	2070.1	2414.9	
ρ/(g/cm^3)	0.669	0.681	0.705	0.725	0.757	0.783	0.826	0.844	

T/K = 423.15 w_B 0.051

P/bar	108.6	139.6	208.6	277.5	346.5	518.8	691.2	1035.9	1380.7
ρ/(g/cm^3)	0.605	0.616	0.635	0.651	0.664	0.692	0.714	0.747	0.774
P/bar	2070.1	2414.9							
ρ/(g/cm^3)	0.818	0.837							

Polymer (B): **poly(ethylene glycol-*b*-propylene glycol-*b*-ethylene glycol)** **2009FRA**
Characterization: M_n/g.mol^{-1} = 2000, M_w/g.mol^{-1} = 2100, 14 wt% PEG, Aldrich Chem. Co., Inc., Milwaukee, WI
Solvent (A): **dimethylsulfoxide** **C_2H_6OS** **67-68-5**

T/K = 298.15

w_B	0.0000	0.0347	0.0671	0.1427	0.2102	0.2927	0.4061	0.4710
ρ/(g/cm^3)	1.09532	1.09268	1.09026	1.08463	1.07968	1.07354	1.06518	1.06030
V^E/(cm^3/g)	0.0	−0.00022	−0.00045	−0.00098	−0.00147	−0.00194	−0.00256	−0.00278

w_B	0.5717	0.6791	0.7731	0.8617	0.9684	1.0000
ρ/(g/cm^3)	1.05262	1.04429	1.03675	1.02943	1.02019	1.01738
V^E/(cm^3/g)	−0.00295	−0.00289	−0.00249	−0.00183	−0.00049	0.0

T/K = 313.15

w_B	0.0000	0.0347	0.0671	0.1427	0.2102	0.2927	0.4061	0.4710
ρ/(g/cm^3)	1.07983	1.07723	1.07493	1.06937	1.06456	1.05861	1.05061	1.04595
V^E/(cm^3/g)	0.0	−0.00015	−0.00039	−0.00075	−0.00117	−0.00156	−0.00215	−0.00237

w_B	0.5717	0.6791	0.7731	0.8617	0.9684	1.0000
ρ/(g/cm^3)	1.03869	1.03073	1.02359	1.01664	1.00789	1.00522
V^E/(cm^3/g)	−0.00261	−0.00256	−0.00226	−0.00166	−0.00046	0.0

Polymer (B): **poly(ethylene glycol-*b*-propylene glycol-*b*-ethylene glycol)** **2009FRA**
Characterization: M_n/g.mol^{-1} = 1240, M_w/g.mol^{-1} = 1290, 15 wt% PEG, Aldrich Chem. Co., Inc., Milwaukee, WI
Solvent (A): **dimethylsulfoxide** **C_2H_6OS** **67-68-5**

T/K = 298.15

w_B	0.0000	0.0248	0.0701	0.1459	0.2142	0.2884	0.4071	0.4902
ρ/(g/cm^3)	1.09532	1.09340	1.09006	1.08452	1.07954	1.07417	1.06552	1.05945
V^E/(cm^3/g)	0.0	−0.00016	−0.00044	−0.00093	−0.00135	−0.00179	−0.00234	−0.00264

w_B	0.5644	0.6861	0.7805	0.8595	0.9534	1.0000
ρ/(g/cm^3)	1.05392	1.04462	1.03724	1.03084	1.02300	1.01899
V^E/(cm^3/g)	−0.00276	−0.00263	−0.00227	−0.00168	−0.00066	0.0

T/K = 313.15

w_B	0.0000	0.0248	0.0701	0.1459	0.2142	0.2883	0.4071	0.4902
ρ/(g/cm^3)	1.07983	1.07794	1.07473	1.06947	1.06468	1.05961	1.05137	1.04560
V^E/(cm^3/g)	0.0	−0.00015	−0.00043	−0.00095	−0.00133	−0.00183	−0.00242	−0.00277

w_B	0.5644	0.6861	0.7806	0.8595	0.9534	1.0000
ρ/(g/cm^3)	1.04033	1.03147	1.02433	1.01815	1.01051	1.00660
V^E/(cm^3/g)	−0.00291	−0.00284	−0.00243	−0.00182	−0.00071	0.0

Polymer (B): **poly(ethylene glycol-*b*-propylene glycol-*b*-ethylene glycol)** **2009FRA**

Characterization: M_n/g.mol^{-1} = 2720, M_w/g.mol^{-1} = 2890, 16 wt% PEG, Aldrich Chem. Co., Inc., Milwaukee, WI

Solvent (A): **dimethylsulfoxide** **C_2H_6OS** **67-68-5**

T/K = 298.15

w_B	0.0000	0.0404	0.0642	0.1405	0.2113	0.2846	0.4078	0.4914
ρ/(g/cm^3)	1.09532	1.09233	1.09055	1.08495	1.07984	1.07453	1.06572	1.05972
V^E /(cm^3/g)	0.0	−0.00016	−0.00026	−0.00062	−0.00099	−0.00131	−0.00186	−0.00213
w_B	0.5629	0.6864	0.7715	0.8721	0.9595	1.0000		
ρ/(g/cm^3)	1.05456	1.04544	1.03899	1.03112	1.02402	1.02065		
V^E /(cm^3/g)	−0.00230	−0.00227	−0.00203	−0.00140	−0.00052	0.0		

T/K = 313.15

w_B	0.0000	0.0404	0.0642	0.1405	0.2113	0.2846	0.4078	0.4914
ρ/(g/cm^3)	1.07983	1.07661	1.07491	1.06959	1.06475	1.05979	1.05146	1.04584
V^E /(cm^3/g)	0.0	−0.00017	−0.00025	−0.00062	−0.00099	−0.00140	−0.00199	−0.00235
w_B	0.5629	0.6864	0.7716	0.8721	0.9595	1.0000		
ρ/(g/cm^3)	1.04097	1.03226	1.02602	1.01848	1.01158	1.00828		
V^E /(cm^3/g)	−0.00256	−0.00253	−0.00220	−0.00156	−0.00058	0.0		

Polymer (B): **poly(ethylene glycol-*b*-propylene glycol-*b*-ethylene glycol)** **2009FRA**

Characterization: M_n/g.mol^{-1} = 2110, M_w/g.mol^{-1} = 2210, 51 wt% PEG, Aldrich Chem. Co., Inc., Milwaukee, WI

Solvent (A): **dimethylsulfoxide** **C_2H_6OS** **67-68-5**

T/K = 298.15

w_B	0.0000	0.0335	0.0642	0.1462	0.2241	0.3081	0.4759	0.4939
ρ/(g/cm^3)	1.09532	1.09440	1.09342	1.09100	1.08870	1.08620	1.08080	1.08014
V^E /(cm^3/g)	0.0	−0.00030	−0.00048	−0.00112	−0.00173	−0.00235	−0.00322	−0.00325
w_B	0.5620	0.6804	0.7839	0.8703	0.9390	1.0000		
ρ/(g/cm^3)	1.07778	1.07320	1.06873	1.06452	1.06101	1.05758		
V^E /(cm^3/g)	−0.00344	−0.00334	−0.00282	−0.00194	−0.00107	0.0		

T/K = 313.15

w_B	0.0000	0.0335	0.0642	0.1462	0.2241	0.3081	0.4759	0.4939
ρ/(g/cm^3)	1.07983	1.07883	1.07806	1.07583	1.07381	1.07155	1.06667	1.06608
V^E /(cm^3/g)	0.0	−0.00027	−0.00056	−0.00116	−0.00180	−0.00242	−0.00330	−0.00333
w_B	0.5620	0.6804	0.7839	0.8703	0.9390	1.0000		
ρ/(g/cm^3)	1.06387	1.05966	1.05553	1.05163	1.04831	1.04504		
V^E /(cm^3/g)	−0.00348	−0.00338	−0.00287	−0.00202	−0.00111	0.0		

Polymer (B): **poly(ethylene glycol-*b*-propylene glycol-*b*-ethylene glycol)** **2008COM**
Characterization: M_n/g.mol^{-1} = 1220, M_w/g.mol^{-1} = 1320, 19 mol% PEG, Aldrich Chem. Co., Inc., Milwaukee, WI
Solvent (A): **2-methyltetrahydrofuran** $C_5H_{10}O$ **96-47-9**

T/K = 298.15

w_B	0.0000	0.0423	0.0938	0.1630	0.2670	0.3525	0.4740	0.5565
ρ/(g/cm^3)	0.84800	0.85434	0.86213	0.87276	0.88911	0.90290	0.92300	0.93708
V^E /(cm^3/g)	0.0	−0.0004	−0.0008	−0.0013	−0.0018	−0.0021	−0.0022	−0.0022
w_B	0.6247	0.7482	0.8195	0.8942	0.9629	1.0000		
ρ/(g/cm^3)	0.94890	0.97104	0.98414	0.99813	1.01132	1.01856		
V^E /(cm^3/g)	−0.0020	−0.0017	−0.0013	−0.0008	−0.0003	0.0		

T/K = 313.15

w_B	0.0000	0.0423	0.0938	0.1630	0.2670	0.3525	0.4740	0.5565
ρ/(g/cm^3)	0.83238	0.83887	0.84690	0.85780	0.87456	0.88875	0.90944	0.92362
V^E /(cm^3/g)	0.0	−0.0005	−0.0011	−0.0017	−0.0024	−0.0029	−0.0032	−0.0032
w_B	0.6247	0.7482	0.8195	0.8942	0.9629	1.0000		
ρ/(g/cm^3)	0.93596	0.95850	0.97183	0.98605	0.99938	1.00667		
V^E /(cm^3/g)	−0.0030	−0.0024	−0.0019	−0.0012	−0.0005	0.0		

Polymer (B): **poly(ethylene glycol-*b*-propylene glycol-*b*-ethylene glycol)** **2008COM**
Characterization: M_n/g.mol^{-1} = 2280, M_w/g.mol^{-1} = 2510, 59 mol% PEG, Aldrich Chem. Co., Inc., Milwaukee, WI
Solvent (A): **2-methyltetrahydrofuran** $C_5H_{10}O$ **96-47-9**

T/K = 298.15

w_B	0.0000	0.0456	0.0981	0.1713	0.2670	0.3527	0.4755	0.5663
ρ/(g/cm^3)	0.84800	0.85613	0.86559	0.87910	0.89723	0.91397	0.93876	0.95779
V^E /(cm^3/g)	0.0	−0.0006	−0.0010	−0.0017	−0.0023	−0.0027	−0.0029	−0.0029
w_B	0.6320	0.7441	0.8145	0.8885	0.9648	1.0000		
ρ/(g/cm^3)	0.97193	0.99683	1.01296	1.03040	1.04880	1.05747		
V^E /(cm^3/g)	−0.0027	−0.0022	−0.0018	−0.0012	−0.0004	0.0		

T/K = 313.15

w_B	0.0000	0.0456	0.0981	0.1713	0.2670	0.3527	0.4755	0.5663
ρ/(g/cm^3)	0.83238	0.84071	0.85042	0.86425	0.88276	0.89989	0.92529	0.94467
V^E /(cm^3/g)	0.0	−0.0007	−0.0014	−0.0023	−0.0031	−0.0037	−0.0041	−0.0040
w_B	0.6320	0.7441	0.8145	0.8885	0.9648	1.0000		
ρ/(g/cm^3)	0.95912	0.98444	1.00080	1.01832	1.03697	1.04567		
V^E /(cm^3/g)	−0.0039	−0.0032	−0.0026	−0.0016	−0.0006	0.0		

Polymer (B): **poly(ethylene oxide-*b*-propylene oxide)** **2004TAB**
Characterization: M_n/g.mol^{-1} = 20700, $EO_{316}PO_{94}$, 77.1 mol% or 71.8 wt% ethylene oxide, synthesized in the laboratory

Solvent (A): **water** **H_2O** **7732-18-5**

T/K = 278.15

c_B/(g/L)	1.003	1.505	2.046	2.529	3.024	3.921	5.038	7.458	9.991
ρ/(g/cm^3)	1.00013	1.00021	1.00030	1.00038	1.00046	1.00060	1.00078	1.00118	1.00158

c_B/(g/L)	12.092	14.991	17.409	19.838
ρ/(g/cm^3)	1.00192	1.00238	1.00277	1.00316

T/K = 283.15

c_B/(g/L)	1.003	1.505	2.046	2.529	3.024	3.921	5.038	7.458	9.991
ρ/(g/cm^3)	0.99986	0.99994	1.00002	1.00010	1.00018	1.00032	1.00049	1.00087	1.00127

c_B/(g/L)	12.092	14.991	17.409	19.838
ρ/(g/cm^3)	1.00159	1.00204	1.00242	1.00279

T/K = 288.15

c_B/(g/L)	1.003	1.505	2.046	2.529	3.024	3.921	5.038	7.458	9.991
ρ/(g/cm^3)	0.99925	0.99933	0.99941	0.99948	0.99956	0.99969	0.99986	1.00023	1.00061

c_B/(g/L)	12.092	14.991	17.409	19.838
ρ/(g/cm^3)	1.00092	1.00135	1.00171	1.00207

T/K = 293.15

c_B/(g/L)	0.989	1.322	1.846	2.453	2.995	4.024	5.305	7.581	10.162
ρ/(g/cm^3)	0.99835	0.99840	0.99848	0.99856	0.99864	0.99879	0.99897	0.99929	0.99965

c_B/(g/L)	12.340	15.132	17.467	19.935
ρ/(g/cm^3)	0.99995	1.00034	1.00067	1.00101

T/K = 298.15

c_B/(g/L)	0.989	1.322	1.846	2.453	2.995	4.024	5.305	7.581	10.162
ρ/(g/cm^3)	0.99718	0.99723	0.99729	0.99737	0.99744	0.99758	0.99774	0.99804	0.99837

c_B/(g/L)	12.340	15.132	17.467	19.935
ρ/(g/cm^3)	0.99864	0.99900	0.99929	0.99960

T/K = 303.15

c_B/(g/L)	0.989	1.322	1.846	2.453	2.995	4.024	5.305	7.581	10.162
ρ/(g/cm^3)	0.99577	0.99581	0.99588	0.99596	0.99602	0.99615	0.99631	0.99660	0.99692

c_B/(g/L)	12.340	15.132	17.467	19.935
ρ/(g/cm^3)	0.99719	0.99753	0.99782	0.99812

continued

continued

T/K = 308.15

c_B/(g/L)	0.989	1.322	1.846	2.453	2.995	4.024	5.305	7.581	10.162
ρ/(g/cm^3)	0.99415	0.99419	0.99426	0.99433	0.99440	0.99452	0.99468	0.99495	0.99527
c_B/(g/L)	12.340	15.132	17.467	19.935					
ρ/(g/cm^3)	0.99553	0.99586	0.99614	0.99644					

T/K = 313.15

c_B/(g/L)	0.989	1.322	1.846	2.453	2.995	4.024	5.305	7.581	10.162
ρ/(g/cm^3)	0.99233	0.99237	0.99244	0.99251	0.99257	0.99269	0.99285	0.99312	0.99342
c_B/(g/L)	12.340	15.132	17.467	19.935					
ρ/(g/cm^3)	0.99368	0.99400	0.99428	0.99456					

T/K = 318.15

c_B/(g/L)	0.989	1.322	1.846	2.453	2.995	4.024	5.305	7.581	10.162
ρ/(g/cm^3)	0.99032	0.99037	0.99043	0.99050	0.99056	0.99068	0.99083	0.99109	0.99139
c_B/(g/L)	12.340	15.132	17.467	19.935					
ρ/(g/cm^3)	0.99164	0.99196	0.99223	0.99251					

T/K = 323.15

c_B/(g/L)	0.989	1.322	1.846	2.453	2.995	4.024	5.305	7.581	10.162
ρ/(g/cm^3)	0.98814	0.98818	0.98824	0.98831	0.98837	0.98849	0.98864	0.98890	0.98919
c_B/(g/L)	12.340	15.132	17.467	19.935					
ρ/(g/cm^3)	0.98944	0.98975	0.99001	0.99029					

Polymer (B): **poly(ethylene oxide-*b*-propylene oxide-*b*-ethylene oxide)** **2006DE1**

Characterization: M_n/g.mol^{-1} = 11400, $EO_{103}PO_{39}EO_{103}$, 15.9 mol% propylene oxide, F88, BASF AG, Germany

Solvent (A): **water** **H_2O** **7732-18-5**

T/K = 298.15

c_B/(mol/kg)	0.00000	0.00126	0.00208	0.00382	0.00393	0.00489	0.00491	0.00559
ρ/(g/cm^3)	0.99705	0.99919	1.00057	1.00339	1.00361	1.00516	1.00515	1.00622
c_B/(mol/kg)	0.00668	0.00739	0.00749	0.00831	0.01021	0.01245	0.01376	0.01450
ρ/(g/cm^3)	1.00789	1.00900	1.00913	1.01036	1.01314	1.01634	1.01813	1.01921
c_B/(mol/kg)	0.01526	0.01667	0.01939	0.02007	0.02126	0.02350		
ρ/(g/cm^3)	1.02018	1.02201	1.02551	1.02627	1.02757	1.02998		

T/K = 308.15

c_B/(mol/kg)	0.00000	0.00126	0.00201	0.00208	0.00292	0.00302	0.00382	0.00393
ρ/(g/cm^3)	0.99404	0.99609	0.99728	0.99739	0.99870	0.99886	1.00009	1.00027

continued

continued

c_B/(mol/kg)	0.00489	0.00559	0.00611	0.00668	0.00713	0.00749	0.00831	0.01021
ρ/(g/cm^3)	1.00175	1.00274	1.00353	1.00428	1.00499	1.00541	1.00653	1.00905
c_B/(mol/kg)	0.01140	0.01245	0.01376	0.01450	0.01667	0.01701	0.01939	0.02126
ρ/(g/cm^3)	1.01058	1.01183	1.01339	1.01441	1.01687	1.01732	1.02010	1.02212
c_B/(mol/kg)	0.02140	0.02342	0.02350					
ρ/(g/cm^3)	1.02226	1.02443	1.02451					

Polymer (B): **poly(ethylene oxide-*b*-propylene oxide-*b*-ethylene oxide)** **2006DE1**

Characterization: M_n/g.mol^{-1} = 14600, $EO_{132}PO_{50}EO_{132}$, 15.9 mol% propylene oxide, F108, BASF AG, Germany

Solvent (A): **water** **H_2O** **7732-18-5**

T/K = 298.15

c_B/(mol/kg)	0.00000	0.00050	0.00067	0.00113	0.00197	0.00264	0.00335	0.00371
ρ/(g/cm^3)	0.99705	0.99816	0.99853	0.99955	1.00136	1.00276	1.00425	1.00498
c_B/(mol/kg)	0.00482	0.00545	0.00606	0.00672	0.00769	0.00828	0.00835	0.00941
ρ/(g/cm^3)	1.00723	1.00849	1.00965	1.01091	1.01265	1.01370	1.01379	1.01571
c_B/(mol/kg)	0.00982	0.01058	0.01127	0.01274	0.01482	0.01636	0.01670	0.01882
ρ/(g/cm^3)	1.01630	1.01761	1.01868	1.02105	1.02432	1.02667	1.02721	1.03023

T/K = 308.15

c_B/(mol/kg)	0.00067	0.00197	0.00264	0.00335	0.00371	0.00482	0.00606	0.00672
ρ/(g/cm^3)	0.99543	0.99803	0.99930	1.00064	1.00133	1.00333	1.00555	1.00675
c_B/(mol/kg)	0.00769	0.00828	0.00835	0.00982	0.01090	0.01127	0.01274	0.01482
ρ/(g/cm^3)	1.00845	1.00947	1.00958	1.01210	1.01388	1.01446	1.01683	1.02012

Polymer (B): **poly(ethylene oxide-*b*-propylene oxide-*b*-ethylene oxide)** **2006DE1**

Characterization: M_n/g.mol^{-1} = 8350, $EO_{76}PO_{29}EO_{76}$, 16.0 mol% propylene oxide, F68, BASF AG, Germany

Solvent (A): **water** **H_2O** **7732-18-5**

T/K = 298.15

c_B/(mol/kg)	0.00000	0.00301	0.00524	0.00712	0.00813	0.00994	0.01197	0.01443
ρ/(g/cm^3)	0.99705	1.00074	1.00339	1.00559	1.00671	1.00881	1.01096	1.01354
c_B/(mol/kg)	0.01647	0.01687	0.01900	0.02007	0.02668	0.03065	0.03905	0.04439
ρ/(g/cm^3)	1.01565	1.01607	1.01821	1.01927	1.02547	1.02895	1.03571	1.03944
c_B/(mol/kg)	0.04950	0.05235	0.05735	0.06981	0.071515			
ρ/(g/cm^3)	1.04272	1.04444	1.04732	1.05356	1.05438			

continued

continued

T/K = 308.15

c_B/(mol/kg)	0.00000	0.00301	0.00524	0.00712	0.00813	0.00994	0.01197	0.01443
ρ/(g/cm^3)	0.99404	0.99755	1.00007	1.00215	1.00322	1.00519	1.00725	1.00967
c_B/(mol/kg)	0.01647	0.01687	0.01900	0.02007	0.02129	0.02523	0.02668	0.03065
ρ/(g/cm^3)	1.01162	1.01204	1.01403	1.01509	1.01603	1.01930	1.02046	1.02352
c_B/(mol/kg)	0.03572	0.03905	0.04439	0.04950	0.05089	0.05235		
ρ/(g/cm^3)	1.02728	1.02939	1.03289	1.03607	1.03708	1.03772		

Polymer (B): **poly(ethylene oxide-*b*-propylene oxide-*b*-ethylene oxide)** **2006DE1**

Characterization: M_n/g.mol^{-1} = 1900, $EO_{11}PO_{16}EO_{11}$, 42.1 mol% propylene oxide, L35, BASF AG, Germany

Solvent (A): **water** $\mathbf{H_2O}$ **7732-18-5**

T/K = 298.15

c_B/(mol/kg)	0.01529	0.04002	0.06275	0.07064	0.07566	0.1076	0.1272	0.1417
ρ/(g/cm^3)	1.00068	1.00619	1.01098	1.01252	1.01350	1.01942	1.02278	1.02518
c_B/(mol/kg)	0.1840	0.2167	0.2503	0.2987	0.3641	0.4400	0.4749	0.5812
ρ/(g/cm^3)	1.03143	1.03573	1.03969	1.04440	1.04944	1.05357	1.05507	1.05866
c_B/(mol/kg)	0.6848	0.7383	0.7981	0.9084	0.9586	1.3439	1.4572	
ρ/(g/cm^3)	1.06117	1.06241	1.06316	1.06455	1.06478	1.06722	1.06707	

T/K = 305.15

c_B/(mol/kg)	0.00000	0.00541	0.00717	0.0153	0.03872	0.06275	0.07409	0.1082
ρ/(g/cm^3)	0.99503	0.99629	0.99668	0.99852	1.00357	1.00833	1.01048	1.01651
c_B/(mol/kg)	0.1417	0.1943	0.2167	0.2503	0.3641	0.4749	0.5812	0.6167
ρ/(g/cm^3)	1.02167	1.02875	1.03134	1.03475	1.04343	0.04878	1.05238	1.05328
c_B/(mol/kg)	0.7383	0.9084	1.3439					
ρ/(g/cm^3)	1.05666	1.05864	1.06144					

T/K = 308.15

c_B/(mol/kg)	0.00000	0.00517	0.01533	0.04002	0.07064	0.07566	0.1076	0.1272
ρ/(g/cm^3)	0.99404	0.99522	0.99748	1.00270	1.00863	1.00954	1.01507	1.01810
c_B/(mol/kg)	0.1840	0.2242	0.2355	0.2503	0.2987	0.3683	0.4400	0.4731
ρ/(g/cm^3)	1.02589	1.03039	1.03152	1.03349	1.03691	1.04161	1.04500	1.04640
c_B/(mol/kg)	0.5801	0.6848	0.7981	0.9586	1.4572			
ρ/(g/cm^3)	1.04995	1.05256	1.05426	1.05644	1.05900			

6.3. References

2000BY1 Byun, H.-S., DiNoia, T.P., and McHugh, M.A., High-pressure densities of ethane, pentane, pentane-d12, 25.5 wt% ethane in pentane-d12, 2.4 wt% deuterated poly(ethylene-*co*-butene) (PEB) in ethane, 5.3 wt% hydrogenated PEB in pentane, 5.1 wt % hydrogenated PEB in pentane-d12, and 4.9 wt % hydrogenated PEB in pentane-d12 + 23.1 wt % ethane, *J. Chem. Eng. Data*, 45, 810, 2000.

2000BY2 Byun, H.-S., Kim, K., and Lee, H.-S., High-pressure phase behavior and mixture density of binary poly(ethylene-*co*-butene)-dimethyl ether system, *Hwahak Konghak*, 38, 826, 2000.

2000CAP Capt, L. and Kamal, M.R., The pressure-volume-temperature behavior of polyethylene melts, *Intern. Polym. Process.*, 15, 83, 2000.

2000SA1 Sato, Y., Inohara, K., Takishima, S., Masuoka, H., Imaizumi, M., Yamamoto, H., and Takasugi, M., Pressure-volume-temperature behavior of polylactide, poly(butylene succinate), and poly(butylene succinate-*co*-adipate), *Polym. Eng. Sci.*, 40, 2602, 2000.

2000SA2 Sato, Y., Tsuboi, A., Sorakubo, A., Takishima, S., Masuoka, H., and Ishikawa, T., Vapor-liquid equilibrium ratios for hexane at infinite dilution in ethylene + impact polypropylene copolymer and propylene + impact polypropylene copolymer, *Fluid Phase Equil.*, 170, 49, 2000.

2000TOM Tomova, D., Kressler, J., and Radusch, H.-J., Phase behaviour in ternary polyamide 6/polyamide 66/elastomer blends (experimental data by D. Tomova), *Polymer*, 41, 7773, 2000.

2001MEK Mekhilef, N., Viscoelastic and pressure-volume-temperature properties of poly(vinylidene fluoride) and poly(vinylidene fluoride)-hexafluoropropylene copolymers (experimental data by N. Mekhilef), *J. Appl. Polym. Sci.*, 80, 230, 2001.

2001PRI Privalko, V.P., Korskanov, V.V., Privalko, E.G., Dolgoshey, V.I., Huhn, W., and Rieger, B., Thermodynamic properties and thermoelastic behavior of the alternating terpolymers of ethene, propene, and carbon monoxide in the melt state, *J. Macromol. Sci.-Phys. B*, 40, 83, 2001.

2004DLU Dlubek, G., Gupta, A.S., Piontek, J., Krause-Rehberg, R., Kaspar, H., and Lochhaas, K.H., Temperature dependence of the free volume in fluoroelastomers from positron lifetime and PVT experiments (experimental data by G. Dlubek), *Macromolecules*, 37, 6606, 2004.

2004TAB Taboada, P., Barbosa, S., and Mosquera, V., Thermodynamic properties of a diblock copolymer of poly(oxyethylene) and poly(oxypropylene) in aqueous solution, *Langmuir*, 20, 8903, 2004.

2005BRI Briatico-Vangosa, F. and Rink, M., Dilatometric behavior and glass transition in a styrene-acrylonitrile copolymer, *J. Polym. Sci.: Part B: Polym. Phys.*, 43, 1904, 2005.

2005KIL Kilburn, D., Dlubek, G., Pionteck, J., Bamford, D., and Alam, M.A., Microstructure of free volume in SMA copolymers I. Free volume from Simha-Somcynsky analysis of *PVT* experiments (experimental data by G. Dlubek), *Polymer*, 46, 859, 2005.

2006BLO Blochowiak, M., Pakula, T., Butt, H.-J., Bruch, M., and Floudas, G., Thermodynamics and rheology of cycloolefin copolymers, *J. Chem. Phys.*, 124, 134903, 2006.

2006DE1 DeLisi, R., Lazzara, G., Lombardo, R., Milioto, S., Muratore, N., and Liveri, M.L.T., Thermodynamic behavior of non-ionic triblock copolymers in water at three temperatures, *J. Solution Chem.*, 35, 659, 2006.

2006DE2 DeLisi, R., Lazzara, G., Milioto, S., and Muratore, N., Volumes of aqueous block copolymers based on poly(propylene oxides) and poly(ethylene oxides) in a large temperature range: A quantitative description, *J. Chem. Thermodyn.*, 38, 1344, 2006.

2007FU2 Funke, Z., Hotani, Y., Ougizawa, T., Kressler, J., and Kammer, H.-W., Equation-of-state properties and surface tension of ethylene-vinyl alcohol random copolymers (experimental data by Z. Funke), *Eur. Polym. J.*, 43, 2371, 2007.

2007SAT Sato, Y., Hashiguchi, H., Inohara, K., Takishima, S., and Masuoka, H., PVT properties of polyethylene copolymer melts, *Fluid Phase Equil.*, 257, 124, 2007.

2008COM Comelli, F., Bigi, A., Vitalini, D., Rubini, K., and Francesconi, R., Densities, viscosities, refractive indices, and heat capacities of poly(propylene glycols) or poly(ethylene glycol)-poly(propylene glycol)-poly(ethylene glycol)-*block*-copolymers + 2-methyltetrahydrofuran at (298.15 and 313.15) K and at atmospheric pressure, *J. Chem. Eng. Data*, 53, 1302, 2008.

2008DLU Dlubek, G., Pionteck, J., Rätzke, K., Kruse, J., and Faupel, F., Temperature dependence of the free volume in amorphous Teflon AF1600 and AF2400: A pressure-volume-temperature and positron lifetime study (experimental data by G. Dlubek), *Macromolecules*, 41, 6125, 2008.

2009FRA Francesconi, R., Bigi, A., Vitalini, D., Rubini, K., and Comelli, F., Densities, viscosities, refractive indices, and heat capacities of four poly(ethylene glycol)-poly(propylene glycol)-poly(ethylene glycol)-*block*-copolymers + dimethyl sulfoxide at (298.15 and 313.15) K and at atmospheric pressure, *J. Chem. Eng. Data*, 54, 956, 2009.

7. SECOND VIRIAL COEFFICIENTS (A_2) OF COPOLYMER SOLUTIONS

7.1. Experimental A_2 data

Copolymer	M_n/ g/mol	M_w/ g/mol	Solvent	T/ K	A_2 10^4/ cm^{3}mol/g^2	Ref.
Carboxymethylcellulose-*g*-poly(*N,N*-dihexylacrylamide)						
(1.1 mol% *N,N*-dihexylacrylamide		730000	water (0.5 M NaCl)	298.15	2.4	2008VID
(5.9 mol% *N,N*-dihexylacrylamide		590000	water (0.5 M NaCl)	298.15	0.71	2008VID
Poly(acrylamide-*co*-*N*-isopropylacrylamide)						
(15 mol% acrylamide)		3100000	water	298.15	3.01	1994MUM
(30 mol% acrylamide)		4500000	water	298.15	3.31	1994MUM
(45 mol% acrylamide)		3900000	water	298.15	3.46	1994MUM
(60 mol% acrylamide)		2200000	water	298.15	3.62	1994MUM
Poly(acrylamide-*co*-maleic anhydride)						
(50 mol% acrylamide)		24400	water	298.15	3.0	1983ION
Poly(acrylamide-*co*-methyl methacrylate)						
(25 mol% methyl methacrylate)		36400	*N*-methylformamide	298.15	6.22	1994OHS
(25 mol% methyl methacrylate)		56200	*N*-methylformamide	298.15	5.69	1994OHS
(25 mol% methyl methacrylate)		88800	*N*-methylformamide	298.15	5.23	1994OHS
(25 mol% methyl methacrylate)		101600	*N*-methylformamide	298.15	5.14	1994OHS
(25 mol% methyl methacrylate)		116700	*N*-methylformamide	298.15	4.84	1994OHS
(25 mol% methyl methacrylate)		138600	*N*-methylformamide	298.15	4.37	1994OHS
(25 mol% methyl methacrylate)		166500	*N*-methylformamide	298.15	4.15	1994OHS
Poly(acrylonitrile-*co*-butyl methacrylate)						
(50 mol% acrylonitrile)		171400	2-butanone	298.15	3.58	1977ARU
(50 mol% acrylonitrile)		277900	2-butanone	298.15	2.51	1977ARU
(50 mol% acrylonitrile)		459200	2-butanone	298.15	2.21	1977ARU
(50 mol% acrylonitrile)		496900	2-butanone	298.15	1.60	1977ARU
(50 mol% acrylonitrile)		570300	2-butanone	298.15	1.37	1977ARU
(50 mol% acrylonitrile)		683100	2-butanone	298.15	1.41	1977ARU
(50 mol% acrylonitrile)		292300	*N,N*-dimethylformamide	298.15	4.13	1977ARU
(50 mol% acrylonitrile)		403200	*N,N*-dimethylformamide	298.15	4.71	1977ARU
(50 mol% acrylonitrile)		463700	*N,N*-dimethylformamide	298.15	4.89	1977ARU

Copolymer	M_n/ g/mol	M_w/ g/mol	**Solvent**	*T*/ K	A_2 10^4/ cm^3mol/g^2	**Ref.**
(50 mol% acrylonitrile)		485700	*N,N*-dimethylformamide	298.15	4.70	1977ARU
(50 mol% acrylonitrile)		598800	*N,N*-dimethylformamide	298.15	4.35	1977ARU
(50 mol% acrylonitrile)		610600	*N,N*-dimethylformamide	298.15	4.86	1977ARU
(50 mol% acrylonitrile)		771100	*N,N*-dimethylformamide	298.15	3.89	1977ARU
Poly(acrylonitrile-*co*-α-methylstyrene)						
(46 mol% acrylonitrile)		45000	tetrahydrofuran	298.15	7.19	1979GL2
(46 mol% acrylonitrile)		54000	tetrahydrofuran	298.15	7.14	1979GL2
(46 mol% acrylonitrile)		65000	tetrahydrofuran	298.15	7.28	1979GL2
(46 mol% acrylonitrile)		72000	tetrahydrofuran	298.15	9.42	1979GL2
(46 mol% acrylonitrile)		75000	tetrahydrofuran	298.15	9.06	1979GL2
(46 mol% acrylonitrile)		79000	tetrahydrofuran	298.15	7.99	1979GL2
(46 mol% acrylonitrile)		93000	tetrahydrofuran	298.15	8.35	1979GL2
(46 mol% acrylonitrile)		104000	tetrahydrofuran	298.15	8.44	1979GL2
(46 mol% acrylonitrile)		105000	tetrahydrofuran	298.15	6.48	1979GL2
(46 mol% acrylonitrile)		137000	tetrahydrofuran	298.15	7.83	1979GL2
(46 mol% acrylonitrile)		160000	tetrahydrofuran	298.15	8.26	1979GL2
(46 mol% acrylonitrile)		170000	tetrahydrofuran	298.15	7.64	1979GL2
(46 mol% acrylonitrile)		208000	tetrahydrofuran	298.15	8.88	1979GL2
(46 mol% acrylonitrile)		218000	tetrahydrofuran	298.15	7.55	1979GL2
(46 mol% acrylonitrile)		226000	tetrahydrofuran	298.15	6.40	1979GL2
(46 mol% acrylonitrile)		268000	tetrahydrofuran	298.15	7.64	1979GL2
(46 mol% acrylonitrile)		313000	tetrahydrofuran	298.15	6.93	1979GL2
(46 mol% acrylonitrile)		332000	tetrahydrofuran	298.15	7.19	1979GL2
(46 mol% acrylonitrile)		417000	tetrahydrofuran	298.15	6.66	1979GL2
(46 mol% acrylonitrile)		45000	tetrahydrofuran	298.15	7.19	1979GL3
(46 mol% acrylonitrile)		54000	tetrahydrofuran	298.15	7.14	1979GL3
(46 mol% acrylonitrile)		65000	tetrahydrofuran	298.15	7.28	1979GL3
(46 mol% acrylonitrile)		72000	tetrahydrofuran	298.15	9.42	1979GL3
(46 mol% acrylonitrile)		75000	tetrahydrofuran	298.15	9.06	1979GL3
(46 mol% acrylonitrile)		79000	tetrahydrofuran	298.15	7.99	1979GL3
(46 mol% acrylonitrile)		93000	tetrahydrofuran	298.15	8.35	1979GL3
(46 mol% acrylonitrile)		104000	tetrahydrofuran	298.15	8.44	1979GL3
(46 mol% acrylonitrile)		105000	tetrahydrofuran	298.15	6.48	1979GL3
(46 mol% acrylonitrile)		137000	tetrahydrofuran	298.15	7.83	1979GL3
(46 mol% acrylonitrile)		160000	tetrahydrofuran	298.15	8.26	1979GL3
(46 mol% acrylonitrile)		170000	tetrahydrofuran	298.15	7.64	1979GL3
(46 mol% acrylonitrile)		208000	tetrahydrofuran	298.15	8.88	1979GL3
(46 mol% acrylonitrile)		218000	tetrahydrofuran	298.15	7.55	1979GL3
(46 mol% acrylonitrile)		226000	tetrahydrofuran	298.15	6.40	1979GL3
(46 mol% acrylonitrile)		268000	tetrahydrofuran	298.15	7.64	1979GL3
(46 mol% acrylonitrile)		313000	tetrahydrofuran	298.15	6.93	1979GL3
(46 mol% acrylonitrile)		332000	tetrahydrofuran	298.15	7.19	1979GL3
(46 mol% acrylonitrile)		417000	tetrahydrofuran	298.15	6.66	1979GL3

Copolymer	M_n/ g/mol	M_w/ g/mol	Solvent	T/ K	A_2 10^4/ cm^3mol/g^2	Ref.
Poly(acrylonitrile-*co*-vinyl acetate)						
(11.0 wt% vinyl acetate)		42200	*N,N*-dimethylformamide	298.15	7.3	1985NIK
(11.6 wt% vinyl acetate)		79400	*N,N*-dimethylformamide	298.15	15.5	1985NIK
(11.8 wt% vinyl acetate)		177800	*N,N*-dimethylformamide	298.15	13.1	1985NIK
(13.0 wt% vinyl acetate)		141300	*N,N*-dimethylformamide	298.15	9.2	1985NIK
(14.5 wt% vinyl acetate)		158500	*N,N*-dimethylformamide	298.15	16.0	1985NIK
(14.5 wt% vinyl acetate)		69100	*N,N*-dimethylformamide	298.15	17.0	1985NIK
(15.5 wt% vinyl acetate)		148600	*N,N*-dimethylformamide	298.15	13.8	1985NIK
(16.9 wt% vinyl acetate)		237100	*N,N*-dimethylformamide	298.15	10.8	1985NIK
(17.0 wt% vinyl acetate)		118900	*N,N*-dimethylformamide	298.15	12.9	1985NIK
(19.0 wt% vinyl acetate)		75000	*N,N*-dimethylformamide	298.15	14.0	1985NIK
(19.6 wt% vinyl acetate)		100000	*N,N*-dimethylformamide	298.15	17.3	1985NIK
(21.0 wt% vinyl acetate)		188400	*N,N*-dimethylformamide	298.15	8.5	1985NIK
(21.0 wt% vinyl acetate)		79400	*N,N*-dimethylformamide	298.15	21.2	1985NIK
(21.7 wt% vinyl acetate)		39800	*N,N*-dimethylformamide	298.15	23.1	1985NIK
(23.5 wt% vinyl acetate)		66800	*N,N*-dimethylformamide	298.15	16.4	1985NIK
(28.0 wt% vinyl acetate)		118900	*N,N*-dimethylformamide	298.15	12.1	1985NIK
(31.7 wt% vinyl acetate)		70800	*N,N*-dimethylformamide	298.15	11.3	1985NIK
Poly(acrylonitrile-*co*-vinylidene chloride)						
(59 wt% acrylonitrile)		204000	γ-butyrolactone	298.15	0.119	1986KAM
(59 wt% acrylonitrile)		42000	*N,N*-dimethylacetamide	298.15	0.132	1986KAM
(59 wt% acrylonitrile)		62000	*N,N*-dimethylacetamide	298.15	0.121	1986KAM
(59 wt% acrylonitrile)		90000	*N,N*-dimethylacetamide	298.15	0.120	1986KAM
(59 wt% acrylonitrile)		140000	*N,N*-dimethylacetamide	298.15	0.120	1986KAM
(59 wt% acrylonitrile)		204000	*N,N*-dimethylacetamide	298.15	0.115	1986KAM
(59 wt% acrylonitrile)		270000	*N,N*-dimethylacetamide	298.15	0.110	1986KAM
(59 wt% acrylonitrile)		348000	*N,N*-dimethylacetamide	298.15	0.095	1986KAM
(59 wt% acrylonitrile)		500000	*N,N*-dimethylacetamide	298.15	0.090	1986KAM
(59 wt% acrylonitrile)		204000	*N,N*-dimethylformamide	298.15	0.137	1986KAM
Poly(benzyl acrylate-*co*-methyl acrylate)						
(12 mol% benzyl acrylate)	218000	371000	2-butanone	293.15	2.2	1970MAE
(30 mol% benzyl acrylate)	260000	370000	2-butanone	293.15	2.4	1970MAE
(50 mol% benzyl acrylate)	177000	343000	2-butanone	293.15	2.3	1970MAE
(71 mol% benzyl acrylate)	125000	289000	2-butanone	293.15	1.9	1970MAE
(12 mol% benzyl acrylate)	260000	370000	*N,N*-dimethylformamide	293.15	5.0	1970MAE
(30 mol% benzyl acrylate)	177000	343000	*N,N*-dimethylformamide	293.15	4.8	1970MAE
(50 mol% benzyl acrylate)	580000	499000	*N,N*-dimethylformamide	293.15	4.8	1970MAE
(71 mol% benzyl acrylate)	125000	289000	*N,N*-dimethylformamide	293.15	3.8	1970MAE

Copolymer	M_n/ g/mol	M_w/ g/mol	**Solvent**	T/ K	$A_2\ 10^4$/ cm^{3}mol/g^2	**Ref.**
Poly(γ-benzyl-L-glutamate-*b*-methyl methacrylate-*b*-γ-benzyl-L-glutamate)						
(33 wt% glutamate)		100000	*N,N*-dimethylformamide	298.15	4.0	1991ULY
(56 wt% glutamate)		060000	*N,N*-dimethylformamide	298.15	3.7	1991ULY
(45 wt% glutamate)		160000	*N,N*-dimethylformamide	298.15	3.1	1991ULY
(56 wt% glutamate)		080000	*N,N*-dimethylformamide	298.15	2.9	1991ULY
(57 wt% glutamate)		055000	*N,N*-dimethylformamide	298.15	3.2	1991ULY
(62 wt% glutamate)		095000	*N,N*-dimethylformamide	298.15	3.2	1991ULY
(60 wt% glutamate)		170000	*N,N*-dimethylformamide	298.15	3.2	1991ULY
(65 wt% glutamate)		320000	*N,N*-dimethylformamide	298.15	2.1	1991ULY
(76 wt% glutamate)		085000	*N,N*-dimethylformamide	298.15	3.1	1991ULY
(76 wt% glutamate)		140000	*N,N*-dimethylformamide	298.15	3.7	1991ULY
(78 wt% glutamate)		140000	*N,N*-dimethylformamide	298.15	1.7	1991ULY
(83 wt% glutamate)		320000	*N,N*-dimethylformamide	298.15	2.2	1991ULY
(86 wt% glutamate)		150000	*N,N*-dimethylformamide	298.15	2.3	1991ULY
Poly[(1,1'-biphenyl)-4-4'diol-*co*-3,3-bis(4-carboxyphenyl)phthalide]						
(1:1)		52000	1,2-dichloroethane	298.15	0.0	1984GLA
(1:1)		52000	trichloromethane	298.15	8.0	1984GLA
Poly(3,3',4,4'-biphenyltetracarboxylic dianhydride-*co*-1,4-phenylenediamine)						
		4800	dimethylsulfoxide	298.15	34	1991SWA
		7200	dimethylsulfoxide	298.15	32.2	1991SWA
		8300	dimethylsulfoxide	298.15	27	1991SWA
		16800	dimethylsulfoxide	298.15	28	1991SWA
		22000	dimethylsulfoxide	298.15	24	1991SWA
		56400	dimethylsulfoxide	298.15	19.6	1991SWA
		5900	1-methyl-2-pyrrolidinone	298.15	28	1991SWA
		9900	1-methyl-2-pyrrolidinone	298.15	24	1991SWA
		18800	1-methyl-2-pyrrolidinone	298.15	22	1991SWA
		75000	1-methyl-2-pyrrolidinone	298.15	15	1991SWA
Poly[9,9'-bis(4-acetylphenyl)fluorene-*co*-4,4'-diamino-3,3'-dibenzoyldiphenyl ether]						
	107000		1,1,2,2-tetrachloroethane	298.15	13.8	1981PAD
	159000		1,1,2,2-tetrachloroethane	298.15	13.6	1981PAD
	314000		1,1,2,2-tetrachloroethane	298.15	13.8	1981PAD
	525000		1,1,2,2-tetrachloroethane	298.15	13.6	1981PAD
Poly[bis(4-aminophenoxy)benzene-*co*-pyromellitic anhydride]						
		20000	*N,N*-dimethylacetamide	294.15	20.0	1985BAR
		50000	*N,N*-dimethylacetamide	294.15	17.0	1985BAR
		80000	*N,N*-dimethylacetamide	294.15	15.0	1985BAR

Copolymer	M_n/ g/mol	M_w/ g/mol	Solvent	T/ K	A_2 10^4/ cm^3mol/g^2	Ref.
Poly[bis(4-aminophenyl) ether-*co*-pyromellitic anhydride]						
		20000	*N,N*-dimethylacetamide	294.15	20.0	1985BAR
		90000	*N,N*-dimethylacetamide	294.15	11.0	1985BAR
		130000	*N,N*-dimethylacetamide	294.15	9.8	1985BAR
		150000	*N,N*-dimethylacetamide	294.15	10.5	1985BAR
		350000	*N,N*-dimethylacetamide	294.15	9.4	1985BAR
Poly[bis(4-aminophenyl) ether-*co*-tricyclodecenetetracarboxylic dianhydride]						
		80000	dimethylsulfoxide	293.15	8.5	1991DON
Poly(bis(4-aminophenyl) sulfone-*co*-isophthaloyl chloride)						
		53000	dimethylsulfoxide	293.15	12.2	2008ZUL
Poly[2,5-bis(carbomethoxy)terephthaloyl chloride-*co*-3,3-bis(4-aminophenyl)phthalide]						
		7800	*N,N*-dimethylformamide	298.15	40.0	1974MOL
		8000	*N,N*-dimethylformamide	298.15	36.0	1974MOL
		12000	*N,N*-dimethylformamide	298.15	35.0	1974MOL
		17000	*N,N*-dimethylformamide	298.15	36.0	1974MOL
		29500	*N,N*-dimethylformamide	298.15	34.0	1974MOL
		34000	*N,N*-dimethylformamide	298.15	34.5	1974MOL
		42000	*N,N*-dimethylformamide	298.15	33.4	1974MOL
		45000	*N,N*-dimethylformamide	298.15	32.1	1974MOL
		47000	*N,N*-dimethylformamide	298.15	29.8	1974MOL
		59000	*N,N*-dimethylformamide	298.15	18.2	1974MOL
		67500	*N,N*-dimethylformamide	298.15	16.2	1974MOL
		75000	*N,N*-dimethylformamide	298.15	14.7	1974MOL
		83000	*N,N*-dimethylformamide	298.15	14.6	1974MOL
		91000	*N,N*-dimethylformamide	298.15	14.5	1974MOL
		111000	*N,N*-dimethylformamide	298.15	15.5	1974MOL
		125000	*N,N*-dimethylformamide	298.15	16.0	1974MOL
Poly[1,4-bis(3,4-dicarboxyphenoxyl)benzenedianhydride-*co*-2,2'-dimethyl-4,4'-methylenedianiline]						
		50700	trichloromethane	298.15	28	1996SID
		86000	trichloromethane	298.15	13.3	2001SID
		90400	trichloromethane	298.15	23	1996SID
		124000	trichloromethane	298.15	16	1996SID
		215000	trichloromethane	298.15	12	1996SID
		345000	trichloromethane	298.15	9.2	1996SID

Copolymer	M_n/ g/mol	M_w/ g/mol	Solvent	T/ K	A_2 10^4/ cm^{3}mol/g^2	Ref.
Poly[2,2'-bis(3,4-dicarboxyphenyl)hexafluoropropanedianhydride-*co*-2,2'-bis(trifluoromethyl)-4,4'-biphenyldiamine]						
		6900	tetrahydrofuran	303.15	53.4	1997KWA
		15400	tetrahydrofuran	303.15	23.0	1997KWA
		59900	tetrahydrofuran	303.15	28.2	1997KWA
		134000	tetrahydrofuran	303.15	11.2	1997KWA
		203000	tetrahydrofuran	303.15	10.3	1997KWA
Poly[2,2'-bis(3,4-dicarboxyphenyl)hexafluoropropane dianhydride-*co*-4,4'-diamino-3,3'-dimethylbiphenyl]						
		14000	cyclopentanone	303.15	2.8	2000HUT
		18000	cyclopentanone	303.15	4.4	2000HUT
		27000	cyclopentanone	303.15	5.2	2000HUT
		32000	cyclopentanone	303.15	4.1	2000HUT
		52000	cyclopentanone	303.15	2.5	2000HUT
		147000	cyclopentanone	303.15	3.5	2000HUT
		160000	cyclopentanone	303.15	3.8	2000HUT
		358000	cyclopentanone	303.15	2.9	2000HUT
Poly{2,2'-bis[4-(3,4-dicarboxyphenoxy)phenyl]propane dianhydride-*co*-4,4'-diamino-3,3'-dimethylbiphenyl}						
		13000	cyclopentanone	303.15	0.1	2000HUT
		22000	cyclopentanone	303.15	−0.5	2000HUT
		23000	cyclopentanone	303.15	−0.1	2000HUT
		32000	cyclopentanone	303.15	0.2	2000HUT
		55000	cyclopentanone	303.15	0.1	2000HUT
		70000	cyclopentanone	303.15	0.1	2000HUT
		146000	cyclopentanone	303.15	0.2	2000HUT
Poly[9,9'-bis(4-hydroxyphenyl)fluorene-*co*-terephthaloyl chloride]						
(1:1)		90000	trichloromethane	298.15	5.0	1977TA1
(1:1)		90000	trichloromethane	298.15	5.0	1977TA3
(1:1)		200000	1,1,2,2-tetrachloroethane	298.15	7.8	1977TA1
(1:1)		200000	1,1,2,2-tetrachloroethane	298.15	7.8	1977TA3
Poly(bisphenol-A diglycidyl ether-*co*-adipic acid)						
		64000	*N,N*-dimethylacetamide	298.15	10.5	2001CAO

Copolymer	M_n/ g/mol	M_w/ g/mol	Solvent	T/ K	A_2 10^4/ cm^3mol/g^2	Ref.
Poly(bisphenol-F-*co*-2,6-dichloro-6-methoxy-1,3,5-triazine)						
	44000	65000	trichloromethane	303.15	8.91	1993KAN
	49400	86000	trichloromethane	303.15	8.41	1993KAN
	53000	110000	trichloromethane	303.15	8.32	1993KAN
	55000	125000	trichloromethane	303.15	7.65	1993KAN
	81400	240000	trichloromethane	303.15	7.09	1993KAN
	87000	319000	trichloromethane	303.15	6.60	1993KAN
	398000	752000	trichloromethane	303.15	1.40	1993KAN
Poly(butadiene-*co*-acrylonitrile)						
(hydrogenated)		150000	trichloromethane	298.15	3.34	1993ROY
Poly(butadiene-*b*-styrene)						
(23.3 mol% styrene)	93200	109000	cyclohexane	298.15	6.59	2009XIO
(23.3 mol% styrene)	93200	109000	cyclohexane	308.15	7.49	2009XIO
(23.3 mol% styrene)	93200	109000	cyclohexane	323.15	5.50	2009XIO
(32.9 wt% styrene)	43100		cyclohexane	307.35	10.0	1968UTR
(17.8 wt% styrene)	105000		cyclohexane	307.35	9.60	1968UTR
(12.3 wt% styrene)	129000		cyclohexane	307.35	10.3	1968UTR
(3.6 wt% styrene)	620000		cyclohexane	307.35	7.55	1968UTR
(32.9 wt% styrene)	43100		1,4-dioxane	307.35	5.80	1968UTR
(17.8 wt% styrene)	105000		1,4-dioxane	307.35	2.72	1968UTR
(12.3 wt% styrene)	129000		1,4-dioxane	307.35	3.41	1968UTR
(3.6 wt% styrene)	620000		1,4-dioxane	307.35	1.64	1968UTR
(23.3 mol% styrene)	93200	109000	tetrahydrofuran	293.15	7.36	2009XIO
(23.3 mol% styrene)	93200	109000	tetrahydrofuran	298.15	9.94	2009XIO
(23.3 mol% styrene)	93200	109000	tetrahydrofuran	313.15	9.88	2009XIO
(32.9 wt% styrene)	43100		toluene	307.35	12.6	1968UTR
(17.8 wt% styrene)	105000		toluene	307.35	11.5	1968UTR
(12.3 wt% styrene)	129000		toluene	307.35	11.9	1968UTR
(3.6 wt% styrene)	620000		toluene	307.35	8.17	1968UTR
Poly(butadiene-*co*-styrene)						
(25.0 mol% styrene)	112000	800000	benzene	301.75	14.2	1961COO
(25.0 mol% styrene)	84100	740000	benzene	301.75	12.0	1961COO
(29.0 mol% styrene)	116000	1700000	benzene	301.75	12.8	1961COO
(30.0 mol% styrene)	58400	2670000	benzene	301.75	14.9	1961COO
(25.0 mol% styrene)	112000	800000	cyclohexane	301.75	0.46	1961COO
(25.0 mol% styrene)	84100	740000	cyclohexane	301.75	1.39	1961COO
(29.0 mol% styrene)	116000	1700000	cyclohexane	301.75	1.62	1961COO
(30.0 mol% styrene)	58400	2670000	cyclohexane	301.75	0.14	1961COO
(24 wt% styrene)	49000		cyclohexane	303.15	1.7	1965HOM
(24 wt% styrene)	76000		cyclohexane	303.15	1.56	1965HOM

Copolymer	M_n/ g/mol	M_w/ g/mol	**Solvent**	*T*/ K	$A_2\ 10^4$/ cm^3mol/g^2	**Ref.**
(24 wt% styrene)	119000		cyclohexane	303.15	1.2	1965HOM
(24 wt% styrene)	225000		cyclohexane	303.15	1.3	1965HOM
(24 wt% styrene)	252000		cyclohexane	303.15	1.16	1965HOM
(24 wt% styrene)	352000		cyclohexane	303.15	1.05	1965HOM
(24 wt% styrene)	514000		cyclohexane	303.15	1.0	1965HOM
(14.6 mol% styrene)	30300	54800	tetrahydrofuran	296.15	14.4	1979STA
(15.5 mol% styrene)	131000	143000	tetrahydrofuran	296.15	11.2	1979STA
(23.9 mol% styrene)	122000	132000	tetrahydrofuran	296.15	10.9	1979STA
(25.0 mol% styrene)	50400	52900	tetrahydrofuran	296.15	13.9	1979STA
(25.0 mol% styrene)	153000	200000	tetrahydrofuran	296.15	9.2	1979STA
(38.7 mol% styrene)	49700	52200	tetrahydrofuran	296.15	12.9	1979STA
(39.4 mol% styrene)	118000	131000	tetrahydrofuran	296.15	8.8	1979STA
(62.4 mol% styrene)	137000	148000	tetrahydrofuran	296.15	7.8	1979STA
(63.6 mol% styrene)	47200	50500	tetrahydrofuran	296.15	10.5	1979STA
Poly(butylene oxide-*b*-ethylene oxide-*b*-butylene oxide)						
($B_6E_{46}B_6$)		2950	water	278.15	−25.1	1997LIU
(79 mol% EO)	2900		water	298.15	−0.3	1998LIU
(90 mol% EO)	4640		water	298.15	+0.3	1998LIU
(91.5 mol% EO)	13190		water	298.15	−1.0	1998LIU
(93 mol% EO)	13430		water	298.15	−0.8	1998LIU
Poly(4-*tert*-butylstyrene-*b*-dimethylsiloxane)						
(29.6 wt% 4-*tert*-BS)	39800	92700	toluene	303.15	4.23	1990KUE
Poly(4-*tert*-butylstyrene-*b*-dimethylsiloxane-*b*-4-*tert*-butylstyrene)						
(29.7 wt% 4-*tert*-BS)	70200	115000	toluene	303.15	3.24	1990KUE
(29.4 wt% 4-*tert*-BS)	151000	291000	toluene	303.15	2.56	1990KUE
(27.3 wt% 4-*tert*-BS)	294000	829000	toluene	303.15	2.10	1990KUE
Poly(*tert*-butylstyrene-*b*-styrene)						
(50 mol% 4-*tert*-BS)		185000	*N,N*-dimethylacetamide	298.15	0.20	1993ZHO
(50 mol% 4-*tert*-BS)		185000	*N,N*-dimethylacetamide	318.15	−0.02	1993ZHO
(50 mol% 4-*tert*-BS)		185000	*N,N*-dimethylacetamide	333.15	0.37	1993ZHO
Poly(4-chlorostyrene-*co*-methyl methacrylate)						
(15 % chlorine)		150000	benzene	295.45	2.5	1967MOH
(15 % chlorine)		1200000	benzene	295.45	1.3	1967MOH
(15 % chlorine)		150000	1,4-dioxane	295.45	3.2	1967MOH
(15 % chlorine)		1200000	1,4-dioxane	295.45	1.6	1967MOH
(15 % chlorine)		150000	trichloromethane	295.45	7.3	1967MOH
(15 % chlorine)		1200000	trichloromethane	295.45	3.3	1967MOH

Copolymer	M_n/ g/mol	M_w/ g/mol	**Solvent**	*T*/ K	$A_2\ 10^4$/ cm^3mol/g^2	**Ref.**
Poly(4-chlorostyrene-*b*-styrene-*b*-4-chlorostyrene)						
(37.2 mol% styrene)	368000		2-butanone	303.15	1.75	1986OGA
(46.7 mol% styrene)	325000		2-butanone	303.15	1.78	1986OGA
(65.0 mol% styrene)	379000		2-butanone	303.15	1.68	1986OGA
(37.2 mol% styrene)	368000		isopropylbenzene	328.15	1.37	1986OGA
(46.7 mol% styrene)	325000		isopropylbenzene	328.15	1.65	1986OGA
(65.0 mol% styrene)	379000		isopropylbenzene	328.15	2.19	1986OGA
(37.2 mol% styrene)	368000		toluene	303.15	2.23	1986OGA
(46.7 mol% styrene)	325000		toluene	303.15	2.62	1986OGA
(65.0 mol% styrene)	379000		toluene	303.15	3.02	1986OGA
Poly(decyl methacrylate-*b*-styrene)						
(81.5 wt% styrene)		37800	methyl acetate	298.15	−0.021	2004PIT
(85.0 wt% styrene)		65100	methyl acetate	298.15	−0.341	2004PIT
(92.2 wt% styrene)		68900	methyl acetate	298.15	−0.655	2004PIT
(89.5 wt% styrene)		134600	methyl acetate	298.15	−0.479	2004PIT
(81.5 wt% styrene)		19900	tetrahydrofuran	298.15	7.50	2004PIT
(79.0 wt% styrene)		40100	tetrahydrofuran	298.15	5.35	2004PIT
(85.0 wt% styrene)		54200	tetrahydrofuran	298.15	3.48	2004PIT
(70.8 wt% styrene)		60700	tetrahydrofuran	298.15	3.72	2004PIT
(92.2 wt% styrene)		62100	tetrahydrofuran	298.15	5.45	2004PIT
(67.0 wt% styrene)		77500	tetrahydrofuran	298.15	6.75	2004PIT
(89.5 wt% styrene)		111400	tetrahydrofuran	298.15	5.10	2004PIT
(79.0 wt% styrene)	37100		toluene	308.15	8.45	2004PIT
(85.0 wt% styrene)	54100		toluene	308.15	3.68	2004PIT
(70.8 wt% styrene)	56900		toluene	308.15	7.13	2004PIT
(92.2 wt% styrene)	60900		toluene	308.15	5.64	2004PIT
(67.0 wt% styrene)	74800		toluene	308.15	5.38	2004PIT
(89.5 wt% styrene)	106200		toluene	308.15	4.76	2004PIT
Poly(4,4'-diaminophenyl-*co*-pyromellitic anhydride]						
		10000	*N,N*-dimethylacetamide	294.15	30.0	1985BAR
		30000	*N,N*-dimethylacetamide	294.15	27.0	1985BAR
		80000	*N,N*-dimethylacetamide	294.15	32.0	1985BAR
Poly(diethyl maleate-*co*-vinyl acetate)						
	30700		toluene	304.15	11.86	1989BEL
	38400		toluene	304.15	9.375	1989BEL
	47200		toluene	304.15	7.753	1989BEL
	51200		toluene	304.15	7.085	1989BEL
	98300		toluene	304.15	4.08	1989BEL
	130000		toluene	304.15	3.01	1989BEL

Copolymer	M_n/ g/mol	M_w/ g/mol	Solvent	T/ K	A_2 10^4/ cm^{3}mol/g^2	Ref.
Poly(*N,N*-dimethylacrylamide–*b*-*N*-isopropylacrylamide-*b*-*N,N*-dimethylacrylamide)						
(72 mol% NIPAM)	41300	42500	water	323.15	3.78	2006CON
(78 mol% NIPAM)	53000	60400	water	323.15	2.58	2006CON
Poly[2,2'-dimethyl-4,4'-methylenedianiline-*co*-3,3',4,4'-oxydi(phthalic anhydride)]						
		11500	trichloromethane	298.15	14	1996SID
		18000	trichloromethane	298.15	12	1996SID
		25000	trichloromethane	298.15	10	1996SID
		30000	trichloromethane	298.15	9.0	1996SID
Poly(*trans*-2,5-dimethylpiperazine-*co*-phthaloyl chloride)						
		6020	1-methyl-2-pyrrolidinone	298.15	82.5	1977MOT
		8200	1-methyl-2-pyrrolidinone	298.15	20.0	1977MOT
		13100	1-methyl-2-pyrrolidinone	298.15	10.1	1977MOT
		20200	1-methyl-2-pyrrolidinone	298.15	11.3	1977MOT
		26800	1-methyl-2-pyrrolidinone	298.15	9.01	1977MOT
		30800	1-methyl-2-pyrrolidinone	298.15	9.77	1977MOT
		43000	1-methyl-2-pyrrolidinone	298.15	7.95	1977MOT
		44600	1-methyl-2-pyrrolidinone	298.15	8.57	1977MOT
		54700	1-methyl-2-pyrrolidinone	298.15	6.53	1977MOT
		66500	1-methyl-2-pyrrolidinone	298.15	8.43	1977MOT
		93500	1-methyl-2-pyrrolidinone	298.15	8.57	1977MOT
		93500	1-methyl-2-pyrrolidinone	298.15	5.58	1977MOT
		128000	1-methyl-2-pyrrolidinone	298.15	6.83	1977MOT
		183000	1-methyl-2-pyrrolidinone	298.15	6.32	1977MOT
		212000	1-methyl-2-pyrrolidinone	298.15	5.97	1977MOT
		293000	1-methyl-2-pyrrolidinone	298.15	5.19	1977MOT
		406000	1-methyl-2-pyrrolidinone	298.15	5.84	1977MOT
		482000	1-methyl-2-pyrrolidinone	298.15	4.78	1977MOT
		105000	trichloromethane	298.15	9.24	1977MOT
		124000	trichloromethane	298.15	8.46	1977MOT
		160000	trichloromethane	298.15	9.83	1977MOT
		165000	trichloromethane	298.15	8.22	1977MOT
		209000	trichloromethane	298.15	6.57	1977MOT
		277000	trichloromethane	298.15	7.80	1977MOT
		406000	trichloromethane	298.15	7.69	1977MOT
		482000	trichloromethane	298.15	6.66	1977MOT
Poly(*trans*-2,5-dimethylpiperazine-*co*-terephthaloyl chloride)						
		5210	2,2,2-trifluoroethanol	298.15	65.7	1978MOT
		6600	2,2,2-trifluoroethanol	298.15	58.6	1978MOT
		9000	2,2,2-trifluoroethanol	298.15	54.2	1978MOT

Copolymer	M_n/ g/mol	M_w/ g/mol	Solvent	T/ K	A_2 10^4/ cm^3mol/g^2	Ref.
		13200	2,2,2-trifluoroethanol	298.15	47.0	1978MOT
		18200	2,2,2-trifluoroethanol	298.15	41.5	1978MOT
		26500	2,2,2-trifluoroethanol	298.15	37.9	1978MOT
		30500	2,2,2-trifluoroethanol	298.15	35.5	1978MOT
	38000	38800	2,2,2-trifluoroethanol	298.15	34.9	1978MOT
		58600	2,2,2-trifluoroethanol	298.15	29.2	1978MOT
	68200	68500	2,2,2-trifluoroethanol	298.15	29.6	1978MOT
		109000	2,2,2-trifluoroethanol	298.15	27.6	1978MOT
	144000	150000	2,2,2-trifluoroethanol	298.15	23.8	1978MOT
		198000	2,2,2-trifluoroethanol	298.15	22.9	1978MOT
		243000	2,2,2-trifluoroethanol	298.15	22.7	1978MOT
		319000	2,2,2-trifluoroethanol	298.15	21.7	1978MOT
Poly(dimethylsiloxane-*b*-ethylene oxide)						
(segmented blocks)		23500	benzene-d6	room	+22	1985HAE
(segmented blocks)		23500	cyclohexane-d12	room	−11	1985HAE
(segmented blocks)		23500	ethanol	room	−19	1985HAE
(segmented blocks)		23500	methanol-d4	room	−13	1985HAE
(segmented blocks)		23500	methanol	room	−14	1985HAE
(segmented blocks)		23500	toluene	room	+1.8	1985HAE
Poly(dimethylsiloxane-*co*-hydromethylsiloxane)						
(0.6 mol% Si-H)	26400	58000	toluene	298.15	4.21	2001LOO
(3 mol% Si-H)	43000	89000	toluene	298.15	4.7	2001LOO
Poly(ethyl acrylate-*co*-methyl methacrylate)						
(16 mol% MMA)	66000	165000	2-butanone	293.15	5.3	1970MAE
(38 mol% MMA)	49000	116000	2-butanone	293.15	5.3	1970MAE
(60 mol% MMA)	45000	100000	2-butanone	293.15	5.5	1970MAE
(80 mol% MMA)	38000	75000	2-butanone	293.15	4.3	1970MAE
(16 mol% MMA)	66000	165000	*N,N*-dimethylformamide	293.15	5.8	1970MAE
(38 mol% MMA)	49000	116000	*N,N*-dimethylformamide	293.15	6.2	1970MAE
(60 mol% MMA)	45000	100000	*N,N*-dimethylformamide	293.15	5.8	1970MAE
(80 mol% MMA)	38000	75000	*N,N*-dimethylformamide	293.15	5.3	1970MAE
(50 wt% MMA)		95000	2-propanone	293.15	5.0	1970WU2
(50 wt% MMA)		140000	2-propanone	293.15	4.3	1970WU2
(50 wt% MMA)		220000	2-propanone	293.15	3.8	1970WU2
(50 wt% MMA)		240000	2-propanone	293.15	3.8	1970WU2
(50 wt% MMA)		500000	2-propanone	293.15	3.5	1970WU2
(50 wt% MMA)		660000	2-propanone	293.15	3.5	1970WU2

Copolymer	M_n/ g/mol	M_w/ g/mol	Solvent	T/ K	A_2 10^4/ cm^3mol/g^2	Ref.
Poly(ethylene-*alt*-1-butene)						
(50 mol% 1-butene)	19500	19600	toluene	310.15	9.5	1984MAY
(50 mol% 1-butene)	54500	56700	toluene	310.15	8.4	1984MAY
(50 mol% 1-butene)	95500	103000	toluene	310.15	7.8	1984MAY
(50 mol% 1-butene)	129000	137000	toluene	310.15	7.55	1984MAY
(50 mol% 1-butene)	209000	228000	toluene	310.15	7.1	1984MAY
Poly(ethylene-*co*-1-butene)						
(4 mol% 1-butene		28000	perdeuterononadecane	423	0.08	2000WES
(21.9 mol% 1-butene)		6400	perdeuterodecane	369.15	67	2002SCH
Poly(ethylene-*alt*-propylene)						
(50 mol% propylene)	8200	8400	toluene	310.15	22.3	1984MAY
(50 mol% propylene)	14500	14900	toluene	310.15	18.4	1984MAY
(50 mol% propylene)	22500	24500	toluene	310.15	15.8	1984MAY
(50 mol% propylene)	34600	36300	toluene	310.15	13.7	1984MAY
(50 mol% propylene)	41600	46400	toluene	310.15	12.85	1984MAY
(50 mol% propylene)	91400	96000	toluene	310.15	9.8	1984MAY
(50 mol% propylene)	267000	301000	toluene	310.15	6.8	1984MAY
(50 mol% propylene)	293000	316000	toluene	310.15	6.6	1984MAY
(50 mol% propylene)	545000	600000	toluene	310.15	5.35	1984MAY
Poly(ethylene-*co*-propylene)						
(27.6 mol% propylene)		356000	1-chloronaphthalene	398.15	7.83	1972MOR
Poly(ethylene-*co*-propylene-*co*-ethylidene norbornene)						
(34.9 mol% propylene)		224000	1-chloronaphthalene	398.15	6.93	1972MOR
(37.1 mol% propylene)		277000	1-chloronaphthalene	398.15	5.43	1972MOR
(40.7 mol% propylene)		455000	1-chloronaphthalene	398.15	1.78	1972MOR
Poly(ethylene-*alt*-tetrafluoroethylene)						
(50 mol% ethylene)		540000	diisobutyl adipate	513.15	1.97	1987CHU
(50 mol% ethylene)	400000	540000	diisobutyl adipate	513.15	1.97	1987WUC
(50 mol% ethylene)		540000	diisobutyl adipate	513.15	1.97	1989CHU
(50 mol% ethylene)	700000	900000	diisobutyl adipate	513.15	1.14	1987WUC
(50 mol% ethylene)		1160000	diisobutyl adipate	513.15	1.02	1987CHU
(50 mol% ethylene)	840000	1160000	diisobutyl adipate	513.15	1.02	1987WUC
(50 mol% ethylene)		1160000	diisobutyl adipate	513.15	1.02	1989CHU
(50 mol% ethylene)	1605000	3210000	diisobutyl adipate	513.15	0.511	1989CHU

Copolymer	M_n/ g/mol	M_w/ g/mol	Solvent	T/ K	A_2 10^4/ cm^{3}mol/g^2	Ref.
Poly(ethylene oxide-*b*-(R)-3-hydroxybutyrate-*b*-ethylene oxide)						
(81.8 wt% EO)	2000-810-2000		water	298.15	−3.2	2006LIX
(92.4 wt% eEO)	5000-780-5000		water	298.15	0.27	2006LIX
Poly(ethylene oxide-*co*-methylene oxide)						
	34000		1H,1H,5H-octafluoro-1-pentanol	383.15	20	1965WAG
	40000		1H,1H,5H-octafluoro-1-pentanol	383.15	13	1965WAG
	44000		1H,1H,5H-octafluoro-1-pentanol	383.15	15	1965WAG
	56000		1H,1H,5H-octafluoro-1-pentanol	383.15	18	1965WAG
		62000	1H,1H,5H-octafluoro-1-pentanol	383.15	24	1965WAG
		72000	1H,1H,5H-octafluoro-1-pentanol	383.15	25	1965WAG
		96000	1H,1H,5H-octafluoro-1-pentanol	383.15	17	1965WAG
		129000	1H,1H,5H-octafluoro-1-pentanol	383.15	17	1965WAG
Poly(ethylene oxide-*b*-methyl methacrylate)						
(9.1 wt% ethylene oxide)		35000	1,4-dioxane	298.15	0.88	2001EDE
(9.1 wt% ethylene oxide)		27000	*N,N*-dimethylformamide	298.15	4.31	2001EDE
(9.1 wt% ethylene oxide)		36600	2,2,2-trifluoroethanol	298.15	15.2	2001EDE
Poly(ethylene oxide-*co*-propylene oxide)						
(72.4 mol% ethylene oxide)		36000	water	298.15	6.0	1991LOU
(79.5 mol% ethylene oxide)		30800	water	298.15	10.5	1991LOU
(79.5 mol% ethylene oxide)		32500	water	298.15	12.5	1991LOU
(86.6 mol% ethylene oxide)		30100	water	298.15	21.5	1991LOU
Poly(ethylene oxide-*b*-propylene oxide-*b*-ethylene oxide)						
(40 wt% ethylene oxide)	3400	3700	1,2-dimethylbenzene	299.35	+45	1993WUG
(70 mol% ethylene oxide)		3600	water	277.15	−190	1998CRO
(70 mol% ethylene oxide)		3600	water	281.25	−210	1998CRO
(70 mol% ethylene oxide)		3600	water	282.55	−230	1998CRO
(70 mol% ethylene oxide)		3600	water	283.55	−240	1998CRO
(70 mol% ethylene oxide)		3600	water	284.65	−170	1998CRO
(70 mol% ethylene oxide)		3600	water	286.75	−220	1998CRO
(70 mol% ethylene oxide)		3600	water	288.65	−180	1998CRO

Copolymer	M_n/ g/mol	M_w/ g/mol	**Solvent**	*T*/ K	$A_2\ 10^4$/ cm^3mol/g^2	**Ref.**
Poly(ethylene oxide-*b*-styrene-*b*-ethylene oxide)						
(18.7 mol% ethylene oxide)		53900	ethyl acetate	298.15	4.0	1989FAN
(51.0 mol% ethylene oxide)		58000	ethyl acetate	298.15	1.8	1989FAN
(61.0 mol% ethylene oxide)		57100	ethyl acetate	298.15	0.8	1989FAN
(18.7 mol% ethylene oxide)		82000	toluene	298.15	6.5	1989FAN
(51.0 mol% ethylene oxide)		114000	toluene	298.15	2.9	1989FAN
(61.0 mol% ethylene oxide)		137000	toluene	298.15	2.6	1989FAN
(18.7 mol% ethylene oxide)		63300	trichloromethane	298.15	7.8	1989FAN
(51.0 mol% ethylene oxide)		100000	trichloromethane	298.15	6.3	1989FAN
(61.0 mol% ethylene oxide)		86200	trichloromethane	298.15	6.2	1989FAN
Poly(2-ethylhexyl methacrylate-*co*-acrylic acid)						
(3.10 mol% acrylic acid)			2-butanone	293.15	0.85	1973BAR
(5.13 mol% acrylic acid)			2-butanone	293.15	0.80	1973BAR
(14.0 mol% acrylic acid)			2-butanone	293.15	1.3	1973BAR
(3.10 mol% acrylic acid)			styrene	293.15	1.4	1973BAR
(5.13 mol% acrylic acid)			styrene	293.15	1.6	1973BAR
(14.0 mol% acrylic acid)			styrene	293.15	0.35	1973BAR
Poly(ethyl methacrylate-*co*-*N*-vinylcarbazole)						
(55 wt% *N*-vinylcarbazole)		114000	benzene	296.15	7.90	1976IOA
(61 wt% *N*-vinylcarbazole)		116000	benzene	296.15	8.48	1976IOA
(64 wt% *N*-vinylcarbazole)		102000	benzene	296.15	10.55	1976IOA
(72 wt% *N*-vinylcarbazole)		82000	benzene	296.15	16.00	1976IOA
(78 wt% *N*-vinylcarbazole)		86000	benzene	296.15	18.30	1976IOA
Poly[2-(*N*-ethylperfluorooctylsulfonamido)ethyl acrylate-*co*-*N*-isopropylacrylamide]						
(0.08 mol% fluorocarbon)		85000	water	298.15	−1.2	1998ZHA
(0.13 mol% fluorocarbon)		110000	water	298.15	−1.1	1998ZHA
(0.26 mol% fluorocarbon)		120000	water	298.15	−3.0	1998ZHA
(0.39 mol% fluorocarbon)		160000	water	298.15	−1.1	1998ZHA
(0.82 mol% fluorocarbon)		130000	water	298.15	−4.1	1998ZHA
Poly[2-(*N*-ethylperfluorooctylsulfonamido)ethyl methacrylate-*co*-*N*-isopropylacrylamide]						
(0.06 mol% fluorocarbon)		73000	water	298.15	−8.6	1998ZHA
(0.10 mol% fluorocarbon)		86000	water	298.15	−2.5	1998ZHA
(0.21 mol% fluorocarbon)		120000	water	298.15	−2.0	1998ZHA
(0.41 mol% fluorocarbon)		150000	water	298.15	−1.7	1998ZHA
(0.88 mol% fluorocarbon)		120000	water	298.15	−3.1	1998ZHA

Copolymer	M_n/ g/mol	M_w/ g/mol	Solvent	T/ K	A_2 10^4/ cm^3mol/g^2	Ref.
Poly(formaldehyde-*co*-phenol) (acetylated)						
		2200	ethyl acetate	298.15	4.0	1992KI1
		3100	ethyl acetate	298.15	3.3	1992KI1
		3900	ethyl acetate	298.15	5.4	1992KI1
		4900	ethyl acetate	298.15	2.1	1992KI1
		6000	ethyl acetate	298.15	4.4	1992KI1
		6900	ethyl acetate	298.15	2.9	1992KI1
		8700	ethyl acetate	298.15	4.9	1992KI1
		8800	ethyl acetate	298.15	4.0	1992KI1
		16100	ethyl acetate	298.15	4.6	1992KI1
		17000	ethyl acetate	298.15	3.8	1992KI1
		20300	ethyl acetate	298.15	6.3	1992KI1
		21700	ethyl acetate	298.15	1.7	1992KI1
		21700	ethyl acetate	298.15	3.2	1992KI1
		22300	ethyl acetate	298.15	2.8	1992KI1
		26000	ethyl acetate	298.15	4.5	1992KI1
		30400	ethyl acetate	298.15	1.6	1992KI1
		31500	ethyl acetate	298.15	3.6	1992KI1
		69300	ethyl acetate	298.15	−6.3	1992KI1
		114000	ethyl acetate	298.15	−6.5	1992KI1
Poly(glycidol-*b*-propylene oxide-*b*-glycidol)						
(20 wt% propylene oxide)			water	298.15	0.38	2007RAN
(20 wt% propylene oxide)			water	313.15	0.23	2007RAN
(20 wt% propylene oxide)			water	323.15	0.57	2007RAN
(20 wt% propylene oxide)			water	333.15	0.08	2007RAN
(30 wt% propylene oxide)			water	298.15	0.07	2007RAN
(30 wt% propylene oxide)			water	313.15	0.45	2007RAN
(30 wt% propylene oxide)			water	323.15	0.27	2007RAN
(30 wt% propylene oxide)			water	333.15	0.28	2007RAN
(70 wt% propylene oxide)			water	298.15	−0.19	2007RAN
(70 wt% propylene oxide)			water	313.15	−0.48	2007RAN
(70 wt% propylene oxide)			water	323.15	−0.87	2007RAN
(70 wt% propylene oxide)			water	333.15	−0.50	2007RAN
Poly(3-hydroxybutanoic acid-*co*-3-hydroxypentanoic acid)						
(12 mol% 3-HPA)		369000	trichlorodeuteromethane	298.15	7.8	2008FOS
Poly(4-iodostyrene-*co*-styrene)						
(57 mol% styrene)		748000	1,4-dioxane	293.15	1.95	1971BRA
(87 mol% styrene)		469000	1,4-dioxane	293.15	2.88	1971BRA

Copolymer	M_n/ g/mol	M_w/ g/mol	**Solvent**	*T*/ K	A_2 10^4/ cm^3mol/g^2	**Ref.**
Poly(isobutyl methacrylate-*co*-2-(*tert*-butylamino)ethyl methacrylate)						
(23 mol% *tert*-BAEMA)		2420000	isopropylamine	303.15	1.22	1990CHU
Poly(isoprene-*b*-styrene-*b*-isoprene)						
(27 wt% styrene)		78300	tetrachloromethane	298.15	5.4	1994PIS
(27 wt% styrene)		69800	tetrahydrofuran	298.15	11.0	1994PIS
(49 wt% styrene)	82500	92000	tetrahydrofuran	298.15	7.6	2000PIS
(27 wt% styrene)	62700		toluene	310.15	10.8	1994PIS
Poly(*N*-isopropylacrylamide) stereoblock polymer (isotactic-atactic-isotactic)						
i2-a28-i2	36700		water	293.15	9.7	2008NUO
i2-a40-i2	47100		water	293.15	2.2	2008NUO
Poly(*N*-isopropylacrylamide-*co*-acrylamide)						
(15 mol% acrylamide)		3100000	water	298.15	3.01	1994MUM
(30 mol% acrylamide)		4500000	water	298.15	3.31	1994MUM
(45 mol% acrylamide)		3900000	water	298.15	3.46	1994MUM
(60 mol% acrylamide)		2200000	water	298.15	3.62	1994MUM
Poly(*N*-isopropylacrylamide-*b*-*N,N*-dimethylacrylamide)						
(33 wt% NIPAM)	35300	39500	water	305.15	0.2	2006LAM
(42 mol% NIPAM)	17900	19700	water	323.15	3.03	2006CON
(64 mol% NIPAM)	29600	34000	water	323.15	1.11	2006CON
(71 mol% NIPAM)	38600	43200	water	323.15	1.14	2006CON
(82 mol% NIPAM)	61900	74900	water	323.15	1.10	2006CON
Poly(*N*-isopropylacrylamide-*b*-L-glutamic acid)						
(NIPAM55-*b*-LGA35)		3370000	water (pH = 3)	298.15	−2.05	2008DEN
Poly{*N*-isopropylacrylamide-*co*-3-[*N*-(3-methacrylamidopropyl)-*N,N*-dimethylammonio] propane sulfate}						
(2 mol% 3-[MAPDMA]PS)		70000	methanol	298.15	13	2005NED
(3 mol% 3-[MAPDMA]PS)		90000	methanol	298.15	12	2005NED
(5 mol% 3-[MAPDMA]PS)		290000	methanol	298.15	12	2005NED
(10 mol% 3-[MAPDMA]PS)		120000	methanol	298.15	43	2005NED
(2 mol% 3-[MAPDMA]PS)		300000	water	298.15	−11	2005NED
(3 mol% 3-[MAPDMA]PS)		240000	water	298.15	−14	2005NED
(5 mol% 3-[MAPDMA]PS)		250000	water	298.15	−11	2005NED
(10 mol% 3-[MAPDMA]PS)		380000	water	298.15	−18	2005NED

Copolymer	M_n/ g/mol	M_w/ g/mol	Solvent	T/ K	A_2 10^4/ cm^3mol/g^2	Ref.
Poly(DL-lactic acid-*co*-glycolic acid)						
(75/25)		93200	trichloromethane	298.15	9.6	1998PEN
Poly(maleic acid-*co*-ethyl vinyl ether)						
	55000		tetrahydrofuran	303.15	3.60	1980SHI
		153000	tetrahydrofuran	298.15	2.03	1980SHI
Poly(maleic anhydride-*co*-ethyl vinyl ether)						
	101000		tetrahydrofuran	303.15	4.06	1980SHI
		95100	tetrahydrofuran	298.15	4.68	1980SHI
		122000	tetrahydrofuran	298.15	4.06	1980SHI
		158000	tetrahydrofuran	298.15	3.56	1980SHI
		209000	tetrahydrofuran	298.15	3.42	1980SHI
		281000	tetrahydrofuran	298.15	2.81	1980SHI
		283000	tetrahydrofuran	298.15	3.10	1980SHI
		470000	tetrahydrofuran	298.15	3.07	1980SHI
		651000	tetrahydrofuran	298.15	2.19	1980SHI
		1320000	tetrahydrofuran	298.15	2.83	1980SHI
Poly(maleic anhydride-*co*-1-octadecene)						
(50 mol% maleic anhydride)		28000	ethyl acetate	298.15	3.0	1985MAT
(50 mol% maleic anhydride)		37000	ethyl acetate	298.15	2.9	1985MAT
(50 mol% maleic anhydride)		51000	ethyl acetate	298.15	2.8	1985MAT
(50 mol% maleic anhydride)		68000	ethyl acetate	298.15	2.5	1985MAT
(50 mol% maleic anhydride)		99000	ethyl acetate	298.15	2.1	1985MAT
Poly(methacrylonitrile-*co*-butyl methacrylate)						
(50 mol% methacrylonitrile)		46340	2-butanone	303.15	8.51	1978RED
(50 mol% methacrylonitrile)		62220	2-butanone	303.15	6.68	1978RED
(50 mol% methacrylonitrile)		82990	2-butanone	303.15	7.16	1978RED
(50 mol% methacrylonitrile)		116880	2-butanone	303.15	7.46	1978RED
(50 mol% methacrylonitrile)		126120	2-butanone	303.15	5.90	1978RED
(50 mol% methacrylonitrile)		129080	2-butanone	303.15	6.96	1978RED
(50 mol% methacrylonitrile)		173930	2-butanone	303.15	4.35	1978RED
Poly(methacrylonitrile-*co*-ethyl methacrylate)						
(50 mol% methacrylonitrile)		78140	2-butanone	303.15	4.65	1978RED
(50 mol% methacrylonitrile)		115570	2-butanone	303.15	4.18	1978RED
(50 mol% methacrylonitrile)		118290	2-butanone	303.15	3.83	1978RED
(50 mol% methacrylonitrile)		138730	2-butanone	303.15	3.60	1978RED

Copolymer	M_n/ g/mol	M_w/ g/mol	**Solvent**	*T*/ K	A_2 10^4/ cm^3mol/g^2	**Ref.**
(50 mol% methacrylonitrile)		153030	2-butanone	303.15	3.49	1978RED
(50 mol% methacrylonitrile)		180960	2-butanone	303.15	5.46	1978RED
(50 mol% methacrylonitrile)		233270	2-butanone	303.15	5.11	1978RED
Poly(methacrylonitrile-*co*-methyl methacrylate)						
(50 mol% methacrylonitrile)		44030	2-butanone	298.15	4.11	1978RED
(50 mol% methacrylonitrile)		67610	2-butanone	298.15	5.18	1978RED
(50 mol% methacrylonitrile)		76450	2-butanone	298.15	5.18	1978RED
(50 mol% methacrylonitrile)		94240	2-butanone	298.15	5.05	1978RED
(50 mol% methacrylonitrile)		104554	2-butanone	298.15	5.05	1978RED
(50 mol% methacrylonitrile)		114110	2-butanone	298.15	4.42	1978RED
Poly(methacrylonitrile-*co*-styrene)						
(50.2 mol% styrene)		227000	2-butanone	303.15	2.16	1977RED
(50.2 mol% styrene)		279000	2-butanone	303.15	2.46	1977RED
(50.2 mol% styrene)		309000	2-butanone	303.15	1.44	1977RED
(50.2 mol% styrene)		329000	2-butanone	303.15	1.88	1977RED
(50.2 mol% styrene)		346000	2-butanone	303.15	3.18	1977RED
(50.2 mol% styrene)		398000	2-butanone	303.15	2.89	1977RED
(50.2 mol% styrene)		223000	*N,N*-dimethylformamide	303.15	4.19	1977RED
(50.2 mol% styrene)		229000	*N,N*-dimethylformamide	303.15	3.37	1977RED
(50.2 mol% styrene)		285000	*N,N*-dimethylformamide	303.15	3.27	1977RED
(50.2 mol% styrene)		300000	*N,N*-dimethylformamide	303.15	3.37	1977RED
(50.2 mol% styrene)		362000	*N,N*-dimethylformamide	303.15	2.76	1977RED
(50.2 mol% styrene)		416000	*N,N*-dimethylformamide	303.15	3.07	1977RED
Poly(methyl acrylate-*co*-methyl methacrylate)						
(11.6 mol% MMA)		560000	2-butanone	298.15	3.5	1965KOT
(11.6 mol% MMA)		940000	2-butanone	298.15	3.4	1965KOT
(11.6 mol% MMA)		1320000	2-butanone	298.15	3.4	1965KOT
(11.6 mol% MMA)		2080000	2-butanone	298.15	3.1	1965KOT
(19.5 mol% MMA)		370000	2-butanone	298.15	3.3	1965KOT
(19.5 mol% MMA)		680000	2-butanone	298.15	3.2	1965KOT
(19.5 mol% MMA)		1180000	2-butanone	298.15	3.1	1965KOT
(19.5 mol% MMA)		1370000	2-butanone	298.15	3.0	1965KOT
(49.4 mol% MMA)		710000	2-butanone	298.15	2.9	1965KOT
(49.4 mol% MMA)		1050000	2-butanone	298.15	2.9	1965KOT
(49.4 mol% MMA)		1300000	2-butanone	298.15	2.7	1965KOT
(49.4 mol% MMA)		1870000	2-butanone	298.15	2.0	1965KOT
(50.0 wt% MMA)		80000	2-propanone	293.15	5.5	1970WU2
(50.0 wt% MMA)		115000	2-propanone	293.15	4.8	1970WU2
(50.0 wt% MMA)		135000	2-propanone	293.15	4.5	1970WU2

Copolymer	M_n/ g/mol	M_w/ g/mol	**Solvent**	T/ K	A_2 10^4/ cm^3mol/g^2	**Ref.**
(50.0 wt% MMA)		155000	2-propanone	293.15	4.2	1970WU2
(50.0 wt% MMA)		200000	2-propanone	293.15	4.0	1970WU2
(50.0 wt% MMA)		240000	2-propanone	293.15	3.9	1970WU2
(50.0 wt% MMA)		420000	2-propanone	293.15	3.4	1970WU2
(50.0 wt% MMA)		525000	2-propanone	293.15	3.5	1970WU2
Poly(methyl methacrylate-*co*-acrylonitrile)						
(23.6 mol% acrylonitrile)		400000	acetonitrile	303.15	1.52	1982MA2
(23.6 mol% acrylonitrile)		440000	acetonitrile	303.15	1.52	1982MA2
(23.6 mol% acrylonitrile)		580000	acetonitrile	303.15	1.46	1982MA2
(23.6 mol% acrylonitrile)		750000	acetonitrile	303.15	1.42	1982MA2
(23.6 mol% acrylonitrile)		1016000	acetonitrile	303.15	1.30	1982MA2
(23.6 mol% acrylonitrile)		1240000	acetonitrile	303.15	1.24	1982MA2
(23.6 mol% acrylonitrile)		1560000	acetonitrile	303.15	0.99	1982MA2
(28.9 mol% acrylonitrile)		2168000	acetonitrile	303.15	0.70	1977KAS
(28.9 mol% acrylonitrile)		2371000	acetonitrile	303.15	0.65	1977KAS
(28.9 mol% acrylonitrile)		3161000	acetonitrile	303.15	0.69	1977KAS
(28.9 mol% acrylonitrile)		4772000	acetonitrile	303.15	0.55	1977KAS
(28.9 mol% acrylonitrile)		6323000	acetonitrile	303.15	0.37	1977KAS
(28.9 mol% acrylonitrile)		7903000	acetonitrile	303.15	0.40	1977KAS
(50.0 mol% acrylonitrile)		270000	acetonitrile	303.15	0.53	1982MA2
(50.0 mol% acrylonitrile)		460000	acetonitrile	303.15	0.51	1982MA2
(50.0 mol% acrylonitrile)		676000	acetonitrile	303.15	0.48	1982MA2
(50.0 mol% acrylonitrile)		758000	acetonitrile	303.15	0.45	1982MA2
(50.0 mol% acrylonitrile)		1150000	acetonitrile	303.15	0.42	1982MA2
(50.0 mol% acrylonitrile)		1200000	acetonitrile	303.15	0.40	1982MA2
(50.0 mol% acrylonitrile)		1830000	acetonitrile	303.15	0.32	1982MA2
(74.0 mol% acrylonitrile)		600000	acetonitrile	303.15	0.29	1982MA2
(74.0 mol% acrylonitrile)		832000	acetonitrile	303.15	0.26	1982MA2
(74.0 mol% acrylonitrile)		871000	acetonitrile	303.15	0.25	1982MA2
(74.0 mol% acrylonitrile)		1000000	acetonitrile	303.15	0.23	1982MA2
(74.0 mol% acrylonitrile)		1047000	acetonitrile	303.15	0.22	1982MA2
(74.0 mol% acrylonitrile)		1175000	acetonitrile	303.15	0.21	1982MA2
(74.0 mol% acrylonitrile)		1352000	acetonitrile	303.15	0.19	1982MA2
(74.0 mol% acrylonitrile)		1530000	acetonitrile	303.15	0.16	1982MA2
(74.0 mol% acrylonitrile)		1700000	acetonitrile	303.15	0.14	1982MA2
(48.0 mol% acrylonitrile)		188000	2-butanone	303.15	2.85	1967SHI
(48.0 mol% acrylonitrile)		337000	2-butanone	303.15	2.40	1967SHI
(48.0 mol% acrylonitrile)		400000	2-butanone	303.15	2.32	1967SHI
(48.0 mol% acrylonitrile)		667000	2-butanone	303.15	2.20	1967SHI
(48.0 mol% acrylonitrile)		769000	2-butanone	303.15	2.12	1967SHI
(48.0 mol% acrylonitrile)		1087000	2-butanone	303.15	1.85	1967SHI
(48.0 mol% acrylonitrile)		667000	*N,N*-dimethylformamide	303.15	4.9	1967SHI

Copolymer	M_n/ g/mol	M_w/ g/mol	Solvent	T/ K	A_2 10^4/ cm^{3}mol/g^2	Ref.
Poly(methyl methacrylate-*co*-butyl acrylate)						
(29.5 wt% methyl methacrylate)		27000	2-propanone	293.15	9.5	1970WU1
(29.5 wt% methyl methacrylate)		46000	2-propanone	293.15	7.8	1970WU1
(29.5 wt% methyl methacrylate)		150000	2-propanone	293.15	5.0	1970WU1
(29.5 wt% methyl methacrylate)		240000	2-propanone	293.15	4.5	1970WU1
(29.5 wt% methyl methacrylate)		320000	2-propanone	293.15	4.3	1970WU1
(29.5 wt% methyl methacrylate)		500000	2-propanone	293.15	3.9	1970WU1
(29.5 wt% methyl methacrylate)		800000	2-propanone	293.15	3.6	1970WU1
(56.0 wt% methyl methacrylate)		24000	2-propanone	293.15	11.0	1970WU1
(56.0 wt% methyl methacrylate)		49000	2-propanone	293.15	8.5	1970WU1
(56.0 wt% methyl methacrylate)		91000	2-propanone	293.15	6.4	1970WU1
(56.0 wt% methyl methacrylate)		160000	2-propanone	293.15	5.3	1970WU1
(56.0 wt% methyl methacrylate)		300000	2-propanone	293.15	4.7	1970WU1
(56.0 wt% methyl methacrylate)		530000	2-propanone	293.15	4.25	1970WU1
(56.0 wt% methyl methacrylate)		710000	2-propanone	293.15	4.0	1970WU1
(79.5 wt% methyl methacrylate)		22000	2-propanone	293.15	8.0	1970WU1
(79.5 wt% methyl methacrylate)		43000	2-propanone	293.15	6.0	1970WU1
(79.5 wt% methyl methacrylate)		87000	2-propanone	293.15	5.2	1970WU1
(79.5 wt% methyl methacrylate)		125000	2-propanone	293.15	4.6	1970WU1
(79.5 wt% methyl methacrylate)		280000	2-propanone	293.15	3.5	1970WU1
(79.5 wt% methyl methacrylate)		530000	2-propanone	293.15	3.2	1970WU1
(79.5 wt% methyl methacrylate)		740000	2-propanone	293.15	2.6	1970WU1
Poly(methyl methacrylate-*co*-butyl methacrylate)						
(50.0 wt% methyl methacrylate)		70000	2-propanone	293.15	3.2	1970WU2
(50.0 wt% methyl methacrylate)		160000	2-propanone	293.15	2.8	1970WU2
(50.0 wt% methyl methacrylate)		210000	2-propanone	293.15	2.5	1970WU2
(50.0 wt% methyl methacrylate)		330000	2-propanone	293.15	2.3	1970WU2
(50.0 wt% methyl methacrylate)		430000	2-propanone	293.15	2.2	1970WU2
Poly(methyl methacrylate-*b*-styrene-*b*-methyl methacrylate)						
(7 vol% styrene)		8200000	2-butanone	room	0.1	1964KRA
(19 vol% styrene)		3830000	2-butanone	room	0.25	1964KRA
(25 vol% styrene)		2170000	2-butanone	room	0.6	1964KRA
(33 vol% styrene)		1460000	2-butanone	room	0.95	1964KRA
(40 vol% styrene)		1580000	2-butanone	room	0.44	1964KRA
(44 vol% styrene)		1270000	2-butanone	room	0.74	1964KRA
(60 vol% styrene)		1150000	2-butanone	room	0.8	1964KRA
(72 vol% styrene)		278000	2-butanone	room	1.8	1964KRA
(73 vol% styrene)		392000	2-butanone	room	1.05	1964KRA
(74 vol% styrene)		1070000	2-butanone	room	0.8	1964KRA
(75 vol% styrene)		810000	2-butanone	room	0.9	1964KRA

Copolymer	M_n/ g/mol	M_w/ g/mol	**Solvent**	*T*/ K	A_2 10^4/ cm^3mol/g^2	**Ref.**
(76 vol% styrene)		529000	2-butanone	room	1.3	1964KRA
(77 vol% styrene)		425000	2-butanone	room	1.3	1964KRA
(77 vol% styrene)		704000	2-butanone	room	0.9	1964KRA
(80 vol% styrene)		318000	2-butanone	room	1.2	1964KRA
(83 vol% styrene)		572000	2-butanone	room	1.3	1964KRA
(31.4 wt% styrene)	37400		toluene	298.15	4.24	1965INA
(46.1 wt% styrene)	36000		toluene	298.15	5.55	1965INA
(48 wt% styrene)	187000		toluene	288.15	3.4	1964URW
(48 wt% styrene)	187000		toluene	298.15	3.9	1964URW
(48 wt% styrene)	187000		toluene	323.15	5.1	1964URW
(48 wt% styrene)		230000	toluene	298.15	3.46	1964URW
(62 wt% styrene)	225000		toluene	298.15	3.6	1964URW
(62 wt% styrene)		266000	toluene	298.15	3.24	1964URW
(70 wt% styrene)	302000		toluene	298.15	3.2	1964URW
(70 wt% styrene)		370000	toluene	298.15	3.17	1964URW
(85 wt% styrene)	390000		toluene	298.15	3.0	1964URW
(85 wt% styrene)		470000	toluene	298.15	3.01	1964URW
Poly(methyl methacrylate-*g*-styrene)						
(6.14 wt% styrene)		6090000	1-ethylnaphthalene	313.15	1.89	1996KIK
(12.8 wt% styrene)		6690000	1-ethylnaphthalene	313.15	1.64	1996KIK
(6.14 wt% styrene)		6090000	trichloromethane	313.15	2.13	1996KIK
(12.8 wt% styrene)		6690000	trichloromethane	313.15	2.07	1996KIK
Poly(4-methyl-1-penten-3-one-*co*-(S)-4-methyl-1-hexen-3-one)						
(10 mol% (S)-MHO)	52000		trichloromethane	309.15	10	1983THI
(22 mol% (S)-MHO)	54000		trichloromethane	309.15	13	1983THI
(50 mol% (S)-MHO)	56000		trichloromethane	309.15	10	1983THI
(72 mol% (S)-MHO)	62000		trichloromethane	309.15	11	1983THI
(90 mol% (S)-MHO)	68000		trichloromethane	309.15	7	1983THI
Poly(α-methylstyrene-*b*-styrene)						
(25-30% α-methylstyrene)		103000	cyclohexane	298	−1.0	1971DON
(25-30% α-methylstyrene)		103000	cyclohexane	305	−0.5	1971DON
(25-30% α-methylstyrene)		103000	cyclohexane	311	0.0	1971DON
(25-30% α-methylstyrene)		103000	cyclohexane	315	0.5	1971DON
(25-30% α-methylstyrene)		103000	cyclohexane	318	1.0	1971DON
(25-30% α-methylstyrene)		103000	cyclohexane	325	1.5	1971DON
(25-30% α-methylstyrene)		103000	tetrahydrofuran	313	0.66	1971DON
(25-30% α-methylstyrene)		103000	tetrahydrofuran	323	0.70	1971DON

Copolymer	M_n/ g/mol	M_w/ g/mol	**Solvent**	T/ K	A_2 10^4/ cm^3mol/g^2	**Ref.**
Poly(1-octadecene-*co*-maleic anhydride)						
(50 mol% maleic anhydride)		28000	ethyl acetate	298.15	3.0	1985MAT
(50 mol% maleic anhydride)		37000	ethyl acetate	298.15	2.9	1985MAT
(50 mol% maleic anhydride)		51000	ethyl acetate	298.15	2.8	1985MAT
(50 mol% maleic anhydride)		68000	ethyl acetate	298.15	2.5	1985MAT
(50 mol% maleic anhydride)		99000	ethyl acetate	298.15	2.1	1985MAT
Poly[4,4'-oxybis(benzoyl chloride)-*co*-phenolphthalein]						
(1:1)		200000	cyclohexanone	298.15	37.5	1977TA3
(1:1)		170000	1,2-dichloroethane	298.15	10.0	1977TA3
(1:1)		170000	*N,N*-dimethylformamide	298.15	50.0	1977TA3
(1:1)		200000	1,4-dioxane	298.15	15.0	1977TA3
(1:1)		160000	1-methyl-2-pyrrolidinone	298.15	9.0	1977TA3
(1:1)		170000	1,1,2,2-tetrachloroethane	298.15	40.0	1977TA3
(1:1)		170000	tetrahydrofuran	298.15	24.5	1977TA3
(1:1)		170000	trichloromethane	298.15	34.5	1977TA3
Poly[4,4'-oxydiphthalic anhydride-*co*-3,3'-bis(p-aminophenyl) phthalate]						
(1:1)		170000	*N,N*-dimethylformamide	298.15	41.0	1977TA3
(1:1)		200000	dimethylsulfoxide	298.15	31.0	1977TA3
(1:1)		170000	1-methyl-2-pyrrolidinone	298.15	30.0	1977TA3
(1:1)		170000	1,1,2,2-tetrachloroethane	298.15	32.0	1977TA3
(1:1)		160000	trichloromethane	298.15	32.0	1977TA3
Poly(4,4'-oxydiphthalic anhydride-*co*-p-phenylenediamine)						
(1:1)		40000	*N,N*-dimethylacetamide	294.15	10.0	1985BAR
Poly(p-phenylenediamine-*co*-pyromellitic dianhydride)						
(1:1)		166000	1-methyl-2-pyrrolidinone	298.15	8.8	1992KI2
Poly(phenolphthalein-*co*-terephthaloyl chloride)						
(1:1)		50000	1,2-dichloroethane	298.15	6.6	1984GLA
(1:1)		50000	tetrahydrofuran	292.15	0.0	1984GLA
(1:1)		50000	trichloromethane	298.15	10.0	1984GLA
Poly(propylene oxide-*b*-ethylene oxide-*b*-propylene oxide)						
(46 mol% EO)	2650	3000	water	298.15	16.5	1994ZHO
(46 mol% EO)	2650	3000	water	313.15	–2.3	1994ZHO

Copolymer	M_n/ g/mol	M_w/ g/mol	**Solvent**	T/ K	$A_2\ 10^4$/ cm^3mol/g^2	**Ref.**
Poly(styrene-*co*-acrylic acid) (12 arm/12 arm)						
(64.7 wt% polystyrene)		278000	1,4-dioxane	298.15	−0.074	2001VOU
Poly(styrene-*co*-acrylonitrile)						
(19.7 mol% acrylonitrile)		120000	2-butanone	303.15	3.60	1966SHI
(38.3 mol% acrylonitrile)		161000	2-butanone	303.15	4.25	1966SHI
(38.3 mol% acrylonitrile)		250000	2-butanone	303.15	4.20	1966SHI
(38.3 mol% acrylonitrile)		816000	2-butanone	303.15	4.10	1966SHI
(38.3 mol% acrylonitrile)		1000000	2-butanone	303.15	4.00	1966SHI
(38.5 mol% acrylonitrile)		96000	2-butanone	298.15	4.72	1977GL1
(38.5 mol% acrylonitrile)		197000	2-butanone	298.15	7.70	1977GL1
(38.5 mol% acrylonitrile)		197000	2-butanone	298.15	7.72	1977GL1
(38.5 mol% acrylonitrile)		238000	2-butanone	298.15	4.15	1977GL1
(38.5 mol% acrylonitrile)		264000	2-butanone	298.15	5.45	1977GL1
(38.5 mol% acrylonitrile)		265000	2-butanone	298.15	5.43	1977GL1
(38.5 mol% acrylonitrile)		323000	2-butanone	298.15	4.04	1977GL1
(38.5 mol% acrylonitrile)		543000	2-butanone	298.15	3.67	1977GL1
(38.5 mol% acrylonitrile)		601000	2-butanone	298.15	5.25	1977GL1
(38.5 mol% acrylonitrile)		602000	2-butanone	298.15	5.23	1977GL1
(38.5 mol% acrylonitrile)		647000	2-butanone	298.15	4.62	1977GL1
(38.5 mol% acrylonitrile)		648000	2-butanone	298.15	4.62	1977GL1
(38.5 mol% acrylonitrile)		756000	2-butanone	298.15	4.80	1977GL1
(38.5 mol% acrylonitrile)		761000	2-butanone	298.15	4.77	1977GL1
(38.5 mol% acrylonitrile)		832000	2-butanone	298.15	5.03	1977GL1
(38.5 mol% acrylonitrile)		836000	2-butanone	298.15	5.01	1977GL1
(38.5 mol% acrylonitrile)		1130000	2-butanone	298.15	4.89	1977GL1
(38.5 mol% acrylonitrile)		1390000	2-butanone	298.15	3.59	1977GL1
(38.5 mol% acrylonitrile)		238000	2-butanone	298.15	4.15	1979GL3
(38.5 mol% acrylonitrile)		264000	2-butanone	298.15	5.45	1979GL3
(38.5 mol% acrylonitrile)		265000	2-butanone	298.15	5.43	1979GL3
(38.5 mol% acrylonitrile)		323000	2-butanone	298.15	4.04	1979GL3
(38.5 mol% acrylonitrile)		543000	2-butanone	298.15	3.67	1979GL3
(38.5 mol% acrylonitrile)		601000	2-butanone	298.15	5.25	1979GL3
(38.5 mol% acrylonitrile)		602000	2-butanone	298.15	5.23	1979GL3
(38.5 mol% acrylonitrile)		647000	2-butanone	298.15	4.62	1979GL3
(38.5 mol% acrylonitrile)		648000	2-butanone	298.15	4.62	1979GL3
(38.5 mol% acrylonitrile)		756000	2-butanone	298.15	4.80	1979GL3
(38.5 mol% acrylonitrile)		761000	2-butanone	298.15	4.77	1979GL3
(38.5 mol% acrylonitrile)		832000	2-butanone	298.15	5.03	1979GL3
(38.5 mol% acrylonitrile)		836000	2-butanone	298.15	5.01	1979GL3
(38.5 mol% acrylonitrile)		1130000	2-butanone	298.15	4.89	1979GL3
(38.5 mol% acrylonitrile)		1390000	2-butanone	298.15	3.59	1979GL3

Copolymer	M_n/ g/mol	M_w/ g/mol	Solvent	T/ K	A_2 10^4/ cm^3mol/g^2	Ref.
(47.5 mol% acrylonitrile)		140000	2-butanone	303.15	5.9	1976RED
(47.5 mol% acrylonitrile)		258000	2-butanone	303.15	4.9	1976RED
(47.5 mol% acrylonitrile)		373000	2-butanone	303.15	5.7	1976RED
(47.5 mol% acrylonitrile)		535000	2-butanone	303.15	5.5	1976RED
(47.5 mol% acrylonitrile)		778000	2-butanone	303.15	5.0	1976RED
(47.5 mol% acrylonitrile)		1555000	2-butanone	303.15	4.1	1976RED
(49.9 mol% acrylonitrile)		493000	2-butanone	303.15	4.46	1977RED
(49.9 mol% acrylonitrile)		614000	2-butanone	303.15	3.63	1977RED
(49.9 mol% acrylonitrile)		673000	2-butanone	303.15	4.74	1977RED
(49.9 mol% acrylonitrile)		736000	2-butanone	303.15	3.62	1977RED
(49.9 mol% acrylonitrile)		813000	2-butanone	303.15	4.74	1977RED
(49.9 mol% acrylonitrile)		1590000	2-butanone	303.15	3.90	1977RED
(51.4 mol% acrylonitrile)		182000	2-butanone	303.15	4.80	1966SHI
(62.6 mol% acrylonitrile)		189000	2-butanone	303.15	1.80	1966SHI
(62.6 mol% acrylonitrile)		556000	2-butanone	303.15	1.80	1966SHI
(72.2 mol% acrylonitrile)		140000	2-butanone	303.15	1.45	1966SHI
(17.2 mol% acrylonitrile)		147000	*N,N*-dimethylformamide	298.15	4.70	1993UCH
(31.2 mol% acrylonitrile)		111000	*N,N*-dimethylformamide	298.15	8.56	1993UCH
(44.0 mol% acrylonitrile)		78000	*N,N*-dimethylformamide	298.15	10.9	1993UCH
(49.9 mol% acrylonitrile)		209000	*N,N*-dimethylformamide	303.15	2.16	1977RED
(49.9 mol% acrylonitrile)		466000	*N,N*-dimethylformamide	303.15	3.31	1977RED
(49.9 mol% acrylonitrile)		719000	*N,N*-dimethylformamide	303.15	4.60	1977RED
(49.9 mol% acrylonitrile)		748000	*N,N*-dimethylformamide	303.15	5.03	1977RED
(49.9 mol% acrylonitrile)		1098000	*N,N*-dimethylformamide	303.15	4.34	1977RED
(49.9 mol% acrylonitrile)		1110000	*N,N*-dimethylformamide	303.15	3.49	1977RED
(55.7 mol% acrylonitrile)		63000	*N,N*-dimethylformamide	298.15	13.3	1993UCH
(63.3 mol% acrylonitrile)		90000	*N,N*-dimethylformamide	298.15	15.8	1993UCH
(68.8 mol% acrylonitrile)		55000	*N,N*-dimethylformamide	298.15	20.3	1993UCH
(80.1 mol% acrylonitrile)		61000	*N,N*-dimethylformamide	298.15	22.2	1993UCH
(27.4 mol% acrylonitrile)		144000	ethyl acetate	303.15	4.8	1976RED
(27.4 mol% acrylonitrile)		203000	ethyl acetate	303.15	3.9	1976RED
(27.4 mol% acrylonitrile)		248000	ethyl acetate	303.15	4.0	1976RED
(27.4 mol% acrylonitrile)		296000	ethyl acetate	303.15	3.8	1976RED
(27.4 mol% acrylonitrile)		379000	ethyl acetate	303.15	2.5	1976RED
(27.4 mol% acrylonitrile)		576000	ethyl acetate	303.15	2.7	1976RED
(38.5 mol% acrylonitrile)		222000	ethyl acetate	303.15	2.8	1976RED
(38.5 mol% acrylonitrile)		352000	ethyl acetate	303.15	3.0	1976RED
(38.5 mol% acrylonitrile)		472000	ethyl acetate	303.15	2.9	1976RED
(38.5 mol% acrylonitrile)		601000	ethyl acetate	303.15	3.0	1976RED
(38.5 mol% acrylonitrile)		647000	ethyl acetate	303.15	3.0	1976RED
(38.5 mol% acrylonitrile)		1060000	ethyl acetate	303.15	2.7	1976RED

Copolymer	M_n/ g/mol	M_w/ g/mol	Solvent	T/ K	A_2 10^4/ cm^3mol/g^2	Ref.
(51 mol% acrylonitrile)		269000	ethyl acetate	303.15	4.28	1982MA1
(51 mol% acrylonitrile)		347000	ethyl acetate	303.15	2.26	1982MA1
(51 mol% acrylonitrile)		457000	ethyl acetate	303.15	1.89	1982MA1
(51 mol% acrylonitrile)		794000	ethyl acetate	303.15	1.23	1982MA1
(51 mol% acrylonitrile)		912000	ethyl acetate	303.15	1.20	1982MA1
(51 mol% acrylonitrile)		1365000	ethyl acetate	303.15	1.14	1982MA1
(51 mol% acrylonitrile)		2240000	ethyl acetate	303.15	1.11	1982MA1
(25.0 wt% acrylonitrile)	90000	147000	tetrahydrofuran	298.15	11.5	1998SCH
(17.2 mol% acrylonitrile)		147000	tetrahydrofuran	298.15	7.32	1993UCH
(31.2 mol% acrylonitrile)		111000	tetrahydrofuran	298.15	7.94	1993UCH
(38.5 mol% acrylonitrile)		144000	tetrahydrofuran	298.15	8.47	1977GL2
(38.5 mol% acrylonitrile)		145000	tetrahydrofuran	298.15	8.29	1977GL2
(38.5 mol% acrylonitrile)		150000	tetrahydrofuran	298.15	8.88	1977GL2
(38.5 mol% acrylonitrile)		230000	tetrahydrofuran	298.15	7.40	1977GL2
(38.5 mol% acrylonitrile)		286000	tetrahydrofuran	298.15	6.11	1977GL2
(38.5 mol% acrylonitrile)		605000	tetrahydrofuran	298.15	7.81	1977GL2
(38.5 mol% acrylonitrile)		645000	tetrahydrofuran	298.15	7.38	1977GL2
(38.5 mol% acrylonitrile)		825000	tetrahydrofuran	298.15	7.55	1977GL2
(38.5 mol% acrylonitrile)		1072000	tetrahydrofuran	298.15	7.74	1977GL2
(38.5 mol% acrylonitrile)		1123000	tetrahydrofuran	298.15	6.83	1977GL2
(38.5 mol% acrylonitrile)		1873000	tetrahydrofuran	298.15	4.30	1977GL2
(38.5 mol% acrylonitrile)		230000	tetrahydrofuran	298.15	7.40	1979GL3
(38.5 mol% acrylonitrile)		286000	tetrahydrofuran	298.15	6.11	1979GL3
(38.5 mol% acrylonitrile)		605000	tetrahydrofuran	298.15	7.81	1979GL3
(38.5 mol% acrylonitrile)		645000	tetrahydrofuran	298.15	7.38	1979GL3
(38.5 mol% acrylonitrile)		825000	tetrahydrofuran	298.15	7.55	1979GL3
(38.5 mol% acrylonitrile)		1072000	tetrahydrofuran	298.15	7.74	1979GL3
(38.5 mol% acrylonitrile)		1123000	tetrahydrofuran	298.15	6.83	1979GL3
(38.5 mol% acrylonitrile)		1873000	tetrahydrofuran	298.15	4.30	1979GL3
(38.5 mol% acrylonitrile)		2644000	tetrahydrofuran	298.15	2.74	1979GL3
(44.0 mol% acrylonitrile)		78000	tetrahydrofuran	298.15	9.10	1993UCH
(55.7 mol% acrylonitrile)		63000	tetrahydrofuran	298.15	8.50	1993UCH
Poly(styrene-*co*-acrylonitrile-*co*-methyl methacrylate)						
(48 wt% styrene, 30 wt% AN)		300000	acetonitrile	293.15	2.3	1973KAM
(48 wt% styrene, 30 wt% AN)		300000	2-butanone	293.15	5.0	1973KAM
(60.8 wt% styrene, 8.9 wt% AN)		450000	2-butanone	293.15	3.8	1973KAM
(48 wt% styrene, 30 wt% AN)		300000	*N,N*-dimethylformamide	293.15	10.8	1973KAM
(60.8 wt% styrene, 8.9 wt% AN)		450000	*N,N*-dimethylformamide	293.15	5.5	1973KAM
(48 wt% styrene, 30 wt% AN)		300000	1,4-dioxane	293.15	1.3	1973KAM
(60.8 wt% styrene, 8.9 wt% AN)		450000	1,4-dioxane	293.15	6.8	1973KAM
(48 wt% styrene, 30 wt% AN)		300000	tetrahydrofuran	293.15	7.3	1973KAM
(60.8 wt% styrene, 8.9 wt% AN)		450000	tetrahydrofuran	293.15	7.8	1973KAM

Copolymer	M_n/ g/mol	M_w/ g/mol	**Solvent**	T/ K	A_2 10^4/ cm^3mol/g^2	**Ref.**
(60.8 wt% styrene, 8.9 wt% AN)		450000	toluene	293.15	3.6	1973KAM
(48 wt% styrene, 30 wt% AN)		300000	trichloromethane	293.15	9.5	1973KAM
(60.8 wt% styrene, 8.9 wt% AN)		450000	trichloromethane	293.15	10.3	1973KAM
Poly(styrene-*alt*-benzyl methacrylate)						
		209000	2-butanone	room	0.37	1982BRA
		404000	2-butanone	room	1.05	1982BRA
		636000	2-butanone	room	0.77	1982BRA
		1052000	2-butanone	room	0.56	1982BRA
		1179000	2-butanone	room	0.79	1982BRA
		1515000	2-butanone	room	0.83	1982BRA
		2067000	2-butanone	room	0.80	1982BRA
		209000	toluene	room	2.86	1982BRA
		404000	toluene	room	1.71	1982BRA
		636000	toluene	room	1.97	1982BRA
		1052000	toluene	room	1.21	1982BRA
		1179000	toluene	room	1.72	1982BRA
		1515000	toluene	room	1.56	1982BRA
		2067000	toluene	room	1.11	1982BRA
		209000	trichloromethane	room	0.37	1982BRA
		404000	trichloromethane	room	3.32	1982BRA
		636000	trichloromethane	room	3.26	1982BRA
		1052000	trichloromethane	room	1.95	1982BRA
		1179000	trichloromethane	room	2.94	1982BRA
		1515000	trichloromethane	room	2.78	1982BRA
		2067000	trichloromethane	room	2.35	1982BRA
Poly(styrene-*co*-benzyl methacrylate)						
(50 mol% styrene)		63000	2-butanone	room	1.42	1982BRA
(50 mol% styrene)		128000	2-butanone	room	1.39	1982BRA
(50 mol% styrene)		185000	2-butanone	room	1.51	1982BRA
(50 mol% styrene)		246000	2-butanone	room	1.28	1982BRA
(50 mol% styrene)		281000	2-butanone	room	1.20	1982BRA
(50 mol% styrene)		351000	2-butanone	room	1.36	1982BRA
(50 mol% styrene)		386000	2-butanone	room	1.04	1982BRA
(50 mol% styrene)		477000	2-butanone	room	1.01	1982BRA
(50 mol% styrene)		612000	2-butanone	room	0.97	1982BRA
(50 mol% styrene)		128000	toluene	room	2.45	1982BRA
(50 mol% styrene)		185000	toluene	room	2.43	1982BRA
(50 mol% styrene)		246000	toluene	room	2.31	1982BRA
(50 mol% styrene)		281000	toluene	room	2.47	1982BRA
(50 mol% styrene)		351000	toluene	room	2.20	1982BRA
(50 mol% styrene)		386000	toluene	room	2.21	1982BRA

Copolymer	M_n/ g/mol	M_w/ g/mol	**Solvent**	*T*/ K	$A_2\ 10^4$/ cm^3mol/g^2	**Ref.**
(50 mol% styrene)		477000	toluene	room	2.05	1982BRA
(50 mol% styrene)		612000	toluene	room	1.53	1982BRA
(50 mol% styrene)		63000	trichloromethane	room	5.50	1982BRA
(50 mol% styrene)		128000	trichloromethane	room	4.89	1982BRA
(50 mol% styrene)		185000	trichloromethane	room	4.78	1982BRA
(50 mol% styrene)		246000	trichloromethane	room	4.30	1982BRA
(50 mol% styrene)		281000	trichloromethane	room	4.28	1982BRA
(50 mol% styrene)		351000	trichloromethane	room	4.03	1982BRA
(50 mol% styrene)		386000	trichloromethane	room	4.01	1982BRA
(50 mol% styrene)		477000	trichloromethane	room	3.49	1982BRA
(50 mol% styrene)		612000	trichloromethane	room	2.46	1982BRA
Poly(styrene-*b*-butadiene) four-arm star-shaped with polybutadiene as inner blocks (SB)4						
(25.6 mol% styrene)	198000	220000	cyclohexane	298.15	7.13	2009XIO
(25.6 mol% styrene)	198000	220000	cyclohexane	308.15	9.21	2009XIO
(25.6 mol% styrene)	198000	220000	cyclohexane	323.15	8.97	2009XIO
(25.6 mol% styrene)	198000	220000	tetrahydrofuran	293.15	13.2	2009XIO
(25.6 mol% styrene)	198000	220000	tetrahydrofuran	298.15	11.4	2009XIO
(25.6 mol% styrene)	198000	220000	tetrahydrofuran	303.15	13.7	2009XIO
(25.6 mol% styrene)	198000	220000	tetrahydrofuran	313.15	14.6	2009XIO
Poly(styrene-*b*-butadiene-*b*-styrene)						
(25.8 mol% styrene)	76300	85500	cyclohexane	298.15	8.93	2009XIO
(25.8 mol% styrene)	76300	85500	cyclohexane	308.15	11.0	2009XIO
(25.8 mol% styrene)	76300	85500	cyclohexane	323.15	11.5	2009XIO
(4.1 wt% styrene)	517000		cyclohexane	307.35	8.69	1968UTR
(11.6 wt% styrene)	170000		cyclohexane	307.35	9.56	1968UTR
(20.7 wt% styrene)	117000		cyclohexane	307.35	9.64	1968UTR
(30.1 wt% styrene)	69400		cyclohexane	307.35	10.2	1968UTR
(36.6 wt% styrene)	54800		cyclohexane	307.35	9.53	1968UTR
(40.2 wt% styrene)	141000		cyclohexane	307.35	8.54	1968UTR
(45.5 wt% styrene)	44800		cyclohexane	307.35	9.40	1968UTR
(45.9 wt% styrene)	34800		cyclohexane	307.35	10.1	1968UTR
(11.6 wt% styrene)	170000		1,4-dioxane	307.35	2.16	1968UTR
(20.7 wt% styrene)	117000		1,4-dioxane	307.35	3.87	1968UTR
(30.1 wt% styrene)	69400		1,4-dioxane	307.35	4.48	1968UTR
(36.6 wt% styrene)	54800		1,4-dioxane	307.35	5.62	1968UTR
(45.5 wt% styrene)	44800		1,4-dioxane	307.35	5.99	1968UTR
(45.9 wt% styrene)	34800		1,4-dioxane	307.35	6.63	1968UTR
(25.8 mol% styrene)	76300	85500	tetrahydrofuran	293.15	13.3	2009XIO
(25.8 mol% styrene)	76300	85500	tetrahydrofuran	298.15	12.6	2009XIO
(25.8 mol% styrene)	76300	85500	tetrahydrofuran	303.15	13.0	2009XIO
(25.8 mol% styrene)	76300	85500	tetrahydrofuran	313.15	15.8	2009XIO

Copolymer	M_n/ g/mol	M_w/ g/mol	**Solvent**	*T*/ K	A_2 10^4/ cm^3mol/g^2	**Ref.**
(4.1 wt% styrene)	517000		toluene	307.35	9.57	1968UTR
(11.6 wt% styrene)	170000		toluene	307.35	11.4	1968UTR
(20.7 wt% styrene)	117000		toluene	307.35	11.7	1968UTR
(30.1 wt% styrene)	69400		toluene	307.35	12.3	1968UTR
(36.6 wt% styrene)	54800		toluene	307.35	12.0	1968UTR
(40.2 wt% styrene)	141000		toluene	307.35	10.2	1968UTR
(45.5 wt% styrene)	44800		toluene	307.35	12.7	1968UTR
(45.9 wt% styrene)	34800		toluene	307.35	12.5	1968UTR
Poly(styrene-*co*-butyl methacrylate)						
(78.9 wt% styrene)			benzene	298.15	5.5	1979GRU
(1:1)		219400	2-butanone	308.15	3.90	1973SRI
(1:1)		337200	2-butanone	308.15	4.04	1973SRI
(1:1)		361900	2-butanone	308.15	3.60	1973SRI
(1:1)		404600	2-butanone	308.15	3.03	1973SRI
(1:1)		469800	2-butanone	308.15	2.99	1973SRI
(1:1)		483800	2-butanone	308.15	2.83	1973SRI
(1:1)		500100	2-butanone	308.15	3.91	1973SRI
(1:1)		556400	2-butanone	308.15	2.64	1973SRI
(20.6 wt% styrene)	185000	311000	2-butanone	298.15	1.94	1987KYO
(25.5 wt% styrene)			2-butanone	298.15	2.4	1979GRU
(67.7 wt% styrene)	121000	395000	2-butanone	298.15	3.22	1987KYO
(78.9 wt% styrene)			2-butanone	298.15	1.7	1979GRU
(80.0 wt% styrene)	193000	249000	2-butanone	298.15	1.08	1987KYO
(85.0 wt% styrene)	176000	308000	2-butanone	298.15	1.94	1987KYO
(25.5 wt% styrene)			butyl acetate	298.15	1.9	1979GRU
(78.9 wt% styrene)			butyl acetate	298.15	1.9	1979GRU
(25.5 wt% styrene)			1-chloronaphthalene	298.15	4.6	1979GRU
(78.9 wt% styrene)			1-chloronaphthalene	298.15	2.9	1979GRU
(25.5 wt% styrene)			cyclohexanone	298.15	2.0	1979GRU
(78.9 wt% styrene)			cyclohexanone	298.15	3.1	1979GRU
(25.5 wt% styrene)			1,2-dichlorobenzene	298.15	4.3	1979GRU
(78.9 wt% styrene)			1,2-dichlorobenzene	298.15	3.5	1979GRU
(25.5 wt% styrene)			1,2-dichloroethane	298.15	3.7	1979GRU
(25.5 wt% styrene)			1,4-dioxane	298.15	1.6	1979GRU
Poly(styrene-*b*-4-chlorostyrene-*b*-styrene)						
(31.1 mol% styrene)	299000		2-butanone	303.15	1.85	1986OGA
(50.9 mol% styrene)	356000		2-butanone	303.15	1.82	1986OGA
(65.5 mol% styrene)	374000		2-butanone	303.15	1.68	1986OGA
(31.1 mol% styrene)	299000		isopropylbenzene	328.15	1.52	1986OGA
(50.9 mol% styrene)	356000		isopropylbenzene	328.15	1.85	1986OGA
(65.5 mol% styrene)	374000		isopropylbenzene	328.15	2.27	1986OGA

Copolymer	M_n/ g/mol	M_w/ g/mol	Solvent	T/ K	A_2 10^4/ cm^3mol/g^2	Ref.
(31.1 mol% styrene)	299000		toluene	303.15	2.14	1986OGA
(50.9 mol% styrene)	356000		toluene	303.15	2.73	1986OGA
(65.5 mol% styrene)	374000		toluene	303.15	3.00	1986OGA
Poly(styrene-*co*-divinylbenzene)						
		233000	cyclohexane	310.05	0.5	1977AMB
		402000	cyclohexane	309.55	−0.2	1977AMB
		402000	cyclohexane	313.45	1.3	1977AMB
		88400	tetrahydrofuran	303.15	7.0	1977AMB
		233000	tetrahydrofuran	303.15	5.4	1977AMB
		402000	tetrahydrofuran	303.15	5.7	1977AMB
		585000	tetrahydrofuran	303.15	3.3	1977AMB
		2230000	tetrahydrofuran	303.15	2.9	1977AMB
		5240000	tetrahydrofuran	303.15	0.7	1977AMB
Poly[styrene-*b*-(ethylene-*co*-propylene)]						
(35 wt% styrene)		105000	2-butanone	298.15	0.042	1993QU2
(26 wt% styrene)		145000	5-methyl-2-hexanone	298.15	0.036	1993QU1
(35 wt% styrene)		105000	5-methyl-2-hexanone	298.15	0.078	1993QU1
(35 wt% styrene)		105000	5-methyl-2-hexanone	298.15	0.078	1993QU2
(35 wt% styrene)		105000	5-methyl-2-hexanone	298.15	0.078	1995QU1
(35 wt% styrene)		105000	4-methyl-2-pentanone	298.15	0.068	1993QU2
(35 wt% styrene)		105000	4-methyl-2-pentanone	298.15	0.058	1995VIL
(26 wt% styrene)		145000	n-octane	298.15	0.89	1993QU1
(35 wt% styrene)		105000	n-octane	298.15	0.36	1993QU1
(35 wt% styrene)		105000	2-pentanone	298.15	0.031	1993QU2
(35 wt% styrene)		105000	3-pentanone	298.15	0.096	1993QU2
Poly[styrene-*b*-(ethylene-*co*-propylene)-*b*-styrene]						
(30 wt% styrene)		60700	5-methyl-2-hexanone	298.15	0.012	1995QU2
(32 wt% styrene)		87300	5-methyl-2-hexanone	298.15	0.006	1995QU2
(30 wt% styrene)		260000	5-methyl-2-hexanone	298.15	0.046	1995QU2
Poly(styrene-*b*-isoprene)						
(13.5 wt% styrene)	102000	108000	cyclohexane	293.15	8.28	1972GIR
(13.5 wt% styrene)	102000	108000	cyclohexane	303.15	8.65	1972GIR
(13.5 wt% styrene)	102000	108000	cyclohexane	318.15	9.10	1972GIR
(13.5 wt% styrene)	102000	108000	cyclohexane	333.15	9.51	1972GIR
(24.7 wt% styrene)	100000	109000	cyclohexane	293.15	7.61	1972GIR
(24.7 wt% styrene)	100000	109000	cyclohexane	303.15	7.74	1972GIR
(24.7 wt% styrene)	100000	109000	cyclohexane	318.15	8.04	1972GIR
(24.7 wt% styrene)	100000	109000	cyclohexane	333.15	8.69	1972GIR

Copolymer	M_n/ g/mol	M_w/ g/mol	Solvent	T/ K	A_2 10^4/ cm^3mol/g^2	Ref.
(33 wt% styrene)	241000		cyclohexane	293.15	6.40	1969CR2
(33 wt% styrene)	241000		cyclohexane	303.15	6.97	1969CR2
(33 wt% styrene)	241000		cyclohexane	313.15	7.19	1969CR2
(33 wt% styrene)		256000	cyclohexane	303.15	7.66	1969CR1
(52.2 wt% styrene)	105000	111000	cyclohexane	293.15	4.65	1972GIR
(52.2 wt% styrene)	105000	111000	cyclohexane	303.15	4.88	1972GIR
(52.2 wt% styrene)	105000	111000	cyclohexane	318.15	5.44	1972GIR
(52.2 wt% styrene)	105000	111000	cyclohexane	333.15	6.09	1972GIR
(56 wt% styrene)	352000		cyclohexane	293.15	3.92	1969CR2
(56 wt% styrene)	352000		cyclohexane	303.15	4.64	1969CR2
(56 wt% styrene)	352000		cyclohexane	313.15	5.40	1969CR2
(56 wt% styrene)		370000	cyclohexane	303.15	4.40	1969CR1
(59 wt% styrene)	211000		cyclohexane	293.15	4.39	1969CR2
(59 wt% styrene)	211000		cyclohexane	303.15	5.39	1969CR2
(59 wt% styrene)	211000		cyclohexane	313.15	5.99	1969CR2
(59 wt% styrene)		221000	cyclohexane	303.15	5.51	1969CR1
(78 wt% styrene)	247000		cyclohexane	293.15	2.19	1969CR2
(78 wt% styrene)	247000		cyclohexane	303.15	3.46	1969CR2
(78 wt% styrene)	247000		cyclohexane	313.15	4.64	1969CR2
(78 wt% styrene)		259000	cyclohexane	303.15	3.48	1969CR1
(78.2 wt% styrene)	104000	112000	cyclohexane	293.15	2.30	1972GIR
(78.2 wt% styrene)	104000	112000	cyclohexane	303.15	2.97	1972GIR
(78.2 wt% styrene)	104000	112000	cyclohexane	318.15	4.15	1972GIR
(78.2 wt% styrene)	104000	112000	cyclohexane	333.15	5.05	1972GIR
(13.5 wt% styrene)	102000	108000	4-methyl-2-pentanone	293.15	1.10	1972GIR
(13.5 wt% styrene)	102000	108000	4-methyl-2-pentanone	303.15	1.20	1972GIR
(13.5 wt% styrene)	102000	108000	4-methyl-2-pentanone	318.15	1.31	1972GIR
(13.5 wt% styrene)	102000	108000	4-methyl-2-pentanone	333.15	1.39	1972GIR
(24.7 wt% styrene)	100000	109000	4-methyl-2-pentanone	293.15	1.45	1972GIR
(24.7 wt% styrene)	100000	109000	4-methyl-2-pentanone	303.15	1.67	1972GIR
(24.7 wt% styrene)	100000	109000	4-methyl-2-pentanone	318.15	1.70	1972GIR
(24.7 wt% styrene)	100000	109000	4-methyl-2-pentanone	333.15	1.77	1972GIR
(33 wt% styrene)	241000		4-methyl-2-pentanone	303.15	1.39	1969CR2
(33 wt% styrene)		256000	4-methyl-2-pentanone	303.15	1.38	1969CR1
(52.2 wt% styrene)	105000	111000	4-methyl-2-pentanone	293.15	1.38	1972GIR
(52.2 wt% styrene)	105000	111000	4-methyl-2-pentanone	303.15	1.56	1972GIR
(52.2 wt% styrene)	105000	111000	4-methyl-2-pentanone	318.15	1.87	1972GIR
(52.2 wt% styrene)	105000	111000	4-methyl-2-pentanone	333.15	2.07	1972GIR
(56 wt% styrene)	352000		4-methyl-2-pentanone	303.15	1.76	1969CR2
(56 wt% styrene)		370000	4-methyl-2-pentanone	303.15	1.74	1969CR1
(59 wt% styrene)	211000		4-methyl-2-pentanone	303.15	1.25	1969CR2
(59 wt% styrene)		221000	4-methyl-2-pentanone	303.15	1.25	1969CR1
(78 wt% styrene)	247000		4-methyl-2-pentanone	303.15	1.59	1969CR2
(78 wt% styrene)		259000	4-methyl-2-pentanone	303.15	1.60	1969CR1

Copolymer	M_n/ g/mol	M_w/ g/mol	Solvent	T/ K	A_2 10^4/ cm^3mol/g^2	Ref.
(78.2 wt% styrene)	104000	112000	4-methyl-2-pentanone	293.15	1.83	1972GIR
(78.2 wt% styrene)	104000	112000	4-methyl-2-pentanone	303.15	1.93	1972GIR
(78.2 wt% styrene)	104000	112000	4-methyl-2-pentanone	318.15	2.04	1972GIR
(78.2 wt% styrene)	104000	112000	4-methyl-2-pentanone	333.15	2.21	1972GIR
(28 wt% styrene)		14600	tetrachloromethane	298.15	12.8	1994PIS
(29 wt% styrene)		20000	tetrachloromethane	298.15	11.8	1994PIS
(30 wt% styrene)		71500	tetrachloromethane	298.15	7.2	1994PIS
(28 wt% styrene)		14100	tetrahydrofuran	298.15	14.6	1994PIS
(28 wt% styrene)		24400	tetrahydrofuran	298.15	13.5	1994PIS
(29 wt% styrene)		17900	tetrahydrofuran	298.15	14.0	1994PIS
(29 wt% styrene)		54900	tetrahydrofuran	298.15	12.0	1994PIS
(30 wt% styrene)		69600	tetrahydrofuran	298.15	9.3	1994PIS
(50 wt% styrene)	94900	98700	tetrahydrofuran	298.15	7.1	2000PIS
(13.5 wt% styrene)	102000	108000	toluene	293.15	9.48	1972GIR
(13.5 wt% styrene)	102000	108000	toluene	303.15	9.73	1972GIR
(13.5 wt% styrene)	102000	108000	toluene	318.15	10.08	1972GIR
(13.5 wt% styrene)	102000	108000	toluene	333.15	9.94	1972GIR
(24.7 wt% styrene)	100000	109000	toluene	293.15	9.13	1972GIR
(24.7 wt% styrene)	100000	109000	toluene	303.15	9.45	1972GIR
(24.7 wt% styrene)	100000	109000	toluene	318.15	9.82	1972GIR
(24.7 wt% styrene)	100000	109000	toluene	333.15	9.95	1972GIR
(28 wt% styrene)	12200		toluene	310.15	12.7	1994PIS
(28 wt% styrene)	22500		toluene	310.15	12.6	1994PIS
(29 wt% styrene)	15700		toluene	310.15	15.1	1994PIS
(29 wt% styrene)	53800		toluene	310.15	10.8	1994PIS
(30 wt% styrene)	61200		toluene	310.15	10.4	1994PIS
(33 wt% styrene)	241000		toluene	293.15	7.70	1969CR2
(33 wt% styrene)	241000		toluene	303.15	8.46	1969CR2
(33 wt% styrene)	241000		toluene	313.15	8.58	1969CR2
(33 wt% styrene)		256000	toluene	303.15	8.10	1969CRA
(52.2 wt% styrene)	105000	111000	toluene	293.15	7.24	1972GIR
(52.2 wt% styrene)	105000	111000	toluene	303.15	7.69	1972GIR
(52.2 wt% styrene)	105000	111000	toluene	318.15	8.00	1972GIR
(52.2 wt% styrene)	105000	111000	toluene	333.15	7.89	1972GIR
(56 wt% styrene)	352000		toluene	293.15	6.97	1969CR2
(56 wt% styrene)	352000		toluene	303.15	7.11	1969CR2
(56 wt% styrene)	352000		toluene	313.15	7.96	1969CR2
(56 wt% styrene)		370000	toluene	303.15	10.2	1969CR1
(59 wt% styrene)	211000		toluene	293.15	8.26	1969CR2
(59 wt% styrene)	211000		toluene	303.15	8.43	1969CR2
(59 wt% styrene)	211000		toluene	313.15	9.27	1969CR2
(59 wt% styrene)		221000	toluene	303.15	9.04	1969CR1
(78 wt% styrene)	247000		toluene	293.15	6.34	1969CR2
(78 wt% styrene)	247000		toluene	303.15	6.87	1969CR2

Copolymer	M_n/ g/mol	M_w/ g/mol	Solvent	T/ K	A_2 10^4/ cm^{3}mol/g^2	Ref.
(78 wt% styrene)	247000		toluene	313.15	7.01	1969CR2
(78 wt% styrene)		259000	toluene	303.15	8.78	1969CR1
(78.2 wt% styrene)	104000	112000	toluene	293.15	7.24	1972GIR
(78.2 wt% styrene)	104000	112000	toluene	303.15	7.69	1972GIR
(78.2 wt% styrene)	104000	112000	toluene	318.15	8.00	1972GIR
(78.2 wt% styrene)	104000	112000	toluene	333.15	7.89	1972GIR
Poly(styrene-*b*-isoprene) (polyblock copolymer)						
(74 wt% styrene)		179000	benzene	293.15	2.17	1998ISH
(74 wt% styrene)		294000	benzene	293.15	0.94	1998ISH
(74 wt% styrene)		332000	benzene	293.15	0.59	1998ISH
(74 wt% styrene)		417000	benzene	293.15	0.48	1998ISH
(79 wt% styrene)		175000	benzene	293.15	0.77	1998ISH
(79 wt% styrene)		261000	benzene	293.15	0.57	1998ISH
(79 wt% styrene)		421000	benzene	293.15	0.34	1998ISH
(79 wt% styrene)		735000	benzene	293.15	0.21	1998ISH
Poly(styrene-*b*-isoprene-*b*-styrene)						
	28100	31000	cyclohexane	305.15	0.0	1976ROO
(36 wt% styrene)		78300	tetrachloromethane	298.15	5.4	1994PIS
(36 wt% styrene)		76300	tetrahydrofuran	298.15	8.1	1994PIS
(48 wt% styrene)	86100	93000	tetrahydrofuran	298.15	8.4	2000PIS
(36 wt% styrene)	70200		toluene	310.15	11.8	1994PIS
Poly(styrene-*co*-methacrylic acid)						
(20 mol% methacrylic acid)		230000	1,4-dioxane	293.15	2.9	1973RAF
(20 mol% methacrylic acid)		355000	1,4-dioxane	293.15	3.0	1973RAF
(20 mol% methacrylic acid)		384000	1,4-dioxane	293.15	2.9	1973RAF
(20 mol% methacrylic acid)		561000	1,4-dioxane	293.15	2.8	1973RAF
(20 mol% methacrylic acid)		602000	1,4-dioxane	293.15	2.7	1973RAF
(20 mol% methacrylic acid)		891000	1,4-dioxane	293.15	2.2	1973RAF
(20 mol% methacrylic acid)		913000	1,4-dioxane	293.15	2.0	1973RAF
(20 mol% methacrylic acid)		1017000	1,4-dioxane	293.15	1.8	1973RAF
(20 mol% methacrylic acid)		1256000	1,4-dioxane	293.15	1.2	1973RAF
Poly(styrene-*co*-methacrylonitrile)						
(50.2 mol% styrene)		227000	2-butanone	303.15	2.16	1977RED
(50.2 mol% styrene)		279000	2-butanone	303.15	2.46	1977RED
(50.2 mol% styrene)		309000	2-butanone	303.15	1.44	1977RED
(50.2 mol% styrene)		329000	2-butanone	303.15	1.88	1977RED
(50.2 mol% styrene)		346000	2-butanone	303.15	3.18	1977RED
(50.2 mol% styrene)		398000	2-butanone	303.15	2.89	1977RED

Copolymer	M_n/ g/mol	M_w/ g/mol	Solvent	T/ K	A_2 10^4/ cm^{3}mol/g^2	Ref.
(50.2 mol% styrene)		223000	*N,N*-dimethylformamide	303.15	4.19	1977RED
(50.2 mol% styrene)		229000	*N,N*-dimethylformamide	303.15	3.37	1977RED
(50.2 mol% styrene)		285000	*N,N*-dimethylformamide	303.15	3.27	1977RED
(50.2 mol% styrene)		300000	*N,N*-dimethylformamide	303.15	3.37	1977RED
(50.2 mol% styrene)		362000	*N,N*-dimethylformamide	303.15	2.76	1977RED
(50.2 mol% styrene)		416000	*N,N*-dimethylformamide	303.15	3.07	1977RED
Poly(styrene-*co*-4-methoxystyrene)						
(26.0 mol% 4-methoxystyrene)		61000	toluene	298.15	5.34	1972PIZ
(26.4 mol% 4-methoxystyrene)		71000	toluene	298.15	5.0	1972PIZ
(26.4 mol% 4-methoxystyrene)		86000	toluene	298.15	5.6	1972PIZ
(26.0 mol% 4-methoxystyrene)		93000	toluene	298.15	5.62	1972PIZ
(26.4 mol% 4-methoxystyrene)		105000	toluene	298.15	4.6	1972PIZ
(26.0 mol% 4-methoxystyrene)		131000	toluene	298.15	5.54	1972PIZ
(26.4 mol% 4-methoxystyrene)		137000	toluene	298.15	4.3	1972PIZ
(26.4 mol% 4-methoxystyrene)		171000	toluene	298.15	4.3	1972PIZ
(26.0 mol% 4-methoxystyrene)		200000	toluene	298.15	4.99	1972PIZ
(26.0 mol% 4-methoxystyrene)		245000	toluene	298.15	4.64	1972PIZ
(26.0 mol% 4-methoxystyrene)		285000	toluene	298.15	4.69	1972PIZ
(26.4 mol% 4-methoxystyrene)		308000	toluene	298.15	3.8	1972PIZ
(26.0 mol% 4-methoxystyrene)		348000	toluene	298.15	4.67	1972PIZ
(26.4 mol% 4-methoxystyrene)		350000	toluene	298.15	3.0	1972PIZ
(26.0 mol% 4-methoxystyrene)		420000	toluene	298.15	4.49	1972PIZ
(26.0 mol% 4-methoxystyrene)		665000	toluene	298.15	4.29	1972PIZ
(53.0 mol% 4-methoxystyrene)		66000	toluene	298.15	5.46	1972PIZ
(53.0 mol% 4-methoxystyrene)		104000	toluene	298.15	5.33	1972PIZ
(53.0 mol% 4-methoxystyrene)		168000	toluene	298.15	4.68	1972PIZ
(53.0 mol% 4-methoxystyrene)		269000	toluene	298.15	4.26	1972PIZ
(53.0 mol% 4-methoxystyrene)		292000	toluene	298.15	4.22	1972PIZ
(53.0 mol% 4-methoxystyrene)		446000	toluene	298.15	3.97	1972PIZ
(53.0 mol% 4-methoxystyrene)		564000	toluene	298.15	3.98	1972PIZ
(53.0 mol% 4-methoxystyrene)		642000	toluene	298.15	3.86	1972PIZ
(53.0 mol% 4-methoxystyrene)		920000	toluene	298.15	3.82	1972PIZ
(53.0 mol% 4-methoxystyrene)		1346000	toluene	298.15	3.81	1972PIZ
(53.0 mol% 4-methoxystyrene)		1783000	toluene	298.15	3.41	1972PIZ
(53.8 mol% 4-methoxystyrene)		35500	toluene	298.15	4.1	1972PIZ
(53.8 mol% 4-methoxystyrene)		50500	toluene	298.15	5.2	1972PIZ
(53.8 mol% 4-methoxystyrene)		60000	toluene	298.15	4.2	1972PIZ
(53.8 mol% 4-methoxystyrene)		72000	toluene	298.15	3.6	1972PIZ
(53.8 mol% 4-methoxystyrene)		94000	toluene	298.15	4.0	1972PIZ
(53.8 mol% 4-methoxystyrene)		100000	toluene	298.15	3.6	1972PIZ
(53.8 mol% 4-methoxystyrene)		121000	toluene	298.15	3.5	1972PIZ
(53.8 mol% 4-methoxystyrene)		154000	toluene	298.15	3.2	1972PIZ

Copolymer	M_n/ g/mol	M_w/ g/mol	Solvent	T/ K	A_2 10^4/ cm^3mol/g^2	Ref.
(53.8 mol% 4-methoxystyrene)		234000	toluene	298.15	4.2	1972PIZ
(53.8 mol% 4-methoxystyrene)		400000	toluene	298.15	3.8	1972PIZ
(53.8 mol% 4-methoxystyrene)		660000	toluene	298.15	3.9	1972PIZ
(53.8 mol% 4-methoxystyrene)		701000	toluene	298.15	3.2	1972PIZ
(75.6 mol% 4-methoxystyrene)		44000	toluene	298.15	5.9	1972PIZ
(75.6 mol% 4-methoxystyrene)		61000	toluene	298.15	5.4	1972PIZ
(75.6 mol% 4-methoxystyrene)		79000	toluene	298.15	4.54	1972PIZ
(75.6 mol% 4-methoxystyrene)		118000	toluene	298.15	4.50	1972PIZ
(75.6 mol% 4-methoxystyrene)		119000	toluene	298.15	4.1	1972PIZ
(75.6 mol% 4-methoxystyrene)		169000	toluene	298.15	3.9	1972PIZ
(75.6 mol% 4-methoxystyrene)		184000	toluene	298.15	3.92	1972PIZ
(75.6 mol% 4-methoxystyrene)		187000	toluene	298.15	4.2	1972PIZ
(75.6 mol% 4-methoxystyrene)		232000	toluene	298.15	3.2	1972PIZ
(75.6 mol% 4-methoxystyrene)		279000	toluene	298.15	3.72	1972PIZ
(75.6 mol% 4-methoxystyrene)		294000	toluene	298.15	3.0	1972PIZ
(75.6 mol% 4-methoxystyrene)		421000	toluene	298.15	3.5	1972PIZ
(75.6 mol% 4-methoxystyrene)		431000	toluene	298.15	3.55	1972PIZ
(75.6 mol% 4-methoxystyrene)		730000	toluene	298.15	3.30	1972PIZ
(75.6 mol% 4-methoxystyrene)		1149000	toluene	298.15	3.05	1972PIZ
(75.6 mol% 4-methoxystyrene)		1717000	toluene	298.15	3.00	1972PIZ
Poly(styrene-*co*-methyl acrylate)						
(15 mol% styrene)	290000	1120000	2-butanone	293.15	2.6	1970MAE
(15 mol% styrene)	290000	1120000	2-butanone	303.15	2.7	1969FIS
(25 mol% styrene)		89100	2-butanone	303.15	4.05	1969MAT
(25 mol% styrene)		146000	2-butanone	303.15	3.46	1969MAT
(25 mol% styrene)		165000	2-butanone	303.15	4.05	1969MAT
(25 mol% styrene)		278000	2-butanone	303.15	3.20	1969MAT
(28 mol% styrene)	77000	254000	2-butanone	293.15	3.2	1970MAE
(28 mol% styrene)	77000	254000	2-butanone	303.15	3.5	1969FIS
(40 mol% styrene)		70400	2-butanone	303.15	4.19	1969MAT
(40 mol% styrene)		97100	2-butanone	303.15	3.70	1969MAT
(40 mol% styrene)		124000	2-butanone	303.15	3.90	1969MAT
(40 mol% styrene)		177000	2-butanone	303.15	3.07	1969MAT
(40 mol% styrene)		183000	2-butanone	303.15	3.68	1969MAT
(40 mol% styrene)		237000	2-butanone	303.15	3.59	1969MAT
(45 mol% styrene)	258000	400000	2-butanone	293.15	3.4	1970MAE
(45 mol% styrene)	258000	400000	2-butanone	303.15	3.7	1969FIS
(62 mol% styrene)	180000	418000	2-butanone	293.15	3.2	1970MAE
(61.7 mol% styrene)	180000	418000	2-butanone	303.15	3.5	1969FIS
(66 mol% styrene)		65700	2-butanone	303.15	3.64	1969MAT
(66 mol% styrene)		141000	2-butanone	303.15	3.18	1969MAT
(66 mol% styrene)		154000	2-butanone	303.15	2.86	1969MAT

Copolymer	M_n/ g/mol	M_w/ g/mol	Solvent	T/ K	A_2 10^4/ cm^3mol/g^2	Ref.
(66 mol% styrene)		183000	2-butanone	303.15	2.94	1969MAT
(66 mol% styrene)		223000	2-butanone	303.15	3.23	1969MAT
(66 mol% styrene)		236000	2-butanone	303.15	2.91	1969MAT
(66 mol% styrene)		355000	2-butanone	303.15	2.50	1969MAT
(77 mol% styrene)		30900	2-butanone	303.15	1.97	1969MAT
(77 mol% styrene)		67400	2-butanone	303.15	3.24	1969MAT
(77 mol% styrene)		118000	2-butanone	303.15	2.79	1969MAT
(77 mol% styrene)		134000	2-butanone	303.15	2.68	1969MAT
(77 mol% styrene)		147000	2-butanone	303.15	2.44	1969MAT
(77 mol% styrene)		172000	2-butanone	303.15	2.70	1969MAT
(77 mol% styrene)		200000	2-butanone	303.15	2.63	1969MAT
(77 mol% styrene)		224000	2-butanone	303.15	2.30	1969MAT
(77 mol% styrene)		332000	2-butanone	303.15	2.05	1969MAT
(77 mol% styrene)		481000	2-butanone	303.15	1.93	1969MAT
(77 mol% styrene)		535000	2-butanone	303.15	1.93	1969MAT
(77 mol% styrene)		568000	2-butanone	303.15	1.91	1969MAT
(77 mol% styrene)		794000	2-butanone	303.15	1.89	1969MAT
(78 mol% styrene)	231000	327000	2-butanone	293.15	2.8	1970MAE
(77.6 mol% styrene)	231000	327000	2-butanone	303.15	2.9	1969FIS
(15 mol% styrene)	290000	1120000	*N,N*-dimethylformamide	293.15	5.2	1970MAE
(15 mol% styrene)	290000	1120000	*N,N*-dimethylformamide	310.15	4.6	1969FIS
(28 mol% styrene)	77000	254000	*N,N*-dimethylformamide	293.15	5.5	1970MAE
(28 mol% styrene)	77000	254000	*N,N*-dimethylformamide	310.15	5.1	1969FIS
(45 mol% styrene)	258000	400000	*N,N*-dimethylformamide	293.15	5.7	1970MAE
(45 mol% styrene)	258000	400000	*N,N*-dimethylformamide	310.15	5.5	1969FIS
(62 mol% styrene)	180000	418000	*N,N*-dimethylformamide	293.15	5.2	1970MAE
(61.7 mol% styrene)	180000	418000	*N,N*-dimethylformamide	310.15	4.9	1969FIS
(78 mol% styrene)	231000	327000	*N,N*-dimethylformamide	293.15	4.1	1970MAE
(77.6 mol% styrene)	231000	327000	*N,N*-dimethylformamide	310.15	4.0	1969FIS
(50 mol% styrene		1000000	ethyl acetate	308.15	<1.0	1968KAR
(25 mol% styrene)	065200		toluene	303.15	4.72	1969MAT
(25 mol% styrene)	093000		toluene	303.15	4.64	1969MAT
(25 mol% styrene)	157000		toluene	303.15	4.13	1969MAT
(25 mol% styrene)	176000		toluene	303.15	4.10	1969MAT
(25 mol% styrene)	208000		toluene	303.15	3.86	1969MAT
(25 mol% styrene)	237000		toluene	303.15	3.86	1969MAT
(40 mol% styrene)	66400		toluene	303.15	5.42	1969MAT
(40 mol% styrene)	94900		toluene	303.15	4.97	1969MAT
(40 mol% styrene)	126000		toluene	303.15	4.76	1969MAT
(40 mol% styrene)	181000		toluene	303.15	4.69	1969MAT
(40 mol% styrene)	236000		toluene	303.15	4.48	1969MAT
(40 mol% styrene)	300000		toluene	303.15	4.35	1969MAT
(50 mol% styrene)	66900		toluene	303.15	5.73	1969MAT
(50 mol% styrene)	80700		toluene	303.15	5.08	1969MAT

Copolymer	M_n/ g/mol	M_w/ g/mol	**Solvent**	T/ K	A_2 10^4/ cm^3mol/g^2	**Ref.**
(50 mol% styrene)	89800		toluene	303.15	4.55	1969MAT
(50 mol% styrene)	125000		toluene	303.15	4.73	1969MAT
(50 mol% styrene)	164000		toluene	303.15	4.10	1969MAT
(50 mol% styrene)	198000		toluene	303.15	4.26	1969MAT
(50 mol% styrene)	244000		toluene	303.15	4.05	1969MAT
(66 mol% styrene)	59200		toluene	303.15	7.47	1969MAT
(66 mol% styrene)	83800		toluene	303.15	5.73	1969MAT
(66 mol% styrene)	100000		toluene	303.15	5.60	1969MAT
(66 mol% styrene)	135000		toluene	303.15	5.50	1969MAT
(66 mol% styrene)	179000		toluene	303.15	3.30	1969MAT
(66 mol% styrene)	216000		toluene	303.15	5.43	1969MAT
(77 mol% styrene)	22200		toluene	303.15	8.30	1969MAT
(77 mol% styrene)	45700		toluene	303.15	7.69	1969MAT
(77 mol% styrene)	64700		toluene	303.15	6.38	1969MAT
(77 mol% styrene)	85300		toluene	303.15	6.20	1969MAT
(77 mol% styrene)	107000		toluene	303.15	5.50	1969MAT
(77 mol% styrene)	127000		toluene	303.15	5.63	1969MAT
(77 mol% styrene)	146000		toluene	303.15	5.50	1969MAT
(77 mol% styrene)	159000		toluene	303.15	5.20	1969MAT
(77 mol% styrene)	207000		toluene	303.15	4.77	1969MAT
(77 mol% styrene)	248000		toluene	303.15	4.37	1969MAT
(77 mol% styrene)	316000		toluene	303.15	5.00	1969MAT
(77 mol% styrene)	349000		toluene	303.15	4.63	1969MAT
(77 mol% styrene)	349000		toluene	303.15	4.63	1969MAT
(77 mol% styrene)	455000		toluene	303.15	4.50	1969MAT
(77 mol% styrene)	909000		toluene	303.15	4.23	1969MAT
Poly(styrene-*b*-methyl methacrylate)						
(38 wt% styrene)		1530000	benzene	303.15	1.48	1974UTI
(38 wt% styrene)		1530000	benzene	303.15	1.25	1974UTI
(27 mol% deuterated styrene)		366000	1,4-dioxane	room	11.0	1983SUN
(33 mol% deuterated styrene)		301000	1,4-dioxane	room	9.98	1983SUN
(46 mol% deuterated styrene)		215000	1,4-dioxane	room	11.8	1983SUN
(56 mol% deuterated styrene)		175000	1,4-dioxane	room	10.9	1983SUN
(38 wt% styrene)		1530000	toluene	303.15	1.65	1974UTI
(38 wt% styrene)		1530000	toluene	342.15	4.92	1974UTI
Poly(styrene-*co*-methyl methacrylate)						
(46.4 wt% methyl methacrylate)		49000	2-butanone	298.15	3.2	1955STO
(46.4 wt% methyl methacrylate)		94000	2-butanone	298.15	2.9	1955STO
(46.4 wt% methyl methacrylate)		205000	2-butanone	298.15	2.35	1955STO
(46.4 wt% methyl methacrylate)		310000	2-butanone	298.15	2.2	1955STO
(46.4 wt% methyl methacrylate)		710000	2-butanone	298.15	1.9	1955STO

Copolymer	M_n/ g/mol	M_w/ g/mol	Solvent	T/ K	A_2 10^4/ cm^3mol/g^2	Ref.
(46.4 wt% methyl methacrylate)		1330000	2-butanone	298.15	1.8	1955STO
(46.4 wt% methyl methacrylate)		1900000	2-butanone	298.15	1.5	1955STO
(29 mol% styrene)	100000	150000	1-chlorobutane	303.15	2.1	1968KOT
(29 mol% styrene)	181000	276000	1-chlorobutane	303.15	1.7	1968KOT
(29 mol% styrene)	276000	447000	1-chlorobutane	303.15	2.3	1968KOT
(29 mol% styrene)	354000	606000	1-chlorobutane	303.15	1.4	1968KOT
(29 mol% styrene)	387000	678000	1-chlorobutane	303.15	1.5	1968KOT
(29 mol% styrene)	440000	774000	1-chlorobutane	303.15	1.3	1968KOT
(29 mol% styrene)	592000	1058000	1-chlorobutane	303.15	1.5	1968KOT
(56 mol% styrene)	34200	47900	1-chlorobutane	303.15	5.6	1968KOT
(56 mol% styrene)	48000	67600	1-chlorobutane	303.15	4.6	1968KOT
(56 mol% styrene)	68900	97100	1-chlorobutane	303.15	4.6	1968KOT
(56 mol% styrene)	96600	137000	1-chlorobutane	303.15	3.4	1968KOT
(56 mol% styrene)	185000	277000	1-chlorobutane	303.15	2.7	1968KOT
(56 mol% styrene)	226000	334000	1-chlorobutane	303.15	2.6	1968KOT
(56 mol% styrene)	264000	396000	1-chlorobutane	303.15	2.3	1968KOT
(56 mol% styrene)	350000	535000	1-chlorobutane	303.15	2.2	1968KOT
(56 mol% styrene)	397000	608000	1-chlorobutane	303.15	2.2	1968KOT
(56 mol% styrene)	460000	745000	1-chlorobutane	303.15	2.1	1968KOT
(56 mol% styrene)	500000	810000	1-chlorobutane	303.15	2.2	1968KOT
(70 mol% styrene)	40000	49600	1-chlorobutane	303.15	5.2	1968KOT
(70 mol% styrene)	47300	58600	1-chlorobutane	303.15	4.1	1968KOT
(70 mol% styrene)	62100	77000	1-chlorobutane	303.15	3.0	1968KOT
(70 mol% styrene)	66700	84000	1-chlorobutane	303.15	5.8	1968KOT
(70 mol% styrene)	145000	183000	1-chlorobutane	303.15	3.2	1968KOT
(70 mol% styrene)	181000	230000	1-chlorobutane	303.15	3.3	1968KOT
(70 mol% styrene)	210000	270000	1-chlorobutane	303.15	3.0	1968KOT
(70 mol% styrene)	274000	352000	1-chlorobutane	303.15	3.0	1968KOT
(70 mol% styrene)	342000	440000	1-chlorobutane	303.15	2.6	1968KOT
(70 mol% styrene)	432000	544000	1-chlorobutane	303.15	2.1	1968KOT
(29.8 mol% methyl methacrylate)		440000	cyclohexanol	336.15	0.0	1966KOT
(70 mol% styrene)	342000	440000	cyclohexanol	336.15	0.0	1968KOT
(30 mol% methyl methacrylate)		440000	cyclohexanol	336.15	0.0	1969KOT
(43.8 mol% methyl methacrylate)		535000	cyclohexanol	334.45	0.0	1966KOT
(56 mol% styrene)	350000	535000	cyclohexanol	334.45	0.0	1968KOT
(44 mol% methyl methacrylate)			cyclohexanol	334.45	0.0	1969KOT
(70 mol% methyl methacrylate)			cyclohexanol	341.35	0.0	1969KOT
(70.7 mol% methyl methacrylate)		606000	cyclohexanol	341.15	0.0	1966KOT
(29 mol% styrene)	354000	606000	cyclohexanol	341.35	0.0	1968KOT
(11.5 wt% methyl methacrylate)		122000	1,4-dioxane	298.15	6.9	1977KRA
(23.5 wt% methyl methacrylate)		119000	1,4-dioxane	298.15	6.2	1977KRA
(36 wt% methyl methacrylate)		131000	1,4-dioxane	298.15	6.6	1977KRA
(46.4 wt% methyl methacrylate)		1330000	1,4-dioxane	298.15	3.3	1955STO
(46.4 wt% methyl methacrylate)		1900000	1,4-dioxane	298.15	3.6	1955STO

Copolymer	M_n/ g/mol	M_w/ g/mol	Solvent	T/ K	A_2 10^4/ cm^{3}mol/g^2	Ref.
(64 wt% methyl methacrylate)		194000	1,4-dioxane	298.15	5.75	1977KRA
(76.5 wt% methyl methacrylate)		196000	1,4-dioxane	298.15	5.35	1977KRA
(88.5 wt% methyl methacrylate)		220000	1,4-dioxane	298.15	5.25	1977KRA
(50 mol% styrene)		7300000	1,4-dioxane	293.15	2.291	1983SIM
(50 mol% styrene)		18400000	1,4-dioxane	293.15	1.910	1983SIM
(50 mol% styrene)		20400000	1,4-dioxane	293.15	1.863	1983SIM
(50 mol% styrene)		36800000	1,4-dioxane	293.15	1.701	1983SIM
(50 mol% styrene)		55300000	1,4-dioxane	293.15	1.556	1983SIM
(50 mol% styrene)		67600000	1,4-dioxane	293.15	1.481	1983SIM
(29.8 mol% methyl methacrylate)		342000	2-ethoxyethanol	345.95	0.0	1966KOT
(70 mol% styrene)	342000	440000	2-ethoxyethanol	345.95	0.0	1968KOT
(30 mol% methyl methacrylate)		440000	2-ethoxyethanol	345.95	0.0	1969KOT
(43.8 mol% methyl methacrylate)		535000	2-ethoxyethanol	331.55	0.0	1966KOT
(44 mol% methyl methacrylate)		535000	2-ethoxyethanol	331.55	0.0	1969KOT
(56 mol% styrene)	350000	535000	2-ethoxyethanol	331.55	0.0	1968KOT
(29 mol% styrene)	354000	606000	2-ethoxyethanol	313.15	0.0	1968KOT
(70 mol% methyl methacrylate)		606000	2-ethoxyethanol	313.15	0.0	1969KOT
(70.7 mol% methyl methacrylate)		606000	2-ethoxyethanol	313.15	0.0	1966KOT
(46.5 mol% methyl methacrylate)		372000	ethyl acetate	303.15	2.3	1976RED
(46.5 mol% methyl methacrylate)		423000	ethyl acetate	303.15	2.1	1976RED
(46.5 mol% methyl methacrylate)		466000	ethyl acetate	303.15	1.8	1976RED
(46.5 mol% methyl methacrylate)		540000	ethyl acetate	303.15	1.7	1976RED
(46.5 mol% methyl methacrylate)		612000	ethyl acetate	303.15	1.6	1976RED
(46.5 mol% methyl methacrylate)		772000	ethyl acetate	303.15	1.5	1976RED
(46.5 mol% methyl methacrylate)		1113000	ethyl acetate	303.15	1.5	1976RED
(46.4 wt% methyl methacrylate)		1330000	nitroethane	298.15	1.5	1955STO
(46.4 wt% methyl methacrylate)		1900000	nitroethane	298.15	1.2	1955STO
(46.4 wt% methyl methacrylate)		1900000	tetrachloromethane	298.15	2.4	1955STO
(46.4 wt% methyl methacrylate)		1330000	tetrachloromethane	298.15	2.3	1955STO
(29 mol% styrene)	46700	70500	toluene	303.15	3.7	1968KOT
(29 mol% styrene)	100000	150000	toluene	303.15	4.3	1968KOT
(29 mol% styrene)	181000	276000	toluene	303.15	3.7	1968KOT
(29 mol% styrene)	276000	447000	toluene	303.15	3.2	1968KOT
(29 mol% styrene)	354000	606000	toluene	303.15	3.0	1968KOT
(29 mol% styrene)	440000	774000	toluene	303.15	2.8	1968KOT
(29 mol% styrene)	592000	1058000	toluene	303.15	2.6	1968KOT
(56 mol% styrene)	96600	137000	toluene	303.15	4.1	1968KOT
(56 mol% styrene)	185000	277000	toluene	303.15	3.9	1968KOT
(56 mol% styrene)	226000	334000	toluene	303.15	3.6	1968KOT
(56 mol% styrene)	264000	396000	toluene	303.15	3.7	1968KOT
(56 mol% styrene)	350000	535000	toluene	303.15	3.6	1968KOT
(56 mol% styrene)	397000	608000	toluene	303.15	3.5	1968KOT
(56 mol% styrene)	460000	745000	toluene	303.15	3.1	1968KOT
(56 mol% styrene)	500000	810000	toluene	303.15	3.3	1968KOT

Copolymer	M_n/ g/mol	M_w/ g/mol	Solvent	T/ K	A_2 10^4/ cm^3mol/g^2	Ref.
(70 mol% styrene)	66700	84000	toluene	303.15	4.1	1968KOT
(70 mol% styrene)	145000	183000	toluene	303.15	4.7	1968KOT
(70 mol% styrene)	210000	270000	toluene	303.15	3.5	1968KOT
(70 mol% styrene)	274000	352000	toluene	303.15	4.0	1968KOT
(70 mol% styrene)	342000	440000	toluene	303.15	3.1	1968KOT
(70 mol% styrene)	432000	544000	toluene	303.15	2.9	1968KOT
(10 wt% styrene)	84800	133000	toluene	323.15	0.12	2008BER
(20 wt% styrene)	80500	103000	toluene	323.15	0.28	2008BER
(50 wt% styrene)	75000	90000	toluene	323.15	2.03	2008BER
(83 wt% styrene)	34200	50700	toluene	323.15	7.31	2008BER
Poly(styrene-*b*-octadecyl methacrylate)						
(67.8 wt% styrene)		43200	ethyl acetate	298.15	1.7	2000PIT
(74.2 wt% styrene)		30500	ethyl acetate	298.15	1.2	2000PIT
(75.6 wt% styrene)		32200	ethyl acetate	298.15	7.4	2000PIT
(86.2 wt% styrene)		467000	ethyl acetate	298.15	1.1	2000PIT
(86.2 wt% styrene)		467000	ethyl acetate	298.15	1.1	2000PIT
(10.8 wt% styrene)		62400	tetrahydrofuran	298.15	3.21	2000PIT
(55.2 wt% styrene)		47300	tetrahydrofuran	298.15	5.00	2000PIT
(67.8 wt% styrene)		42600	tetrahydrofuran	298.15	7.05	2000PIT
(74.2 wt% styrene)		30200	tetrahydrofuran	298.15	9.15	2000PIT
(75.6 wt% styrene)		32000	tetrahydrofuran	298.15	8.95	2000PIT
(83.1 wt% styrene)		88800	tetrahydrofuran	298.15	5.07	2000PIT
(86.2 wt% styrene)		464500	tetrahydrofuran	298.15	3.24	2000PIT
(10.8 wt% styrene)	51700		toluene	308.15	4.88	2000PIT
(43.6 wt% styrene)	15300		toluene	308.15	9.1	2000PIT
(55.2 wt% styrene)	43500		toluene	308.15	5.79	2000PIT
(67.8 wt% styrene)	38500		toluene	308.15	6.14	2000PIT
(74.2 wt% styrene)	28600		toluene	308.15	10.0	2000PIT
(75.6 wt% styrene)	30200		toluene	308.15	6.97	2000PIT
(83.1 wt% styrene)	88100		toluene	308.15	4.66	2000PIT
Poly[styrene-*b*-(styrene-*co*-acrylonitrile)] [28 wt%-*b*-72 wt%(71 wt%-*co*-29 wt%)]						
	145000	260000	tetrahydrofuran	298.15	6.7	2009GRO
Poly(styrene-*co*-4-vinylbenzophenone)						
(0.5 mol% VBP)		288500	benzene	293.15	4.85	1974BRA
(1.0 mol% VBP)		307000	benzene	293.15	4.45	1974BRA
(2.0 mol% VBP)		286000	benzene	293.15	4.09	1974BRA
(5.0 mol% VBP)		324000	benzene	293.15	4.52	1974BRA
(10.0 mol% VBP)		351000	benzene	293.15	2.93	1974BRA
(20.0 mol% VBP)		251000	benzene	293.15	7.68	1974BRA

Copolymer	M_n/ g/mol	M_w/ g/mol	**Solvent**	T/ K	A_2 10^4/ cm^3mol/g^2	**Ref.**
(0.5 mol% VBP)	132000		*N,N*-dimethylformamide	293.15	2.69	1974BRA
(1.0 mol% VBP)	134000		*N,N*-dimethylformamide	293.15	1.95	1974BRA
(2.0 mol% VBP)	123000		*N,N*-dimethylformamide	293.15	2.37	1974BRA
(5.0 mol% VBP)	129000		*N,N*-dimethylformamide	293.15	2.26	1974BRA
(10.0 mol% VBP)	182000		*N,N*-dimethylformamide	293.15	1.43	1974BRA
(20.0 mol% VBP)	124000		*N,N*-dimethylformamide	293.15	1.89	1974BRA
Poly(styrene-*b*-1-vinyl-2-pyrrolidinone-*b*-styrene)						
		360000	*N,N*-dimethylformamide	294.15	2.0	1978ESK
		600000	1,4-dioxane	294.15	1.5	1978ESK
		270000	trichloromethane	303.15	0.0	1978ESK
		1200000	trichloromethane	294.15	2.0	1978ESK
Poly(terephthalic acid-*co*-4-aminobenzohydrazide)						
		3780	dimethylsulfoxide	296.15	78.7	1988KRI
		5670	dimethylsulfoxide	296.15	52.9	1988KRI
		7060	dimethylsulfoxide	296.15	63.6	1988KRI
		9160	dimethylsulfoxide	296.15	57.2	1988KRI
		9790	dimethylsulfoxide	296.15	40.6	1988KRI
		16600	dimethylsulfoxide	296.15	44.2	1988KRI
		28500	dimethylsulfoxide	296.15	59.4	1988KRI
		47100	dimethylsulfoxide	296.15	55.9	1988KRI
		83900	dimethylsulfoxide	296.15	49.8	1988KRI
		1550000	dimethylsulfoxide	296.15	17.1	1988KRI
		2600000	dimethylsulfoxide	296.15	32.7	1988KRI
Poly(terephthaloyl chloride-*co*-4-aminobenzohydrazide)						
		9370	*N,N*-dimethylacetamide	298.15	48.4	1984SAK
		21500	*N,N*-dimethylacetamide	298.15	48.0	1984SAK
		37800	*N,N*-dimethylacetamide	298.15	38.6	1984SAK
		1980	dimethylsulfoxide	298.15	78.0	1984SAK
		2740	dimethylsulfoxide	298.15	76.8	1984SAK
		3440	dimethylsulfoxide	298.15	74.0	1984SAK
		4670	dimethylsulfoxide	298.15	64.5	1984SAK
		5680	dimethylsulfoxide	298.15	70.4	1984SAK
		6330	dimethylsulfoxide	298.15	63.4	1984SAK
		12200	dimethylsulfoxide	298.15	64.7	1984SAK
		22200	dimethylsulfoxide	298.15	54.0	1984SAK
		6300	dimethylsulfoxide	298.15	57.8	1984SAK
		9350	dimethylsulfoxide	298.15	43.5	1984SAK
		12300	dimethylsulfoxide	298.15	51.1	1984SAK

Copolymer	M_n/ g/mol	M_w/ g/mol	**Solvent**	*T*/ K	A_2 10^4/ cm^3mol/g^2	**Ref.**
		15300	dimethylsulfoxide	298.15	48.6	1984SAK
		18200	dimethylsulfoxide	298.15	42.5	1984SAK
		22100	dimethylsulfoxide	298.15	49.4	1984SAK
		28800	dimethylsulfoxide	298.15	48.2	1984SAK
		36600	dimethylsulfoxide	298.15	43.3	1984SAK
		48600	dimethylsulfoxide	298.15	48.2	1984SAK
		57800	dimethylsulfoxide	298.15	34.2	1984SAK
Poly(vinyl acetate-*co*-vinyl alcohol)						
(0.32 wt% vinyl acetate)		73000	water	293.15	2.2	1974AND
(3.95 wt% vinyl acetate)		87000	water	293.15	3.5	1974AND
(6.08 wt% vinyl acetate)		77500	water	293.15	2.9	1974AND
(9.82 wt% vinyl acetate)		87000	water	293.15	2.9	1974AND
(15.6 wt% vinyl acetate)		93800	water	293.15	2.7	1974AND
Poly(vinyl acetate-*co*-vinyl chloride)						
(11.4 mol% vinyl acetate)		29500	tetrahydrofuran	298.15	11.9	1985MIN
(11.4 mol% vinyl acetate)		42500	tetrahydrofuran	298.15	11.8	1985MIN
(11.4 mol% vinyl acetate)		96700	tetrahydrofuran	298.15	8.9	1985MIN
(25.0 mol% vinyl acetate)		21000	tetrahydrofuran	298.15	14.8	1985MIN
(25.0 mol% vinyl acetate)		49900	tetrahydrofuran	298.15	8.2	1985MIN
(25.0 mol% vinyl acetate)		185000	tetrahydrofuran	298.15	5.0	1985MIN
(40.0 mol% vinyl acetate)		43000	tetrahydrofuran	298.15	11.8	1985MIN
(40.0 mol% vinyl acetate)		79000	tetrahydrofuran	298.15	9.2	1985MIN
(40.0 mol% vinyl acetate)		179000	tetrahydrofuran	298.15	6.4	1985MIN
(66.0 mol% vinyl acetate)		100000	tetrahydrofuran	298.15	2.8	1985MIN
(66.0 mol% vinyl acetate)		248000	tetrahydrofuran	298.15	2.5	1985MIN
(66.0 mol% vinyl acetate)		708000	tetrahydrofuran	298.15	1.2	1985MIN
Poly(*N*-vinylcaprolactam-*co*-1-vinylimidazole)						
(26.0 mol% 1-vinylimidazole)		254000	water	293.15	−4.0	2006LOZ
(26.0 mol% 1-vinylimidazole)		254000	water	323.15	−12.0	2006LOZ
(27.0 mol% 1-vinylimidazole)		30000	water	293.15	−18.0	2006LOZ
(27.0 mol% 1-vinylimidazole)		30000	water	323.15	9.0	2006LOZ
(38.0 mol% 1-vinylimidazole)		66000	water	293.15	9.0	2006LOZ
(38.0 mol% 1-vinylimidazole)		66000	water	323.15	5.0	2006LOZ
(38.5 mol% 1-vinylimidazole)		28000	water	293.15	19.0	2006LOZ
(38.5 mol% 1-vinylimidazole)		28000	water	323.15	14.0	2006LOZ

Copolymer	M_n/ g/mol	M_w/ g/mol	**Solvent**	T/ K	A_2 10^4/ cm^{3}mol/g^2	**Ref.**
Poly(vinyl chloride-*co*-butyl acrylate)						
(16.6 mol% butyl acrylate)		173000	tetrahydrofuran	298.15	5.18	1985MIN
(16.6 mol% butyl acrylate)		242000	tetrahydrofuran	298.15	5.50	1985MIN
(16.6 mol% butyl acrylate)		302000	tetrahydrofuran	298.15	6.49	1985MIN
(40.0 mol% butyl acrylate)		205000	tetrahydrofuran	298.15	2.44	1985MIN
(40.0 mol% butyl acrylate)		632000	tetrahydrofuran	298.15	1.83	1985MIN
Poly(vinyl chloride-*co*-methyl acrylate)						
(25.5 mol% methyl acrylate)		221000	tetrahydrofuran	298.15	5.70	1985MIN
(25.5 mol% methyl acrylate)		476000	tetrahydrofuran	298.15	5.20	1985MIN
(41.6 mol% methyl acrylate)		121000	tetrahydrofuran	298.15	8.03	1985MIN
(41.6 mol% methyl acrylate)		320000	tetrahydrofuran	298.15	4.72	1985MIN
(41.6 mol% methyl acrylate)		891000	tetrahydrofuran	298.15	3.16	1985MIN
(64.0 mol% methyl acrylate)		379000	tetrahydrofuran	298.15	1.4	1985MIN
(64.0 mol% methyl acrylate)		1060000	tetrahydrofuran	298.15	2.2	1985MIN
(64.0 mol% methyl acrylate)		1340000	tetrahydrofuran	298.15	3.3	1985MIN
(64.0 mol% methyl acrylate)		2510000	tetrahydrofuran	298.15	2.2	1985MIN
Poly(vinyl chloride-*co*-vinylidene chloride)						
(15.0 mol% vinylidene chloride)		39000	tetrahydrofuran	298.15	6.20	1984PAN
(15.0 mol% vinylidene chloride)		50000	tetrahydrofuran	298.15	9.10	1984PAN
(15.0 mol% vinylidene chloride)		99000	tetrahydrofuran	298.15	6.00	1984PAN
(15.0 mol% vinylidene chloride)		121000	tetrahydrofuran	298.15	7.04	1984PAN
(15.0 mol% vinylidene chloride)		223000	tetrahydrofuran	298.15	4.49	1984PAN
(15.0 mol% vinylidene chloride)		50000	tetrahydrofuran	298.15	9.10	1985MIN
(15.0 mol% vinylidene chloride)		121000	tetrahydrofuran	298.15	7.04	1985MIN
(15.0 mol% vinylidene chloride)		223000	tetrahydrofuran	298.15	4.49	1985MIN
(42.0 mol% vinylidene chloride)		79000	tetrahydrofuran	298.15	2.60	1984PAN
(42.0 mol% vinylidene chloride)		93000	tetrahydrofuran	298.15	2.59	1984PAN
(42.0 mol% vinylidene chloride)		107000	tetrahydrofuran	298.15	2.80	1984PAN
(42.0 mol% vinylidene chloride)		133000	tetrahydrofuran	298.15	2.68	1984PAN
(42.0 mol% vinylidene chloride)		156000	tetrahydrofuran	298.15	4.01	1984PAN
(42.0 mol% vinylidene chloride)		187000	tetrahydrofuran	298.15	3.96	1984PAN
(42.0 mol% vinylidene chloride)		79000	tetrahydrofuran	298.15	2.60	1985MIN
(42.0 mol% vinylidene chloride)		133000	tetrahydrofuran	298.15	2.68	1985MIN
(42.0 mol% vinylidene chloride)		187000	tetrahydrofuran	298.15	3.96	1985MIN
Poly[vinylidene fluoride-*co*-perfluoro(methyl vinyl ether)]						
(75 mol% vinylidene fluoride)			ethyl acetate	293.15	2.2	1978ERE

7.2. References

1955STO Stockmayer, W.H., Moore, L.D., Fixman, M., and Epstein, B.N., Copolymers in dilute solution. I. Preliminary results for styrene-methyl methacrylate, *J. Polym. Sci.*, 16, 517, 1955.

1961COO Cooper, W., Vaughan, G., Eaves, D.E., and Madden, R.W., Molecular weight distribution and branching in butadiene polymers and copolymers, *J. Polym. Sci.*, 50, 159, 1961.

1961MOR Morneau, G.A., Roth, P.I., and Shultz, A.R., Trifluoronitrosomethane/ tetrafluoroethylene elastomers dilute solution properties and molecular weight, *J. Polym. Sci.*, 55, 609, 1961.

1964KRA Krause, S., Dilute solution properties of a styrene-methyl methacrylate block copolymer, *J. Phys. Chem.*, 68, 1948, 1964.

1964URW Urwin, J.R. and Stearne, J.M., Solution properties of block copolymers of styrene and methyl methacrylate. Part. II. Viscosity, osmotic pressure, and light scattering studies, *Makromol. Chem.*, 78, 204, 1964.

1965HOM Homma, T. and Fujita, H., Further sedimentation analysis of styrene-butadiene copolymer rubber, *J. Appl. Polym. Sci.*, 9, 1701, 1965.

1965INA Inagaki, H. and Miyamoto, T., Preparation of block copolymers. of A-B-A type and its behaviour in dilute solution, *Makromol. Chem.*, 87, 166, 1965.

1965KOT Kotera, A., Saito, T., Watanabe, Y., and Ohama, M., Studies on the solution properties of copolymers of methyl acrylate and methyl methacrylate, *Makromol. Chem.*, 87, 195, 1965.

1965WAG Wagner, H.L. and Wissbrun, K.F, Molecular weight and rheology of acetal copolymers, *Makromol. Chem.*, 81, 14, 1965.

1966KOT Kotaka, T., Ohnuma, H., and Murakami, S., The theta-condition for random and block copolymers of styrene and methyl methacrylate, Y, *J. Phys. Chem.*, 70, 4099, 1966.

1966SHI Shimura, Y., Effects of composition on the solution properties of styrene-acrylonitrile copolymers, *J. Polym. Sci.: Part A-2*, 4, 423, 1966.

1967MOH Mohite, R.B., Gundiah, S., and Kapur, S.L., Solution properties of copolymer of p-chlorostyrene and methyl methacrylate, *Makromol. Chem.*, 108, 52, 1967.

1967SHI Shimura, Y., Solution properties of methyl methacrylate-acrylonitrile copolymer, *Bull. Chem. Soc. Japan*, 40, 273, 1967.

1968KAR Karunakaran, K. and Santappa, M., Solution properties of poly(methyl acrylate) and (1:1) poly(styrene-*co*-methyl acrylate), *J. Polym. Sci.: Part A-2*, 6, 713, 196.

1968KOT Kotaka, T., Murakami, Y., and Inagaki, H., Dilute solution properties of styrene-methyl methacrylate random copolymers, *J. Phys. Chem.*, 72, 829, 1968.

1968UTR Utracki, L.W., Simha, R., and Fetters, L.J., Solution properties of polystyrene-polybutadiene block copolymers, *J. Polym. Sci.: Part A-2*, 6, 2051, 1968.

1969CR1 Cramond, D.N. and Urwin, J.R., Solution properties of block copolymers of poly(isoprene-styrene): I. Molecular dimensions by viscosity and light scattering methods, *Eur. Polym. J.*, 5, 35, 1969.

1969CR2 Cramond, D.N. and Urwin, J.R., Solution properties of block copolymers of poly(isoprene-styrene): II. Thermodynamic parameters from osmotic data, *Eur. Polym. J.*, 5, 45, 1969.

1969FIS Fischer, H. and Mächtle, W., Untersuchungen an hochverdünnten Lösungen von statistischen Styrol-Acrylsäuremethylester-Copolymeren, *Koll. Z.-Z. Polym.*, 230, 221, 1969.

1969KOT Kotaka, T., Ohnuma, H., and Inagaki, H., Thermodynamic and conformational properties of styrene-methyl methacrylate block copolymers in dilute solution. II-Behavior in theta solvents, *Polymer*, 10, 517, 1969.

1969MAT Matsuda, H., Yamano, K., and Inagaki, H., Styrene-methyl acrylate copolymers and acrylate homopolymers in solution, *J. Polym. Sci.: Part A-2*, 7, 609, 1969.

1970MAE Mächtle, W., Untersuchungen an hochverdunnten Lösungen von einigen statistischen Copolymeren, *Angew. Makromol. Chem.*, 10, 1, 1970.

1970WU1 Wunderlich, W., Lösungseigenschaften von statistischen Copolymeren aus Methylmethacrylat und n-Butylacrylat, *Angew. Makromol. Chem.*, 11, 73, 1970.

1970WU2 Wunderlich, W., Flexibilität und thermodynamische Eigenschaften von Polyacrylaten, Polymethacrylaten und statistischen Acrylat/Methacrylat-Copolymeren, *Angew. Makromol. Chem.*, 11, 189, 1970.

1971BRA Braun, D., Chaudhari, D., and Mächtle, W., Lösungseigenschaften und spezifische Berechnungsinkremente von p-Jodstyrol/Styrol-Copolymeren, *Angew. Makromol. Chem.*, 15, 83, 1971.

1971DON Dondos, A., Investigations on conformations of styrene-α-methylstyrene block copolymers in dilute solution, *J Polym. Sci.: Polym. Lett.*, 9, 871, 1971.

1972GIR Girolamo, M. and Urwin, J.R., Thermodynamic parameters from osmotic studies on solutions of block copolymers of polyisoprene and polystyrene, *Eur. Polym. J.*, 8, 299, 1972.

1972MOR Morimoto, M. and Okamoto, Y., Solution properties and molecular structures of EPM and EPDM, *J. Appl. Polym. Sci.*, 16, 2795, 1972.

1972PIZ Pizzoli, M. and Ceccorulli, G., Solution properties of styrene-p-methoxystyrene random copolymers II, *Eur. Polym. J.*, 8, 769, 1972.

1973BAR Baranovskaya, I.A. and Eskin, V.E., Conformation of molecules of a copolymer of 2-ethylhexyl methacrylate and acrylic acid in different solvents (Russ.), *Vysokomol. Soedin., Ser. B*, 15, 794, 1973.

1973KAM Kambe, H., Kambe, Y., and Honda, C., Light scattering from terpolymer solutions, *Polymer*, 14, 460, 1973.

1973RAF Rafikov, S.R., Monakov, Yu.B., Duvakina, N.V., Marina, N.G., Budtov, V.P., Minchenkova, N.Kh., and Sharikova, A.M., Viscosity, diffusion, and light scattering of copolymers of styrene and methacrylic acid (Russ.), *Vysokomol. Soedin., Ser. B*, 15, 807, 1973.

1973SRI Srinivasan, K.S.V. and Santappa, M., Dilute solution properties and molecular characterization of poly(butyl methacrylate-*co*-styrene), *J. Polym. Sci.: Polym. Phys. Ed.*, 11, 331, 1973.

1974AND Andreeva, V.M., Anikeeva, A.A., Tager, A.A., and Kosareva, L.P., Phase equilibrium of aqueous solutions of poly(vinyl alcohol) and products of its acetylation (Russ.), *Vysokomol. Soedin., Ser. B*, 16, 277, 1974.

1974BRA Braun, D. and Traser, G., Über eine polymeranaloge Umsetzung zur Darstellung von 1-Phenyl-1-(4-vinylphenyl)äthylen/Styrol-Copolymeren, *Makromol. Chem.*, 175, 2275, 1974.

1974MOL Molodtsova, E.D., Pavlova, S.A., Timofeeva, G.I., Vygodskii, Ya.S., Vinogradova, S.V., and Korshak, V.V., Molecular weight characteristics of cardo poly(amido esters) produced by the low temperature polycondensation (Russ.), *Vysokomol. Soedin., Ser. A*, 16, 2183, 1974.

1974UTI Utiyama, H., Takenaka, H., Mizumori, M., and Fukuda, M., Light-scattering studies of a polystyrene-poly(methyl methacrylate) two-block copolymer in dilute solutions, *Macromolecules*, 7, 28, 1974.

1976IOA Ioan, S. and Benedek, J., Study of *N*-vinylcarbazole-ethyl methacrylate statistical coplymer by light-scattering, *Rev. Roum. Chim.*, 21, 145, 1976.

1976RED Reddy, C.R. and Kalpagam, V., Dilute solution properties of styrene-acrylonitrile and styrene-methyl methacrylate copolymers. I. Intrinsic viscosity-molecular weight relations, *J. Polym. Sci.: Polym. Phys. Ed.*, 14, 749, 1976.

1976ROO Roovers, J.E.L. and Bywater, S., Modification of the solution properties of polystyrene in cyclohexane by the incorporation of small amounts of foreign groups, *Macromolecules*, 9, 873, 1976.

1977AMB Ambler, M.R. and McIntyre, D., Randomly branched styrene/divinylbenzene copolymers. 11. Solution properties and structure, *J. Appl. Polym. Sci.*, 21, 2269, 1977.

1977ARU Arulsamy, S.M. and Santappa, M., Solution properties of poly(butyl methacrylate-*co*-acrylonitrile), *Makromol. Chem.*, 178, 2451, 1977.

1977BIR Birshtein, T.M., Zubkov, V.A., Milevskaya, I.S., Eskin, V.E., and Baranovskaya, I.A., Flexibility of aromatic polyimides and polyamidoacids, *Eur. Polym. J.*, 13, 375, 1977.

1977GL1 Glöckner, G. and Mauksch, D., Streulichtmessungen an azeotropen Styrol-Acrylnitril-Copolymeren in Butanon, *Z. Phys. Chem., Leipzig*, 258, 1142, 1977.

1977GL2 Glöckner, G., Über das Verhalten von Styrol-Acrylnitril-Copolymeren in Lösung III, *Faserforsch. Textiltechn.*, 28, 111, 1977.

1977KAS Kashyap, A.K., Kalpagam, V., and Reddy, C.R., Dilute solution properties of methyl methacrylate-acrylonitrile copolymer (MA1), *Polymer*, 18, 878, 1977.

1977KRA Kratochvil, P., Strakova, D., and Tuzar, Z., Interaction between polymers in dilute solutions of two polymers in a single solvent measured by light scattering and the compatibility of polymers, *Br. Polym. J.*, 9, 217, 1977.

1977MOT Motowka, M., Norisuya, T., and Fujita, H., Stiff chain behavior of poly(phthaloyl-*trans*-2,5-dimethylpiperazine) in dilute solution, *Polym. J.*, 9, 613, 1977.

1977RED Reddy, G.V., Srinivasan, K.S.V., and Santappa, M., Dilute solution properties of poly(methacrylonitrile-*co*-styrene) and poly(acrylonitrile-*co*-styrene), *J. Macromol. Sci.-Chem.*, A11, 2123, 1977.

1977TA1 Tager, A.A., Kolmakova, L.K., Anufriev, V.A., Bessonov, Yu.S., Zhigunova, O.A., Vinogradova, S.V., Salazkin, S.N., and Tsilipotkina, M.V., Thermodynamics of the dissolution of cardo polyarylates in chloroform and tetrachloroethane (Russ.), *Vysokomol. Soedin., Ser. A*, 19, 2367, 1977.

1977TA2 Tager, A.A. and Ikanina, T.V., Thermodynamic study of dilute and concentrated solutions of poly(p-chlorostyrene) (Russ.), *Vysokomol. Soedin., Ser. B*, 19, 192, 1977.

1977TA3 Tager, A.A., Kolmakova, L.K., Vshivkov, S.A., Kremlyakova, E.V., Vygodskii, Ya.S., Salazkin, S.N., and Vinogradova, S.V., Second virial coefficients of solutions of some cardo polymers (Russ.), *Vysokomol. Soedin., Ser. B*, 19, 738, 1977.

1978ERE Erenburg, E.G., Pavlova, L.V., Osipchuk, E.O., Dolgopolskii, I.M., Konshin, A.I., Rabinovich, R.L., and Poddubnyi, I.Ya., Flexibility of molecular chains and molecular-weight distribution of vinylidene fluoride-perfluoro(methyl vinyl ether) copolymers, *Vysokomol. Soedin., Ser. A*, 20, 382, 1978.

1978ESK Eskin, V.E., Grigorev, A.I., Baranovskaya, I.A., and Rudkovskaya, G.D., Conformational properties of styrene-vinylpyrrolidone block copolymers in solution, *Vysokomol. Soedin., Ser. A*, 20, 55, 1978.

1978MOT Motowoka, M., Fujita, H., and Norisuye, T., Stiff chain behavior of poly(terephthaloyl-*trans*-2,5-dimethylpiperazine) in dilute solution, *Polym. J.*, 10, 331, 1978.

1978RED Reddy, G.V. and Santappa, M., Dilute solution properties of poly(methacrylonitrile) and copolymers of methacrylonitrile with methacrylates, *Indian J. Chem.*, 16A, 99, 1978.

1979GL1 Glöckner, G., Francuskiewicz, F., and Reichardt, H.-U., On the behaviour of styrene-acrylonitrile copolymers in solution IV. Light scattering investigation of azeotropic copolymers in tetrahydrofuran, *Acta Polym.*, 30, 551, 1979.

1979GL2 Glöckner, G., Francuskiewicz, F., and Reichardt, H.-U., On the behaviour of styrene-acrylonitrile copolymers in solution V. Light scattering investigation of α-methylstyrene copolymers with about 46 mol% acrylonitrile in tetrahydrofuran as a solvent, *Acta Polym.*, 30, 628, 1979.

1979GL3 Glöckner, G., On the behavior of styrene-acrylonitrile copolymers in solution. VI. Presentation of light-scattering results according to excluded-volume theory, *Eur. Polym. J.*, 15, 727, 1979.

1979GRU Gruber, E. and Knell, W.L., Zur Solvatation von Copolymeren: Untersuchung an Poly(styrol-*co*-butylacrylat)en und Poly(styrol-*co*-butylmethacrylat)en, *Progr. Colloid Polym. Sci.*, 66, 393, 1979.

1979STA Stacy, C.J. and Kraus, G., Second virial coefficients of homopolymers and copolymers of butadiene and styrene in tetrahydrofuran, *J. Polym. Sci., Polym. Phys. Ed.*, 17, 2007, 1979.

1980SHI Shimizu, T., Minakata, A., and Tomiyama, T., The solution properties of maleic anhydride and maleic acid copolymers: 1. Light scattering, osmotic pressure and viscosity measurements, *Polymer*, 21, 1427, 1980.

1981PAD Padaki, S.M. and Stille, J.K., Dilute-solution parameters for a fluorene-containing "cardo" polyquinoline, *Macromolecules*, 14, 888, 1981.

1982BRA Braun, D. and Menzel, W., Lösungseigenschaften von äquimolaren statistischen und alternierenden Copolymeren aus Styrol und Benzylmethacrylat, *Colloid Polym. Sci.*, 260, 1011, 1982.

1982MA1 Mangalam, P.V. and Kalpagam, V., Styrene-acrylonitrile random copolymer in ethyl acetate, *J. Polym. Sci., Polym. Phys. Ed.*, 20, 773, 1982.

1982MA2 Mangalam, P.V. and Kalpagam, V., Behaviour of methylmethacrylate-acrylonitrile random copolymers in dilute solution, *Polymer*, 23, 991, 1982.

1983ION Ionescu, L.M., Petrea, I., Bujor, I.I., and Demetrescu, I., Light scattering from aqueous solutions of poly(acrylamide-*co*-maleic anhydride), *Makromol. Chem.*, 184, 1005, 1983.

1983SIM Simionescu, C.I. and Simionescu, B.C., Solution properties of ultrahigh molecular weight polymers, 1. Methyl methacrylate-styrene 50:50 random copolymers, *Makromol. Chem.*, 184, 829, 1983.

1983SUN Sun, S.F., Light scattering studies of deuterated polystyrene-poly(methyl methacrylate) diblock copolymers in 1,4-dioxane solutions, *Makromol. Chem., Rapid Commun.*, 4, 203, 1983.

1983THI Thien, N.-T, Suter, U.W., and Pino, P., Optically active vinyl polymers, 24. Copolymers of 4-methyl-1-penten-3-one and (S)-4-methyl-l-hexen-3-one, *Makromol. Chem.*, 184, 2335, 1983.

1984GLA Gladkova, E.A., Pavlova, S.-S.A., Dubrovina, L.V., and Korshak, V.V., Thermodynamics of dilute solutions of polyarylates (Russ.), *Vysokomol. Soed., Ser. A*, 26, 53, 1984.

1984MAY Mays, J., Hadjichristidis, N., and Fetters, L.J., Characteristic ratios of model polydienes and polyolefins, *Macromolecules*, 17, 2723, 1984.

1984PAN Pancheshnikova, R.B., Brodko, L.S., Monakov, Yu.B., Petrenko, P.I., Yanovskii, D.M., and Minsker, K.S., Hydrodynamic and conformational properties of vinyl chloride-vinylidene chloride copolymers in solutions (Russ.), *Vysokomol. Soed., Ser. B*, 26, 276, 1984.

1984SAK Sakurai, K., Ochi, K., Norisuye, T., and Fujita, H., Stiff-chain behavior of poly(terephthalamide-p-benzohydrazide) in dimethyl sulfoxide, *Polym. J.*, 16, 559, 1984.

1985BAR Baranovskaya, I.A., Kudryavtsev, V.V., Dyakonova, N.V., Sklizkova, V.P., Eskin, V.E., and Koton, M.M., On equilibrium flexibility of polyamide acids (Russ.), *Vysokomol. Soed., Ser. A*, 27, 604, 1985.

1985HAE Haesslin, H.-W., Dimethylsiloxane-ethylene oxide block copolymers. 2. Preliminary results on dilute solution properties, *Makromol. Chem.*, 186, 357, 1985.

1985MAT Matsuo, K., Stockmayer, W.H., and Bangerter, F., Conformational properties of poly(1-octadecene/maleic anhydride) in solution, *Macromolecules*, 18, 1346, 1985.

1985MIN Minsker, K.S., Panchesnikova, R.B., Monakov, Yu.B., and Zaikov, G.E., Thermodynamic and hydrodynamic properties of chlorine-containing carbochain polymers in solutions, *Eur. Polym. J.*, 21, 981, 1985.

1985NIK1 Nikitina, N.P., Nekrasov, I.K., and Glazkovskii, Yu.B., Hydrodynamic properties of dilute solutions and molecular characteristics of copolymers of acrylonitrile and vinyl acetate, *Vysokomol. Soed., Ser. A*, 27, 1364, 1985.

1986KAM Kamide, K., Miyazaki, Y., and Yamazaki, H., Dilute solution properties of acrylonitrile/ vinylidene chloride copolymer, *Polym. J.*, 18, 645, 1986.

1986OGA Ogawa, E., Yamaguchi, N., and Shima, M., Estimation of the interaction parameter between polystyrene and poly(p-chlorostyrene) from osmotic pressure measurements, *Polym. J.*, 18, 903, 1986.

1987KYO Kyohmen, M., Inoue, K., Baba, Y., Kagemoto, A., and Beatty, C.L., Heats of dilution of poly[styrene-*ran*-(butyl methacrylate)] solutions measured with an automatic flow microcalorimeter, *Makromol. Chem.*, 188, 2721, 1987.

1987WUC Wu, C., Buck, W., and Chu, B., Light scattering characterization of an alternating copolymer of ethylene and tetrafluoroethylene. 2. Molecular weight distributions, *Macromolecules*, 20, 98, 1987.

1988KRI Krigbaum, W.R. and Brelsford, G., Static and dynamic light scattering studies of poly(terephthalic acid-*co*-4-aminobenzohydrazide) in dimethyl sulfoxide, *Macromolecules*, 21, 2502, 1988.

1989BEL Beldie, C., Nemtoi, G., Dumitriu, E., and Barboiu, V., The behaviour of diethyl maleate-vinyl acetate copolymers in solution, *Rev. Roum. Chim.*, 34, 51, 1989.

1989EGO Egorochkin, G.A., Semchikov, Yu.D., Smirnova, L.A., Knyazeva, T.E., Tokhonova, Z.A., Karyakin, N.V., and Sveshnikova, T.G., Thermodynamic analysis of the copolymerization of styrene with acrylonitrile and *N*-vinylpyrrolidone with vinyl acetate (Russ.), *Vysokomol. Soedin., Ser. B*, 31, 46, 1989.

1989CHU Chu, B., Wu, C., and Buek, W., Light scattering characterization of an alternating copolymer of ethylene and tetrafluoroethylene. 3. Temperature dependence of polymer size, *Macromolecules*, 22, 371, 1989.

1989FAN Fang, T., Buo, S., Xie, H., Zhang, W., and Yu, L., Water-soluble and amphiphilic polymers. 4. Light scattering studies of some dilute solutions of PEO-PSt-PEO triblock polymers, *Polym. Bull.*, 22, 311, 1989.

1990CHU Chu, B., Wang, J., and Shuely, W.J., Solution behavior of a random copolymer of poly(isobutyl methacrylate-*co*-(*tert*-butylamino)ethyl methacrylate). 1. Laser light scattering studies, *Macromolecules*, 23, 2252, 1990.

1990KUE Kücükyavuz, S., Kücükyavuz, Z., and Erdogan, G., Characterization of p-*tert*-butylstyrene-dimethylsiloxane triblock copolymers in solution, *Polymer*, 31, 379, 1990.

1991DON Donenov, B.K. and Zhubanov, B.A., Conformational properties and equilibrium rigidity of alicyclic polyimide macromolecules (Russ.), *Vysokomol. Soedin., Ser. A*, 33, 2056, 1991.

1991LOU Louai, A., Sarazin, D., Pollet, G., Francois, J., and Moreaux, F., Properties of ethylene oxide-propylene oxide statistical copolymers in aqueous solution, *Polymer*, 32, 703, 1991.

1991SWA Swanson, S.A., Cotts, P.M., Siemens, R., and Kim, S.H., Synthesis and solution properties of poly[*N,N'*-(p-phenylene)-3,3',4,4'-biphenyltetracarboxylic acid diamide], *Macromolecules*, 24, 1352, 1991.

1991ULY Ulyanova, N.N., Baranovskaya, I.A., Liubina, S.Ya., Bezrukova, M.A., Rudkovskaya, G.D., Shabsels, B.M., Vlasov, G.P., and Eskin, V.E., Investigation of macromolecules exhibiting the structure of a once-broken rod by molecular optics. 2. Synthesis and investigation of three-block copolymers: Poly(γ-benzyl-L-glutamate)-poly(methyl methacrylate)-poly(γ-benzyl-L-glutamate), *Macromolecules*, 24, 3324, 1991.

1992KI1 Kim, M.G., Nieh, W.L., Sellers, T., Wilson, W.W., and Mays, J.W., Polymer solution properties of a phenol-formaldehyde resol resin by gel permeation chromatography, intrinsic viscosity, static light scattering, and vapor pressure osmometric methods, *Ind. Eng. Chem. Res.*, 31, 973, 1992.

1992KI2 Kim, S., Cotts, P.M., and Volksen, W., On-line measurement of the rms radius of gyration and molecular weight of polyimide precursor fractions eluting from a size-exclusion chromatograph, *J. Polym. Sci.: Part B: Polym. Phys.*, 30, 177, 1992.

1993KAN Kansara, S., Patel, N.K., and Patel, C.K., Synthesis and characterization of poly(2-methoxycyanurate) of bisphenol-F, *Polymer*, 34, 1303, 1993.

1993QU1 Quintana, J.R., Villacampa, M., and Katime, I., Influence of the block length on the structure of micelles formed by polystyrene-*block*-poly(ethylene-*co*-propene) in octane and 5-methyl-2-hexanone, *Macromol. Chem. Phys.*, 194, 983, 1993.

1993QU2 Quintana, J.R., Villacampa, M., and Katime, I., Micellization of a polystyrene-*block*-poly(ethylene/propylene) block copolymer in ketones, *Polymer*, 34, 2380, 1993.

1993ROY Roy, S., Bhattacharjee, S., Bhowmick, A.K., Gupta, B.R., and Kulkarni, R.A., Solution properties of epoxidized natural rubber and hydrogenated nitrile rubber, *J. Macromol. Sci., Macromol. Rep.*, A30, 310, 1993.

1993UCH Uchida, H., Aoki, Y., and Kato, R., Characterization of poly(styrene-*co*-acrylonitrile)s by light scattering, *Kobunshi Ronbunshu*, 50, 941, 1993.

1993WUG Wu, G., Zhou, Z., and Chu, B., Water-induced micelle formation of block copoly(oxyethylene-oxypropylene-oxyethylene) in o-xylene, *Macromolecules*, 26, 2117, 1993.

1993ZHO Zhou, Z., Chu, B., and Peiffer, D.G., Temperature-induced micelle formation of a diblock copolymer of styrene and *tert*-butylstyrene in *N,N*-dimethylacetamide, *Macromolecules*, 26, 1876, 1993.

1994MUM Mumick, P.S., and McCormick, C.L., Water soluble copolymers. 54: *N*-isopropyl-acrylamide-*co*-acrylamide copolymers in drag reduction: Synthesis, characterization, and dilute solution behavior, *Polym. Eng. Sci.*, 34, 1419, 1994.

1994OHS Oh, S.Y. and Siegel, R.A., Study of the dilute spöution properties of poly(acrylamide-*co*-methyl methacrylate), *J. Macromol. Sci.-Pure Appl. Chem.*, A31, 231, 1994.

1994PIS Pispas, S. and Hadjichristidis, N., End-functionalized block copolymers of styrene and isoprene: Synthesis and association behavior in dilute solution, *Macromolecules*, 27, 1891, 1994.

1994ZHO Zhou, Z. and Chu, B., Phase behavior and association properties of poly(oxypropylene)-poly(oxyethy1ene)-poly(oxypropylene) triblock copolymer in aqueous solution, *Macromolecules*, 27, 2025, 1994.

1995QU1 Quintana, J.R., Janez, M.D, Villacampa, M., and Katime, I., Diblock copolymer micelles in solvent binary mixtures. 1. Selective solvent/precipitant, *Macromolecules*, 28, 4139, 1995.

1995QU2 Quintana, J.R., Salazar, R.A, and Katime, I., Micelle formation and polyisobutylene solubilization by polystyrene-*block*-poly(ethylene-*co*-butylene)-*block*-polystyrene block copolymers, *Macromol. Chem. Phys.*, 196, 1625, 1995.

1995VIL Villacampa, M., Apodaca, E.D. de, Quintana, J.R., and Katime, I., Diblock copolymer micelles in solvent binary mixtures. 2. Selective solvent/good solvent, *Macromolecules*, 28, 4144, 1995.

1996KIK Kikuchi, A. and Nose, T., Unimolecular-micelle formation of poly(methyl methacrylate)-*graft*-polystyrene in isoamyl acetate, *Polymer*, 37, 5889, 1996.

1996SID Siddiq, M., Hu, H., Ding, M., Li, B., and Wu, C., Laser light scattering studies of soluble high-performance polyimides: Solution properties and molar mass distributions, *Macromolecules*, 29, 7426, 1996.

1997KWA Kwan, S.C.M., Wu, C., Li, F., Savitski, E.P., Harris, F.W., and Cheng, S.Z.D., Laser light scattering studies of soluble high performance fluorine-containing polyimides, *Macromol. Chem. Phys.*, 198, 3605, 1997.

1997LIU Liu, T., Zhou, Z., Wu, C., Chu, B., Schneider, D.K., and Nace, V.M., Self-assembly of poly(oxybutylene)-*b*-poly(oxyethylene)-*b*-poly(oxybutylene) ($B_6E_{46}B_6$) triblock copolymer in aqueous solution, *J. Phys. Chem. B*, 101, 8808, 1997.

1998CRO Crowther, N.J. and Eagland, D., Studies of aggregation in dilute solutions of the triblock copolymer POE-POP-POE, Synperonic F87-I, *Eur. Polym. J.*, 34, 605, 1998.

1998ISH Ishizu, K., Tsubaki, K.-I., and Ono, T., Synthesis and dilute solution properties of poly(diblock macromonomer)s, *Polymer*, 39, 2935, 1998.

1998LIU Liu, T., Zhou, Z., Wu, C., Nace, V.M., Chu, B., Dominant factors on the micellization of $B_nE_mB_n$-type triblock copolymers in aqueous solution, *J. Phys. Chem. B*, 102, 2875, 1998.

1998PEN Penco, M., Donetti, R., Mendichi, R., and Ferrutti, P., New poly(ester-carbonate) multi-block copolymers based on poly(lactic-glycolic acid) and poly(ε-caprolactone) segments, *Macromol. Chem. Phys.*, 199, 1737, 1998.

1998SCH Schneider, A., Homopolymer- und Copolymerlösungen im Vergleich: Wechselwirkungsparameter und Grenzflächenspannung, *Dissertation*, Johannes-Gutenberg Universität Mainz, 1998.

1998ZHA Zhang, Y., Li, M., Fang, Q., Zhang, Y.-X., Jiang, M., and Wu, C., Effect of incorporating a trace amount of fluorocarbon into poly(*N*-isopropylacrylamide) on its association in water, *Macromolecules*, 31, 2527, 1998.

2000HUT Hu, T., Wang, C., Li, F., Harris, F.W., Cheng, S.Z.D., and Wu, C., Scalings of fluorine-containing polyimides in cyclopentanone, *J. Polym. Sci.: Part B: Polym. Phys.*, 38, 2077, 2000.

2000PIS Pispas, S., Hadjichristidis, N., Potemkin, I., and Khokhlov, A., Effect of architecture on the micellization properties of block copolymers: A2B miktoarm stars vs. AB diblocks, *Macromolecules*, 33, 1741, 2000.

2000PIT Pitsikalis, M., Siakali-Kioulafa, E., and Hadjichristidis, N., Block copolymers of styrene and stearyl methacrylate. Synthesis and micellization properties in selective solvents, *Macromolecules*, 33, 5460, 2000.

2000WES Westermann, S., Willner, L., Richter, D., and Fetters, L.J., The evaluation of polyethylene chain dimensions as a function of concentration in nonadecane, *Macromol. Chem. Phys.*, 201, 500, 2000.

2001CAO Cao, X., Sessa, D.J., Wolf, W.J., and Willett, J.L., Light-scattering study of poly(hydroxy ester ether) in *N,N*-dimethylacetamide solution, *J. Appl. Polym. Sci.*, 80, 1737, 2001.

2001EDE Edelmann, K., Janich, M., Pyckhout-Hintzen, W., and Höring, S., The aggregation behavior of poly(ethylene oxide)-poly(methyl methacrylate) diblock copolymers in organic solvents, *Macromol. Chem. Phys.*, 202, 1638, 2001.

2001LOO Loos, K., Jonas, G., and Stadler, R., Carbohydrate modified polysiloxanes. 3. Solution properties of carbohydrate-polysiloxane conjugates in toluene, *Macromol. Chem. Phys.*, 202, 3210, 2001.

2001SID Siddiq, M. and Wu, C., Dynamic light-scattering characterization of the molecular weight distribution of unfractionated polyimide, *J. Appl. Polym. Sci.*, 81, 1670, 2001.

2001VOU Voulgaris, D. and Tsitsilianis, C., Aggregation behavior of polystyrene/poly(acrylic acid) heteroarm star copolymers in 1,4-dioxane and aqueous media, *Macromol. Chem. Phys.*, 202, 3284, 2001.

2002SCH Schwahn, D., Richter, D., Wight, P.J., Symon, C., Fetters, L.J., and Lin, M., Self-assembling behavior in decane solution of potential wax crystal nucleators based on poly(co-olefins), *Macromolecules*, 35, 861, 2002.

2004PIT Pitsikalis, M., Siakali-Kioulafa, E., and Hadjichristidis, N., Block copolymers of styrene and n-alkyl methacrylates with long alkyl groups. Micellization behavior in selective solvents, *J. Polym. Sci.: Part A: Polym. Chem.*, 42, 4177, 2004.

2005NED Nedelcheva, A.N., Novakov, C.P., Miloshev, S.M., and Berlinova, I.V., Electrostatic self-assembly of thermally responsive zwitterionic poly(*N*-isopropylacrylamide) and poly(ethylene oxide) modified with ionic groups, *Polymer*, 46, 2059, 2005.

2006CON Convertine, A.J., Lokitz, B.S., Vasileva, Y., Myrick, L.J., Scales, C.W., Lowe, A.B., and McCormick, C.L., Direct synthesis of thermally responsive DMA/NIPAM diblock and DMA/NIPAM/DMA triblock copolymers via aqueous, room temperature RAFT polymerization, *Macromolecules*, 39, 1724, 2006.

2006LAM Lambeth, R.H., Ramakrishnan, S., Mueller, R., Poziemski, J.P., Miguel, G.S., Markoski, L.J., Zukoski, C.F., and Moore, J.S., Synthesis and aggregation behavior of thermally responsive star polymers, *Langmuir*, 22, 6352, 2006.

2006LIX Li, X., Mya, K.Y., Ni, X., He, C., Leong, K.W., and Li, J., Dynamic and static light scattering studies on self-aggregation behavior of biodegradable amphiphilic poly(ethylene oxide)-poly[(R)-3-hydroxybutyrate]-poly(ethylene oxide) triblock copolymers in aqueous solution, *J. Phys. Chem. B*, 110, 5920, 2006.

2006LOZ Lozinskii, V.I., Simenel, I.A., Semenova, M.G., Belyakova, L.E., Ilin, M.M., Grinberg, V. Ya., Dubovik, A.S., and Khokhlov, A.R., Behavior of protein-like *N*-vinylcaprolactam and *N*-vinylimidazole copolymers in aqueous solutions, *Polym. Sci., Ser. A*, 48, 435, 2006.

2007RAN Rangelov, S., Almgren, M., Halacheva, S., and Tsvetanov, C., Polyglycidol-based analogues of pluronic block copolymers. Light scattering and cryogenic transmission electron microscopy studies, *J. Phys. Chem. C*, 111, 13185, 2007.

2008BER Bercea, M., Eckelt, J., and Wolf, B.A., Random copolymers: Their solution thermodynamics as compared with that of the corresponding homopolymers (experimental data by M. Bercea), *Ind. Eng. Chem. Res.*, 47, 2434, 2008.

2008DEN Deng, L., Shi, K., Zhang, Y., Wang, H., Zeng, J., Guo, X., Du, Z., and Zhang, B., Synthesis of well-defined poly(*N*-isopropylacrylamide)-*b*-poly(L-glutamic acid) by a versatile approach and micellization, *J. Colloid Interface Sci.*, 323, 169, 2008.

2008FOS Foster, L.J.R., Schwahn, D., Pipich, V., Holden, P.J., and Richter, D., Small-angle neutron scattering characterization of polyhydroxyalkanoates and their bioPEGylated hybrids in solution, *Biomacromolecules*, 9, 314, 2008.

2008NUO Nuopponen, M., Kalliomaeki, K., Aseyev, V., and Tenhu, H., Spontaneous and thermally induced self-organization of A-B-A stereoblock polymers of *N*-isopropylacrylamide in aqueous solutions, *Macromolecules*, 41, 4881, 2008.

2008VID Vidal, R.R.L., Balaban, R., and Borsali, R., Amphiphilic derivatives of carboxymethylcellulose: Evidence for intra- and intermolecular hydrophobic associations in aqueous solutions, *Polym. Eng. Sci.*, 48, 2011, 2008.

2008ZUL Zulfiqar, S., Ishaq, M., Ahmad, Z., and Sarwar, M.I., Synthesis, static, and dynamic light scattering studies of soluble aromatic polyamide, *Polym. Adv. Technol.*, 19, 1250, 2008.

2009GRO Gromadzki, D., Lokaj, J., Slouf, M., and Stepanek, P., Dilute solutions and phase behavior of polydisperse A-*b*-(A-*co*-B) diblock copolymers, *Polymer*, 50, 2451, 2009.

2009XIO Xiong, X., Eckelt, J., Zhang, L., and Wolf, B.A., Thermodynamics of block copolymer solutions as compared with the corresponding homopolymer solutions: Experiment and theory (exp. data by J. Eckelt and B.A. Wolf), *Macromolecules*, 42, 8398, 2009.

APPENDICES

Appendix 1 List of copolymers in alphabetical order

Polymer	**Page(s)**
Poly(butylene succinate-*co*-butylene adipate)	22-23, 379
Poly(butyl methacrylate-*co*-*N,N*-dimethylaminoethyl methacrylate)	23-25
Poly(butyl methacrylate-*b*-perfluoroalkylethyl acrylate)	25-26
Poly(4-*tert*-butylstyrene-*b*-dimethylsiloxane)	426
Poly(4-*tert*-butylstyrene-*b*-dimethylsiloxane-*b*-4-*tert*-butylstyrene)	426
Poly(*tert*-butylstyrene-*b*-styrene)	426
Poly(ε-caprolactone-*b*-ethylene glycol-*b*-ε-caprolactone)	152
Poly(ε-caprolactone-*b*-*N*-isopropylacrylamide-*b*-ε-caprolactone)	152
Poly(4-chlorostyrene-*co*-methyl methacrylate)	426
Poly(4-chlorostyrene-*b*-styrene-*b*-4-chlorostyrene)	427
Poly(cyclohexene carbonate-*b*-ethylene oxide-*b*-cyclohexene carbonate)	310
Poly(cyclohexene oxide-*co*-carbon dioxide)	259, 315-316
Poly[*N*-cyclopropylacrylamide-*co*-4-(2-phenyl-diazenyl)benzamido-*N*-(2-aminoethyl)acrylamide]	152
Poly(decyl methacrylate-*b*-styrene)	427
Poly(diacetone acrylamide-*co*-acrylamide)	152, 219
Poly(diacetone acrylamide-*co*-hydroxyethyl acrylate)	152, 230
Poly(diallyaminoethanoate-*co*-dimethylsulfoxide)	219
Poly(*N,N*-diallylammonioethanoic acid-*co*-sulfur dioxide)	174
Poly(4,4'-diaminophenyl-*co*-pyromellitic anhydride]	427
Poly(3,4-dichlorobenzyl methacrylate-*co*-ethyl methacrylate)	73-78, 363-366
Poly(*N,N*-diethylacrylamide-*co*-acrylamide)	153
Poly(*N,N*-diethylacrylamide-*co*-acrylic acid)	111, 153, 219, 230
Poly(*N,N*-diethylacrylamide-*co*-*N,N*-dimethylacrylamide)	153
Poly(*N,N*-diethylacrylamide-*co*-*N*-ethylacrylamide)	153, 230
Poly(*N,N*-diethylacrylamide-*co*-methacrylic acid)	153
Poly[*N,N*-diethylacrylamide-*co*-4-(2-phenyldiazenyl) benzamido-*N*-(2-aminoethyl)acrylamide]	153
Poly(*N,N*-diethylaminoethyl methacrylate-*co*-methyl methacrylate)	26
Poly[di(ethylene glycol) methyl ether methacrylate-*b*-tri(ethylene glycol) methyl ether methacrylate]	153
Poly[di(ethylene glycol) methyl ether methacrylate-*co*-tri(ethylene glycol) methyl ether methacrylate]	153
Poly(diethyl maleate-*co*-vinyl acetate)	427
Poly(1,1-dihydroperfluorooctyl acrylate-*b*-styrene)	310
Poly(1,1-dihydroperfluorooctyl acrylate-*b*-vinyl acetate)	310
Poly(*N,N*-dimethylacrylamide-*co*-allyl methacrylate)	111-112
Poly(*N,N*-dimethylacrylamide-*co*-*tert*-butylacrylamide)	26-27, 174-175, 220
Poly(*N,N*-dimethylacrylamide-*co*-glycidyl methacrylate)	153
Poly(*N,N*-dimethylacrylamide-*co*-2-hydroxyethyl methacrylate)	220
Poly(*N,N*-dimethylacrylamide-*b*-*N*-isopropylacrylamide)	153
Poly(*N,N*-dimethylacrylamide-*b*-*N*-isopropylacrylamide-*b*-*N*-acryloylpyrrolidine)	154

Polymer	**Page(s)**
Poly(ethyl ethylene phosphate-*b*-propylene oxide-*b*-ethyl ethylene phosphate)	158
Poly(2-ethylhexyl methacrylate-*co*-acrylic acid)	432
Poly(2-ethylhexylacrylate-*co*-acrylic acid-*co*-vinyl acetate)	223
Poly(ethyl methacrylate-*co*-*N*-vinylcarbazole)	432
Poly[*N*-ethyl-*N*-methylacrylamide-*co*-4-(2-phenyldiazenyl) benzamido-*N*-(2-aminoethyl)acrylamide]	158
Poly(2-ethyl-2-oxazoline-*b*-ε-caprolactone)	158
Poly(2-ethyl-2-oxazoline-*b*-2-nonyl-2-oxazoline)	158, 224
Poly(2-ethyl-2-oxazoline-*co*-2-nonyl-2-oxazoline)	158, 224
Poly(2-ethyl-2-oxazoline-*co*-2-phenyl-2-oxazoline)	224
Poly(2-ethyl-2-oxazoline-*co*-2-propyl-2-oxazoline)	231
Poly[2-(*N*-ethylperfluorooctylsulfonamido)ethyl acrylate-*co*-*N*-isopropylacrylamide]	432
Poly[2-(*N*-ethylperfluorooctylsulfonamido)ethyl methacrylate-*co*-*N*-isopropylacrylamide]	432
Poly(3-ethyl-1-vinyl-2-pyrrolidinone-*co*-1-vinyl-2-pyrrolidinone)	158
Poly(formaldehyde-*co*-phenol) (acetylated)	433
Poly(L-glutamic acid-*g*-*N*-isopropylacrylamide)	159
Poly(glycidol-*b*-propylene oxide-*b*-glycidol)	159, 433
Poly(glycidyl methacrylate-*co*-butyl methacrylate)	88-89, 367
Poly(glycidyl methacrylate-*co*-ethyl methacrylate)	90, 368
Poly(glycidyl methacrylate-*co*-methyl methacrylate)	91, 368
Poly(1H,1H,2H,2H-heptadecafluorodecyl methacrylate-*co*-methyl methacrylate)	312
Poly(hexafluoroacetone-*co*-vinylidene fluoride)	224
Poly(3-hydroxybutanoic acid-*co*-3-hydroxypentanoic acid)	224, 433
Poly(3-hydroxybutyrate-*co*-3-hydroxyvalerate)	58
Poly(2-hydroxyethyl acrylate-*co*-butyl acrylate)	159, 224
Poly(2-hydroxyethyl acrylate-*co*-2-hydroxyethyl methacrylate)	159
Poly(2-hydroxyethyl acrylate-*co*-hydroxypropyl acrylate)	159, 231
Poly(2-hydroxyethyl acrylate-*co*-vinyl butyl ether)	159
Poly(2-hydroxyethyl methacrylate-*g*-ethylene glycol)	159
Poly(2-hydroxyethyl methacrylate-*b*-*N*-isopropylacrylamide)	159
Poly(2-hydroxyisopropyl acrylate-*co*-aminoethyl methacrylate)	159
Poly(2-hydroxypropyl acrylate-*co*-*N*-acryloylmorpholine)	126
Poly(2-hydroxypropyl acrylate-*co*-aminoethyl methacrylate)	159
Poly(2-hydroxypropyl acrylate-*co*-*N,N*-dimethylacrylamide)	127
Poly[*N*-(2-hydroxypropyl)methacrylamide-*b*-*N*-isopropylacrylamide]	159
Poly[*N*-(2-hydroxypropyl)methacrylamide monolactate-*co*-*N*-(2-hydroxypropyl) methacrylamide dilactate]	160, 231
Poly(hydroxystearic acid-*b*-ethylene oxide-*b*-hydroxystearic acid)	224
Poly(4-iodostyrene-*co*-styrene)	433
Poly(isobutyl methacrylate-*co*-2-(*tert*-butylamino)ethyl methacrylate)	434
Poly(isobutyl vinyl ether-*co*-2-hydroxyethyl vinyl ether)	160, 231

Polymer	**Page(s)**
Poly(2-isopropyl-2-oxazoline-*co*-2-nonyl-2-oxazoline)	168, 233
Poly(2-isopropyl-2-oxazoline-*co*-2-propyl-2-oxazoline)	168, 233
Poly(DL-lactic acid-*co*-glycolic acid)	41-44, 59, 312, 352, 435
Poly(L-lactic acid-*co*-glycolic acid)	91-92
Poly(DL-lactic acid-*b*-perfluoropropylene oxide	313
Poly(DL-lactide-*co*-glycolide)	303-306
Poly(L-lactide-*co*-diglycidyl ether of bisphenol A-*co*-4,4'-hexafluoroisopropylidenediphenol)	306-307
Poly(L-lysine-*g*-*N*-isopropylacrylamide)	169
Poly(maleic acid-*co*-acrylic acid)	226
Poly(maleic acid-*co*-ethyl vinyl ether)	435
Poly(maleic acid-*co*-styrene)	226
Poly(maleic acid-*co*-vinyl acetate)	226
Poly(maleic anhydride-*alt*-*tert*-butylstyrene)-*g*-poly(ethylene glycol) monomethyl ether	169
Poly(maleic anhydride-*co*-ethyl vinyl ether)	435
Poly(maleic anhydride-*co*-1-octadecene)	435
Poly(maleic anhydride-*alt*-styrene)-*g*-poly(ethylene glycol) monomethyl ether	169
Poly(3-mesityl-2-hydroxypropyl methacrylate-*co*-1-vinyl-2-pyrrolidinone)	92-94, 369
Poly[methacrylic acid-*co*-butyl methacrylate-*co*-poly(ethylene glycol) monomethyl ether methacrylate]	169
Poly(methacrylic acid-*co*-glycidyl methacrylate-*co*-poly(ethylene glycol) monomethyl ether methacrylate]	169
Poly(methacrylic acid-*co*-lauryl methacrylate-*co*-poly(ethylene glycol) monomethyl ether methacrylate]	169
Poly(methacrylonitrile-*co*-butyl methacrylate)	435
Poly(methacrylonitrile-*co*-ethyl methacrylate)	435-436
Poly(methacrylonitrile-*co*-methyl methacrylate)	436
Poly(methacrylonitrile-*co*-styrene)	436
Poly(methoxydiethylene glycol methacrylate-*co*-dodecyl methacrylate)	143
Poly(methoxydiethylene glycol methacrylate-*co*-methoxyoligo(ethylene glycol) methacrylate)	143-145
Poly[2-(2-methoxyethoxy)ethyl methacrylate-*co*-oligo(ethylene glycol) methacrylate]	169, 227
Poly[2-(2-methoxyethoxy)ethyl methacrylate-*co*-oligo(ethylene glycol) methyl ether methacrylate]	169
Poly[methoxypoly(ethylene glycol)-*b*-ε-caprolactone]	169
Poly[methoxytri(ethylene glycol) acrylate-*b*-4-vinylbenzyl methoxytris(oxyethylene) ether]	170
Poly(methyl acrylate-*co*-methyl methacrylate)	436-437
Poly(methylhydrosiloxane-*co*-dimethylsiloxane)	369
Poly(methyl methacrylate-*co*-acrylonitrile)	437

Appendix 2 List of systems and properties in order of the copolymers

Copolymer	**Solvent**	**Property**	**Page(s)**
Poly(acrylonitrile-*co*-α-methylstyrene)			
	tetrahydrofuran	A_2	420
Poly(acrylonitrile-*co*-vinyl acetate)			
	N,N-dimethylformamide	A_2	421
Poly(acrylonitrile-*co*-vinylidene chloride)			
	γ-butyrolactone	A_2	421
	N,N-dimethylacetamide	A_2	421
	N,N-dimethylformamide	A_2	421
Poly(benzyl acrylate-*co*-methyl acrylate)			
	2-butanone	A_2	421
	N,N-dimethylformamide	A_2	421
Poly(γ-benzyl-L-glutamate-*b*-methyl methacrylate-*b*-γ-benzyl-L-glutamate)			
	N,N-dimethylformamide	A_2	422
Poly[(1,1'-biphenyl)-4-4'diol-*co*-3,3-bis(4-carboxyphenyl)-phthalide]			
	1,2-dichloroethane	A_2	422
	trichloromethane	A_2	422
Poly(3,3',4,4'-biphenyltetracarb-oxylic dianhydride-*co*-1,4-phenylenediamine)			
	dimethylsulfoxide	A_2	422
	1-methyl-2-pyrrolidinone	A_2	422
Poly[9,9'-bis(4-acetylphenyl)-fluorene-*co*-4,4'-diamino-3,3'-dibenzoyldiphenyl ether]			
	1,1,2,2-terachloroethane	A_2	422
	trichloromethane	A_2	422

Copolymer	Solvent	Property	Page(s)
Poly[bis(4-aminophenoxy)-benzene-*co*-pyromellitic anhydride]	*N,N*-dimethylacetamide	A_2	422
Poly[bis(4-aminophenyl) ether-*co*-pyromellitic anhydride]	*N,N*-dimethylacetamide	A_2	423
Poly[bis(4-aminophenyl) ether-*co*-tricyclodecenetetracarboxylic dianhydride]	dimethylsulfoxide	A_2	423
Poly(bis(4-aminophenyl) sulfone-*co*-isophthaloyl chloride)	dimethylsulfoxide	A_2	423
Poly[2,5-bis(carbomethoxy)-terephthaloyl chloride-*co*-3,3-bis(4-aminophenyl)phthalide]	*N,N*-dimethylformamide	A_2	423
Poly[1,4-bis(3,4-dicarboxyphen-oxyl)benzenedianhydride-*co*-2,2'-dimethyl-4,4'-methylenedianiline]	trichloromethane	A_2	423
Poly[2,2'-bis(3,4-dicarboxy-phenyl)hexafluoropropane-dianhydride-*co*-2,2'-bis(trifluoro-methyl)-4,4'-biphenyldiamine]	tetrahydrofuran	A_2	424
Poly[2,2'-bis(3,4-dicarboxy-phenyl)hexafluoropropane-dianhydride-*co*-4,4'-diamino-3,3'-dimethylbiphenyl]	cyclopentanone	A_2	424

Copolymer	**Solvent**	**Property**	**Page(s)**
Poly{2,2'-bis[4-(3,4-dicarboxy-phenoxy)phenyl]propane dianhydride-*co*-4,4'-diamino-3,3'-dimethylbiphenyl}			
	cyclopentanone	A_2	424
Poly[9,9'-bis(4-hydroxyphenyl)-fluorene-*co*-terephthaloyl chloride]			
	1,1,2,2-tetrachloroethane	A_2	424
	trichloromethane	A_2	424
Poly(bisphenol-A diglycidyl ether-*co*-adipic acid)			
	N,N-dimethylacetamide	A_2	424
Poly(bisphenol-F-*co*-2,6-dichloro-6-methoxy-1,3,5-triazine)			
	trichloromethane	A_2	425
Poly(butadiene-*co*-acrylonitrile)			
	trichloromethane	A_2	425
Poly(butadiene-*co*-α-methyl-styrene)			
	ethyl acetate	UCST/LCST	230
Poly(butadiene-*b*-styrene)			
	cyclohexane	A_2	425
	1,4-dioxane	A_2	425
	tetrahydrofuran	VLE	21-22
	tetrahydrofuran	A_2	425
	toluene	A_2	425
Poly(butadiene-*co*-styrene)			
	benzene	A_2	425
	cyclohexane	A_2	425-426
	tetrahydrofuran	A_2	426

Copolymer	Solvent	Property	Page(s)
Poly(*tert*-butyl acrylate-*b*-methyl methacrylate)			
	benzene	Henry	71
	benzene	$\Delta_M H_A^\infty$, $\Delta_{sol} H_{A(vap)}^\infty$	362
	butyl acetate	Henry	72
	butyl acetate	$\Delta_M H_A^\infty$, $\Delta_{sol} H_{A(vap)}^\infty$	362
	tert-butyl acetate	Henry	71, 72
	tert-butyl acetate	$\Delta_M H_A^\infty$, $\Delta_{sol} H_{A(vap)}^\infty$	362
	chlorobenzene	Henry	71, 72, 73
	chlorobenzene	$\Delta_M H_A^\infty$, $\Delta_{sol} H_{A(vap)}^\infty$	362-363
	cumene	Henry	72
	n-decane	Henry	71, 72, 73
	n-decane	$\Delta_M H_A^\infty$, $\Delta_{sol} H_{A(vap)}^\infty$	362-363
	ethyl acetate	Henry	71, 72
	ethyl acetate	$\Delta_M H_A^\infty$, $\Delta_{sol} H_{A(vap)}^\infty$	362-363
	ethylbenzene	Henry	71, 72, 73
	ethylbenzene	$\Delta_M H_A^\infty$, $\Delta_{sol} H_{A(vap)}^\infty$	362-363
	isobutyl acetate	Henry	71-72
	isobutyl acetate	$\Delta_M H_A^\infty$, $\Delta_{sol} H_{A(vap)}^\infty$	362
	isopropylbenzene	Henry	71, 73
	isopropylbenzene	$\Delta_M H_A^\infty$, $\Delta_{sol} H_{A(vap)}^\infty$	362-363
	3-methylbutyl acetate	Henry	71, 72, 73
	3-methylbutyl acetate	$\Delta_M H_A^\infty$, $\Delta_{sol} H_{A(vap)}^\infty$	362-363
	n-nonane	Henry	71, 73
	n-nonane	$\Delta_M H_A^\infty$, $\Delta_{sol} H_{A(vap)}^\infty$	362-363
	n-octane	Henry	71, 73
	n-octane	$\Delta_M H_A^\infty$, $\Delta_{sol} H_{A(vap)}^\infty$	362-363
	propyl acetate	Henry	71, 72
	propyl acetate	$\Delta_M H_A^\infty$, $\Delta_{sol} H_{A(vap)}^\infty$	362-363
	2-propyl acetate	Henry	72
	2-propyl acetate	$\Delta_M H_A^\infty$, $\Delta_{sol} H_{A(vap)}^\infty$	362
	propylbenzene	Henry	71, 72, 73
	propylbenzene	$\Delta_M H_A^\infty$, $\Delta_{sol} H_{A(vap)}^\infty$	362-363
	toluene	Henry	71, 72
	toluene	$\Delta_M H_A^\infty$, $\Delta_{sol} H_{A(vap)}^\infty$	362-363
Poly(butylene oxide-*b*-ethylene oxide-*b*-butylene oxide)			
	water	A_2	426
Poly(butylene succinate-*co*-butylene adipate)			
	carbon dioxide	gas solubility	22-23
	–	PVT	379

Copolymer	Solvent	Property	Page(s)
Poly(butyl methacrylate-*co*-*N,N*-dimethylaminoethyl methacrylate)			
	benzene	VLE	23
	carbon dioxide	gas solubility	23-24
	toluene	VLE	24-25
Poly(butyl methacrylate-*b*-perfluoroalkylethyl acrylate)			
	carbon dioxide	gas solubility	25-26
Poly(4-*tert*-butylstyrene-*b*-dimethylsiloxane)			
	toluene	A_2	426
Poly(4-*tert*-butylstyrene-*b*-dimethylsiloxane-*b*-4-*tert*-butylstyrene)			
	toluene	A_2	426
Poly(*tert*-butylstyrene-*b*-styrene)			
	N,N-dimethylacetamide	A_2	426
Poly(4-chlorostyrene-*co*-methyl methacrylate)			
	benzene	A_2	426
	1,4-dioxane	A_2	426
	trichloromethane	A_2	426
Poly(4-chlorostyrene-*b*-styrene-*b*-4-chlorostyrene)			
	2-butanone	A_2	427
	isopropylbenzene	A_2	427
	toluene	A_2	427
Poly(cyclohexene oxide-*co*-carbon dioxide)			
	carbon dioxide	HPPE	259
	carbon dioxide + cyclohexene oxide	HPPE	315-316

Copolymer	Solvent	Property	Page(s)
Poly(decyl methacrylate-*b*-styrene)			
	methyl acetate	A_2	427
	tetrahydrofuran	A_2	427
Poly(diacetone acrylamide-*co*-hydroxyethyl acrylate)			
	water	LLE	152
	water	UCST/LCST	230
Poly(*N,N*-diallylammonioethanoic acid-*co*-sulfur dioxide)			
	water + poly(ethylene glycol)	LLE	174
Poly(4,4'-diaminophenyl-*co*-pyromellitic anhydride]			
	N,N-dimethylacetamide	A_2	427
Poly(3,4-dichlorobenzyl methacrylate-*co*-ethyl methacrylate)			
	benzene	Henry	73-78
	benzene	$\Delta_M H_A^\infty$	363-366
	2-butanone	Henry	74-78
	2-butanone	$\Delta_M H_A^\infty$	363-366
	n-decane	Henry	74-78
	n-decane	$\Delta_M H_A^\infty$	363-366
	1,2-dimethylbenzene	Henry	73-78
	1,2-dimethylbenzene	$\Delta_M H_A^\infty$	363-366
	n-dodecane	Henry	74-78
	n-dodecane	$\Delta_M H_A^\infty$	363-366
	ethanol	Henry	74-78
	ethanol	$\Delta_M H_A^\infty$	363-366
	ethyl acetate	Henry	74-78
	ethyl acetate	$\Delta_M H_A^\infty$	363-366
	methanol	Henry	73-78
	methanol	$\Delta_M H_A^\infty$	363-366
	methyl acetate	Henry	74-78
	methyl acetate	$\Delta_M H_A^\infty$	363-366
	n-nonane	Henry	74-78
	n-nonane	$\Delta_M H_A^\infty$	363-366
	n-octane	Henry	74-78
	n-octane	$\Delta_M H_A^\infty$	363-366
	2-propanone	Henry	74-78

Copolymer	Solvent	Property	Page(s)
	2-propanone	$\Delta_M H_A^\infty$	363-366
	toluene	Henry	73-78
	toluene	$\Delta_M H_A^\infty$	363-366
	n-undecane	Henry	74-78
	n-undecane	$\Delta_M H_A^\infty$	363-366
Poly(*N,N*-diethylacrylamide-*co*-acrylic acid)			
	water	LLE	111
	water	UCST/LCST	230
Poly(*N,N*-diethylacrylamide-*co*-*N*-ethylacrylamide			
	water	UCST/LCST	230
Poly(diethyl maleate-*co*-vinyl acetate)			
	toluene	A_2	427
Poly(*N,N*-dimethylacrylamide-*co*-allyl methacrylate)			
	water	LLE	111-112
Poly(*N,N*-dimethylacrylamide-*co*-*tert*-butylacrylamide)			
	water	VLE	26-27
	water + monoammonium phosphate	LLE	174
	water + monopotassium phosphate	LLE	175
	water + monosodium carbonate	LLE	175
	water + sodium chloride	LLE	175
Poly(*N,N*-dimethylacrylamide-*b*-*N*-isopropylacrylamide-*b*-*N,N*-dimethylacrylamide)			
	water	A_2	428

Copolymer	**Solvent**	**Property**	**Page(s)**
Poly(*N,N*-dimethylacrylamide-*co*-*N*-phenylacrylamide)			
	water	LLE	113-114
Poly[2-(*N,N*-dimethylamino)-ethyl methacrylate]-*b*-2-(*N*-morpholino)ethyl methacrylate]			
	benzene	Henry	78
	benzene	$\Delta_M H_A^{\infty}$	366
	cycloheptane	Henry	78
	cycloheptane	$\Delta_M H_A^{\infty}$	366
	cyclohexane	Henry	78
	cyclohexane	$\Delta_M H_A^{\infty}$	366
	cyclopentane	Henry	78
	cyclopentane	$\Delta_M H_A^{\infty}$	366
	n-decane	Henry	78
	n-decane	$\Delta_M H_A^{\infty}$	366
	n-heptane	Henry	78
	n-heptane	$\Delta_M H_A^{\infty}$	366
	n-hexane	Henry	78
	n-hexane	$\Delta_M H_A^{\infty}$	366
	n-nonane	Henry	78
	n-nonane	$\Delta_M H_A^{\infty}$	366
	n-octane	Henry	78
	n-octane	$\Delta_M H_A^{\infty}$	366
	toluene	Henry	78
	toluene	$\Delta_M H_A^{\infty}$	366
Poly[2,2'-dimethyl-4,4'-methylenedianiline-*co*-3,3',4,4'-oxydi(phthalic anhydride)]			
	trichloromethane	A_2	428
Poly(*trans*-2,5-dimethylpiperazine-*co*-phthaloyl chloride)			
	1-methyl-2-pyrrolidinone	A_2	428
	trichloromethane	A_2	428
Poly(*trans*-2,5-dimethylpiperazine-*co*-terephthaloyl chloride)			
	2,2,2-trifluoroethanol	A_2	428-429

Copolymer	Solvent	Property	Page(s)
Poly(dimethylsiloxane-*b*-ethylene oxide)			
	benzene-d6	A_2	429
	cyclohexane-d12	A_2	429
	ethanol	A_2	429
	methanol	A_2	429
	methanol-d4	A_2	429
	toluene	A_2	429
Poly(dimethylsiloxane-*co*-hydromethylsiloxane)			
	toluene	A_2	429
Poly[dimethylsiloxane-*co*-methyl(4-cyanobiphenoxy)-butylsiloxane]			
	benzene	$\Delta_{sol}H_{A(vap)}^{\infty}$	367
	cyclohexane	$\Delta_{sol}H_{A(vap)}^{\infty}$	367
	1,2-dimethylbenzene	$\Delta_{sol}H_{A(vap)}^{\infty}$	367
	1,3-dimethylbenzene	$\Delta_{sol}H_{A(vap)}^{\infty}$	367
	1,4-dimethylbenzene	$\Delta_{sol}H_{A(vap)}^{\infty}$	367
	2,3-dimethylpentane	$\Delta_{sol}H_{A(vap)}^{\infty}$	367
	2,4-dimethylpentane	$\Delta_{sol}H_{A(vap)}^{\infty}$	367
	ethylbenzene	$\Delta_{sol}H_{A(vap)}^{\infty}$	367
	n-heptane	$\Delta_{sol}H_{A(vap)}^{\infty}$	367
	n-hexane	$\Delta_{sol}H_{A(vap)}^{\infty}$	367
	2-methylhexane	$\Delta_{sol}H_{A(vap)}^{\infty}$	367
	3-methylhexane	$\Delta_{sol}H_{A(vap)}^{\infty}$	367
	n-nonane	$\Delta_{sol}H_{A(vap)}^{\infty}$	367
	n-octane	$\Delta_{sol}H_{A(vap)}^{\infty}$	367
	n-pentane	$\Delta_{sol}H_{A(vap)}^{\infty}$	367
	toluene	$\Delta_{sol}H_{A(vap)}^{\infty}$	367
	2,2,3-trimethylbutane	$\Delta_{sol}H_{A(vap)}^{\infty}$	367
Poly(*N*-ethylacrylamide-*co*-*N*-isopropylacrylamide)			
	water	UCST/LCST	230
Poly(ethyl acrylate-*co*-methyl methacrylate)			
	2-butanone	A_2	429
	N,N-dimethylformamide	A_2	429
	2-propanone	A_2	429

Copolymer	Solvent	Property	Page(s)
Poly(ethylene-*co*-1-butene) (deuterated)			
	2,2-dimethylpropane	LLE	120
	2,2-dimethylpropane	HPPE	265
	ethane	density	407
	ethane + n-pentane-d12	density	408
	2-methylbutane	LLE	120-121
	2-methylbutane	HPPE	268
	n-pentane	LLE	121
	n-pentane	HPPE	269
Poly(ethylene-*co*-butyl acrylate)			
	ethene	HPPE	271-273
	ethene + butyl acrylate	HPPE	325-327
	ethene + methyl acrylate	HPPE	327
Poly(ethylene-*co*-butyl meth-acrylate)			
	ethene	HPPE	273-276
Poly(ethylene-*co*-butyl meth-acrylate-*co*-methacrylic acid)			
	ethene	HPPE	276-279
Poly(ethylene-*co*-ethyl acrylate)			
	ethene	HPPE	279-280
	carbon dioxide	$\Delta_{sol}H_{A(vap)}^{\infty}$	376
Poly(ethylene-*co*-2-ethylhexyl acrylate)			
	ethene	HPPE	281
	ethene + 2-ethylhexyl acrylate	HPPE	328
Poly(ethylene-*co*-ethyl methacrylate)			
	ethene + ethyl methacrylate	HPPE	328-329

Copolymer	Solvent	Property	Page(s)
Poly(ethylene-*co*-1-hexene)			
	–	PVT	381-382
	ethene	HPPE	281-282
	ethene + n-butane	HPPE	329-330
	ethene + carbon dioxide	HPPE	330-331
	ethene + ethane	HPPE	331
	ethene + helium	HPPE	331-332
	ethene + 1-hexene	HPPE	332-333
	ethene + 1-hexene + nitrogen	HPPE	333-334
	ethene + methane	HPPE	334
	ethene + nitrogen	HPPE	335
	ethene + propane	HPPE	335-336
	ethene + 2-methylpropane + 1-hexene	HPPE	336
	1-hexene	vapor solubility	29-33
	2-methylpropane	HPPE	282-283
	propane	HPPE	283-286
Poly(ethylene-*co*-methacrylic acid)			
	ethene	HPPE	286-288
Poly(ethylene-*co*-methyl acrylate)			
	ethene	HPPE	288-290
	ethene + butyl acrylate	HPPE	336-337
	ethene + 2-ethylhexyl acrylate	HPPE	337
	ethene + n-heptane	HPPE	337
	ethene + methyl acrylate	HPPE	338
	ethene + vinyl acetate	HPPE	338
Poly(ethylene-*co*-methyl acrylate-*co*-vinyl acetate)			
	ethene	HPPE	290-291
	ethene + vinyl acetate	HPPE	338-339
Poly(ethylene-*co*-methyl methacrylate)			
	ethene	HPPE	291-293
Poly(ethylene-*co*-norbornene)			
	–	PVT	382
	ethene + toluene + bicyclo[2,2,1]-2-heptene	VLE	61-63

Copolymer	Solvent	Property	Page(s)
Poly(ethylene-*co*-1-octadecene)			
	1-hexene	vapor solubility	33-35
Poly(ethylene-*co*-1-octene)			
	–	PVT	383-384
	benzene	Henry	79-88
	cyclohexane	Henry	79-88
	cyclohexane	VLE	35-36
	cyclopentane	VLE	36
	n-dodecane	Henry	79-88
	ethene	HPPE	293-294
	ethene + 1-octene	HPPE	339
	n-heptane	Henry	79-88
	n-heptane	VLE	36
	n-hexane	Henry	79-88
	n-hexane	VLE	36
	1-hexene	Henry	79-88
	1-hexene	vapor solubility	36-37
	n-nonane	Henry	79
	n-octane	Henry	79-88
	n-octane	VLE	37
	1-octene	Henry	79-88
	n-pentadecane	Henry	79-88
	n-pentane	VLE	37-38
	n-pentane	LLE	121
	propane	HPPE	294
	toluene	Henry	79-88
Poly(ethylene-*alt*-propylene)			
	toluene	A_2	430
Poly(ethylene-*b*-propylene)			
	–	PVT	384-385
	ethene	gas solubility	38
	ethene + n-hexane	K^{∞}	64
	propene	gas solubility	38-39
	propene + n-hexane	K^{∞}	64
Poly(ethylene-*co*-propylene)			
	1-chloronaphthalene	A_2	430
	ethene	HPPE	294-296
	ethene + carbon dioxide	HPPE	340

Copolymer	**Solvent**	**Property**	**Page(s)**
	ethene + vinyl acetate + helium	HPPE	346
	ethene + vinyl acetate + methane	HPPE	347-348
	ethene + vinyl acetate + 2-methyl-2-propanol	HPPE	348-349
	ethene + vinyl acetate + nitrogen	HPPE	350
	styrene + polystyrene	LLE	176
Poly(ethylene-*co*-vinyl alcohol)	–	PVT	387-392
Poly(ethylene glycol-*b*-propylene glycol-*b*-ethylene glycol)	dimethylsulfoxide	density, V^E	410-411
	2-methyltetrahydrofuran	density, V^E	412
Poly[(ethylene glycol) mono-dodecyl ether-*b*-propylene glycol]	carbon dioxide	HPPE	300-301
Poly[(ethylene glycol) monometh-acrylate-*co*-methyl methacrylate]	water	UCST/LCST	231
Poly(ethylene oxide-*b*-dimethylsiloxane)	1,2,3,4-tetrahydronaphtha-lene + poly(ethylene oxide)	LLE	176
	toluene + poly(ethylene oxide)	LLE	177-178
Poly(ethylene oxide-*b*-(R)-3-hydroxybutyrate-*b*-ethylene oxide)	water	A_2	431
Poly(ethylene oxide-*co*-methylene oxide)	1H,1H,5H-octafluoro-1-pentanol	A_2	431

Copolymer	Solvent	Property	Page(s)
	1,2-dimethylbenzene	A_2	431
	water	LLE	125-126
	water	UCST/LCST	231
	water	$\Delta_M H$	376
	water	density	414-416
	water	A_2	431
	water + dextran	LLE	186-194
	water + dipotassium phosphate	LLE	194-201
	water + disodium tartrate	LLE	201-202
	water + lithium sulfate	LLE	202-203
	water + magnesium sulfate	LLE	203-206
	water + sodium acetate	LLE	206
	water + sodium bromide	LLE	206
	water + sodium chloride	LLE	206-207
	water + sodium citrate	LLE	207
	water + sodium nitrate	LLE	208
	water + sodium perchlorate	LLE	208
	water + sodium sulfate	LLE	209-210
	water + surfactant	ΔH	376
	water + zinc sulfate	LLE	210-211
Poly(ethylene oxide-*b*-styrene-*b*-ethylene oxide)			
	ethyl acetate	A_2	432
	toluene	A_2	432
	trichloromethane	A_2	432
Poly(ethylene oxide-*co*-tetrahydrofuran)			
	2-propanol	$\Delta_M H$	376
	tetrachloromethane	$\Delta_M H$	376
	trichloromethane	$\Delta_M H$	376
Poly(ethylene oxide-*b*-1,1,2,2-tetrahydroperfluorodecyl acrylate			
	carbon dioxide	HPPE	302
	carbon dioxide + 2-hydroxyethyl methacrylate	HPPE	350
Poly(2-ethylhexyl methacrylate-*co*-acrylic acid)			
	2-butanone	A_2	432
	styrene	A_2	432

Copolymer	Solvent	Property	Page(s)
Poly(ethyl methacrylate-*co*-*N*-vinylcarbazole)			
	benzene	A_2	432
Poly[2-(*N*-ethylperfluorooctyl-sulfonamido)ethyl acrylate-*co*-*N*-isopropylacrylamide]			
	water	A_2	432
Poly[2-(*N*-ethylperfluorooctyl-sulfonamido)ethyl methacrylate-*co*-*N*-isopropylacrylamide]			
	water	A_2	432
Poly(2-ethyl-2-oxazoline-*co*-2-propyl-2-oxazoline)			
	water	UCST/LCST	231-232
Poly(formaldehyde-*co*-phenol) (acetylated)			
	ethyl acetate	A_2	433
Poly(glycidol-*b*-propylene oxide-*b*-glycidol)			
	water	A_2	433
Poly(glycidyl methacrylate-*co*-butyl methacrylate)			
	benzene	Henry	88
	benzene	$\Delta_M H_A^\infty$	367
	1-chlorobutane	Henry	89
	1-chlorobutane	$\Delta_M H_A^\infty$	367
	1-chloropropane	Henry	89
	1-chloropropane	$\Delta_M H_A^\infty$	367
	n-decane	Henry	88-89
	n-decane	$\Delta_M H_A^\infty$	367
	1,4-dimethylbenzene	Henry	88-89
	1,4-dimethylbenzene	$\Delta_M H_A^\infty$	367
	n-heptane	Henry	89
	n-heptane	$\Delta_M H_A^\infty$	367
	n-hexane	Henry	89
	n-hexane	$\Delta_M H_A^\infty$	367

Copolymer	Solvent	Property	Page(s)
Poly(4-iodostyrene-*co*-styrene)	1,4-dioxane	A_2	433
Poly(isobutyl methacrylate-*co*-2-(*tert*-butylamino)ethyl methacrylate)	isopropylamine	A_2	434
Poly(isobutyl vinyl ether-*co*-2-hydroxyethyl vinyl ether)	water	UCST/LCST	231
Poly(isoprene-*b*-styrene-*b*-isoprene)	tetrachloromethane	A_2	434
	tetrahydrofuran	A_2	434
	toluene	A_2	434
Poly(*N*-isopropylacrylamide) stereoblock polymer (isotactic-atactic-isotactic)	water	A_2	434
Poly(*N*-isopropylacrylamide-*co*-acrylamide)	water	LLE	127-128
	water	A_2	434
Poly(*N*-isopropylacrylamide-*co*-acrylic acid)	water	LLE	128-129
	water	ΔH	376
Poly(*N*-isopropylacrylamide-*b*-ε-caprolactone-*b*-*N*-isopropylacrylamide)	water	LLE	129-130
Poly(*N*-isopropylacrylamide-*co*-1-deoxy-1-methacrylamido-D-glucitol)	water	LLE	130-131

Copolymer	Solvent	Property	Page(s)
Poly{*N*-isopropylacrylamide-*co*-3-[*N*-(3-methacrylamidopropyl)-*N,N*-dimethylammonio] propane sulfate}			
	methanol	A_2	434
	water	A_2	434
Poly(*N*-isopropylacrylamide-*co*-2-methacryloamidohistidine)			
	water	LLE	136-137
Poly(*N*-isopropylacrylamide-*co*-8-methacryloyloxyoctanoic acid)			
	water	LLE	137-138
	water	LLE	166
Poly(*N*-isopropylacrylamide-*co*-5-methacryloyloxypentanoic acid)			
	water	LLE	138-139
	water	LLE	166
Poly(*N*-isopropylacrylamide-*co*-11-methacryloyloxyundecanoic acid)			
	water	LLE	139-140
	water	LLE	166
Poly[*N*-isopropylacrylamide-*co*-oligo(ethylene glycol) mono-methacrylate]			
	deuterium oxide	UCST/LCST	233
	water	UCST/LCST	233
Poly(*N*-isopropylacrylamide-*b*-propylene glycol-*b*-*N*-isopropyl-acrylamide)			
	water	UCST/LCST	233
Poly(*N*-isopropylacrylamide-*co*-4-pentenoic acid)			
	water	LLE	140-142
	water	LLE	167

Copolymer	Solvent	Property	Page(s)
Poly(DL-lactide-*co*-glycolide)	carbon dioxide	HPPE	303-304
	chlorodifluoromethane	HPPE	304-305
	dimethyl ether	HPPE	305-306
	trifluoromethane	HPPE	306
Poly(L-lactide-*co*-diglycidyl ether of bisphenol A-*co*-4,4'-hexa-fluoroisopropylidenediphenol)	dimethyl ether	HPPE	306-307
Poly(maleic acid-*co*-ethyl vinyl ether)	tetrahydrofuran	A_2	435
Poly(maleic anhydride-*co*-ethyl vinyl ether)	tetrahydrofuran	A_2	435
Poly(maleic anhydride-*co*-1-octadecene)	ethyl acetate	A_2	435
Poly(3-mesityl-2-hydroxypropyl methacrylate-*co*-1-vinyl-2-pyrrolidinone)	1-butanol	Henry	93
	1-butanol	$\Delta_M H_A^\infty$, $\Delta_{sol} H_{A(vap)}^\infty$	369
	n-decane	Henry	93
	n-decane	$\Delta_M H_A^\infty$, $\Delta_{sol} H_{A(vap)}^\infty$	369
	ethanol	Henry	92-93
	ethanol	$\Delta_M H_A^\infty$, $\Delta_{sol} H_{A(vap)}^\infty$	369
	n-heptane	Henry	92-93
	n-heptane	$\Delta_M H_A^\infty$, $\Delta_{sol} H_{A(vap)}^\infty$	369
	n-hexane	Henry	92-93
	n-hexane	$\Delta_M H_A^\infty$, $\Delta_{sol} H_{A(vap)}^\infty$	369
	methanol	Henry	92-94
	methanol	$\Delta_M H_A^\infty$, $\Delta_{sol} H_{A(vap)}^\infty$	369
	n-octane	Henry	93
	n-octane	$\Delta_M H_A^\infty$, $\Delta_{sol} H_{A(vap)}^\infty$	369
	1-pentanol	Henry	93-94

Copolymer	Solvent	Property	Page(s)
	1-pentanol	$\Delta_M H_A^\infty$, $\Delta_{sol} H_{A(vap)}^\infty$	369
	1-propanol	Henry	93
	1-propanol	$\Delta_M H_A^\infty$, $\Delta_{sol} H_{A(vap)}^\infty$	369
Poly(methacrylonitrile-*co*-butyl methacrylate)			
	2-butanone	A_2	435
Poly(methacrylonitrile-*co*-ethyl methacrylate)			
	2-butanone	A_2	435-436
Poly(methacrylonitrile-*co*-methyl methacrylate)			
	2-butanone	A_2	436
Poly(methacrylonitrile-*co*-styrene)			
	2-butanone	A_2	436
	N,N-dimethylformamide	A_2	436
Poly(methoxydiethylene glycol methacrylate-*co*-dodecyl methacrylate)			
	water	LLE	143
Poly(methoxydiethylene glycol methacrylate-*co*-methoxyoligo-(ethylene glycol) methacrylate)			
	water	LLE	143-145
Poly(methyl acrylate-*co*-methyl methacrylate)			
	2-butanone	A_2	436
	2-propanone	A_2	436-437
Poly(methyl methacrylate-*co*-acrylonitrile)			
	acetonitrile	A_2	437
	2-butanone	A_2	437
	N,N-dimethylformamide	A_2	437

Copolymer	Solvent	Property	Page(s)
	ethyl acetate	$\Delta_M H_A^\infty$, $\Delta_{sol} H_{A(vap)}^\infty$	369
	n-heptane	$\Delta_M H_A^\infty$, $\Delta_{sol} H_{A(vap)}^\infty$	369
	n-hexane	$\Delta_M H_A^\infty$, $\Delta_{sol} H_{A(vap)}^\infty$	369
	methyl acetate	$\Delta_M H_A^\infty$, $\Delta_{sol} H_{A(vap)}^\infty$	369
	n-pentane	$\Delta_M H_A^\infty$, $\Delta_{sol} H_{A(vap)}^\infty$	369
Poly(4-methyl-1-penten-3-one-*co*-(S)-4-methyl-1-hexen-3-one)			
	trichloromethane	A_2	439
Poly(α-methylstyrene-*b*-styrene)			
	cyclohexane	A_2	439
	tetrahydrofuran	A_2	439
Poly(1-octadecene-*co*-maleic anhydride)			
	ethyl acetate	A_2	440
Poly[oligo(ethylene glycol) methylacrylate-*co*-oligo-(propylene glycol) methacrylate]			
	water	LLE	145
Poly[4,4'-oxybis(benzoyl chloride)-*co*-phenolphthalein]			
	cyclohexanone	A_2	440
	1,2-dichloroethane	A_2	440
	N,N-dimethylformamide	A_2	440
	1,4-dioxane	A_2	440
	1-methyl-2-pyrrolidinone	A_2	440
	1,1,2,2-tetrachloroethane	A_2	440
	tetrahydrofuran	A_2	440
	trichloromethane	A_2	440
Poly[4,4'-oxydiphthalic anhydride-*co*-3,3'-bis(p-aminophenyl) phthalate]			
	N,N-dimethylformamide	A_2	440
	dimethylsulfoxide	A_2	440
	1-methyl-2-pyrrolidinone	A_2	440
	1,1,2,2-tetrachloroethane	A_2	440
	trichloromethane	A_2	440

Copolymer	**Solvent**	**Property**	**Page(s)**
Poly(4,4'-oxydiphthalic anhydride-*co*-p-phenylenediamine)			
	N,N-dimethylacetamide	A_2	440
Poly[(2-phenyl-1,3-dioxolane-4-yl)methyl methacrylate-*co*-butyl methacrylate]			
	1-butanol	Henry	95
	1-butanol	$\Delta_M H_A^\infty$	370
	ethanol	Henry	95
	ethanol	$\Delta_M H_A^\infty$	370
	methanol	Henry	95
	methanol	$\Delta_M H_A^\infty$	370
	1-pentanol	Henry	95
	1-pentanol	$\Delta_M H_A^\infty$	370
	1-propanol	Henry	95
	1-propanol	$\Delta_M H_A^\infty$	370
Poly[(2-phenyl-1,3-dioxolane-4-yl)methyl methacrylate-*co*-ethyl methacrylate]			
	1-butanol	Henry	96
	1-butanol	$\Delta_{sol} H_{A(vap)}^\infty$	370
	ethanol	Henry	96
	ethanol	$\Delta_{sol} H_{A(vap)}^\infty$	370
	n-heptane	Henry	96
	n-heptane	$\Delta_{sol} H_{A(vap)}^\infty$	370
	n-hexane	Henry	96
	n-hexane	$\Delta_{sol} H_{A(vap)}^\infty$	370
	n-octane	Henry	96
	n-octane	$\Delta_{sol} H_{A(vap)}^\infty$	370
	1-propanol	Henry	96
	1-propanol	$\Delta_{sol} H_{A(vap)}^\infty$	370
Poly[(2-phenyl-1,3-dioxolane-4-yl)methyl methacrylate-*co*-glycidyl methacrylate]			
	1-butanol	Henry	97
	1-butanol	$\Delta_M H_A^\infty$	371
	n-decane	Henry	97
	n-decane	$\Delta_M H_A^\infty$	371
	ethanol	Henry	96

Copolymer	Solvent	Property	Page(s)
	2-butanone	Henry	98
	2-butanone	$\Delta_M H_A^\infty$	372
	1,2-dimethylbenzene	Henry	98
	1,2-dimethylbenzene	$\Delta_M H_A^\infty$	372
	n-dodecane	Henry	98
	n-dodecane	$\Delta_M H_A^\infty$	372
	ethanol	Henry	98
	ethanol	$\Delta_M H_A^\infty$	372
	ethyl acetate	Henry	98
	ethyl acetate	$\Delta_M H_A^\infty$	372
	methanol	Henry	98
	methanol	$\Delta_M H_A^\infty$	372
	methyl acetate	Henry	98
	methyl acetate	$\Delta_M H_A^\infty$	372
	2-propanone	Henry	98
	2-propanone	$\Delta_M H_A^\infty$	372
	toluene	Henry	98
	toluene	$\Delta_M H_A^\infty$	372
Poly[propylene-*b*-(ethylene-*co*-propylene)]	–	PVT	392
Poly(propylene oxide-*b*-ethylene oxide-*b*-propylene oxide)	water	A_2	440
Poly(styrene-*co*-acrylic acid)	1,4-dioxane	A_2	441
Poly(styrene-*co*-acrylonitrile)	–	PVT	392
	2-butanone	A_2	441-442
	N,N-dimethylacetamide + polyacrylonitrile	LLE	211-212
	N,N-dimethylformamide	A_2	442
	ethyl acetate	A_2	442-443
	tetrahydrofuran	A_2	443
	toluene	UCST/LCST	233
	trichloromethane	$\Delta_{sol}H_B^\infty$, $\Delta_M H_B^\infty$	359-361

Copolymer	Solvent	Property	Page(s)
Poly(styrene-*b*-butadiene-*b*-styrene-*b*-butadiene-*b*-styrene)			
	n-heptane	LLE	145
Poly[styrene-*b*-(1-butene-*co*-ethylene)-*b*-styrene]			
	benzene	Henry	98
	benzene	$\Delta_M H_A^\infty$, $\Delta_{sol} H_{A(vap)}^\infty$	372
	cyclohexane	Henry	98-99
	cyclohexane	$\Delta_M H_A^\infty$, $\Delta_{sol} H_{A(vap)}^\infty$	372
	cyclopentane	Henry	99
	cyclopentane	$\Delta_M H_A^\infty$, $\Delta_{sol} H_{A(vap)}^\infty$	372
	1,2-dimethylbenzene	Henry	99
	1,2-dimethylbenzene	$\Delta_M H_A^\infty$, $\Delta_{sol} H_{A(vap)}^\infty$	372
	ethylbenzene	Henry	99
	ethylbenzene	$\Delta_M H_A^\infty$, $\Delta_{sol} H_{A(vap)}^\infty$	372
	n-heptane	Henry	99
	n-heptane	$\Delta_M H_A^\infty$, $\Delta_{sol} H_{A(vap)}^\infty$	372
	n-hexane	Henry	98-99
	n-hexane	$\Delta_M H_A^\infty$, $\Delta_{sol} H_{A(vap)}^\infty$	372
	1-hexene	Henry	98
	1-hexene	$\Delta_M H_A^\infty$, $\Delta_{sol} H_{A(vap)}^\infty$	372
	methylcyclohexane	Henry	99
	methylcyclohexane	$\Delta_M H_A^\infty$, $\Delta_{sol} H_{A(vap)}^\infty$	372
	n-octane	Henry	99
	n-octane	$\Delta_M H_A^\infty$, $\Delta_{sol} H_{A(vap)}^\infty$	372
	n-pentane	Henry	99
	n-pentane	$\Delta_M H_A^\infty$, $\Delta_{sol} H_{A(vap)}^\infty$	372
	tetrahydrofuran	Henry	99
	tetrahydrofuran	$\Delta_M H_A^\infty$, $\Delta_{sol} H_{A(vap)}^\infty$	372
	toluene	Henry	99
	toluene	$\Delta_M H_A^\infty$, $\Delta_{sol} H_{A(vap)}^\infty$	372
Poly(styrene-*co*-butyl methacrylate)			
	benzene	A_2	446
	2-butanone	A_2	446
	1-chloronaphthalene	A_2	446
	cyclohexanone	A_2	446
	1,2-dichlorobenzene	A_2	446
	1,2-dichloroethane	A_2	446
	1,4-dioxane	A_2	446

Copolymer	Solvent	Property	Page(s)
Poly(styrene-*b*-4-chlorostyrene-*b*-styrene)			
	2-butanone	A_2	446
	isopropylbenzene	A_2	446
	toluene	A_2	447
Poly(styrene-*co*-divinylbenzene)			
	cyclohexane	A_2	447
	tetrahydrofuran	A_2	447
Poly[styrene-*b*-(ethylene-*co*-propylene)]			
	2-butanone	A_2	447
	5-methyl-2-hexanone	A_2	447
	4-methyl-2-pentanone	A_2	447
	n-octane	A_2	447
	2-pentanone	A_2	447
	3-pentanone	A_2	447
Poly[styrene-*b*-(ethylene-*co*-propylene)-*b*-styrene]			
	5-methyl-2-hexanone	A_2	447
Poly(styrene-*b*-ethylene oxide-*b*-styrene)			
	1-butanol	Henry	100-102
	1-butanol	$\Delta_M H_A^\infty$, $\Delta_{sol} H_{A(vap)}^\infty$	373
	butyl acetate	Henry	99-102
	butyl acetate	$\Delta_M H_A^\infty$, $\Delta_{sol} H_{A(vap)}^\infty$	373
	n-decane	Henry	100-102
	n-decane	$\Delta_M H_A^\infty$, $\Delta_{sol} H_{A(vap)}^\infty$	373
	ethanol	Henry	100-102
	ethanol	$\Delta_M H_A^\infty$, $\Delta_{sol} H_{A(vap)}^\infty$	373
	ethyl acetate	Henry	100-102
	ethyl acetate	$\Delta_M H_A^\infty$, $\Delta_{sol} H_{A(vap)}^\infty$	373
	n-heptane	Henry	99-102
	n-heptane	$\Delta_M H_A^\infty$, $\Delta_{sol} H_{A(vap)}^\infty$	373
	n-hexane	Henry	99-102
	n-hexane	$\Delta_M H_A^\infty$, $\Delta_{sol} H_{A(vap)}^\infty$	373
	methanol	Henry	100-102
	methanol	$\Delta_M H_A^\infty$, $\Delta_{sol} H_{A(vap)}^\infty$	373
	methyl acetate	Henry	100-102

Copolymer	Solvent	Property	Page(s)
	methyl acetate	$\Delta_M H_A^\infty$, $\Delta_{sol} H_{A(vap)}^\infty$	373
	n-nonane	Henry	100-102
	n-nonane	$\Delta_M H_A^\infty$, $\Delta_{sol} H_{A(vap)}^\infty$	373
	n-octane	Henry	100-102
	n-octane	$\Delta_M H_A^\infty$, $\Delta_{sol} H_{A(vap)}^\infty$	373
	pentyl acetate	Henry	99-102
	pentyl acetate	$\Delta_M H_A^\infty$, $\Delta_{sol} H_{A(vap)}^\infty$	373
	1-propanol	Henry	100-102
	1-propanol	$\Delta_M H_A^\infty$, $\Delta_{sol} H_{A(vap)}^\infty$	373
	propyl acetate	Henry	99-102
	propyl acetate	$\Delta_M H_A^\infty$, $\Delta_{sol} H_{A(vap)}^\infty$	373
	1-pentanol	Henry	100-102
	1-pentanol	$\Delta_M H_A^\infty$, $\Delta_{sol} H_{A(vap)}^\infty$	373
Poly(styrene-*g*-ethyl methacrylate)			
	1-butanol	Henry	102-103
	1-butanol	$\Delta_M H_A^\infty$	374
	n-decane	Henry	103
	n-decane	$\Delta_M H_A^\infty$	374
	ethanol	Henry	102
	ethanol	$\Delta_M H_A^\infty$	374
	n-heptane	Henry	102
	n-heptane	$\Delta_M H_A^\infty$	374
	n-hexane	Henry	102-103
	n-hexane	$\Delta_M H_A^\infty$	374
	methanol	Henry	102
	methanol	$\Delta_M H_A^\infty$	374
	n-nonane	Henry	102
	n-nonane	$\Delta_M H_A^\infty$	374
	n-octane	Henry	102
	n-octane	$\Delta_M H_A^\infty$	374
	1-propanol	Henry	102
	1-propanol	$\Delta_M H_A^\infty$	374
	1-pentanol	Henry	103
	1-pentanol	$\Delta_M H_A^\infty$	374
	tetrachlormethane	Henry	103
	tetrachlormethane	$\Delta_M H_A^\infty$	374
Poly(styrene-*b*-isoprene)			
	carbon dioxide	gas solubility	47
	cyclohexane	A_2	447-448
	4-methyl-2-pentanone	A_2	448-449

Copolymer	Solvent	Property	Page(s)
Poly[tetrafluoroethylene-*co*-perfluoro(methyl vinyl ether)]			
	–	PVT	398-400
	benzene	$\Delta_M H_A^\infty$, $\Delta_{sol} H_{A(vap)}^\infty$	374
	n-decane	$\Delta_M H_A^\infty$, $\Delta_{sol} H_{A(vap)}^\infty$	374
	ethane	$\Delta_M H_A^\infty$, $\Delta_{sol} H_{A(vap)}^\infty$	374
	4-fluorotoluene	$\Delta_M H_A^\infty$, $\Delta_{sol} H_{A(vap)}^\infty$	374
	n-heptane	$\Delta_M H_A^\infty$, $\Delta_{sol} H_{A(vap)}^\infty$	374
	hexafluoroethane	$\Delta_M H_A^\infty$, $\Delta_{sol} H_{A(vap)}^\infty$	374
	n-hexane	$\Delta_M H_A^\infty$, $\Delta_{sol} H_{A(vap)}^\infty$	374
	methylcyclohexane	$\Delta_M H_A^\infty$, $\Delta_{sol} H_{A(vap)}^\infty$	374
	n-nonane	$\Delta_M H_A^\infty$, $\Delta_{sol} H_{A(vap)}^\infty$	374
	n-octane	$\Delta_M H_A^\infty$, $\Delta_{sol} H_{A(vap)}^\infty$	374
	2,3,4,5,6-pentafluorotoluene	$\Delta_M H_A^\infty$, $\Delta_{sol} H_{A(vap)}^\infty$	374
	perfluorobenzene	$\Delta_M H_A^\infty$, $\Delta_{sol} H_{A(vap)}^\infty$	374
	n-perfluorohexane	$\Delta_M H_A^\infty$, $\Delta_{sol} H_{A(vap)}^\infty$	374
	perfluoromethylcyclohexane	$\Delta_M H_A^\infty$, $\Delta_{sol} H_{A(vap)}^\infty$	374
	n-perfluorooctane	$\Delta_M H_A^\infty$, $\Delta_{sol} H_{A(vap)}^\infty$	374
	perfluorotoluene	$\Delta_M H_A^\infty$, $\Delta_{sol} H_{A(vap)}^\infty$	374
	toluene	$\Delta_M H_A^\infty$, $\Delta_{sol} H_{A(vap)}^\infty$	374
Poly(*N*-vinylacetamide-*co*-vinyl acetate)			
	water	UCST/LCST	234
Poly(vinyl acetate-*co*-vinyl alcohol)			
	methanol	VLE	51
	methyl acetate	VLE	51-52
	water	VLE	52
	water	LLE	146
	water	A_2	459
	water + ethanol	VLE	66
	water + poly(ethylene oxide-*co*-propylene oxide)	LLE	186
	water + poly(ethylene glycol)	LLE	214
	water + cesium chloride	LLE	215
	water + lithium chloride	LLE	215
	water + potassium chloride	LLE	215
	water + sodium bromide	LLE	215-216
	water + sodium chloride	LLE	216
	water + sodium fluoride	LLE	216
	water + sodium iodide	LLE	216

Copolymer	Solvent	Property	Page(s)
Poly(*N*-vinylcaprolactam-*co*-1-vinylimidazole)			
	water	LLE	148
	water	UCST/LCST	234
	water	A_2	459
Poly(vinyl chloride-*co*-butyl acrylate)			
	tetrahydrofuran	A_2	460
Poly(vinyl chloride-*co*-methyl acrylate)			
	tetrahydrofuran	A_2	460
Poly(vinyl chloride-*co*-vinylidene chloride)			
	tetrahydrofuran	A_2	460
Poly(vinylidene fluoride-*co*-chlorotrifluoroethylene)			
	carbon dioxide + poly(vinylidene fluoride)	gas solubility	68-69
	methane + poly(vinylidene fluoride)	gas solubility	69
Poly(vinylidene fluoride-*co*-hexafluoropropylene)			
	–	PVT	400-405
	carbon dioxide	HPPE	309, 314
Poly[vinylidene fluoride-*co*-perfluoro(methyl vinyl ether)]			
	ethyl acetate	A_2	460
Poly(*N*-vinylisobutyramide-*co*-*N*-vinylamine)			
	water	LLE	148-149
Poly(*N*-vinylformamide-*co*-vinyl acetate)			
	water	UCST/LCST	234

Appendix 3 List of solvents in alphabetical order

Name	Formula	CAS-RN	Page(s)
acetonitrile	C_2H_3N	75-05-8	437, 443
benzene	C_6H_6	71-43-2	23, 44, 48-49, 52, 71, 74-88, 90-91, 98, 103, 362-368, 372, 375, 425-426, 432, 446, 450, 454, 457
benzene-d6	C_6D_6	1076-43-3	429
benzyl methacrylate	$C_{11}H_{12}O_2$	2495-37-6	320-321
bicyclo[2,2,1]-2-heptene	C_7H_{10}	498-66-8	61-62
bisphenol-A diglycidyl ether	$C_{21}H_{24}O_4$	1675-54-3	212-213
n-butane	C_4H_{10}	106-97-8	263, 329-330, 342, 345
1-butanol	$C_4H_{10}O$	71-36-3	93-97, 100-103, 369-371, 373-374
2-butanone	C_4H_8O	78-93-3	53, 74-77, 98, 363-366, 372, 419, 421, 427, 429, 432, 435-439, 441-444, 446-447, 450, 452-455
1-butene	C_4H_8	106-98-9	263-264, 321
butyl acetate	$C_6H_{12}O_2$	123-86-4	72, 99, 101, 362, 373, 446
tert-butyl acetate	$C_6H_{12}O_2$	540-88-5	71-72, 362, 369
butyl acrylate	$C_7H_{12}O_2$	141-32-2	325-326, 336
butyl methacrylate	$C_8H_{14}O_2$	97-88-1	94
γ-butyrolactone	$C_4H_6O_2$	96-48-0	421
carbon dioxide	CO_2	124-38-9	22-26, 41-44, 46-48, 50, 68, 259, 300-303, 308-309, 315-316, 330-331, 340, 342, 346, 350
chlorobenzene	C_6H_5Cl	108-90-7	71-73, 362-363
1-chlorobutane	C_4H_9Cl	109-69-3	89-91, 367-368, 455
chlorodifluoromethane	$CHClF_2$	75-45-6	304-305
1-chloronaphthalene	$C_{10}H_7Cl$	90-13-1	430, 446
1-chloropropane	C_3H_7Cl	540-54-5	89-91, 367-368

Name	Formula	CAS-RN	Page(s)
cycloheptane	C_7H_{14}	291-64-5	78, 366
cyclohexane	C_6H_{12}	110-82-7	35, 78-79, 81, 84, 87-88, 98-99, 366-367, 372, 425-426, 439, 445, 447-448, 450
cyclohexane-d12	C_6D_{12}	1735-17-7	429
cyclohexanol	$C_6H_{12}O$	108-93-0	146, 455
cyclohexanone	$C_6H_{10}O$	108-94-1	440, 446
cyclohexene oxide	$C_6H_{10}O$	286-20-4	315-316
cyclopentane	C_5H_{10}	287-92-3	36, 78, 99, 121, 366, 372
cyclopentanone	C_5H_8O	120-92-3	359, 424
cyclopentene	C_5H_8	142-29-0	122
n-decane	$C_{10}H_{22}$	124-18-5	71-78, 88-91, 93, 97, 100-101, 103, 316, 362-369, 371, 373-375
deuterium oxide	D_2O	7789-20-0	233
1,2-dichlorobenzene	$C_6H_4Cl_2$	95-50-1	446
1,2-dichloroethane	$C_2H_4Cl_2$	107-06-2	422, 440, 446
dichloromethane	CH_2Cl_2	75-09-2	91, 94
diisobutyl adipate	$C_{14}H_{26}O_2$	141-04-8	430
1,1-difluoroethane	$C_2H_4F_2$	75-37-6	103
N,N-dimethylacetamide	C_4H_9NO	127-19-5	211-212, 421, 423-424, 426-427, 440, 458
1,2-dimethylbenzene	C_8H_{10}	95-47-6	73-77, 98-99, 363-367, 372, 431
1,3-dimethylbenzene	C_8H_{10}	108-38-3	367
1,4-dimethylbenzene	C_8H_{10}	106-42-3	88-91, 367-368
dimethyl ether	C_2H_6O	115-10-6	265, 305-306, 406
dimethyl ether-d6	C_2D_6O	17222-37-6	406-407
N,N-dimethylformamide	C_3H_7NO	68-12-2	419-423, 429, 431, 436-437, 440, 442-443, 451, 453, 458
2,3-dimethylpentane	C_7H_{16}	565-59-3	367
2,4-dimethylpentane	C_7H_{16}	108-08-7	367
2,2-dimethylpropane	C_5H_{12}	463-82-1	120, 265
dimethylsulfoxide	C_2H_6OS	67-68-5	410-411, 422-423, 440, 458-459
1,4-dioxane	$C_4H_8O_2$	123-91-1	91, 368, 425-426, 431, 433, 440-441, 443, 445-446, 450, 454-456, 458

Name	Formula	CAS-RN	Page(s)
n-dodecane	$C_{12}H_{26}$	112-40-3	74-79, 81, 84, 87-88, 98, 341, 363-366, 372
ethane-d6	C_2D_6	1632-99-1	266
ethane	C_2H_6	74-84-0	266, 331, 374, 407-408
ethanol	C_2H_6O	64-17-5	53, 66, 73-77, 92-98, 100-102, 316-317, 363-366, 369-374, 429
ethene	C_2H_4	74-85-1	38, 61-62, 64-65, 259-263, 266-268, 271-281, 286-295, 298-299, 316-340, 342-350
2-ethoxyethanol	$C_4H_{10}O_2$	110-80-5	456
ethyl acetate	$C_4H_8O_2$	141-78-6	53, 71-72, 74-78, 91-92, 94, 98, 100, 102, 230, 318, 362-366, 369, 372-373, 432-433, 435, 440, 442-443, 453, 456-457, 460
ethylbenzene	C_8H_{10}	100-41-4	71-73, 99, 362-363, 367, 372
2-ethylhexyl acrylate	$C_{11}H_{20}O_2$	103-11-7	328, 337
ethyl methacrylate	$C_6H_{10}O_2$	97-63-2	328-329
1-ethylnaphthalene	$C_{12}H_{12}$	1127-76-0	439
4-fluorotoluene	C_7H_7F	352-32-9	375
helium	He	7440-59-7	331-332, 342, 346
n-heptane	C_7H_{16}	142-82-5	36, 78-79, 81-85, 87-93, 96-97, 99-102, 145, 318, 337, 366-375
hexafluoroethane	C_2F_6	76-16-4	104, 374
n-hexane	C_6H_{14}	110-54-3	21, 27, 36, 44, 53, 61, 64, 78-92, 96-99, 101-103, 114, 321, 366-374
1-hexene	C_6H_{12}	592-41-6	28, 30-36, 79-88, 98, 332-333, 336, 372
2-hydroxyethyl methacrylate	$C_6H_{10}O_3$	868-77-9	350
isobutyl acetate	$C_6H_{12}O_2$	110-19-0	71, 362
isopropylamine	C_3H_9N	75-31-0	434
isopropylbenzene	C_9H_{12}	98-82-8	71-73, 362, 427, 446
methane	CH_4	74-82-8	69, 334, 343, 347
methanol	CH_4O	67-56-1	51, 55-56, 73-77, 92, 94-95, 98, 100-102, 363-366, 369-370, 372-374, 429, 434
methanol-d4	CD_4O	811-98-3	429

Name	Formula	CAS-RN	Page(s)
methyl acetate	$C_3H_6O_2$	79-20-9	51, 74-78, 94, 98, 100-101, 363-366, 369, 372-373, 427
methyl acrylate	$C_4H_6O_2$	96-33-3	327, 338
2-methylbutane	C_5H_{12}	78-78-4	120, 268
3-methylbutyl acetate	$C_7H_{14}O_2$	123-92-2	71-73, 362-363
methylcyclohexane	C_7H_{14}	108-87-2	99, 372, 375
N-methylformamide	C_2H_5NO	123-39-7	419
2-methylhexane	C_7H_{16}	591-76-4	367
3-methylhexane	C_7H_{16}	589-34-4	367
5-methyl-2-hexanone	$C_7H_{14}O$	110-12-3	447
methyl methacrylate	$C_5H_8O_2$	80-62-6	94
2-methylpentane	C_6H_{14}	107-83-5	29, 117
4-methyl-2-pentanone	$C_6H_{12}O$	108-10-1	447-449
2-methylpropane	C_4H_{10}	75-28-5	282-283, 336
2-methyl-2-propanol	$C_4H_{10}O$	75-65-0	348-349
1-methyl-2-pyrrolidinone	C_5H_9NO	872-50-4	412, 428, 440
2-methyltetrahydrofuran	$C_5H_{10}O$	96-47-9	412
nitroethane	$C_2H_5NO_2$	79-24-3	456
nitrogen	N_2	7727-37-9	333, 335, 344, 350
n-nonane	C_9H_{20}	111-84-2	71, 73-77, 79, 81, 84, 87-91, 100-102, 362-368, 373-375
norbornene	C_7H_{10}	498-66-8	61-62
octafluorocyclobutane	C_4F_8	115-25-3	103
1H,1H,5H-octafluoro-1-pentanol	$C_5H_4F_8O$	355-80-6	431
n-octane	C_8H_{18}	111-65-9	37, 71, 73-91, 93, 96-97, 99-102, 362-375, 447
octanoic acid	$C_8H_{16}O_2$	124-07-2	319
1-octene	C_8H_{16}	111-66-0	79-88, 339
n-pentadecane	$C_{15}H_{32}$	629-62-9	79, 82, 85, 87-88
n-pentane	C_5H_{12}	109-66-0	37, 88, 90-91, 99, 121, 269, 367-367, 369, 372, 408
n-pentane-d12	C_5D_{12}	2031-90-5	408-409
1-pentanol	$C_5H_{12}O$	71-41-0	93-95, 97, 100, 102-103, 369-371, 373-374
2-pentanone	$C_5H_{10}O$	107-87-9	447
3-pentanone	$C_5H_{10}O$	96-22-0	447
pentyl acetate	$C_7H_{14}O_2$	628-63-7	99-101, 373

Name	**Formula**	**CAS-RN**	**Page(s)**
perfluorobenzene	C_6F_6	392-56-3	375
n-perfluorohexane	C_6F_{14}	355-42-0	375
perfluoromethylcyclohexane	C_7F_{14}	355-02-2	375
n-perfluorooctane	C_8F_{18}	307-34-6	375
perfluorotoluene	C_7F_8	434-64-0	375
propane	C_3H_8	74-98-6	269-271, 283-285, 294, 296, 307-308, 335
1-propanol	C_3H_8O	71-23-8	67, 93-97, 100-102, 369-371, 373-374
2-propanol	C_3H_8O	67-63-0	56, 67
2-propanone	C_3H_6O	67-64-1	74-77, 91-92, 94, 98, 363-366, 372, 429, 436-438
propene	C_3H_6	115-07-1	38, 64, 297, 341
propyl acetate	$C_5H_{10}O_2$	109-60-4	71-73, 94, 99, 101-102, 362-363, 373
2-propyl acetate	$C_5H_{10}O_2$	108-21-4	72, 362
propylbenzene	C_9H_{12}	103-65-1	71-73, 362-363
styrene	C_8H_8	100-42-5	176, 432
1,1,2,2-tetrachloroethane	$C_2H_2Cl_4$	79-34-5	422, 424, 440
tetrachloromethane	CCl_4	56-23-5	89-91, 103, 367-368, 374, 434, 449-450, 456
tetrahydrofuran	C_4H_8O	109-99-9	21, 45-46, 91-92, 99, 372, 420, 424-427, 434-435, 439-440, 443, 445, 447, 449-450, 457, 459-460
1,2,3,4-tetrahydronaphthalene	$C_{10}H_{12}$	119-64-2	176
toluene	C_7H_8	108-88-3	24-25, 45, 49-50, 54, 61-62, 65, 71-72, 74-79, 82, 85, 87-91, 98-99, 103, 177, 233, 362-366-368, 372, 375, 425-427, 429-430, 432, 434, 439, 444-447, 449-452-454, 456-457
trichlorodeuteromethane	CCl_3D	865-49-6	433
trichloromethane	$CHCl_3$	67-66-3	92, 95, 359-361, 422-426, 428, 432, 435, 439-440, 444-445, 458
1,1,2-trichloro-1,2,2-trifluoroethane	$C_2Cl_3F_3$	76-13-1	104
2,2,2-trifluoroethanol	$C_2H_3F_3O$	75-89-9	428-429, 431
trifluoromethane	CHF_3	75-46-7	104, 306
2,2,3-trimethylbutane	C_7H_{16}	464-06-2	367

Name	Formula	CAS-RN	Page(s)
2,2,4-trimethylpentane	C_8H_{18}	540-84-1	104, 319
n-undecane	$C_{11}H_{24}$	1120-21-4	74-78, 363, 366
vinyl acetate	$C_4H_6O_2$	108-05-4	54, 338-339, 345-350
water	H_2O	7732-18-5	26-27, 39-41, 52, 54-55, 65-68, 92, 111-114, 122-149, 174-175, 178-211, 214-217, 230-234, 413-416, 419, 426, 428, 431-434, 440, 459

Appendix 4 List of solvents in order of their molecular formulas

Formula	Name	CAS-RN	Page(s)
CCl_3D	trichlorodeuteromethane	865-49-6	433
CCl_4	tetrachloromethane	56-23-5	89-91, 103, 367-368, 374, 434, 449-450, 456
CD_4O	methanol-d4	811-98-3	429
$CHClF_2$	chlorodifluoromethane	75-45-6	304-305
$CHCl_3$	trichloromethane	67-66-3	92, 95, 359-361, 422-426, 428, 432, 435, 439-440, 444-445, 458
CHF_3	trifluoromethane	75-46-7	104, 306
CH_2Cl_2	dichloromethane	75-09-2	91, 94
CH_4	methane	74-82-8	69, 334, 343, 347
CH_4O	methanol	67-56-1	51, 55-56, 73-77, 92, 94-95, 98, 100-102, 363-366, 369-370, 372-374, 429, 434
CO_2	carbon dioxide	124-38-9	22-26, 41-44, 46-48, 50, 68, 259, 300-303, 308-309, 315-316, 330-331, 340, 342, 346, 350
$C_2Cl_3F_3$	1,1,2-trichloro-1,2,2-trifluoroethane	76-13-1	104
C_2D_6	ethane-d6	1632-99-1	266
C_2D_6O	dimethyl ether-d6	17222-37-6	406-407
C_2F_6	hexafluoroethane	76-16-4	104, 374
$C_2H_2Cl_4$	1,1,2,2-tetrachloroethane	79-34-5	422, 424, 440
$C_2H_3F_3O$	2,2,2-trifluoroethanol	75-89-9	428-429, 431
C_2H_3N	acetonitrile	75-05-8	437, 443
C_2H_4	ethene	74-85-1	38, 61-62, 64-65, 259-263, 266-268, 271-281, 286-295, 298-299, 316-340, 342-350
$C_2H_4Cl_2$	1,2-dichloroethane	107-06-2	422, 440, 446
$C_2H_4F_2$	1,1-difluoroethane	75-37-6	103
C_2H_5NO	*N*-methylformamide	123-39-7	419

Formula	Name	CAS-RN	Page(s)
$C_2H_5NO_2$	nitroethane	79-24-3	456
C_2H_6	ethane	74-84-0	266, 331, 374, 407-408
C_2H_6O	dimethyl ether	115-10-6	265, 305-306, 406
C_2H_6O	ethanol	64-17-5	53, 66, 73-77, 92-98, 100-102, 316-317, 363-366, 369-374, 429
C_2H_6OS	dimethylsulfoxide	67-68-5	410-411, 422-423, 440, 458-459
C_3H_6	propene	115-07-1	38, 64, 297, 341
C_3H_6O	2-propanone	67-64-1	74-77, 91-92, 94, 98, 363-366, 372, 429, 436-438
$C_3H_6O_2$	methyl acetate	79-20-9	51, 74-78, 94, 98, 100-101, 363-366, 369, 372-373, 427
C_3H_7Cl	1-chloropropane	540-54-5	89-91, 367-368
C_3H_7NO	*N,N*-dimethylformamide	68-12-2	419-423, 429, 431, 436-437, 440, 442-443, 451, 453, 458
C_3H_8	propane	74-98-6	269-271, 283-285, 294, 296, 307-308, 335
C_3H_8O	1-propanol	71-23-8	67, 93-97, 100-102, 369-371, 373-374
C_3H_8O	2-propanol	67-63-0	56, 67
C_3H_9N	isopropylamine	75-31-0	434
C_4F_8	octafluorocyclobutane	115-25-3	103
$C_4H_6O_2$	γ-butyrolactone	96-48-0	421
$C_4H_6O_2$	methyl acrylate	96-33-3	327, 338
$C_4H_6O_2$	vinyl acetate	108-05-4	54, 338-339, 345-350
C_4H_8	1-butene	106-98-9	263-264, 321
C_4H_8O	tetrahydrofuran	109-99-9	21, 45-46, 91-92, 99, 372, 420, 424-427, 434-435, 439-440, 443, 445, 447, 449-450, 457, 459-460
C_4H_8O	2-butanone	78-93-3	53, 74-77, 98, 363-366, 372, 419, 421, 427, 429, 432, 435-439, 441-444, 446-447, 450, 452-455
$C_4H_8O_2$	1,4-dioxane	123-91-1	91, 368, 425-426, 431, 433, 440-441, 443, 445-446, 450, 454-456, 458

Formula	Name	CAS-RN	Page(s)
$C_4H_8O_2$	ethyl acetate	141-78-6	53, 71-72, 74-78, 91-92, 94, 98, 100, 102, 230, 318, 362-366, 369, 372-373, 432-433, 435, 440, 442-443, 453, 456-457, 460
C_4H_9Cl	1-chlorobutane	109-69-3	89-91, 367-368, 455
C_4H_9NO	*N,N*-dimethylacetamide	127-19-5	211-212, 421, 423-424, 426-427, 440, 458
C_4H_{10}	n-butane	106-97-8	263, 329-330, 342, 345
C_4H_{10}	2-methylpropane	75-28-5	282-283, 336
$C_4H_{10}O$	1-butanol	71-36-3	93-97, 100-103, 369-371, 373-374
$C_4H_{10}O$	2-methyl-2-propanol	75-65-0	348-349
$C_4H_{10}O_2$	2-ethoxyethanol	110-80-5	456
C_5H_{12}	n-pentane	109-66-0	37, 88, 90-91, 99, 121, 269, 367-367, 369, 372, 408
C_5D_{12}	n-pentane-d12	2031-90-5	408-409
$C_5H_4F_8O$	1H,1H,5H-octafluoro-1-pentanol	355-80-6	431
C_5H_8	cyclopentene	142-29-0	122
C_5H_8O	cyclopentanone	120-92-3	359, 424
$C_5H_8O_2$	methyl methacrylate	80-62-6	94
C_5H_9NO	1-methyl-2-pyrrolidinone	872-50-4	412, 428, 440
C_5H_{10}	cyclopentane	287-92-3	36, 78, 99, 121, 366, 372
$C_5H_{10}O$	2-methyltetrahydrofuran	96-47-9	412
$C_5H_{10}O$	2-pentanone	107-87-9	447
$C_5H_{10}O$	3-pentanone	96-22-0	447
$C_5H_{10}O_2$	propyl acetate	109-60-4	71-73, 94, 99, 101-102, 362-363, 373
$C_5H_{10}O_2$	2-propyl acetate	108-21-4	72, 362
C_5H_{12}	2,2-dimethylpropane	463-82-1	120, 265
C_5H_{12}	2-methylbutane	78-78-4	120, 268
$C_5H_{12}O$	1-pentanol	71-41-0	93-95, 97, 100, 102-103, 369-371, 373-374
C_6D_6	benzene-d6	1076-43-3	429
C_6D_{12}	cyclohexane-d12	1735-17-7	429
C_6F_6	perfluorobenzene	392-56-3	375
C_6F_{14}	n-perfluorohexane	355-42-0	375
$C_6H_4Cl_2$	1,2-dichlorobenzene	95-50-1	446
C_6H_5Cl	chlorobenzene	108-90-7	71-73, 362-363

Formula	Name	CAS-RN	Page(s)
C_6H_6	benzene	71-43-2	23, 44, 48-49, 52, 71, 74-88, 90-91, 98, 103, 362-368, 372, 375, 425-426, 432, 446, 450, 454, 457
$C_6H_{10}O$	cyclohexanone	108-94-1	440, 446
$C_6H_{10}O$	cyclohexene oxide	286-20-4	315-316
$C_6H_{12}O$	4-methyl-2-pentanone	108-10-1	447-449
$C_6H_{10}O_2$	ethyl methacrylate	97-63-2	328-329
$C_6H_{10}O_3$	2-hydroxyethyl methacrylate	868-77-9	350
C_6H_{12}	cyclohexane	110-82-7	35, 78-79, 81, 84, 87-88, 98-99, 366-367, 372, 425-426, 439, 445, 447-448, 450
C_6H_{12}	1-hexene	592-41-6	28, 30-36, 79-88, 98, 332-333, 336, 372
$C_6H_{12}O$	cyclohexanol	108-93-0	146, 455
$C_6H_{12}O_2$	butyl acetate	123-86-4	72, 99, 101, 362, 373, 446
$C_6H_{12}O_2$	*tert*-butyl acetate	540-88-5	71-72, 362, 369
$C_6H_{12}O_2$	isobutyl acetate	110-19-0	71, 362
C_6H_{14}	n-hexane	110-54-3	21, 27, 36, 44, 53, 61, 64, 78-92, 96-99, 101-103, 114, 321, 366-374
C_6H_{14}	2-methylpentane	107-83-5	29, 117
C_7F_8	perfluorotoluene	434-64-0	375
C_7F_{14}	perfluoromethylcyclohexane	355-02-2	375
C_7H_7F	4-fluorotoluene	352-32-9	375
C_7H_8	toluene	108-88-3	24-25, 45, 49-50, 54, 61-62, 65, 71-72, 74-79, 82, 85, 87-91, 98-99, 103, 177, 233, 362-366-368, 372, 375, 425-427, 429-430, 432, 434, 439, 444-447, 449-452-454, 456-457
C_7H_{10}	bicyclo[2,2,1]-2-heptene	498-66-8	61-62
C_7H_{10}	norbornene	498-66-8	61-62
$C_7H_{12}O_2$	butyl acrylate	141-32-2	325-326, 336
C_7H_{14}	cycloheptane	291-64-5	78, 366
C_7H_{14}	methylcyclohexane	108-87-2	99, 372, 375
$C_7H_{14}O$	5-methyl-2-hexanone	110-12-3	447
$C_7H_{14}O_2$	3-methylbutyl acetate	123-92-2	71-73, 362-363
$C_7H_{14}O_2$	pentyl acetate	628-63-7	99-101, 373

Formula	Name	CAS-RN	Page(s)
C_7H_{16}	2,3-dimethylpentane	565-59-3	367
C_7H_{16}	2,4-dimethylpentane	108-08-7	367
C_7H_{16}	n-heptane	142-82-5	36, 78-79, 81-85, 87-93, 96-97, 99-102, 145, 318, 337, 366-375
C_7H_{16}	2-methylhexane	591-76-4	367
C_7H_{16}	3-methylhexane	589-34-4	367
C_7H_{16}	2,2,3-trimethylbutane	464-06-2	367
C_8F_{18}	n-perfluorooctane	307-34-6	375
C_8H_8	styrene	100-42-5	176, 432
C_8H_{10}	1,2-dimethylbenzene	95-47-6	73-77, 98-99, 363-367, 372, 431
C_8H_{10}	1,3-dimethylbenzene	108-38-3	367
C_8H_{10}	1,4-dimethylbenzene	106-42-3	88-91, 367-368
C_8H_{10}	ethylbenzene	100-41-4	71-73, 99, 362-363, 367, 372
$C_8H_{14}O_2$	butyl methacrylate	97-88-1	94
C_8H_{16}	1-octene	111-66-0	79-88, 339
$C_8H_{16}O_2$	octanoic acid	124-07-2	319
C_8H_{18}	n-octane	111-65-9	37, 71, 73-91, 93, 96-97, 99-102, 362-375, 447
C_8H_{18}	2,2,4-trimethylpentane	540-84-1	104, 319
C_9H_{12}	isopropylbenzene	98-82-8	71-73, 362, 427, 446
C_9H_{12}	propylbenzene	103-65-1	71-73, 362-363
C_9H_{20}	n-nonane	111-84-2	71, 73-77, 79, 81, 84, 87-91, 100-102, 362-368, 373-375
$C_{10}H_7Cl$	1-chloronaphthalene	90-13-1	430, 446
$C_{10}H_{12}$	1,2,3,4-tetrahydronaphthalene	119-64-2	176
$C_{10}H_{22}$	n-decane	124-18-5	71-78, 88-91, 93, 97, 100-101, 103, 316, 362-369, 371, 373-375
$C_{11}H_{12}O_2$	benzyl methacrylate	2495-37-6	320-321
$C_{11}H_{20}O_2$	2-ethylhexyl acrylate	103-11-7	328, 337
$C_{11}H_{24}$	n-undecane	1120-21-4	74-78, 363, 366
$C_{12}H_{12}$	1-ethylnaphthalene	1127-76-0	439
$C_{12}H_{26}$	n-dodecane	112-40-3	74-79, 81, 84, 87-88, 98, 341, 363-366, 372
$C_{14}H_{26}O_2$	diisobutyl adipate	141-04-8	430
$C_{15}H_{32}$	n-pentadecane	629-62-9	79, 82, 85, 87-88
$C_{21}H_{24}O_4$	bisphenol-A diglycidyl ether	1675-54-3	212-213

Formula	Name	CAS-RN	Page(s)
D_2O	deuterium oxide	7789-20-0	233
H_2O	water	7732-18-5	26-27, 39-41, 52, 54-55, 65-68, 92, 111-114, 122-149, 174-175, 178-211, 214-217, 230-234, 413-416, 419, 426, 428, 431-434, 440, 459
He	helium	7440-59-7	331-332, 342, 346
N_2	nitrogen	7727-37-9	333, 335, 344, 350

INDEX

Milton Keynes UK
Ingram Content Group UK Ltd.
UKHW052026071024
449327UK00027B/2445

9 780367 383312